Boundary Value Problems of Heat Conduction

M. NECATI ÖZIŞIK

Professor, Mechanical and Aerospace Engineering
North Carolina State University

DOVER PUBLICATIONS, INC., New York

to Gül and Hakan

Published in Canada by General Publishing Company, Ltd., 30 Lesmill Road, Don Mills, Toronto, Ontario.
Published in the United Kingdom by Constable and Company, Ltd., 10 Orange Street, London WC2H 7EG.

This Dover edition, first published in 1989, is an unabridged, corrected republication of the work first published by the International Textbook Company, Scranton, Pa., 1968, in the series "International Textbooks in Mechanical Engineering."

Manufactured in the United States of America
Dover Publications, Inc., 31 East 2nd Street, Mineola, N.Y. 11501

Library of Congress Cataloging-in-Publication Data

Özışık, M. Necati.
Boundary value problems of heat conduction / by M. Necati Özışık.
p. cm.
Reprint. Originally published: Scranton : International Textbook Co., 1968. Originally published in series: International textbooks in mechanical engineering.
Includes bibliographies and index.
ISBN 0-486-65990-9
1. Heat—Conduction. 2. Boundary value problems. I. Title.
QC321.033 1989
536′.23—dc20 89-1460
CIP

Preface

With the advent of nuclear reactors and space technology there have been striking advances in the field of thermal sciences. In parallel with these developments the importance of thermal sciences has been emphasized in engineering schools, and three-semester graduate level sequence courses on conduction, convection, and radiation heat transfer have been developed to meet the need of advanced heat transfer work. This text is evolved to a large extent from the one-semester graduate level course on heat conduction given to the students from the mechanical and aerospace and nuclear engineering departments at this institution as the first of the sequence courses in heat transfer. It is concerned primarily with the determination of temperature distribution in solids by means of analytical methods.

The aim of the first part of this book is to provide a unified treatment of the solution of linear boundary value problems of heat conduction. With this objective in mind Chapters 1 through 4 are devoted to the solution of linear boundary value problems of heat conduction with systematic application of the integral transform technique. In the opinion of the author, among several different approaches that are available for the solution of boundary value problems of heat conduction, the integral transform technique offers the most straight forward and elegant approach, provided that the transforms, the inversions, and the kernels are readily available.

Chapter 1 presents the basic theory. Chapters 2, 3, and 4 are devoted to systematic solution of boundary value problems of heat conduction in the Cartesian, cylindrical, and spherical coordinate systems respectively. The transforms, the inversions, and the kernels are tabulated, and at the end of each chapter a summary of the transform of the Laplacian operator is presented. With the aid of these transform tables the solution of linear boundary value problems of heat conduction becomes a relatively simple and straightforward matter. The classical method of separation of variables is also included in the text to solve the homogeneous problems. Although this procedure gives rise to a small amount of repetition, it will help the reader to envision how the transform technique derives its basis from the classical method of separation of variables. It will also enable the reader to see how it differs from this method and provides an orderly approach to the solution of heat conduction problems. Chapter 5 deals with the use of Duhamel's method and Green's functions. Chapter 6 presents the method of solution of heat conduction problems in one-dimensional composite slabs, cylinders, and spheres with the straight forward approach of orthogonal expansion of functions over multilayer regions. Various approximate analytical methods treated in chapter 7 provide a relatively simple means for handling heat conduction problems involving nonlinear boundary conditions, variable thermal properties, a change of phase, or irregular geometries. Chapter 8 presents the application of a number of useful transformations in the solution of nonlinear boundary value problems of heat conduction. Chapter 9 is included to introduce the reader to the numerical techniques such as the finite differences and the Monte Carlo methods in the solution of heat conduction problems. Finally, chapter 10 deals with heat conduction in anisotropic solids and presents an orderly compilation of the useful material on this subject which had been scattered in the literature.

The transform tables that are presented in Chapters 2 through 4 are unique in that they are applicable not only to the solution of boundary value problems of heat conduction but to the solutions of similar partial differential equations that are encountered in various branches of science and engineering.

Whenever possible an attempt is made to develop all the pertinent relations from the fundamentals in a level that a reader with a background in engineering curriculum could follow the derivations with little difficulty; otherwise the results are stated and the original references are cited. A working knowledge in differential equations and advanced calculus is needed as a background to follow the material presented in this book.

M. Necati Özişik

Raleigh, North Carolina
March, 1968

Contents

1

Basic Relations

In this chapter we investigate the basic laws and definitions, the differential equation of heat conduction, the formulation of the boundary-value problem of heat conduction, general methods of solution by separation of variables and by finite integral-transform technique, splitting up of boundary-value problems of heat conduction into simpler ones, and thermal properties of solids which are important in the process of heat conduction.

1-1. THE HEAT FLUX

Temperature and heat flow are two important quantities in the problems of heat conduction. Temperature at any point in the solid is completely defined by its numerical value because it is a scalar quantity, whereas heat flow is defined by its value and direction.

When temperature distribution is not uniform at all points within a solid body, experience has shown that there is heat flow in the solid, the magnitude and direction of which depends on the distribution of temperature, and that heat flow is always in the direction of decreasing temperature. We introduce a vector quantity $\bar{\mathbf{q}}(\bar{r}, t)$, called the *heat-flux vector*, to denote heat flow at a spacial position $\bar{r}$ in a solid body, at any instant t. The magnitude of the heat-flux vector is equal to the quantity of heat crossing a unit area, normal to the direction of heat flow, at the position under consideration, per unit time. The basic law which gives the relationship between the heat flow and the temperature gradient is due to the French mathematician Jean Baptiste Joseph Fourier [1][1] (1768–1830); for a stationary, homogeneous, isotropic solid (i.e., material in which thermal conductivity is independent of direction) it is given in the form

$$\bar{\mathbf{q}}(\bar{r}, t) = -k \quad \nabla T(\bar{r}, t) \tag{1-1}$$

In this relationship the temperature gradient vector $\nabla T(\bar{r}, t)$, by definition, points in the direction in which temperature increases at the highest rate and its magnitude represents the maximum rate of increase of temperature at the point considered. Since the heat-flux vector $\bar{\mathbf{q}}(\bar{r}, t)$ points in the direction of decreasing temperature, the minus sign is included in Eq. 1-1 in order to make the heat flow a positive quantity. The propor-

[1] Bracketed numbers refer to references at end of chapter.

tionality factor k is called the *thermal conductivity* of the material; it is a scalar quantity and a property of the material of the solid. The dimension of thermal conductivity depends on the dimensions chosen for the heat flux and the temperature gradient. When heat flux is in Btu/hr ft^2 and temperature gradient in °F/ft, the dimension of thermal conductivity is in Btu/hr ft^2 (°F/ft).

Figure 1-1 shows a set of isothermal surfaces each differing in temperature by a small constant amount δT, and a unit direction vector $\hat{\mathbf{s}}$,

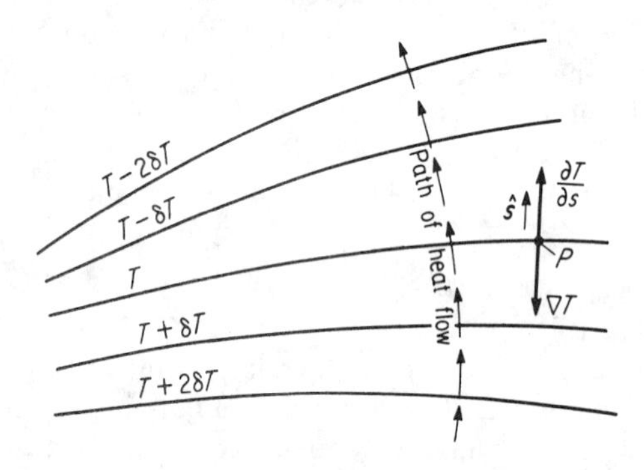

FIG. 1-1. An outward-drawn normal to an isothermal surface.

drawn normal to the isothermal surface T at the point P, and pointing in the direction of decreasing temperature. Based on the hypothesis that heat crosses from the inside to the outside of the isothermal surface, the unit direction vector $\hat{\mathbf{s}}$ is referred to as *outward-drawn* normal to the isothermal surface. Denoting differentiation along the $\hat{\mathbf{s}}$ direction by $\partial/\partial s$, the derivative of temperature in the $\hat{\mathbf{s}}$ direction, $\partial T/\partial s$, represents the maximum rate of decrease of temperature. Since $\partial T/\partial s$ and ∇T are of equal magnitudes but point in the opposite directions, the magnitude of the heat-flux vector in Eq. 1-1 is given as

$$|\bar{\mathbf{q}}| = -k \frac{\partial T}{\partial s} \tag{1-2}$$

where $\partial/\partial s$ denotes differentiation along the outward-drawn normal to the isothermal surface.

Equation 1-2 gives the magnitude of the heat-flux vector across an isothermal surface. In most problems the magnitude of the heat flux across a surface which is not necessarily isothermal is required. As a special case we first examine the magnitude of heat flux across a surface that fits a coordinate surface of an orthogonal coordinate system. Let $\hat{\mathbf{s}}$ and $\hat{\mathbf{n}}$ be the unit direction vectors drawn normal at the point P to an

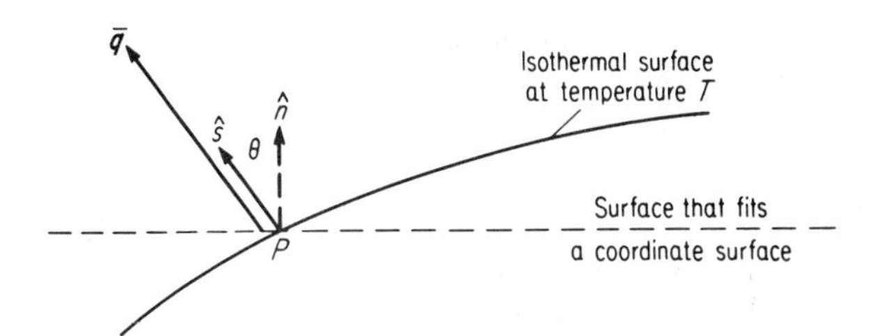

FIG. 1-2. Outward-drawn normals to an isothermal surface and to a coordinate surface.

isothermal surface and a coordinate surface respectively as shown in Fig. 1-2. The unit vectors $\hat{\mathbf{s}}$ and $\hat{\mathbf{n}}$ both point in the direction of decreasing temperature (i.e., in the direction of heat flow) and θ is the angle between them. If $\bar{\mathbf{q}}$ is the heat-flux vector across the isothermal surface at the point P, the magnitude of the heat flow per unit area per unit time across the coordinate surface is given as

$$q_n = \bar{\mathbf{q}} \cdot \hat{\mathbf{n}} = |\bar{\mathbf{q}}| \hat{\mathbf{s}} \cdot \hat{\mathbf{n}} = |\bar{\mathbf{q}}| \cos\theta = -k \frac{\partial T}{\partial s} \cos\theta \tag{1-3}$$

It is apparent from this relation that the heat flow per unit area across the nonisothermal surface that fits the coordinate surface of an orthogonal coordinate system is diminished by a factor of $\cos\theta$ as compared with that across the isothermal surface. Since, by definition, $\cos\theta \cdot \dfrac{\partial}{\partial s}$ is the derivative taken along the normal $\hat{\mathbf{n}}$, Eq. 1-3 is written in the form

$$q_n = -k \frac{\partial T}{\partial n} \tag{1-4}$$

where $\partial/\partial n$ denotes differentiation along the normal drawn to the coordinate surface in the direction of decreasing temperature. In the cartesian coordinate system, for example, the three components of the heat-flux vector, $\bar{\mathbf{q}}$, across the three coordinate surfaces are

$$q_x = -k \frac{\partial T}{\partial x} \qquad q_y = -k \frac{\partial T}{\partial y} \qquad q_z = -k \frac{\partial T}{\partial z} \tag{1-5}$$

The magnitude of the heat flow per unit area across an arbitrary surface passing through the point P can be related to the three components of the flux vector, given by Eq. 1-5. Let $\hat{\mathbf{u}}$ be the unit direction vector drawn normal to the arbitrary surface at the point P, and pointing in the direction of decreasing temperature, and c_{xu}, c_{yu}, c_{zu} be the direction cosines of vector $\hat{\mathbf{u}}$ (i.e., cosines of the angles between x, y, z coordinate axes and the $\hat{\mathbf{u}}$ vector). Then the magnitude of the heat flowing per unit area per unit time across the arbitrary surface at the point P is given by

$$q_u = c_{xu} \; q_x + c_{yu} \; q_y + c_{zu} \; q_z \tag{1-6}$$

If the surface under consideration fits one of the coordinate surfaces, two of the direction cosines vanish and Eq. 1-6 reduces to that given by Eq. 1-4.

1-2. TOTAL HEAT FLOW

Many engineering problems are concerned with the evaluation of the amount of heat transferred through the bounding surfaces of a solid in a given time interval. Consider a differential area dA on the boundary surface of an arbitrary solid as shown in Fig. 1-3. Let $\hat{\mathbf{u}}$ be the outward-

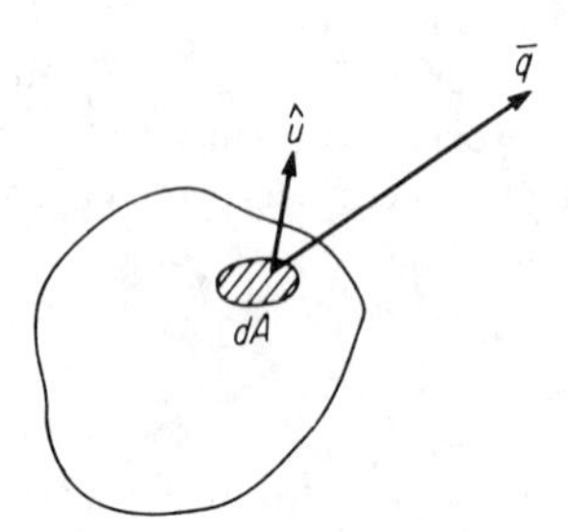

FIG. 1-3. Heat flow through the bounding surfaces of a solid.

drawn normal unit vector and $\bar{\mathbf{q}}$ be the heat-flux vector at the surface element dA. The rate of heat flowing out of the solid through the surface element dA per unit time is

$$dQ = \bar{\mathbf{q}} \cdot \hat{\mathbf{u}} dA = q_u dA \tag{1-7}$$

where q_u is the component of the heat-flux vector in the $\hat{\mathbf{u}}$ direction. The rate of heat flowing out of the entire bounding surface of the solid per unit time is given by

$$Q = \int_A q_u \; dA \qquad \text{Btu/hr} \tag{1-8}$$

where the integration is performed over the boundary surface of the solid.

The total amount of heat flowing out of the solid in a time interval t_1 to t_2 is obtained by integrating Eq. 1-8 with respect to time.

$$\underset{t_1 - t_2}{Q} = \int_{t_1}^{t_2} Q \; dt = \int_{t_1}^{t_2} \int_A q_u dA \, dt \qquad \text{Btu} \tag{1-9}$$

For a region having continuous bounding surfaces A_i, $i = 1, 2, \ldots, s$ in number,[2] that fit the coordinate surfaces of an orthogonal coordinate system, Eq. 1-9 is given in the form

$$Q_{t_1-t_2} = -\int_{t_1}^{t_2}\left(\sum_{i=1}^{s}\int_{A_i} k\frac{\partial T}{\partial n_i}\,dA\right)dt \tag{1-10}$$

where $\partial/\partial n_i$ denote differentiation along the outward-drawn normal to the boundary surface A_i.

It is apparent from the above relations that to evaluate the heat flow across a given surface the temperature gradient across that surface is needed, and that can be evaluated if the temperature distribution within the solid is known. Temperature distribution is determined from the solution of the differential equation of heat conduction which will be given in the next section.

1-3. THE DIFFERENTIAL EQUATION OF HEAT CONDUCTION

In this section we derive the differential equation of heat conduction for a stationary, homogeneous, isotropic solid with heat generation within the region. Heat generation may be due to nuclear, electrical, chemical, or infrared, sources which may be a function of both position and time. It is denoted by $g(\bar{r}, t)$, in Btu/hr ft^3. A heat source with a negative sign denotes the heat sink.

Consider the energy-balance equation for a small control volume V.

$$\begin{pmatrix}\text{Rate of energy}\\ \text{storage in } V\end{pmatrix} = \begin{pmatrix}\text{rate of heat entering } V\\ \text{through its bounding surfaces}\end{pmatrix} + \begin{pmatrix}\text{rate of heat}\\ \text{generation in } V\end{pmatrix} \tag{1-11}$$

Various terms in this relation are evaluated as follows.

The energy storage term, assuming density and specific heat are time-independent, is

$$\begin{pmatrix}\text{Rate of energy}\\ \text{storage in } V\end{pmatrix} = \int_V \rho c_p \frac{\partial T(\bar{r}, t)}{\partial t}\,dV \tag{1-12}$$

Heat entering V through a small area dA on the bounding surface is

$$-\bar{\mathbf{q}}(\bar{r}, t)\cdot\hat{\mathbf{u}}$$

where $\hat{\mathbf{u}}$ is the outward-drawn normal unit direction vector and $\bar{\mathbf{q}}$ the heat-flux vector at dA. Minus sign is used to indicate that heat flows into the volume element V. Heat entering through the entire boundary of V is

$$\begin{pmatrix}\text{Rate of heat entering } V\\ \text{through its bounding surfaces}\end{pmatrix} = -\int_A \bar{\mathbf{q}}\cdot\hat{\mathbf{u}}\,dA = -\int_V \operatorname{div}\bar{\mathbf{q}}\,dV \tag{1-13}$$

[2]For example, a rectangular parallelepiped with one corner at the origin and three edges coinciding with the ox-, oy-, oz-axes has six continuous boundary surfaces, hence $s = 6$.

where divergence theorem is used to convert the surface integral to volume integral.

Heat generation in the volume V is

$$\text{(Rate of heat generation in } V) = \int_V g(\bar{r}, t) \; dV \tag{1-14}$$

Substituting Eqs. 1-12, 1-13, and 1-14 into Eq. 1-11, we obtain

$$\int_V \left[\rho c_p \frac{\partial T(\bar{r}, t)}{\partial t} + \operatorname{div} \bar{\mathbf{q}}(\bar{r}, t) - g(\bar{r}, t) \right] dV = 0 \tag{1-15}$$

Equation 1-15 is derived for an arbitrary small-volume element V within the solid, and volume element V may be chosen so small as to remove the integral. We obtain

$$\rho c_p \frac{\partial T(\bar{r}, t)}{\partial t} = -\operatorname{div} \bar{\mathbf{q}}(\bar{r}, t) + g(\bar{r}, t) \tag{1-16}$$

Substituting $\bar{\mathbf{q}}(\bar{r}, t)$ from Eq. 1-1 into Eq. 1-16, we obtain

$$\rho c_p \frac{\partial T(\bar{r}, t)}{\partial t} = \nabla \cdot [k \nabla T(\bar{r}, t)] + g(\bar{r}, t) \tag{1-17}$$

Equation 1-17 is called *differential equation of heat conduction* for a stationary, homogeneous, isotropic solid with heat generation within the solid.

We now consider the following special cases of Eq. 1-17.

(1) Thermal conductivity is uniform, i.e., independent of position and temperature:

$$\frac{1}{\alpha} \frac{\partial T(\bar{r}, t)}{\partial t} = \nabla^2 T(\bar{r}, t) + \frac{g(\bar{r}, t)}{k} \tag{1-18}$$

where the constant α is called the *thermal diffusivity* of the medium and is defined as

$$\alpha = \frac{k}{\rho c_p} \tag{1-19}$$

It combines three physical properties (ρ, c_p, k) of the solid into a single constant and has dimensions in ft^2/hr.

(2) Uniform thermal conductivity, no heat sources.

$$\frac{1}{\alpha} \frac{\partial T(\bar{r}, t)}{\partial t} = \nabla^2 T(\bar{r}, t) \tag{1-20}$$

which is called the *Fourier equation* of heat conduction or the *diffusion equation*.

(3) Uniform thermal conductivity, steady state, heat generation within the solid.

$$\nabla^2 T(\bar{r}) + \frac{g(\bar{r})}{k} = 0 \tag{1-21}$$

which is *Poisson's equation.*

(4) Steady state, no heat generation.

$$\nabla^2 T(\bar{r}) = 0 \tag{1-22}$$

is the *Laplace equation.*

1-4. THE BOUNDARY CONDITIONS

The differential equation of heat conduction will have numerous solutions unless a set of boundary conditions and an initial condition (for the time-dependent problem) are prescribed. The boundary conditions that prescribe the conditions at the boundary surfaces of the region may be linear or nonlinear. In this book we shall be concerned primarily with problems involving linear boundary conditions; some nonlinear cases will be treated separately in a later chapter. For convenience the linear boundary conditions will be separated into the following three groups:

(1) *Boundary condition of the first kind.* Temperature is prescribed along the boundary surface and for the general case it is a function of both time and position, i.e.,

$$T = f_i(\bar{r}_s, t) \qquad \text{on the boundary surface } s_i \tag{1-23}$$

Special cases include temperature at the boundary surface is a function of position only $f_i(\bar{r}_s)$, or a function of time only $f_i(t)$, or a constant.

If the temperature at the boundary surface vanishes, we have

$$T = 0 \qquad \text{on the boundary surface } s_i \tag{1-24}$$

This special case is called the *homogeneous boundary condition of the first kind.* A boundary surface which is kept at zero temperature satisfies homogeneous boundary conditions of the first kind. A boundary surface which is kept at a constant temperature T_0, also satisfies the homogeneous boundary condition of the first kind if the temperature is measured in excess of T_0.

(2) *Boundary condition of the second kind.* The normal derivative of temperature is prescribed at the boundary surface and it may be a function of both time and position. For a boundary surface that fits the coordinate surface of an orthogonal coordinate system it is given in the form

$$\frac{\partial T}{\partial n_i} = f_i(\bar{r}_s, t) \qquad \text{on the boundary surface } s_i \tag{1-25}$$

where $\partial/\partial n_i$ denotes differentiation along the outward-drawn normal at the boundary surface s_i. This boundary condition is equivalent to that of prescribing the magnitude of the heat flux along the boundary surface, since the left-hand side of Eq. 1-25 becomes the magnitude of the heat flux at the surface s_i when both sides of Eq. 1-25 are multiplied by the thermal conductivity of the material.

Special cases of Eq. 1-25 include the normal derivative of temperature at the boundary surface to be a function of position only $f_i(\bar{r}_s)$, or a function of time only $f_i(t)$, or a constant. If the normal derivative of temperature at the boundary surface vanishes, we have

$$\frac{\partial T}{\partial n_i} = 0 \qquad \text{on the boundary } s_i \tag{1-26}$$

This special case is called the *homogeneous boundary condition of the second kind.* An insulated boundary condition satisfies this condition.

(3) *Boundary condition of the third kind.* A linear combination of the temperature and its normal derivative is prescribed at the boundary surface. For a boundary surface that fits the coordinate surface of an orthogonal coordinate system it is given as

$$k_i \frac{\partial T}{\partial n_i} + h_i T = f_i(\bar{r}_s, t) \qquad \text{on the boundary surface } s_i \tag{1-27}$$

The boundary conditions of the first and second kinds that are discussed above are obtainable by choosing k_i and h_i equal to zero respectively in Eq. 1-27. The physical significance of Eq. 1-27 is that the boundary surface under consideration dissipates heat by convection according to Newton's law of cooling (i.e., heat transfer is proportional to temperature difference) to a surrounding temperature which varies both with time and position along the boundary surface. By writing an energy balance for the boundary surface s_i shown in Fig. 1-4, we have

$$-k_i \frac{\partial T}{\partial n_i} = h_i(T - T_a) \tag{1-28}$$

or

$$k_i \frac{\partial T}{\partial n_i} + h_i T = h_i \cdot T_a \equiv f_i(\bar{r}, t) \tag{1-29}$$

which is in the same form as Eq. 1-27.

A special case of Eq. 1-27 is

$$k_i \frac{\partial T}{\partial n_i} + h_i T = 0 \tag{1-30}$$

which is called the *homogeneous boundary condition of the third kind.* Physical situation described by Eq. 1-30 is that of heat dissipation by convection from the boundary surface into a surrounding at zero temperature.

Three types of boundary conditions described above cover most cases of practical interest. There are also thermal radiation boundary condi-

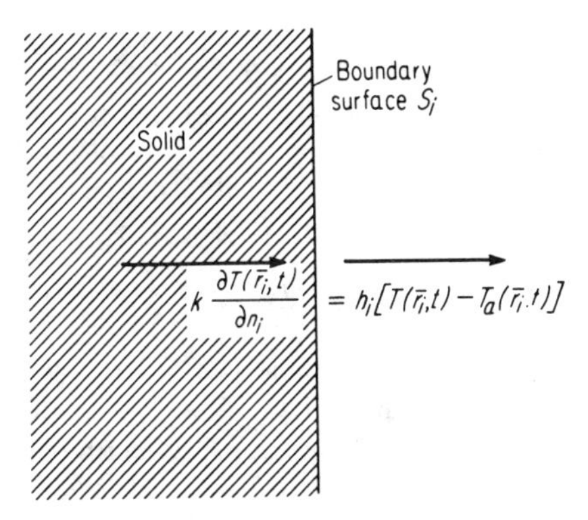

FIG. 1-4. Heat dissipation at the boundary surface according to Newton's law of cooling.

tions with heat transfer obeying the fourth-power temperature law, or the natural convection boundary condition with heat transfer proportional to the $\frac{5}{4}$ power of temperature difference. Such boundary conditions are nonlinear because a power of temperature enters the boundary condition. Provided that the temperature differences are not great the nonlinearities associated with these boundary conditions are avoided by approximating them with the boundary condition of the third kind.

Boundary conditions associated with change of phase (e.e., melting, solidification, ablation) are also nonlinear boundary conditions. In this book we shall be concerned primarily with the solution of linear differential equations of heat conduction subject to the linear boundary conditions discussed above. Solutions of nonlinear problems will be treated later in this book in a separate chapter.

1-5. USE OF DIMENSIONLESS PARAMETERS IN HEAT CONDUCTION PROBLEMS

The number of variables in the solution of heat-conduction problems may be reduced by introducing dimensionless variables. We shall illustrate this with the following boundary-value problem of heat conduction.

A stationary, homogeneous, isotropic solid is initially at a constant temperature T_0. For times $t > 0$, heat is generated within the solid and dissipated by convection from the bounding surfaces into a medium at constant temperature T_∞. Assuming a rectangular geometry and a finite region, the boundary-value problem of heat conduction is

$$\frac{\partial^2 T}{\partial x^2} + \frac{\partial^2 T}{\partial y^2} + \frac{\partial^2 T}{\partial z^2} + \frac{g}{k} = \frac{1}{\alpha}\frac{\partial T}{\partial t} \quad \text{in region } R,\ t > 0 \tag{1-31a}$$

$$-k_i \frac{\partial T}{\partial n_i} = h_i(T_i - T_\infty) \qquad \text{on } s_i, \text{ boundary of } R, \ t > 0 \tag{1-31b}$$

$i = 1, 2, \ldots,$ = number of continuous bounding surfaces of the solid

$$T = T_0 \qquad \text{in region } R, \ t = 0 \tag{1-31c}$$

where $\partial / \partial n_i$ denotes differentiation along the outward-drawn normal to the bounding surface s_i.

To express Eqs. 1-31 in dimensionless form, we divide all length variables by a reference length L (i.e., a characteristic dimension of the solid), and define the following dimensionless length variables.

$$\xi = \frac{x}{L} \qquad \eta = \frac{y}{L} \qquad \zeta = \frac{z}{L} \tag{1-32}$$

and $\partial / \partial N \equiv$ differentiation along outward-drawn normal in the new dimensionless coordinate system (ξ, η, ζ).

We measure all temperatures from an arbitrary reference value and divide them by an arbitrary reference temperature difference. For this particular problem we choose T_∞ as the reference temperature, $(T_0 - T_\infty)$ as the reference temperature difference, and define the following dimensionless excess temperature

$$\theta = \frac{T - T_\infty}{T_0 - T_\infty} \tag{1-33}$$

Substituting Eqs. 1-32 and 1-33 into Eqs. 1-31 we obtain

$$\frac{T_0 - T_\infty}{L^2} \left(\frac{\partial^2 \theta}{\partial \xi^2} + \frac{\partial^2 \theta}{\partial \eta^2} + \frac{\partial^2 \theta}{\partial \zeta^2} \right) + \frac{g}{k} = \frac{T_0 - T_\infty}{\alpha} \frac{\partial \theta}{\partial t} \qquad \text{in } R, \ t > 0 \tag{1-34a}$$

$$-\frac{k_i}{L} \frac{\partial \theta}{\partial N_i} = h_i \theta \qquad \text{on } s_i, \ t > 0 \tag{1-34b}$$

$i = 1, 2, \ldots,$ = number of continuous bounding surfaces

$$\theta = 1 \qquad \text{in } R, \ t = 0 \tag{1-34c}$$

Equation 1-34 is made dimensionless by rearranging the parameters.

$$\left(\frac{\partial^2 \theta}{\partial \xi^2} + \frac{\partial^2 \theta}{\partial \eta^2} + \frac{\partial^2 \theta}{\partial \zeta^2} \right) + \Phi = \frac{\partial \theta}{\partial \tau} \qquad \text{in } R, \ \tau > 0 \tag{1-35a}$$

$$\frac{\partial \theta}{\partial N_i} + B_i \theta = 0 \qquad \text{on } s_i, \ \tau > 0 \tag{1-35b}$$

$$\theta = 1 \qquad \text{in } R, \ \tau = 0 \tag{1-35c}$$

where the dimensionless parameters are defined as

$$\tau = \frac{\alpha t}{L^2} = \text{dimensionless time variable} \tag{1-36a}$$

$$\Phi = \frac{gL^2}{k(T_0 - T_\infty)} = \text{dimensionless heat-generation variable} \tag{1-36b}$$

$$B = \frac{hL}{k} = \text{dimensionless heat-conduction variable} \tag{1-36c}$$

When the dimensionless system 1-35 is compared with the original system 1-31, we notice that the number of variables is less in the dimensionless system. Physical significance of some of the dimensionless parameters defined above are immediately apparent, but parameters τ, Φ, and B may require further examination.

The dimensionless time variable τ, which is sometimes called the *Fourier number*, is the ratio of the rate of heat transferred by conduction to rate of energy stored. This becomes apparent if we rearrange τ as follows:

$$\tau = \frac{\alpha t}{L^2} = \frac{\left(\frac{k}{L}\right)\cdot L^2}{(\rho c_p L^3/t)} = \frac{\begin{pmatrix}\text{rate of heat conducted across } L \text{ in the} \\ \text{reference volume } L^3\text{, in Btu/hr °F}\end{pmatrix}}{\begin{pmatrix}\text{rate of energy storage in reference} \\ \text{volume } L^3\text{, Btu/hr °F}\end{pmatrix}}$$

The Fourier number is a useful parameter in comparing the rate of change of temperature of solids during transient heat-conduction problems. The larger the Fourier number, the larger the amount of heat conducted through the solid as compared with that stored in the solid, or the deeper the penetration of temperature into the solid over a given period of time, or the faster the temperature change at a given depth in the solid.

The dimensionless heat-generation variable Φ is the ratio of the rate of heat generation to the rate of heat conduction in a reference volume L^3. This becomes apparent if we rearrange Φ in the form

$$\Phi = \frac{gL^2}{k(T_0 - T_\infty)} = \frac{gL^3}{\left(\frac{k}{L}\right)L^2(T_0 - T_\infty)} = \frac{\begin{pmatrix}\text{rate of heat generation in} \\ \text{reference volume } L^3\text{, in Btu/hr}\end{pmatrix}}{\begin{pmatrix}\text{rate of heat conduction across} \\ L \text{ in reference volume } L^3\text{, with} \\ \text{temperature difference} \\ (T_0 - T_\infty)\text{, Btu/hr}\end{pmatrix}}$$

Dimensionless heat-conduction parameter, B, which is called the *Biot number*, is similar to the Nusselt number $\text{Nu} = hL/k_f$ for forced-convection heat transfer, except for the definition of thermal conductivity used; thermal conductivity of fluid k_f is used in the Nusselt number, whereas thermal conductivity of the solid k is used in the Biot number. If we rearrange the terms in the Biot number,

$$B = \frac{hL}{k} = \frac{h}{k/L} = \frac{\begin{pmatrix}\text{heat-transfer coefficient at the}\\ \text{boundary surface of the solid}\end{pmatrix}}{\begin{pmatrix}\text{unit conductance of the solid}\\ \text{across reference thickness } L\end{pmatrix}}$$

the Biot number is interpreted as the ratio of the heat-transfer coefficient at the boundary surface of the solid to the internal unit conductance of the solid.

1-6. HOMOGENEOUS AND NONHOMOGENEOUS BOUNDARY-VALUE PROBLEMS OF HEAT CONDUCTION

In this book we shall be concerned primarily with the solution of linear boundary-value problems of heat conduction. For convenience in the analysis, the time-dependent boundary-value problems of heat conduction will be considered in two different groups: *homogeneous problems and nonhomogeneous problems.*

The time-dependent boundary-value problem of heat conduction will be referred to as a *homogeneous problem* when both the differential equation and the boundary conditions are homogeneous. The problem in the form

$$\nabla^2 T = \frac{1}{\alpha}\frac{\partial T}{\partial t} \qquad \text{in region } R,\ t > 0 \tag{1-37a}$$

$$k_i \frac{\partial T}{\partial n_i} + h_i T = 0 \qquad \text{on boundary } s_i,\ t > 0 \tag{1-37b}$$

$$T = F(\bar{r}) \qquad \text{in region } R,\ t = 0 \tag{1-37c}$$

will be referred to as the *homogeneous problem* because both the differential equation and the boundary condition are homogeneous. The boundary condition in Eq. 1-37 could be a homogeneous boundary condition of the first or second kind.

The boundary-value problem of heat conduction will be referred to as *nonhomogeneous* if the differential equation, or the boundary conditions or both are nonhomogeneous. For example, the boundary-value problem of heat conduction in the form

$$\nabla^2 T + \frac{g(\bar{r},t)}{k} = \frac{1}{\alpha}\frac{\partial T}{\partial t} \qquad \text{in region } R,\ t > 0 \tag{1-38a}$$

$$k_i \frac{\partial T}{\partial n_i} + h_i T = f_i(\bar{r},t) \qquad \text{on boundary } s_i,\ t > 0 \tag{1-38b}$$

$$T = F(\bar{r}) \qquad \text{in region } R,\ t = 0 \tag{1-38c}$$

is nonhomogeneous because the differential equation and the boundary condition are nonhomogeneous (i.e., functions $g(\bar{r}, t)$ and $f_i(\bar{r}, t)$ do not include T as product).

The boundary-value problem of heat conduction in the form

$$\nabla^2 T + \frac{g(\bar{r}, t)}{k} = \frac{1}{\alpha}\frac{\partial T}{\partial t} \qquad \text{in region } R, \quad t > 0$$

$$k_i \frac{\partial T}{\partial n_i} + h_i T = 0 \qquad \text{on boundary } s_i, \quad t > 0 \tag{1-39}$$

$$T = F(\bar{r}) \qquad \text{in region } R, \quad t = 0$$

is a nonhomogeneous problem because the differential equation is nonhomogeneous.

1-7. ON THE SOLUTION OF HOMOGENEOUS PROBLEM BY SEPARATION OF VARIABLES

Among various methods that are available for the solution of the homogeneous boundary-value problems of heat conduction the method of separation of variables is most effective and straightforward to apply when separation is possible. Consider a stationary, isotropic, homogeneous solid with constant thermal properties. We choose an orthogonal coordinate system and assume that the solid has continuous bounding surfaces, s_i, $i = 1, 2, \ldots, s$ in number, that fit the coordinate surfaces. The homogeneous boundary-value problem of heat conduction is given as

$$\nabla^2 T(\bar{r}, t) = \frac{1}{\alpha}\frac{\partial T(\bar{r}, t)}{\partial t} \qquad \text{in a finite region } R, \quad t > 0 \tag{1-40a}$$

$$k_i \frac{\partial T(\bar{r}_i, t)}{\partial n_i} + h_i T(\bar{r}_i, t) = 0 \qquad \text{on } s_i, \text{ boundary of } R, \quad t > 0 \tag{1-40b}$$

$$T(\bar{r}, t) = F(\bar{r}) \qquad \text{in region } R, \quad t = 0 \tag{1-40c}$$

where $\partial/\partial n_i$ denotes differentiation along the outward-drawn normal.

In this problem for generality we choose a boundary condition of the third kind at all boundary surfaces. Setting $k_i = 0$ or $h_i = 0$ boundary conditions of the first or second kind are obtained.

We first separate the time and space variables by assuming separation of $T(\bar{r}, t)$ in the form

$$T(\bar{r}, t) = \psi(\bar{r}) \cdot \Gamma(t) \tag{1-41}$$

Substituting this into the homogeneous differential equation 1-40a we immediately obtain

$$\frac{\nabla^2 \psi(\bar{r})}{\psi(\bar{r})} = \frac{1}{\alpha}\frac{\dot{\Gamma}(t)}{\Gamma(t)} \tag{1-42a}$$

where $\dot{\Gamma}(t)$ denotes differentiation with respect to time. In Eq. 1-42a the term on the left is a function of space coordinates $\bar{r}$, alone, and the one on the right of time t, alone, and the only way this relation holds is if each side is equal to the same constant, say $-\lambda^2$. Then we have

$$\frac{\nabla^2\psi(\bar{r})}{\psi(\bar{r})} = \frac{1}{\alpha}\frac{\dot{\Gamma}(t)}{\Gamma(t)} = -\lambda^2 \tag{1-42b}$$

The separated equation for the time variable becomes

$$\dot{\Gamma}(t) + \alpha\lambda^2\Gamma(t) = 0 \tag{1-43}$$

which has a solution in the form

$$\Gamma(t): e^{-\alpha\lambda^2 t}$$

The negative sign chosen for λ^2 and the positive nature of λ^2 implies that the solution asymptotically approaches to zero as time increases indefinitely. This result is expected from the physical nature of the problem, that is, for a homogeneous boundary condition of the third kind it implies that a solid dissipates heat from its bounding surfaces by convection into a surrounding at zero temperature. Since the solid initially has a prescribed temperature distribution and for times $t > 0$ it looses heat to the surrounding continuously but it has no gains to make up for the losses, then the temperature of the solid will eventually approach to the temperature of the surrounding (i.e., zero) as time goes to infinity. For finite regions this argument holds when all h_i's do not simultaneously vanish—that is, there is at least one bounding surface of the solid which is kept at zero temperature or which dissipates heat by convection into a surrounding at zero temperature. For finite regions when all h_i's vanish simultaneously, the temperature transients will pass out with time, and temperature will be uniform across the cross section since there are no heat losses or gains at the boundary surfaces.

The space-variable function $\psi(\bar{r})$ satisfies the following eigenvalue problem.

$$\nabla^2\psi(\bar{r}) + \lambda^2\psi(\bar{r}) = 0 \qquad \text{in region } R \tag{1-44a}$$

$$k_i\frac{\partial\psi(\bar{r})}{\partial n_i} + h_i\psi(\bar{r}_i) = 0 \qquad \text{on boundary } s_i \tag{1-44b}$$

Equation 1-44a is the *Helmholtz* equation subject to homogeneous boundary conditions. System 1-44 has nontrivial solutions only for certain values of the separation variable $\lambda \equiv \lambda_m$, called *eigenvalues*, and the corresponding nontrivial solutions $\psi(\lambda_m, \bar{r}) \equiv \psi_m(\bar{r})$ are called *eigenfunctions*. If the eigenvalue problem 1-44 involves more than one space variable, the separation of variables is an effective method for solving the Helmholtz equation provided that the coordinate system used permits separation of variables. It is shown that there are eleven orthogonal coordinate systems in which the Helmholtz equation separates into ordinary differential equations [2, 3, 4]. Table 1-1 shows these eleven coordinate systems and the type of functions that will be the solutions. Therefore, separation is an important method for solving the Helmholtz equation, since for many problems of practical significance one of these eleven coordinate systems may be suitable in fitting the boundaries of the region under consideration.

TABLE 1-1

ORTHOGONAL COORDINATE SYSTEMS IN WHICH SEPARATION OF HELMHOLTZ EQUATION IS POSSIBLE

Coordinate System	Functions that Appear in Solution
1. Rectangular	Exponential, circular, hyperbolic
2. Circular-cylinder	Bessel, exponential, circular
3. Elliptic-cylinder	Mathieu, circular
4. Parabolic-cylinder	Weber, circular
5. Spherical	Legendre, power, circular
6. Prolate spheroidal	Legendre, circular
7. Oblate spheroidal	Legendre, circular
8. Parabolic	Bessel, circular
9. Conical	Lamé, power
10. Ellipsoidal	Lamé
11. Paraboloidal	Baer

Assuming the eigenfunctions $\psi_m(\bar{r})$ and the eigenvalues λ_m are determined, the complete solution of the temperature function $T(\bar{r}, t)$ is given in the form [5]

$$T(\bar{r}, t) = \sum_{m=1}^{\infty} c_m \cdot \psi_m(\bar{r}) \cdot e^{-\alpha\lambda_m^2 t} \tag{1-45}$$

The summation is taken over all discrete spectrum of eigenvalues λ_m for the problem under consideration. It is to be noted that for three-dimensional problems (in finite regions) the summation in Eq. 1-45 is a triple infinite series.

The solution of Eq. 1-45 contains the unknown coefficients c_m. Since the solution of Eq. 1-45 should satisfy the initial condition of the problem, by substituting $t = 0$ in Eq. 1-45 we obtain

$$F(\bar{r}) = \sum_{m=1}^{\infty} c_m \cdot \psi_m(\bar{r}) \tag{1-46}$$

If the eigenfunctions $\psi_m(\bar{r})$ constitute an orthogonal set in the region considered, the unknown coefficients c_m are determined by making use of the orthogonality property of eigenfunctions $\psi_m(\bar{r})$; that is,

$$\int_R \psi_m(\bar{r}) \cdot \psi_n(\bar{r}) \cdot d\bar{r} = 0 \quad \text{for } m \neq n$$

A discussion of orthogonality of eigenfunctions will be given later in this chapter. If we multiply both sides of Eq. 1-46 by $\psi_n(\bar{r})$, integrate it over the region and make use of the orthogonality condition, we obtain

$$c_m = \frac{\int_R \psi_m(\bar{r}) \cdot F(\bar{r}) \cdot d\bar{r}}{N} \tag{1-47a}$$

$$N = \int_R \psi_m^2(\bar{r}) \cdot d\bar{r} \tag{1-47b}$$

where N is called the *norm* of the eigenfunction $\psi_m(\bar{r})$.

Having determined the coefficients c_m, the complete solution of the homogeneous boundary-value problem of heat conduction in Eq. 1-40 is given in the form

$$T(\bar{r}, t) = \sum_{m=1}^{\infty} \frac{e^{-\alpha\lambda_m^2 t}}{N} \psi_m(\bar{r}) \cdot \int_R \psi_m(\bar{r}') \cdot F(\bar{r}') \cdot d\bar{r}' \tag{1-48a}$$

where

$$N = \int_R \psi_m^2(\bar{r}')d\bar{r}'$$

Sometimes the eigenfunctions are so adjusted that the norm becomes unity. This is done if we define the normalized eigenfunctions $K(\lambda_m, \bar{r})$ as

$$K(\lambda_m, \bar{r}) \equiv \frac{\psi_m(\bar{r})}{\sqrt{N}}$$

The solution, Eq. 1-48a, then becomes

$$T(\bar{r}, t) = \sum_{m=1}^{\infty} e^{-\alpha\lambda_m^2 t} \cdot K(\lambda_m, \bar{r}) \cdot \int_R K(\lambda_m, \bar{r}') \cdot F(\bar{r}') \cdot d\bar{r}' \tag{1-48b}$$

where the summation is taken over all eigenvalues λ_m of the problem.

For finite regions if all h_i's simultaneously vanish, that is when the entire bounding surfaces of the region are insulated, the above solution should include the following term

$$\frac{1}{\text{region}} \cdot \int_R F(\bar{r}') \cdot d\bar{r}'$$

The "region" in the denominator refers to the volume of the region for three-dimensional problems, to the surface of the region for two-dimensional problems, and to the linear dimension if it is one-dimensional. The physical significance of this term is that after the temperature transients have passed, the initial temperature distribution will tend to reach an average value over the region, since boundaries are insulated and there are no heat losses or gains.

If the region R is not finite but extends to infinity, say, in the x direction, the boundary conditions for the x-separation will not be sufficient to establish a relation for determining a set of discrete values for the eigenvalues for the x-separation. In such cases eigenvalues for the x-separation assume all values from 0 to ∞, and the summation for the x-separation is replaced by an integration. Such cases will be illustrated with examples in later chapters.

1-8. THE INTEGRAL-TRANSFORM TECHNIQUE (GENERAL CONSIDERATIONS)

In solving the transient heat-conduction problems it is common to use a Laplace transform to remove time variable from the partial differential equations. However, in many problems it is more convenient to apply an integral transformation that removes the space variable from the partial differential equation. The integral-transform technique is especially attractive for transient and steady-state heat-conduction problems in that it treats all space variables in the same manner and has no inversion difficulties as in the case of the Laplace transformation because both the integral transform and the inversion formula are defined at the onset of the problem. For a given problem, however, the type of integral transform and the corresponding inversion formula depend on the range of the space variable (i.e., finite, semi-infinite or infinite extend) and on the type of the boundary conditions. In the next three chapters the integral transforms and the corresponding inversion formulas are systematically tabulated for the finite, semi-infinite and infinite regions, for various combinations of boundary conditions in the Cartesian, cylindrical, and spherical coordinate systems. With the aid of integral-transform tables presented in this book the solution of linear heat-conduction problems involving more than one space variable become a relatively easy and straightforward matter as compared with the classical method of solution in which experience and ingenuity are required to find the correct form of solution. With the integral-transform technique the solutions for finite regions are in the form of infinite series, but with present-day computing facilities evaluation of such series presents no serious difficulty.

As the integral transform and the inversion formula are essentially based on the technique of rewriting in two parts the classical expansion of a function, such technique is utilized in this book in deriving various integral transforms and inversion formulas. Applications are restricted to the solution of heat-conduction problems, but the transforms that are tabulated are applicable to the solution of partial differential equations with similar boundary conditions. Readers interested in the more fundamental treatment of the theory of integral transform and its application to the solution of partial differential equations of physical problems in several other fields should refer to the excellent book on this subject by

Sneddon [6]. A summary of various integral transforms is given in the texts by Tranter [7] and by Ditkin and Prudnikov [8]. Mathematical theory of integral transforms is given by Titchmarsh [9].

Before examining a general treatment of the solution of the heat-conduction equation with the integral-transform technique we present the following simple example in order to illustrate the basic concepts associated with the integral-transform technique.

Consider a slab $0 \le x \le L$ initially at temperature $F(x)$, and for times $t > 0$ the boundary surface at $x = 0$ is kept insulated and that at $x = L$ is kept at zero temperature. Temperature history of the solid for times $t > 0$ will be determined by solving this problem with the integral-transform technique.

The boundary-value problem of heat conduction is given as

$$\frac{\partial^2 T}{\partial x^2} = \frac{1}{\alpha}\frac{\partial T}{\partial t} \quad \text{in } 0 \le x \le L, \quad t > 0 \tag{1-49a}$$

$$\frac{\partial T}{\partial x} = 0 \quad \text{at } x = 0, \quad t > 0 \tag{1-49b}$$

$$T = 0 \quad \text{at } x = L, \quad t > 0 \tag{1-49c}$$

$$T = F(x) \quad \text{in } 0 \le x \le L, \quad t = 0 \tag{1-49d}$$

where $T \equiv T(x, t)$.

In the above system the range of the space variable is finite and the boundary conditions associated with it are of the second kind at $x = 0$ and of the first kind at $x = L$. The appropriate integral transform and the inversion formula with respect to the x-variable of the temperature function $T(x, t)$ are immediately obtained from the tabulations presented in the next chapter (see Table 2-1) as

Integral transform:

$$\bar{T}(\beta_m, t) = \int_{x'=0}^{L} K(\beta_m, x') \cdot T(x', t) \cdot dx' \tag{1-50a}$$

Inversion formula:

$$T(x, t) = \sum_{m=1}^{\infty} K(\beta_m, x) \cdot \bar{T}(\beta_m, t) \tag{1-50b}$$

where the kernel $K(\beta_m, x)$ is

$$K(\beta_m, x) = \sqrt{\frac{2}{L}} \cos \beta_m x \tag{1-50c}$$

and the summation is taken over all positive roots of the transcendental equation

$$\cos \beta L = 0 \tag{1-50d}$$

It will be shown in the next chapter that the integral transform and the inversion formula as defined by Eq. 1-50 are obtained from the expansion of an arbitrary function in a finite interval $0 \leq x \leq L$ in an infinite series of the normalized eigenfunctions of the following eigenvalue problem.

$$\frac{d^2X}{dx^2} + \beta^2 X = 0 \qquad \text{in } 0 \leq x \leq L \tag{1-51a}$$

$$\frac{dX}{dx} = 0 \qquad \text{at } x = 0 \tag{1-51b}$$

$$X = 0 \qquad \text{at } x = L \tag{1-51c}$$

We now take the integral transform of system 1-49 by applying the transform 1-50a; that is, we operate both sides of the differential equation 1-49a by

$$\int_{x'=0}^{L} K(\beta_m, x') \cdot dx'$$

which yields

$$\int_0^L \frac{d^2T}{dx^2} \cdot K(\beta_m, x) \cdot dx = \frac{1}{\alpha} \frac{d\overline{T}(\beta_m, t)}{dt}$$

The quantity with a bar on the right-hand side denotes the integral transform of the temperature function with respect to the space variable x according to the transform 1-50a. The integral on the left-hand side is the integral transform of the partial derivative $\partial^2 T/\partial x^2$ according to the transform 1-50a. Evaluation of the integral transform of the partial derivatives will be discussed in greater detail in the next section. Here it is suffice to say that the integral transform of the partial derivative on the left-hand side is evaluated by integrating by parts and by making use of the eigenvalue Prob. 1-51, the boundary conditions in Eqs. 1-49b and 1-49c and is given as

$$\int_0^L \frac{d^2T}{dx^2} \cdot K(\beta_m, x) \cdot dx = -\beta_m^2 \; \overline{T}$$

Then, the system of partial differential equation 1-49 reduces to the following ordinary differential equation with respect to the time variable

$$\frac{d\overline{T}(\beta_m, t)}{dt} + \alpha\beta_m^2 \overline{T}(\beta_m, t) = 0 \tag{1-52a}$$

$$\overline{T}(\beta_m, t) = \overline{F}(\beta_m) \qquad \text{for } t = 0 \tag{1-52b}$$

where

$$\overline{F}(\beta_m) = \int_{x'=0}^{L} K(\beta_m, x') \cdot F(x') \cdot dx' \tag{1-52c}$$

Thus we removed from the partial differential equation of heat conduction the partial derivative with respect to the space variable and reduced it to an ordinary differential equation. The solution of system 1-52 is straightforward and yields

$$\bar{T}(\beta_m, t) = \bar{F}(\beta_m) \cdot e^{-\alpha\beta_m^2 t}$$

When the transform of temperature thus determined is inverted by the inversion formula 1-50b, the solution of the boundary-value problem of heat conduction, Eq. 1-49, becomes

$$T(x, t) = \sum_{m=1}^{\infty} K(\beta_m, x) \cdot e^{-\alpha\beta_m^2 t} \cdot \int_{x'=0}^{L} K(\beta_m, x') \cdot F(x') \cdot dx'$$

where all the quantities have been defined previously.

It is to be noted that when the differential equation involves more than one space variable, the partial derivatives with respect to the space variables are removed by the repeated application of one-dimensional integral transform and the problem is again reduced to an ordinary differential equation. For problems involving heat generation and/or nonhomogeneous boundary conditions, the technique essentially remains unaltered but the resulting ordinary differential equation 1-52a would include some additional terms—i.e., the transform of heat-generation and/or boundary-condition functions.

1-9. THE USE OF INTEGRAL TRANSFORMS IN THE SOLUTION OF BOUNDARY-VALUE PROBLEMS OF HEAT CONDUCTION IN FINITE REGIONS

In this section we limit our discussion to the solution with the integral-transform technique of the transient boundary-value problem of heat conduction in finite regions. The treatment of problems involving semi-infinite and infinite regions will be discussed in the following chapters.

A general application of the finite integral-transform technique to the solution of heat-conduction problems in finite regions with time-dependent heat sources and boundary conditions is given by Olcer. [10, 11] Here we present the work by Olcer [10] for the solution of heat-conduction equation in infinite regions with the integral-transform technique.

Time-Dependent Problem. Consider a solid with continuous bounding surfaces s_i, $i = 1, 2, \ldots, s$ in number, that fit the coordinate surfaces of an orthogonal coordinate system. Assume that solid is stationary, homogeneous, isotropic, and has constant thermal properties. The boundary-value problem of heat conduction is given as

$$\nabla^2 T(\bar{r}, t) + \frac{g(\bar{r}, t)}{k} = \frac{1}{\alpha}\frac{\partial T(\bar{r}, t)}{\partial t} \qquad \text{in } R,\ t > 0 \qquad (1\text{-}53a)$$

$$k_i \frac{\partial T(\bar{r}_i, t)}{\partial n_i} + h_i T(\bar{r}_i, t) = f_i(\bar{r}, t) \qquad \text{on } s_i,\ t > 0 \qquad (1\text{-}53b)$$

$$T(\bar{r}, t) = F(\bar{r}) \qquad \text{in } R,\ t = 0 \qquad (1\text{-}53c)$$

where $i = 1, 2, \ldots, s$ number of continuous bounding surfaces of the solid

$\dfrac{\partial}{\partial n_i}$ = differentiation along the outward-drawn normal to the bounding surface s_i

As a first step to solving the system 1-53, we consider the following *auxiliary* homogeneous eigenvalue problem for the space variable $\psi(\bar{r})$, in the same region R, subject to homogeneous boundary conditions.

$$\nabla^2\psi(\bar{r}) + \lambda^2\psi(\bar{r}) = 0 \qquad \text{in } R \tag{1-54a}$$

$$k_i \frac{\partial\psi(\bar{r}_i)}{\partial n_i} + h_i\psi(\bar{r}_i) = 0 \qquad \text{on } s_i \tag{1-54b}$$

It is to be noted that the homogeneous system 1-54 is similar to the eigenvalue problem 1-44 considered in the previous section. Let $\psi_m(\bar{r}) \equiv \psi(\lambda_m, \bar{r})$ be the eigenfunctions and λ_m the eigenvalues of the homogeneous system 1-54.

We take the eigenfunction $\psi_m(\bar{r})$ as the *kernel* and define a three-dimensional *finite integral transform* of the temperature function $T(\bar{r}, t)$ as

$$\overline{T}(\lambda_m, t) = \int_R \psi_m(\bar{r}') \cdot T(\bar{r}', t) \cdot d\bar{r}' \tag{1-55a}$$

and the *inversion formula* as

$$T(\bar{r}, t) = \sum_{m=1}^{\infty} B_m \cdot \psi_m(\bar{r}) \cdot \overline{T}(\lambda_m, t) \tag{1-55b}$$

where the summation is taken over all eigenvalues; for three-dimensional problems the summation is a triple infinite series.

The unknown coefficient B_m in Eq. 1-55b can be determined if the eigenfunctions $\psi_m(\bar{r})$ constitute an orthogonal set in the region under consideration, i.e.,

$$\int_R \psi_m(\bar{r}) \cdot \psi_n(\bar{r}) \cdot d\bar{r} = 0 \qquad \text{when } m \neq n \tag{1-56}$$

Multiplying both sides of Eq. 1-55b by $\psi_n(\bar{r})$, integrating it over the region, and by making use of the orthogonality of the eigenfunctions, we obtain

$$B_m = \frac{1}{\displaystyle\int_R \psi_m^2(\bar{r}) \cdot d\bar{r}} \equiv \frac{1}{N} \tag{1-57}$$

The unknown coefficient B_m is equal to the reciprocal of the *norm* N.

Sometimes it is convenient to introduce *normalized eigenfunctions* in defining the integral transform and the inversion formula. Let the kernel

$K(\lambda_m, \bar{r})$ denote the normalized eigenfunctions of the eigenvalue problem 1-54 defined as

$$K(\lambda_m, \bar{r}) = \frac{\psi_m(\bar{r})}{\sqrt{N}} \tag{1-58}$$

Using the normalized eigenfunctions $K(\lambda_m, \bar{r})$ as the kernel, the finite integral-transform and inversion formula of temperature is defined as

$$\begin{pmatrix}\text{Integral}\\ \text{transform}\end{pmatrix} \quad \bar{T}(\lambda_m, t) = \int_R K(\lambda_m, \bar{r}') \cdot T(\bar{r}', t) \cdot d\bar{r}' \tag{1-59a}$$

$$\begin{pmatrix}\text{Inversion}\\ \text{formula}\end{pmatrix} \quad T(\bar{r}, t) = \sum_{m=1}^{\infty} K(\lambda_m, \bar{r}) \cdot \bar{T}(\lambda_m, t) \tag{1-59b}$$

In the present analysis we shall define the integral transform and the inversion formula using the normalized eigenfunctions as given by Eqs. 1-59a and 1-59b.

We now proceed with the solution of the boundary-value problem of heat conduction, Eq. 1-53.

We take the integral transform of the differential equation of heat conduction, Eq. 1-53a, by applying the integral transform, Eq. 1-59a.

$$\int_R K(\lambda_m, \bar{r}) \cdot \nabla^2 T(\bar{r}, t) \cdot d\bar{r} + \frac{1}{k} \int_R K(\lambda_m, \bar{r}) \cdot g(\bar{r}, t) \cdot d\bar{r}$$
$$= \frac{1}{\alpha} \int_R K(\lambda_m, \bar{r}) \cdot \frac{\partial T(\bar{r}, t)}{\partial t} \cdot d\bar{r}$$

which is written in the form

$$\int_R K(\lambda_m, \bar{r}) \cdot \nabla^2 T(\bar{r}, t) \cdot d\bar{r} + \frac{1}{k} \bar{g}(\lambda_m, t) = \frac{1}{\alpha} \frac{d\bar{T}(\lambda_m, t)}{dt} \tag{1-60}$$

where quantities with bars (i.e., $\bar{g}, \bar{T}$) refer to the integral transform according to the transform 1-59a.

The first integral on the left-hand side of Eq. 1-60 is the integral transform of $\nabla^2 T(\bar{r}, t)$. This integral can be evaluated by making use of Green's theorem, which may be written in the form

$$\int_R K_m \nabla^2 T \, d\bar{r} = \int_R T \ \nabla^2 K_m \, d\bar{r} + \sum_{i=1}^{s} \int_{s_i} \left[K_m \frac{\partial T}{\partial n_i} - T \frac{\partial K_m}{\partial n_i} \right] ds_i \tag{1-61}$$

where $K_m \equiv K(\lambda_m, \bar{r})$ and the summation is taken over all continuous bounding surfaces, $i = 1, 2, \ldots, s$ in number, of the finite region under consideration.

Various terms on the right-hand side of Eq. 1-61 are evaluated as follows.

The first term is obtained by multiplying the auxiliary differential equation 1-54a by T and integrating it over the region R.

$$\int_R T\nabla^2 K_m\, d\bar{r} = -\lambda_m^2 \int_R K_m T\, d\bar{r} \equiv -\lambda_m^2 \cdot \bar{T}(\lambda_m, t) \tag{1-62}$$

The second term on the right-hand side of Eq. 1-61 is evaluated by making use of the boundary conditions 1-53b and 1-54b. We obtain

$$\left(K_m \frac{\partial T}{\partial n_i} - T \frac{\partial K_m}{\partial n_i}\right) = \frac{K(\lambda_m, \bar{r}_i)}{k_i} \cdot f_i(\bar{r}, t) \tag{1-63}$$

Substituting Eqs. 1-62 and 1-63 into Eq. 1-61,

$$\int_R K_m \nabla^2 T\, d\bar{r} = -\lambda_m^2 \cdot \bar{T}(\lambda_m, t) + \sum_{i=1}^{s} \int_{s_i} \frac{K(\lambda_m, \bar{r}_i)}{k_i} \cdot f_i(\bar{r}, t) \cdot ds_i \tag{1-64}$$

Substituting Eq. 1-64 into Eq. 1-60,

$$\frac{d\bar{T}(\lambda_m, t)}{dt} + \alpha\lambda_m^2 \cdot \bar{T}(\lambda_m, t) = A(\lambda_m, t) \tag{1-65a}$$

where

$$A(\lambda_m, t) = \frac{\alpha}{k} \bar{g}(\lambda_m, t) + \alpha \sum_{i=1}^{s} \int_{s_i} \frac{K(\lambda_m, \bar{r}_i')}{k_i} \cdot f_i(\bar{r}', t) \cdot ds_i \tag{1-65b}$$

$$\bar{g}(\lambda_m, t) = \int_R K(\lambda_m, \bar{r}') \cdot g(\bar{r}', t) \cdot d\bar{r}' \tag{1-65c}$$

Thus, by the integral-transform technique we removed from the partial differential equation of heat conduction, Eq. 1-53a, the second partial derivatives with respect to the space variables, and reduced it to a first-order, linear, ordinary differential equation for the integral transform of temperature $\bar{T}(\lambda_m, t)$ as given by Eq. 1-65. In order to solve the differential equation 1-65 an initial condition is needed, which is obtained by taking the integral transform of the initial condition, Eq. 1-53c, of the original boundary-value problem. That is,

$$\bar{T}(\lambda_m, t) = \int_R K(\lambda_m, \bar{r}) \cdot F(\bar{r}) \cdot d\bar{r} \equiv \bar{F}(\lambda_m) \qquad \text{for } t = 0 \tag{1-66}$$

The solution of the differential equation 1-65 subject to the initial condition 1-66 is

$$\bar{T}(\lambda_m, t) = e^{-\alpha\lambda_m^2 t}\left[\bar{F}(\lambda_m) + \int_0^t e^{\alpha\lambda_m^2 t'} \cdot A(\lambda_m, t') \cdot dt'\right] \tag{1-67}$$

Substituting the integral transform of temperature given by Eq. 1-67 into the inversion formula 1-59b, we obtain the solution of the boundary-value problem of heat conduction, Eq. 1-53, in the form

$$T(\bar{r}, t) = \sum_{m=1}^{\infty} e^{-\alpha\lambda_m^2 t} \cdot K(\lambda_m, \bar{r}) \cdot \left[\bar{F}(\lambda_m) + \int_{t'=0}^{t} e^{\alpha\lambda_m^2 t'} \cdot A(\lambda_m, t') \cdot dt' \right] \tag{1-68a}$$

where

$$A(\lambda_m, t') = \frac{\alpha}{k} \bar{g}(\lambda_m, t') + \alpha \sum_{i=1}^{s} \int_{s_i} \frac{K(\lambda_m, \bar{r}_i')}{k_i} \cdot f_i(\bar{r}', t') \cdot ds_i \tag{1-68b}$$

$$\bar{F}(\lambda_m) = \int_R K(\lambda_m, \bar{r}') \cdot F(\bar{r}') \cdot d\bar{r}' \tag{1-68c}$$

$$\bar{g}(\lambda_m, t') = \int_R K(\lambda_m, t') \cdot g(\bar{r}', t') \cdot d\bar{r}' \tag{1-68d}$$

$$K(\lambda_m, \bar{r}) = \frac{\psi_m(\bar{r})}{\sqrt{N}} \tag{1-68e}$$

$$N = \int_R \psi_m^2(\bar{r}') \cdot d\bar{r}' \tag{1-68f}$$

where $\psi_m(\bar{r})$'s are the eigenfunctions of the eigenvalue problem 1-54, and summation in Eq. 1-68 is taken over all eigenvalues.

Solution (1-68) is derived for a boundary condition of the third kind, i.e.,

$$k_i \frac{\partial T}{\partial n_i} + h_i T = f_i(\bar{r}, t)$$

on boundary surface i for all boundary surfaces of the region. When the temperature is prescribed as a function of position and time on the boundary surface i, we set k_i equal to zero, then the above boundary condition reduces to the boundary condition of the first kind, i.e.,

$$T = \frac{f_i(\bar{r}, t)}{h_i}$$

on boundary surface i. When the heat flux is prescribed on the boundary surface i, we set h_i equal to zero and the boundary condition reduces to the boundary condition of the second kind, i.e.,

$$k_i \frac{\partial T}{\partial n_i} = f_i(\bar{r}, t)$$

on boundary surface i.

In the solution 1-68 the parameter k_i is in the denominator in Eq. 1-68b. To avoid difficulty for the cases when $k_i = 0$ (i.e., boundary condition of the first kind) the following change should be made in Eq. 1-68b. When $k_i = 0$, replace

$$\frac{K(\lambda_m, r_i)}{k_i}$$

by

$$-\frac{1}{h_i}\frac{\partial K(\lambda_m, \bar{r}_i)}{\partial n_i}$$

The validity of this relation is apparent from the boundary condition 1-54b of the auxiliary equation.

An Alternate Solution. The solution of the boundary-value problem of heat conduction as given above by Eq. 1-68 is not always uniformly convergent. Olcer [9] presented another form of Eq. 1-68 which is composed of a quasi-steady state and transient parts, and converges uniformly and more rapidly. To obtain the alternate solution we consider quasi-steady-state temperature functions $T_{0j}(\bar{r}, t)$ satisfying the following system.

$$\nabla^2 T_{0j}(\bar{r}, t) + \delta_{0j} \cdot \frac{g(\bar{r}, t)}{k} = 0 \qquad \text{in the region } R \tag{1-69a}$$

$$k_i \frac{\partial T_{0j}(\bar{r}, t)}{\partial n_i} + h_i T_{0j}(\bar{r}, t) = \delta_{ij} \cdot f_i(\bar{r}, t) \qquad \text{on boundary } s_i \tag{1-69b}$$

where δ_{ij} = Kroneker delta $= \begin{cases} 0 \text{ for } i \neq j \\ 1 \text{ for } i = j \end{cases}$

$i = 1, 2, \ldots, s$ the number of continuous bounding surfaces of the solid

$j = 0, 1, 2, \ldots, s$

Here we assume further that all h_i's do not simultaneously vanish.[3]

Consider the integral transform and the inversion formulas defined as

$$\begin{pmatrix}\text{Integral}\\ \text{transform}\end{pmatrix} \quad \bar{T}_{0j}(\lambda_m, t) = \int_R K(\lambda_m, \bar{r}') \cdot T_{0j}(\bar{r}', t) \cdot dr' \tag{1-70a}$$

$$\begin{pmatrix}\text{Inversion}\\ \text{formula}\end{pmatrix} \quad T_{0j}(\bar{r}, t) = \sum_{m=1}^{\infty} K(\lambda_m, \bar{r}) \cdot \bar{T}_{0j}(\bar{r}, t) \tag{1-70b}$$

Integral transform of system 1-69 by applying the transform 1-70a yields

$$\lambda_m^2 \bar{T}_{0j}(\lambda_m, t) = \frac{\delta_{0j} \cdot \bar{g}(\lambda_m, t)}{k} + \sum_{i=1}^{s} \int_{s_i} \frac{K(\lambda_m, \bar{r}_i')}{k_i} \, \delta_{ij} \cdot f_i(\bar{r}_i', t) \cdot ds_i$$

$$j = 0, 1, 2, \ldots, s \tag{1-71}$$

[3]Otherwise the solutions $T_{0j}(\bar{r}, t)$ satisfying Eq. 1-69 do not exist, unless

$$\delta_{0j} \int_R g(\bar{r}, t) \cdot d\bar{r} + k \cdot \delta_{ij} \int_{\text{boundary}} \frac{f_i(\bar{r}_s, t)}{k_i} \, ds = 0$$

which is written in the form

$$\bar{T}_{00}(\lambda_m, t) = \frac{\bar{g}(\lambda_m, t)}{\lambda_m^2 k} \tag{1-72a}$$

$$\bar{T}_{0j}(\lambda_m, t) = \int_{s_j} \frac{K(\lambda_m, \bar{r}_j')}{\lambda_m^2\ k_j} \cdot f_j(\bar{r}_j', t) \cdot ds_j \qquad j = 1, 2, 3, \ldots, s \tag{1-72b}$$

Inverting the transforms 1-72 by the inversion formula 1-70b, the solution of the quasi-steady-state system (1-69) is given in the form

$$T_{00}(\bar{r}, t) = \frac{1}{k} \sum_{m=1}^{\infty} \frac{1}{\lambda_m^2} K(\lambda_m, \bar{r}) \cdot \bar{g}(\lambda_m, t) \tag{1-73a}$$

$$T_{0j}(\bar{r}, t) = \sum_{m=1}^{\infty} \frac{K(\lambda_m, \bar{r})}{\lambda_m^2} \cdot \int_{s_j} \frac{K(\lambda_m, \bar{r}_j')}{k_j} \cdot f_j(\bar{r}_j', t) \cdot ds_j \tag{1-73b}$$

We shall now express the alternate solution 1-68 of system 1-53 in terms of $T_{0j}(\bar{r}, t)$ functions; the procedure is as follows.

From the transform 1-71 we obtain

$$\lambda_m^2 \sum_{j=0}^{s} \bar{T}_{0j}(\lambda_m, t) = \frac{\bar{g}(\lambda_m, t)}{k} + \sum_{i=1}^{s} \int_{s_i} \frac{K(\lambda_m, \bar{r}_i')}{k_i} f_i(\bar{r}_i', t) ds_i \tag{1-74}$$

By comparing the right-hand side of Eq. 1-74 with Eq. 1-65b, we note that it is equal to $A(\lambda_m, t)/\alpha$, hence Eq. 1-74 is written in the form

$$\alpha\lambda_m^2 \sum_{j=0}^{s} \bar{T}_{0j}(\lambda_m, t) = A(\lambda_m, t) \tag{1-75}$$

Substituting Eq. 1-75 into Eq. 1-67,

$$\bar{T}(\lambda_m, t) = e^{-\alpha\lambda_m^2 t}\left[\bar{F}(\lambda_m) + \sum_{j=0}^{s} \int_{t'=0}^{t} \alpha\lambda_m^2 \cdot e^{\alpha\lambda_m^2 t'} \cdot \bar{T}_{0j}(\lambda_m, t') \cdot dt'\right] \tag{1-76}$$

The integral on the right-hand side of Eq. 1-76 is evaluated by parts[4]

---

[4]
$$\begin{aligned}\int_0^t \bar{T}_{0j}(\lambda_m, t') \cdot (\alpha\lambda_m^2 e^{\alpha\lambda_m^2 t'}) dt' &= \int_0^t \bar{T}_{0j}(\lambda_m, t') \cdot (d e^{\alpha\lambda_m^2 t'}) \\ &= \bar{T}_{0j}(\lambda_m, t') \cdot e^{\alpha\lambda_m^2 t'}\Big|_0^t - \int_0^t e^{\alpha\lambda_m^2 t'} \cdot \dot{\bar{T}}_{0j}(\lambda_m, t') \cdot dt' \\ &= e^{\alpha\lambda_m^2 t} \cdot \bar{T}_{0j}(\lambda_m, t) - \bar{T}_{0j}(\lambda_m, 0) - \int_0^t e^{\alpha\lambda_m^2 t'} \cdot \dot{\bar{T}}_{0j}(\lambda_m, t') \cdot dt'\end{aligned}$$

$$\bar{T}(\lambda_m, t) = e^{-\alpha\lambda_m^2 t}\left[\bar{F}(\lambda_m) + \sum_{j=0}^{s} e^{\alpha\lambda_m^2 t} \cdot \bar{T}_{0j}(\lambda_m, t) - \sum_{j=0}^{s} \bar{T}_{0j}(\lambda_m, 0) - \sum_{j=0}^{s} \int_{t'=0}^{t} e^{\alpha\lambda_m^2 t'} \cdot \dot{\bar{T}}_{0j}(\lambda_m, t') \cdot dt'\right]$$

After rearranging,

$$\bar{T}(\lambda_m, t) - \sum_{j=0}^{s} \bar{T}_{0j}(\lambda_m, t) = e^{-\alpha\lambda_m^2 t}\left[\bar{F}(\lambda_m) - \sum_{j=0}^{s} \bar{T}_{0j}(\lambda_m, 0) - \sum_{j=0}^{s} \int_{t'=0}^{t} e^{\alpha\lambda_m^2 t'} \cdot \dot{\bar{T}}_{0j}(\lambda_m, t') \cdot dt'\right] \quad (1\text{-}77)$$

where dots denote derivative with respect to the t'-variable. The inversion formulas 1-59b and 1-70b are combined to give the following inversion formula.

$$T(\bar{r}, t) = \sum_{j=0}^{s} T_{0j}(\bar{r}, t) + \sum_{m=1}^{\infty} K(\lambda_m, \bar{r}) \cdot \left[\bar{T}(\lambda_m, t) - \sum_{j=0}^{s} \bar{T}_{0j}(\lambda_m, t)\right] \quad (1\text{-}78)$$

Inverting Eq. 1-77 by the inversion formula 1-78 we obtain

$$T(\bar{r}, t) = \sum_{j=0}^{s} T_{0j}(\bar{r}, t) + \sum_{m=1}^{\infty} e^{-\alpha\lambda_m^2 t} \cdot K(\lambda_m, \bar{r})\left[\bar{F}(\lambda_m) - \sum_{j=0}^{s} \bar{T}_{0j}(\lambda_m, 0)\right] - \sum_{j=0}^{s}\left[\sum_{m=1}^{\infty} e^{-\alpha\lambda_m^2 t} K(\lambda_m, \bar{r}) \cdot \int_{t'=0}^{t} e^{\alpha\lambda_m^2 t'} \cdot \dot{\bar{T}}_{0j}(\lambda_m, t') \cdot dt'\right] \quad (1\text{-}79)$$

where

$$\bar{F}(\lambda_m) = \int_R K(\lambda_m, \bar{r}') \cdot F(\bar{r}') \cdot d\bar{r}'$$

$$\bar{T}_{0j}(\lambda_m, 0) = \int_R K(\lambda_m, \bar{r}') \cdot T_{0j}(\bar{r}', 0) \cdot d\bar{r}'$$

$$\dot{\bar{T}}_{0j}(\lambda_m, t') = \int_R K(\lambda_m, \bar{r}') \cdot \dot{T}_{0j}(\bar{r}', t') \cdot dr'$$

Equation 1-79 is the alternate form of the solution 1-68 of the boundary-value problem of heat conduction, Eq. 1-53.

Equation 1-79 can be expressed in another form that includes explicitly the heat-generation and the boundary-condition functions. Such a result is obtained by substituting from Eq. 1-71 into Eq. 1-79 the equivalent expressions for the terms

$$\sum_{j=0}^{s} \bar{T}_{0j}(\lambda_m, 0) \qquad \text{and} \qquad \sum_{j=0}^{s} \dot{\bar{T}}_{0j}(\lambda_m, t')$$

$$T(\bar{r},t) = \sum_{j=0}^{s} T_{0j}(\bar{r},t) + \sum_{m=1}^{\infty} e^{-\alpha\lambda_m^2 t}\cdot K(\lambda_m,\bar{r})\left\{\bar{F}(\lambda_m) - \frac{1}{\lambda_m^2}\right.$$

$$\left[\frac{\bar{g}(\lambda_m,0)}{k} + \sum_{i=1}^{s}\int_{s_i}\frac{K(\lambda_m,\bar{r}_i')}{k_i} f_i(\bar{r}',0)\cdot ds_i\right]$$

$$\left. - \frac{1}{\lambda_m^2}\int_0^t e^{\alpha\lambda_m^2 t'}\left[\frac{\dot{\bar{g}}(\lambda_m,t')}{k} + \sum_{i=1}^{s}\int_{s_i}\frac{K(\lambda_m,\bar{r}_i')}{k_i}\cdot \dot{f}_i(\bar{r}_s,t')ds'\right]dt'\right\} \quad (1\text{-}80)$$

Equation 1-80 may be preferred to Eq. 1-79 whenever evaluation of the surface integrals are easier than that of the volume integrals.

The alternate solutions of boundary-value problem of heat conduction Eq. (1-53) given by Eqs. 1-79 and 1-80 are uniformly convergent provided that the functions $F(\bar{r})$, $g(\bar{r},t)$ and $f_i(\bar{r}_s,t)$ possess continuous first and second partial derivatives in the space variables, and that $g(\bar{r},t)$ and $f_i(\bar{r},t)$ functions possess continuous first-order partial derivatives with respect to t.

Extension to the Case of Space-Dependent Thermal Conductivity. The finite integral-transform technique discussed above can be extended to the case of space-dependent thermal conductivity [12, 13]. Consider the boundary-value problem of heat conduction for a finite region R in which the thermal conductivity is a function of position:

$$\nabla\cdot[k(\bar{r})\nabla T(\bar{r},t)] + g(\bar{r},t) = \rho C_p \frac{\partial T(\bar{r},t)}{\partial t} \qquad \text{in region } R,\ t>0 \qquad (1\text{-}81a)$$

$$k_i(\bar{r})\frac{\partial T(\bar{r}_i)}{\partial n_i} + h_i T(\bar{r}_i) = f_i(\bar{r}_i) \qquad \text{on boundary } s_i,\ t>0 \qquad (1\text{-}81b)$$

$$T(\bar{r},t) = F(\bar{r}) \qquad \text{in region } R,\ t=0 \qquad (1\text{-}81c)$$

We consider the following *auxiliary* eigenvalue problem for the space variable $\psi(\bar{r})$ in the region R.

$$\nabla\cdot[k(\bar{r})\nabla\psi(\bar{r})] + \lambda^2\psi(\bar{r}) = 0 \qquad \text{in region } R \qquad (1\text{-}82a)$$

$$k_i(\bar{r})\frac{\partial\psi(\bar{r}_i)}{\partial n_i} + h_i\psi(\bar{r}) = 0 \qquad \text{on boundary } s_i \qquad (1\text{-}82b)$$

Let $K(\lambda_m,\bar{r})$ be the normalized eigenfunctions and λ_m's the eigenvalues of the eigenvalue problem 1-82. Taking $K(\lambda_m,\bar{r})$ as the kernel we define the integral transform and the inversion formula with respect to the space variable $\bar{r}$ of the temperature function $T(\bar{r},t)$ as given by Eqs. 1-59a and 1-59b. We take the integral transform of the heat conduction problem given by Eq. 1-81 by applying the integral transform just defined. The procedure of taking the integral transform is similar to that described previously, except the transform to the term

$$\nabla \cdot [k(\bar{r}) \nabla T(\bar{r})]$$

The integral transform of $\nabla \cdot k(\bar{r}) \nabla T(\bar{r})$ is involved for the problem 1-81 is given as:

$$\int_R K_m \nabla \cdot [k(\bar{r}) \nabla T d\bar{r}] = -\lambda_m^2 \bar{T}(\lambda_m, t) + \sum_{i=1}^{s} \int_{s_i} k(\bar{r}_i) \left[\frac{K(\lambda_m, \bar{r}_i)}{k_i}\right] \cdot f_i(\bar{r}_i, t) ds_i \qquad (1\text{-}83)$$

When the boundary condition for the surface i is of the first kind, in Eq. 1-83 $\left[\frac{K(\lambda_m, \bar{r}_i)}{k_i}\right]$ should be replaced by

$$-\frac{1}{h_i} \frac{\partial K(\lambda_m, \bar{r}_i)}{\partial n_i}$$

In deriving the above results it is assumed that T, K_m, $k(\bar{r}) \nabla T$ and $k(\bar{r}) \nabla K_n$ are continuous and differentiable.

1-10. SPLITTING UP OF HEAT-CONDUCTION PROBLEM INTO SIMPLER PROBLEMS

Since it is generally easier to solve a simpler problem, splitting up of a boundary-value problem of heat conduction has been frequently applied. Consider the boundary-value problem of heat conduction in the form

$$\nabla^2 T(\bar{r}, t) + \frac{g(\bar{r}, t)}{k} = \frac{1}{\alpha} \frac{\partial T(\bar{r}, t)}{\partial t} \quad \text{in } R,\ t > 0 \qquad (1\text{-}84a)$$

$$k_i \frac{\partial T(\bar{r}_i, t)}{\partial n_i} + h_i T(\bar{r}_i, t) = f_i(\bar{r}, t) \quad \text{on } S_i,\ t > 0 \qquad (1\text{-}84b)$$

$$T(\bar{r}, t) = F(\bar{r}) \quad \text{in } R,\ t = 0 \qquad (1\text{-}84c)$$

where $i = 1, 2, \ldots, s$ number of continuous bounding surfaces of the solid

$\frac{\partial}{\partial n_i}$ = differentiation along the outward-drawn normal to the boundary surface s_i

The alternate solution of this system, given by Eq. 1-79, is written in the form

$$T(\bar{r}, t) = \sum_{j=0}^{s} T_{0j}(\bar{r}, t) + T_1(\bar{r}, t) - \sum_{j=0}^{s} T_{2j}(\bar{r}, t) \qquad (1\text{-}85)$$

where

$$T_1(\bar{r}, t) = \sum_{m=1}^{\infty} e^{-\alpha\lambda_m^2 t} \cdot K(\lambda_m, \bar{r}) \cdot \left[\bar{F}(\lambda_m) - \sum_{j=0}^{s} \bar{T}_{0j}(\lambda_m, 0)\right] \tag{1-86}$$

$$T_{2j}(\bar{r}, t) = \sum_{m=1}^{\infty} e^{-\alpha\lambda_m^2 t} \cdot K(\lambda_m, \bar{r}) \cdot \int_{t'=0}^{t} e^{\alpha\lambda_m^2 t'} \cdot \dot{\bar{T}}_{0j}(\lambda_m, t') \cdot dt' \tag{1-87}$$

Equation 1-85 implies that the solution of system 1-84 consists of three simpler solutions described by the temperature functions $T_{0j}(\bar{r}, t)$, $T_1(\bar{r}, t)$ and $T_{2j}(\bar{r}, t)$. We examine the heat-conduction problems satisfied by these three simpler solutions.

We have seen that $T_{0j}(\bar{r}, t)$ functions are given by the Eq. 1-73 and they satisfy the quasi-steady-state problem 1-69.

It is easily seen that the $T_1(\bar{r}, t)$ function as defined by Eq. 1-86 is the solution of the following homogeneous system.

$$\nabla^2 T_1(\bar{r}, t) = \frac{1}{\alpha} \frac{\partial T_1(\bar{r}, t)}{\partial t} \quad \text{in region } R,\ t > 0 \tag{1-88a}$$

$$k_i \frac{\partial T_1(\bar{r}, t)}{\partial n_i} + h_i T_1(\bar{r}, t) = 0 \quad \text{on boundary } s_i,\ t > 0 \tag{1-88b}$$

$$T_1(\bar{r}, t) = F(\bar{r}) - \sum_{j=0}^{s} T_{0j}(\bar{r}, 0) \quad \text{in region } R,\ t = 0 \tag{1-88c}$$

$T_{2j}(\bar{r}, t)$ functions as defined by Eq. 1-87 will be related to the solution of the following homogeneous system:

$$\nabla^2 \theta_j(\bar{r}, \tau, t) = \frac{1}{\alpha} \frac{\partial \theta_j(\bar{r}, \tau, t)}{\partial t} \quad \text{in region } R,\ t > 0 \tag{1-89a}$$

$$k_i \frac{\partial \theta_j(\bar{r}, \tau, t)}{\partial n_i} + h_i \theta_j(\bar{r}, \tau, t) = 0 \quad \text{on boundary } S_i,\ t > 0 \tag{1-89b}$$

$$\theta_j(\bar{r}, \tau, t) = T_{0j}(\bar{r}, \tau) \quad \text{in region } R,\ t = 0 \tag{1-89c}$$

The solution of system 1-89 is given in the form

$$\theta_j(\bar{r}, \tau, t) = \sum_{m=1}^{\infty} e^{-\alpha\lambda_m^2 t} \cdot K(\lambda_m, \bar{r}) \cdot \overline{T}_{0j}(\bar{r}, \tau) \tag{1-90}$$

The solution 1-90 is related to the $T_{2j}(\bar{r}, t)$ functions with the following considerations.

$$\left.\frac{\partial \theta_j(\bar{r}, \tau', t - \tau)}{\partial \tau'}\right|_{\tau' = \tau} = \sum_{m=1}^{\infty} e^{-\alpha\lambda_m^2 (t-\tau)} \cdot K(\lambda_m, \bar{r}) \cdot \dot{\bar{T}}_{0j}(\bar{r}, \tau)$$

or

$$\int_{\tau=0}^{t} \left. \frac{\partial \theta_j(\bar{r}, \tau', t - \tau)}{\partial \tau'} \right|_{\tau'=\tau} \cdot d\tau = \sum_{m=1}^{\infty} e^{-\alpha \lambda_m^2 t} \cdot K(\lambda_m, \bar{r}) \int_{\tau=0}^{t} e^{\alpha \lambda_m^2 \tau} \cdot \dot{\bar{T}}_{0j}(\bar{r}, \tau) \cdot d\tau$$

$$= T_{2j}(\bar{r}, t) \tag{1-91}$$

Hence, $\theta_j(\bar{r}, \tau, t)$ functions are related to $T_{2j}(\bar{r}, t)$ functions. The solution is now written in the form

$$T(\bar{r}, t) = T_1(\bar{r}, t) + \sum_{j=0}^{s} T_{0j}(\bar{r}, t) - \sum_{j=0}^{s} \int_{\tau=0}^{t} \left[\frac{\partial \theta_j(\bar{r}, \tau', t - \tau)}{\partial \tau'} \right] \cdot d\tau \tag{1-92}$$

where $T_{0j}(\bar{r}, t)$, $T_1(\bar{r}, t)$, and $\theta_j(\bar{r}, \tau, t)$ functions are respectively the solutions of the simpler heat-conduction problems described by the systems 1-69, 1-88, and 1-89.

The Special Case of Time-Independent Heat-Generation and Boundary Conditions. We now examine the splitting up of boundary-value problems of heat conduction for the special case of time-independent heat-generation and boundary conditions.

The problem under consideration is given in the form

$$\nabla^2 T(\bar{r}, t) + \frac{g(\bar{r})}{k} = \frac{1}{\alpha} \frac{\partial T(\bar{r}, t)}{\partial t} \qquad \text{in region } R,\ t > 0 \tag{1-93a}$$

$$k_i \frac{\partial T(\bar{r}_i, t)}{\partial n_i} + h_i T(\bar{r}_i, t) = f_i(\bar{r}_i) \qquad \text{on the boundary } s_i,\ t > 0 \tag{1-93b}$$

$$T(\bar{r}, t) = F(\bar{r}) \qquad \text{in region } R,\ t = 0 \tag{1-93c}$$

where $i = 1, 2, \ldots, s$ number of continuous bounding surfaces of the solid;

$\frac{\partial}{\partial n_i}$ = differentiation along the outward-drawn normal to the bounding surface s_i

In this case $T_{2j}(\bar{r}, t)$ functions vanish because they are related to T_{oj} functions by Eq. 1-87, and for time independent heat generation and boundary condition functions, T_{oj} functions are time independent (See: Eq. 1-69) and their derivative with respect to the t'-variable is zero. Hence the solution of system 1-93 from Eq. 1-85 is written in the form

$$T(\bar{r}, t) = \sum_{j=0}^{s} T_{0j}(\bar{r}) + T_1(\bar{r}, t) \tag{1-94}$$

where $T_{0j}(\bar{r})$ functions are the solution of the following steady-state problem.

$$\nabla^2 T_{0j}(\bar{r}) + \delta_{0j} \frac{g(\bar{r})}{k} = 0 \qquad \text{in region } R \tag{1-95a}$$

$$k_i \frac{\partial T_{0j}(\bar{r})}{\partial n_i} + h_i T_{0j}(\bar{r}) = \delta_{ij} f_i(\bar{r}) \qquad \text{on the boundary } s_i \tag{1-95b}$$

where $i = 1, 2, s$ the number of continuous bounding surfaces of the solid

$j = 0, 1, 2, \ldots, s$

$$\delta_{ij} = \text{Kroneker delta} = \begin{cases} 0 \text{ for } i \neq j \\ 1 \text{ for } i = j \end{cases}$$

Here we assume further that all h_i's do not simultaneously vanish.

$T_1(\bar{r}, t)$ functions satisfy the following homogeneous problem.

$$\nabla^2 T_1(\bar{r}, t) = \frac{1}{\alpha} \frac{\partial T_1(\bar{r}, t)}{\partial t} \qquad \text{in region } R,\ t > 0 \tag{1-96a}$$

$$k_i \frac{\partial T_1(\bar{r}_i, t)}{\partial n_i} + h_i T_1(\bar{r}_i, t) = 0 \qquad \text{on boundary } s_i,\ t > 0 \tag{1-96b}$$

$$T_1(\bar{r}, t) = F(\bar{r}) - \sum_{j=0}^{s} T_{0j}(\bar{r}) \qquad \text{in region } R,\ t = 0 \tag{1-96c}$$

Summarizing, we split up the nonhomogeneous boundary-value problem of heat conduction, 1-93, into (a) a set of steady-state equations given by Eq. 1-95, that consists of a Poisson's equation with homogeneous boundary conditions for $j = 0$, and a set of Laplace's equations for $j = 1, 2, \ldots, s$; (b) and a time-dependent homogeneous boundary-value problem of heat conduction as given by Eq. 1-96.

1-11. ORTHOGONALITY OF EIGENFUNCTIONS

In the process of solving the boundary value of heat conduction, a given function is expanded in an infinite series of eigenfunctions of a homogeneous problem. In determining the unknown coefficients in the series expansion, use is made of the orthogonality property of these eigenfunctions. The subject of orthogonality was investigated by *Sturm* and *Liouville* in 1836, and the homogeneous boundary-value problem investigated by them is called the *Sturm-Liouville* system.[14]

Consider the following Sturm-Liouville system:

$$\frac{d}{dx}\left[p(x) \frac{d\psi(\lambda, x)}{dx}\right] + [q(x) + \lambda \ w(x)]\psi(\lambda, x) = 0 \qquad \text{in } a \le x \le b \tag{1-97a}$$

$$A_1 \frac{d\psi(\lambda, x)}{dx} + A_2\psi(\lambda, x) = 0 \qquad \text{at } x = a$$

(1-97b)

$$B_1 \frac{d\psi(\lambda, x)}{dx} + B_2\psi(\lambda, x) = 0 \qquad \text{at } x = b$$

(1-97c)

where the functions $p(x)$, $q(x)$, $w(x)$ and $\frac{dp(x)}{dx}$ are assumed to be real valued, and continuous, and $p(x) > 0$ and $w(x) > 0$ over the interval (a, b). The constants A_1, A_2, B_1, B_2 are real and independent of the parameter λ.

The homogeneous boundary-value problem 1-97 has solutions $\psi_n(x) = \psi(\lambda_n, x)$ only for certain values of the parameter $\lambda = \lambda_n$, and for other values of λ it has trivial solutions, i.e., $\psi(\lambda, x) = 0$. The nontrivial solutions $\psi_n(x)$ are the eigenfunctions and the corresponding parameters λ_n are the eigenvalues.

The eigenfunctions $\psi_n(x)$ of the Sturm-Liouville system constitute an orthogonal set with respect to the weighting function $w(x)$, in the interval (a, b), and this orthogonality condition is given as

$$\int_a^b w(x) \cdot \psi_m(x) \cdot \psi_n(x) dx = 0 \qquad \text{for } m \neq n \text{ (i.e., } \lambda_m \neq \lambda_n) \qquad (1\text{-}98)$$

To prove this result, let

$$L[\psi(\lambda, x)] \equiv \frac{d}{dx}\left[p(x) \frac{d\psi(\lambda, x)}{dx}\right] + q(x) \cdot \psi(\lambda, x) \qquad (1\text{-}99)$$

We then write Eq. 1-97a for any two eigenfunctions $\psi_n(x)$ and $\psi_m(x)$

$$L[\psi_m(x)] + \lambda_m w(x) \cdot \psi_m(x) = 0 \qquad (1\text{-}100a)$$

$$L[\psi_n(x)] + \lambda_n w(x) \cdot \psi_n(x) = 0 \qquad (1\text{-}100b)$$

We multiply Eq. 1-100a by $\psi_n(x)$ and Eq. 1-100b by $\psi_m(x)$, subtract the resulting equations, and obtain

$$\psi_n \cdot L(\psi_m) - \psi_m \cdot L(\psi_n) = (\lambda_n - \lambda_m) \cdot w \cdot \psi_m \cdot \psi_n \qquad (1\text{-}101)$$

Replacing $L(\psi)$'s by the expression as given by Eq. 1-99, we obtain

$$\frac{d}{dx}[p(\psi_n \cdot \psi_m' - \psi_m \cdot \psi_n')] = (\lambda_n - \lambda_m) \cdot w \cdot \psi_m \cdot \psi_n \qquad (1\text{-}102)$$

When both sides of Eq. 1-102 are integrated with respect to x from $x = a$ to $x = b$, the left-hand side vanishes in view of the homogeneous boundary conditions Eqs. 1-97b and 1-97c, and obtain

$$(\lambda_n - \lambda_m) \int_a^b w(x) \cdot \psi_m(x) \cdot \psi_n(x) \cdot dx = 0 \qquad (1\text{-}103)$$

For $\lambda_n \neq \lambda_m$, Eq. 1-103 is satisfied if

$$\int_a^b w(x) \cdot \psi_m(x) \cdot \psi_n(x) \cdot dx = 0 \qquad \text{for } \lambda_n \neq \lambda_m \tag{1-104}$$

which proves that eigenfunctions of the Sturm-Liouville system are orthogonal with respect to the weighting function $w(x)$ in the interval (a,b).

1-12. TRANSFORMATION OF COORDINATES

In the solution of heat-conduction problems we choose a coordinate system that fits the boundary surfaces of the solid. Rectangular bodies require rectangular coordinate systems; circular cylinder, cylindrical coordinate systems; and so on. When solving the heat-conduction problem in a coordinate system other than the cartesian coordinates, the differential equation of heat conduction should be transformed to the coordinate system under consideration. In this respect we limit ourselves to the transformation from the cartesian to an orthogonal curvilinear coordinate system, since skew coordinate systems do not permit the separation of the heat-conduction equation.

Consider a cartesian coordinate system (x_1, x_2, x_3) and an orthogonal curvilinear coordinate system (u_1, u_2, u_3). A differential length ds in the cartesian coordinate system is given as

$$ds^2 = dx_1^2 + dx_2^2 + dx_3^2 \tag{1-105}$$

Let, the functional relationship between the two systems be

$$x_i = x_i(u_1, u_2, u_3) \qquad i = 1, 2, 3 \tag{1-106}$$

A differential length dx_i along the x_i axis is related to the coordinates (u_1, u_2, u_3) by

$$dx_i = \sum_{k=1}^{3} \frac{\partial x_i}{\partial u_k} du_k \tag{1-107}$$

Substituting Eq. 1-107 into Eq. 1-105 and rearranging, we obtain

$$ds^2 = a_1^2 du_1^2 + a_2^2 du_2^2 + a_3^2 du_3^2 \tag{1-108}$$

where

$$a_i^2 = \sum_{k=1}^{3} \left(\frac{\partial x_k}{\partial u_i} \right)^2 \qquad i = 1, 2, 3 \tag{1-109}$$

The coefficients a_i are called *scale factors*, which may be constants or functions of the coordinates. When the functional relationship between the coordinates of the cartesian and the orthogonal curvilinear coordinate system is given, the scale factors a_i are evaluated using Eq. 1-109. Once the scale factors are known, transformation of various operators such as gradient, divergence, and Laplacian from the cartesian coordinate system to an orthogonal coordinate system is made using the following relationships. For proof of these relationships the reader may refer to any standard text on mathematics.

Gradient. Consider a scalar quantity, say temperature function T. The gradient of T in the orthogonal curvilinear coordinate system (u_1, u_2, u_3) is given by the following relationship

$$\nabla T(u_1, u_2, u_3) = \sum_{i=1}^{3} \hat{\mathbf{u}}_i \cdot \frac{1}{a_i} \frac{\partial T(u_1, u_2, u_3)}{\partial u_i} \tag{1-110}$$

where $\hat{\mathbf{u}}_i$'s are the unit direction vectors in the (u_1, u_2, u_3) system in the u_1, u_2, u_3 directions.

Divergence. Consider a vector quantity, say, heat-flux vector $\bar{\mathbf{q}}$. Divergence of $\bar{\mathbf{q}}$ in the (u_1, u_2, u_3) system is given

$$\nabla \cdot \bar{\mathbf{q}}(u_1, u_2, u_3) = \frac{1}{a_1 a_2 a_3} \sum_{i=1}^{3} \frac{\partial}{\partial u_i}\left(\frac{a_1 a_2 a_3}{a_i} q_i\right) \tag{1-111}$$

where q_i's are the components of the heat-flux vector along $\hat{\mathbf{u}}_i$ directions in the (u_1, u_2, u_3) system.

Laplacian. Consider the Laplacian of a scalar quantity, say, of the temperature function T. In the (u_1, u_2, u_3) coordinate system it is given by

$$\nabla^2 T(u_1, u_2, u_3) = \frac{1}{a_1 a_2 a_3} \sum_{i=1}^{3} \frac{\partial}{\partial u_i}\left(\frac{a_1 a_2 a_3}{a_i^2} \frac{\partial T(u_1, u_2, u_3)}{\partial u_i}\right) \tag{1-112}$$

EXAMPLE 1-1. Consider transformation from the cartesian coordinate system (x, y, z) to the cylindrical coordinate system (r, ϕ, z), shown in Fig. 1-5. The functional relationship between the coordinates is given by

$$x = r\cos\phi \qquad y = r\sin\phi \qquad z = z \tag{1-113}$$

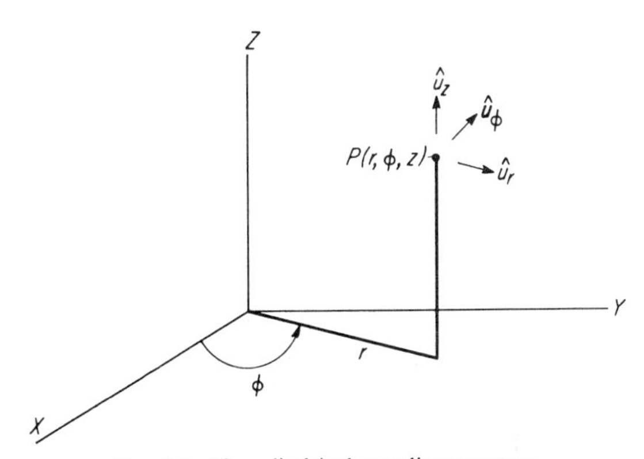

FIG. 1-5. The cylindrical coordinate system.

Let a_r, a_ϕ, a_z be the scale factors for the (r, ϕ, z) system. They are evaluated by using Eq. 1-109 as follows:

$$a_r^2 = \left(\frac{\partial x}{\partial r}\right)^2 + \left(\frac{\partial y}{\partial r}\right)^2 + \left(\frac{\partial z}{\partial r}\right)^2 = \cos^2\phi + \sin^2\phi + 0 = 1$$

$$a_\phi^2 = \left(\frac{\partial x}{\partial \phi}\right)^2 + \left(\frac{\partial y}{\partial \phi}\right)^2 + \left(\frac{\partial z}{\partial \phi}\right)^2 = (-r\sin\phi)^2 + (r\cos\phi)^2 + 0 = r^2$$

$$a_z^2 = \left(\frac{\partial x}{\partial z}\right)^2 + \left(\frac{\partial y}{\partial z}\right)^2 + \left(\frac{\partial z}{\partial z}\right)^2 = 0 + 0 + 1 = 1$$

Hence the scale factors for the cylindrical coordinate system are

$$a_r = 1 \qquad a_\phi = r \qquad a_z = 1 \tag{1-114}$$

The Laplacian in the cylindrical coordinate system, from Eq. 1-112, is given by

$$\nabla^2 T(r, \phi, z) = \frac{1}{r}\frac{\partial}{\partial r}\left(r\frac{\partial T}{\partial r}\right) + \frac{1}{r^2}\frac{\partial^2 T}{\partial \phi^2} + \frac{\partial^2 T}{\partial z^2} \tag{1-115}$$

EXAMPLE 1-2. Figure 1-6 shows a spherical coordinate system (r, ϕ, ψ). The functional relationship between the coordinates is

$$x = r\sin\psi\cdot\cos\phi \qquad y = r\sin\psi\cdot\sin\phi \qquad z = r\cos\psi \tag{1-116}$$

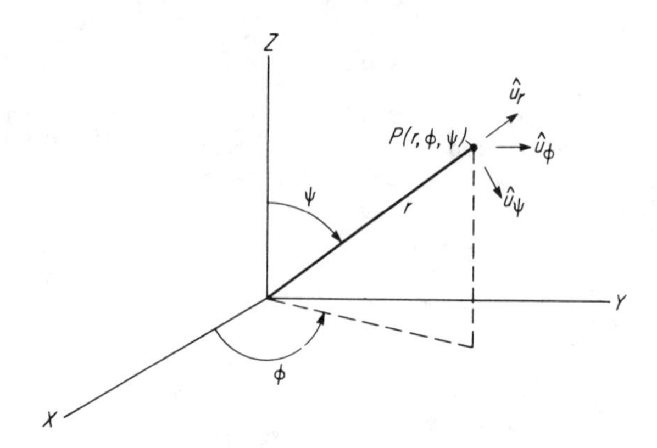

FIG. 1-6. The spherical coordinate system.

The scale factors become

$$a_r = 1 \qquad a_\phi = r\sin\psi \qquad a_\psi = r \tag{1-117}$$

and the Laplacian of temperature in the spherical coordinate system is given by

$$\nabla^2 T(r, \phi, \psi) = \frac{1}{r^2}\frac{\partial}{\partial r}\left(r^2\frac{\partial T}{\partial r}\right) + \frac{1}{r^2\sin\psi}\frac{\partial}{\partial\psi}\left(\sin\psi\frac{\partial T}{\partial\psi}\right) + \frac{1}{r^2\sin^2\psi}\frac{\partial^2 T}{\partial\phi^2} \tag{1-118}$$

1-13. THERMAL PROPERTIES

Thermal diffusivity and thermal conductivity are two important thermal properties that enter the differential equation of heat conduction. Accuracy of the values chosen for these properties affects the accuracy of the results in the heat-conduction problems. There is a tremendously wide difference in the thermal conductivities of various engineering materials, as shown in Fig. 1-7. The highest value is given by pure metals and the lowest value by gases and vapors; the amorphous insulating materials and inorganic liquids have thermal conductivities that lie between them. Thermal conductivity varies by an order of magnitude in each of these groups. For metals it varies from about 240 Btu/hr ft^2

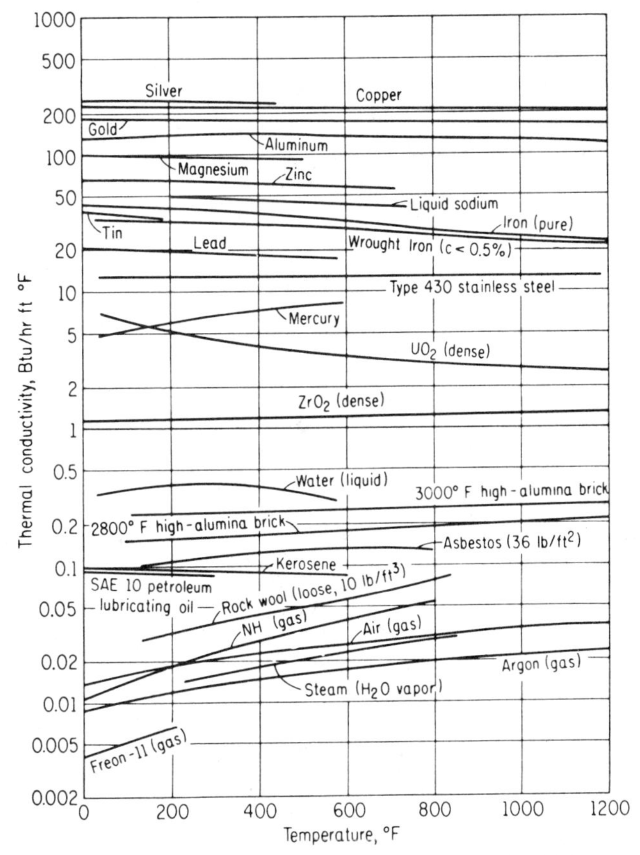

FIG. 1-7. Effects of temperature on the thermal conductivities of engineering materials. (Data from a number of sources, including Refs. 16, 17, and 18.)

(°F/ft) for pure copper to about 10 Btu/hr ft^2 (°F/ft) for chrome-nickel steel; for nonmetallic liquids from about 0.4 Btu/hr ft^2 (°F/ft) for water to about 0.04 Btu/hr ft^2 (°F/ft) for Freon-114; for gases and vapors it may be as low as 0.004 Btu/hr ft^2 (°F/ft) for Freon vapor.

Thermal conductivity of metals varies with temperature. In most heat-conduction problems, however, to avoid nonlinearity, thermal conductivity is usually assumed temperature independent. This assumption is considered reasonable provided that the variation of thermal conductivity with temperature over the temperature range under consideration is not great. When the variation is large, a mean thermal conductivity defined as

$$k_m = \frac{1}{T_2 - T_1} \int_{T_1}^{T_2} k\, dT \tag{1-119}$$

should be used, if T_1 and T_2 are the temperature limits.

For pure metals thermal conductivity decreases with temperature, for gases and insulating materials it increases with temperature. For water and for some other liquids it first increases, then decreases with temperature. At low temperatures variation of thermal conductivity with temperature is very large. Figure 1-8 shows thermal conductivity of metals at low temperatures.

Structure of the material also affects thermal conductivity. Metallic single crystals may have much higher thermal conductivities; for instance, with copper crystals, values of about 5000 Btu/hr ft^2 (°F/ft) and even higher are possible [15].

Thermal conductivity of metals depends very much on their purity; small amounts of impurity usually cause a large reduction in the thermal conductivity of pure metals. Exposure to fast-neutron radiation may reduce thermal conductivity of metals and ceramics by a factor of 2 or more [16].

Table 1-2 gives thermal diffusivities of various engineering materials; there are wide differences in the thermal diffusivities between materials. The physical significance of thermal diffusivity is associated with the speed of propagation of temperature into the solid in transient heat-conduction problems. The higher the thermal diffusivity, the faster the propagation of temperature into the solid. This statement is better illustrated if we refer to a specific example, say, transient cooling of a semi-infinite solid (in region $x > 0$), which is initially at a constant uniform temperature T_0, and for times $t > 0$ the boundary surface at $x = 0$ is kept at zero temperature. The temperature within the solid will decrease with time and is given by the relation (See: Eq. 2-116)

$$\frac{T(x,t)}{T_0} = erf\left(\frac{x}{\sqrt{4\alpha t}}\right) \tag{1-120}$$

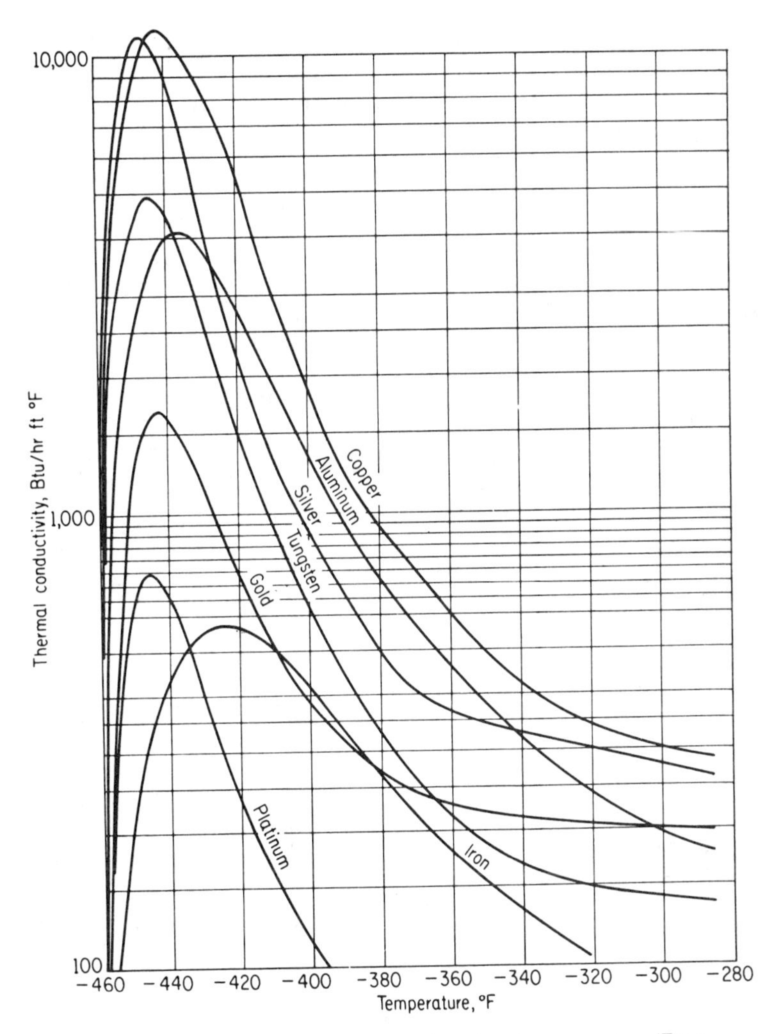

FIG. 1-8. Thermal conductivity of metals at low temperatures. (From Ref. 18.)

Consider a region at a depth, say, one foot from the boundary surface (i.e. $x = 1$ ft). The time required for the temperature to decrease to one-half of the initial temperature (i.e. $T = \frac{1}{2}T_0$) is given for different materials as

	Silver	Copper	Steel	Glass
α ft^2/hr	6.6	4.0	0.5	0.023
Time	9.5 min	16.5 min	2.2 hr	2.00 days

It is apparent from these results that the larger the thermal diffusivity, the shorter the time required for the temperature to penetrate a given distance in the solid. Conversely, $1/\alpha$ is a measure of the time required for the temperature to penetrate into the solid.

TABLE 1-2

THERMAL PROPERTIES OF ENGINEERING MATERIALS

Material	Average Temperature, °F	k, Btu/hr ft °F	c_p, Btu/lb °F	ρ, lb/ft³	α, ft²/hr
Metals					
Aluminum	32	117	0.208	169	3.33
Copper	32	224	0.091	558	4.42
Gold	68	169	0.030	1204	4.68
Iron, pure	32	36	0.104	491	0.70
Cast iron ($c \simeq 4\%$)	68	30	0.10	454	0.66
Lead	70	20	0.030	705	0.95
Mercury	32	4.83	0.033	849	0.172
Nickel	32	34.4	0.103	555	0.60
Silver	32	242	0.056	655	6.60
Steel, mild	32	26	0.11	490	0.48
Tungsten	32	92	0.032	1204	2.39
Zinc	32	65	0.091	446	1.60
Nonmetals					
Asbestos	32	0.087	0.25	36	0.010
Brick, fire clay	400	0.58	0.20	144	0.020
Cork, ground	100	0.024	0.48	8	0.006
Glass, Pyrex		0.68	0.20	150	0.023
Granite	32	1.6	0.19	168	0.050
Ice	32	1.28	0.49	57	0.046
Oak, across grain	85	0.111	0.41	44	0.0062
Pine, across grain	85	0.092	0.42	37	0.0059
Quartz sand, dry		0.15	0.19	103	0.008
Rubber, soft		0.10	0.45	69	0.003
Water	32	0.32	1.01	62.4	0.005

REFERENCES

1. J. B. Fourier, *Theorie analytique de la chaleur*, Paris, 1822. (English translation by A. Freeman, Dover Publications, Inc., New York, 1955.)

2. John W. Dettman, *Mathematical Methods in Physics and Engineering*, McGraw-Hill Book Company, New York, 1962, p. 137.

3. Parry Moon and Domina Eberle Spencer, *Field Theory for Engineers*, D. Van Nostrand Company, Inc., Princeton, N.J., 1961, p. 158.

4. M. Philip Morse and H. Feshbach, *Methods of Theoretical Physics*, Part I, McGraw-Hill Book Company, New York, 1953, p. 513.

5. R. Courant and D. Hilbert, *Methods of Mathematical Physics*, Vol. I, Interscience Publishers, Inc., New York, 1953, p. 312.

6. Ian N. Sneddon, *Fourier Transforms*, McGraw-Hill Book Company, New York, 1951.

———. *Introduction to the Use of Integral Transforms*, McGraw-Hill Book Company, New York, (In press).

7. C. J. Tranter, *Integral Transforms in Mathematical Physics*, John Wiley and Sons, Inc., New York, 1962.

8. V. A. Ditkin and A. P. Prudnikov, *Integral Transforms and Operational Calculus*, Pergamon Press, New York, 1965.

9. E. C. Titchmarsh, *Fourier Integrals*, Clarendon Press, Oxford, 1962.

10. Nurettin Y. Ölçer, "On the Theory of Conductive Heat Transfer in Finite Regions," *Int. J. Heat Mass Transfer*, Vol. 7, 1964, pp. 307–314.

11. ———. "On the Theory of Conductive Heat Transfer in Finite Regions with Boundary Conditions of the Second Kind," *Int. J. Heat Mass Transfer*, Vol. 8, 1965, pp. 529–556.

12. Nurettin Y. Ölçer, "On a Class of Boundary-Initial Value Problems," *Osterr. Ing.-Arch*, Vol. 18, 1964, pp. 104–113.

13. S. Kaplan and G. Sonnemann, "The Helmholtz Transformation Theory and Application," Research Report No. 2 of the Department of Mechanical Engineering, University of Pittsburgh, 1961.

14. Ruel V. Churchill, *Operational Mathematics*, McGraw-Hill Book Company, New York, 1958, p. 264.

15. M. Jakob, *Heat Transfer*, Vol. I, John Wiley and Sons, Inc., New York, 1949, p. 111.

16. A. P. Fraas and M. N. Özışık, *Heat Exchanger Design*, John Wiley and Sons, Inc., New York, 1965.

17. E. R. G. Eckert and R. M. Drake, Jr., *Heat and Mass Transfer*, McGraw-Hill Book Company, New York, 1959, pp. 496–507.

18. R. W. Powell, C. Y. Ho, and P. E. Liley, "Thermal Conductivity of Selected Materials," NSRDS-NBS 8, U.S. Department of Commerce, National Bureau of Standards, 1966.

PROBLEMS

1. A rectangular region ($0 \le x \le a$, $0 \le y \le b$) is initially at a constant temperature T_0. For times $t > 0$ the sides at $x = 0$ and $y = 0$ are kept at zero temperatures while the sides $x = a$ and $y = b$ are subjected to convection (i.e., boundary condition of the third kind) with a medium at constant temperature T_∞. Thermal properties of the region are assumed to be uniform. (a) Formulate the boundary-value problem of heat conduction. (b) Split up the problem into simpler ones.

2. A right-angle solid cylinder with bounding surfaces at $r = a$, $z = 0$ and $z = b$, and with axis coinciding with the z-axis is initially at zero temperature. For times $t > 0$ heat is generated within the solid at a rate of $g(r)$ Btu/hr ft^3 and it is dissipated by convection from the bounding surfaces into a medium at zero temperature. Assume constant properties. (a) Formulate the boundary-value problem of heat conduction. (b) Split up the problem into simpler ones.

3. A solid hemisphere of radius $r = a$ has its flat surface coinciding with the xy plane and its center is situated at the origin. Initially the sphere is at zero temperature. For times $t > 0$ heat is generated within the solid at a constant rate of g Btu/hr ft^3, the plane surface of the sphere is kept at zero temperature and the hemispherical surface $r = a$ is subjected to convection with a medium at constant temperature T_∞. Assume constant properties. (a) Formulate the boundary-value problem of heat conduction. (b) Split up the problem into simpler ones.

4. Consider one-dimensional steady-state heat-conduction problem in a slab $0 \leq x \leq L$. The boundary surfaces at $x = 0$ and $x = L$ are kept at constant temperatures T_1 and T_2 respectively. Heat is generated within the solid at a rate of

$$g(x) = \left[1 - \left(\frac{x}{L}\right)^2\right] g_0 \quad \text{Btu/hr ft}^3$$

where g_0 = constant. Determine the steady-state temperature distribution within the solid. (Assume constant thermal properties.)

5. Determine the one-dimensional, steady-state temperature distribution in a long, solid cylinder of radius $r = b$, having an axis that coincides with the z-axis, for the following conditions. Heat is generated within the cylinder at a rate of

$$g(r) = \left[1 - \left(\frac{r}{b}\right)^2\right] g_0 \quad \text{Btu/hr ft}^3$$

where g_0 = constant and the boundary surface at $r = b$ is subjected to convection into a medium at zero temperature.

6. Consider the following one-dimensional, time-dependent boundary-value problem of heat conduction for a slab $0 \leq x \leq L$.

$$\frac{\partial^2 T}{\partial x^2} = \frac{1}{\alpha}\frac{\partial T}{\partial t} \quad \text{in} \quad 0 \leq x \leq L, \quad t > 0$$

$$T = 0 \quad \text{at} \quad x = 0, \quad t > 0$$

$$T = 0 \quad \text{at} \quad x = L, \quad t > 0$$

$$T = F(x) \quad \text{in} \quad 0 \leq x \leq L, \quad t = 0$$

Determine the transient temperature distribution in the slab: (a) by the method of separation of variables; (b) directly from the generalized solution given by Eq. 1-68. Determine the kernel $K(\lambda_m, x)$ and the eigenvalues λ_m by comparing the solution obtained from Eq. 1-68 with that obtained by the method of separation.

7. Discuss the physical significance of the dimensionless parameters *Biot number* and *Fourier number* in heat-conduction problems.

8. Show that the integral transform with respect to the space variable x of system 1-49 is as given by Eq. 1-52.

2

Heat Conduction in The Cartesian Coordinate System

In this chapter we examine the solution of boundary-value problems of heat conduction in the cartesian coordinate system for the one-, two-, and three-dimensional finite, semi-infinite and infinite regions. Separation of variables will be used in the solution of homogeneous boundary-value problems of heat conduction. The integral-transform technique will be applied for the solution of nonhomogeneous problems.

2-1. INTEGRAL TRANSFORM (FOURIER TRANSFORM) AND INVERSION FORMULA

The integral transforms for use in the cartesian coordinate system are usually called *Fourier* transforms because they are derived with Fourier expansion of an arbitrary function in a given interval. In this section we examine the Fourier transform and the corresponding inversion formula for a finite, semi-infinite, and infinite region.

Finite region, $0 \leq x \leq L$. The integral-transform and the inversion formula for a function $F(x)$ in the finite interval $0 \leq x \leq L$ is defined as

$$\begin{pmatrix}\text{Inversion}\\ \text{formula}\end{pmatrix} F(x) = \sum_{m=1}^{\infty} K(\beta_m, x) \cdot \bar{F}(\beta_m) \tag{2-1a}$$

$$\begin{pmatrix}\text{Integral}\\ \text{transform}\end{pmatrix} \bar{F}(\beta_m) = \int_{x'=0}^{L} K(\beta_m, x') \cdot F(x') \cdot dx' \tag{2-1b}$$

The function $K(\beta_m, x)$ is called the *transform kernel* and β_m's are called the *eigenvalues*; the type of kernels and the eigenvalues for use in Eq. 2-1 depends on the combination of boundary conditions at $x = 0$ and $x = L$. We now evaluate the kernels and the eigenvalues, for generality, for boundary conditions of the third kind at both boundaries. These general relations will be used to evaluate the kernels and eigenvalues for other combinations of boundary conditions.

Consider the following eigenvalue problem for a finite region $0 \leq x \leq L$ with homogeneous boundary conditions of the third kind at both ends. (See Fig. 2-1.)

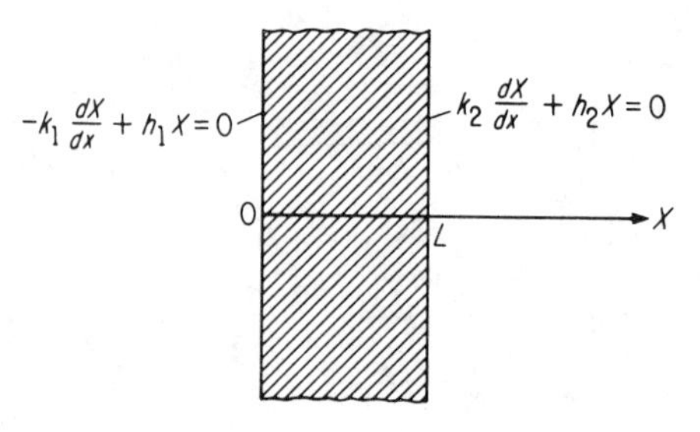

FIG. 2-1. Boundary condition for the auxiliary problem in the region $0 \le x \le L$.

$$\frac{d^2X(x)}{dx^2} + \beta^2 X(x) = 0 \qquad \text{in } 0 \le x \le L \tag{2-2a}$$

$$-k_1 \frac{dX(x)}{dx} + h_1 X(x) = 0 \qquad \text{at } x = 0 \tag{2-2b}$$

$$k_2 \frac{dX(x)}{dx} + h_2 X(x) = 0 \qquad \text{at } x = L \tag{2-2c}$$

This system is a special case of the Sturm-Liouville system with

$$p(x) = 1 \qquad w(x) = 1 \qquad q(x) = 0 \qquad \lambda = \beta^2$$

the eigenfunctions $X(\beta_m, x)$ constitute an orthogonal set in the interval $0 \le x \le L$ with respect to a weighting function $w(x) = 1$; that is,

$$\int_0^L X(\beta_m, x) \cdot X(\beta_n, x) \cdot dx = 0 \qquad \text{for} \qquad m \ne n$$

An arbitrary function $F(x)$ may be represented in the interval $0 \le x \le L$ in an infinite series of the eigenfunctions $X(\beta_m, x)$ in the form

$$F(x) = \sum_{m=1}^{\infty} c_m \cdot X(\beta_m, x) \tag{2-3}$$

where the summation is taken over all eigenvalues β_m. The unknown coefficient c_m is determined by making use of the orthogonality property of eigenfunctions and immediately is given as

$$c_m = \frac{\int_{x'=0}^{L} F(x') \cdot X(\beta_m, x') \cdot dx'}{N} \tag{2-4a}$$

where the norm N is

$$N = \int_0^L X^2(\beta_m, x') \cdot dx' \tag{2-4b}$$

Then the expansion 2-3 is given in the form

$$F(x) = \sum_{m=1}^{\infty} \frac{X(\beta_m, x)}{\sqrt{N}} \int_{x'=0}^{L} \frac{X(\beta_m, x')}{\sqrt{N}} \cdot F(x') \cdot dx' \tag{2-5}$$

We define a kernel $K(\beta_m, x)$ to denote the normalized eigenfunctions as

$$K(\beta_m, x) = \frac{X(\beta_m, x)}{\sqrt{N}} \tag{2-6}$$

and rearrange the expansion 2-5 as the *inversion formula* for the *integral transform* (i.e., the Fourier transform) of function $F(x)$ in the finite interval $0 \leq x \leq L$, in the form as

$$\begin{pmatrix}\text{Inversion}\\ \text{formula}\end{pmatrix} F(x) = \sum_{m=1}^{\infty} K(\beta_m, x) \cdot \bar{F}(\beta_m) \tag{2-7a}$$

$$\begin{pmatrix}\text{Integral}\\ \text{transform}\end{pmatrix} \bar{F}(\beta_m) = \int_{x'=0}^{L} K(\beta_m, x') \cdot F(x') \cdot dx' \tag{2-7b}$$

It is now apparent that the integral transform and the corresponding inversion formula as defined by Eq. 2-1 are merely the Fourier expansions of an arbitrary function in the finite interval $0 \leq x \leq L$.

The kernel $K(\beta_m, x)$ and the eigenvalues β_m can now be evaluated.

A particular solution of the differential Eq. 2-2a is in the form

$$X(\beta_m, x) \colon a \cos \beta_m x + b \sin \beta_m x \tag{2-8}$$

A relationship is established between the unknown coefficients a and b, and the eigenvalues β_m are determined by making use of the boundary conditions of the eigenvalue problem 2-2. If Eq. 2-8 should satisfy the boundary condition at $x = 0$, we have

$$\frac{b}{a} = \frac{h_1}{k_1 \beta_m} \tag{2-9}$$

It also satisfies the boundary condition at $x = L$, if

$$\tan \beta L = \frac{\beta(H_1 + H_2)}{\beta^2 - H_1 \cdot H_2} \tag{2-10}$$

where

$$H_1 \equiv \frac{h_1}{k_1} \qquad H_2 \equiv \frac{h_2}{k_2}$$

Equation 2-10 is the relation for determining the eigenvalues β_m; that is, the positive roots of the transcendental equation 2-10 give the eigenvalues β_m.

We choose the eigenfunctions $X(\beta_m, x)$ in the form

$$X(\beta_m, x) = \cos \beta_m x + \frac{H_1}{\beta_m} \sin \beta_m x \tag{2-11}$$

The norm is given as

$$N = \int_0^L X^2(\beta_m, x) \cdot dx$$

The integral is evaluated by following Carslaw and Jaeger [1], p. 116.
From the differential equation 2-2a

$$\begin{aligned}\int_0^L X_m^2 \cdot dx &= -\frac{1}{\beta_m^2} \int_0^L X_m \cdot X_m'' \cdot dx \\ &= -\frac{1}{\beta_m^2} [X_m \cdot X_m']_0^L + \frac{1}{\beta_m^2} \int_0^L (X_m')^2\, dx\end{aligned} \tag{2-12}$$

where primes denote differentiation with respect to x and $X_m \equiv X(\beta_m, x)$.
From Eq. 2-11,

$$\frac{1}{\beta_m} X_m' = -\sin \beta_m x + \frac{H_1}{\beta_m} \cos \beta_m x \tag{2-13}$$

From Eqs. 2-11 and 2-13,

$$X_m^2 + \frac{1}{\beta_m^2} (X_m')^2 = 1 + \left(\frac{H_1}{\beta_m}\right)^2 \tag{2-14}$$

Integrating both sides of Eq. 2-14,

$$\int_0^L X_m^2 dx = \left[1 + \left(\frac{H_1}{\beta_m}\right)^2\right] \cdot L - \frac{1}{\beta_m^2} \int_0^L (X_m')^2\, dx \tag{2-15}$$

Adding Eqs. 2-12 and 2-15, in view of Eq. 2-4b, we have

$$2N = \left[1 + \left(\frac{H_1}{\beta_m}\right)^2\right] L - \frac{1}{\beta_m^2} [X_m \cdot X_m']_0^L \tag{2-16}$$

The last term on the right-hand side of Eq. 2-16 is evaluated as follows. From the boundary conditions 2-2b and 2-2c,

$$\frac{1}{\beta_m^2} (X_m')^2 \big|_{x=0} = \left(\frac{H_1}{\beta_m}\right)^2 \cdot X_m^2 \big|_{x=0} \tag{2-17a}$$

$$\frac{1}{\beta_m^2} (X_m')^2 \big|_{x=L} = \left(\frac{H_2}{\beta_m}\right)^2 \cdot X_m^2 \big|_{x=L} \tag{2-17b}$$

Evaluating Eq. 2-14 at $x = 0$, and combining with Eq. 2-17a,

$$X_m^2 \big|_{x=0} = 1 \tag{2-18}$$

Evaluating Eq. 2-14 at $x = L$ and combining with Eq. 2-17b,

$$X_m^2 \big|_{x=L} = \frac{1 + (H_1/\beta_m)^2}{1 + (H_2/\beta_m)^2} \tag{2-19}$$

From the boundary conditions 2-2a and 2-2b we also have

$$[X_m \cdot X_m']_{x=0} = H_1 \cdot X_m^2 \big|_{x=0} \tag{2-20a}$$

$$[X_m \cdot X'_m]_{x=L} = -H_2 \cdot X_m^2 |_{x=L} \tag{2-20b}$$

Combining Eqs. 2-20a and 2-20b

$$[X_m \cdot X'_m]_0^L = -H_2 \cdot X_m^2 |_{x=L} - H_1 \cdot X_m^2 |_{x=0} \tag{2-21}$$

In view of the relations 2-18 and 2-19, Eq. 2-21 becomes

$$[X_m \cdot X'_m]_0^L = -H_2 \cdot \frac{1 + (H_1/\beta_m)^2}{1 + (H_2/\beta_m)^2} - H_1 \tag{2-22}$$

Substituting Eq. 2-22 into Eq. 2-16, the norm N becomes

$$N = \frac{1}{2}\left[\frac{\beta_m^2 + H_1^2}{\beta_m^2}\left(L + \frac{H_2}{\beta_m^2 + H_2^2}\right) + \frac{H_1}{\beta_m^2}\right] \tag{2-23}$$

We summarize the above results as follows:

The integral transform and the inversion formula with respect to the space variable x of a function $F(x)$ in the finite interval $0 \leq x \leq L$, subject to the boundary condition of the third kind at both ends are given as

$$\begin{pmatrix}\text{Inversion}\\ \text{formula}\end{pmatrix} F(x) = \sum_{m=1}^{\infty} K(\beta_m, x) \cdot \bar{F}(\beta_m) \tag{2-24a}$$

$$\begin{pmatrix}\text{Integral}\\ \text{transform}\end{pmatrix} \bar{F}(\beta_m) = \int_{x'=0}^{L} K(\beta_m, x') \cdot F(x') \cdot dx' \tag{2-24b}$$

where the summation is taken over all eigenvalues β_m, which are the positive roots of the transcendental equation

$$\tan \beta L = \frac{\beta(H_1 + H_2)}{\beta^2 - H_1 \cdot H_2} \tag{2-24c}$$

The kernel $K(\beta_m, x)$ is given as

$$K(\beta_m, x) \equiv \frac{X(\beta_m, x)}{\sqrt{N}} = \sqrt{2}\frac{\beta_m \cos \beta_m x + H_1 \sin \beta_m x}{\left[(\beta_m^2 + H_1^2)\left(L + \frac{H_2}{\beta_m^2 + H_2^2}\right) + H_1\right]^{1/2}} \tag{2-24d}$$

where

$$H_1 = \frac{h_1}{k_1} \qquad H_2 = \frac{h_2}{k_2}$$

If all combinations of the boundary conditions of the first, second and third kinds are considered, for a finite region $0 \leq x \leq L$ there exists nine combinations of boundary conditions. The kernel $K(\beta_m, x)$ and the eigenvalues β_m for the remaining eight combinations of boundary conditions are easily obtained from Eqs. 2-24c and 2-24d by choosing the values of

H_1 and H_2 as zero, finite or infinite.[1] Table 2-1 gives a summary of the kernels $K(\beta_m, x)$ and the corresponding eigenvalues β_m for the nine combinations of boundary conditions for a finite region $0 \le x \le L$ in the cartesian coordinate system.

The representation of an arbitrary function $F(x)$ as given above is valid if function $F(x)$ satisfies Dirichlet conditions at each point of the interval $(0, L)$ at which $F(x)$ is continuous [2]. We say that the function $F(x)$ satisfies Dirichlet's conditions in the interval $(0, L)$ if

(a) $F(x)$ has only a finite number of maxima and minima in $(0, L)$;

(b) $F(x)$ has only a finite number of discontinuities in $(0, L)$ and no infinite discontinuities.

Semi-infinite region, $0 \le x < \infty$. The integral transform and inversion formula of a function $F(x)$ in the semi-infinite interval $0 \le x < \infty$ is defined as

$$\begin{pmatrix}\text{Inversion}\\ \text{formula}\end{pmatrix} F(x) = \int_{\beta=0}^{\infty} K(\beta, x) \cdot \bar{F}(\beta) \cdot d\beta \tag{2-25a}$$

$$\begin{pmatrix}\text{Integral}\\ \text{transform}\end{pmatrix} \bar{F}(\beta) = \int_{x'=0}^{\infty} K(\beta, x') \cdot F(x') \cdot dx' \tag{2-25b}$$

The function $K(\beta, x)$ is the transform kernel and will now be evaluated.

Consider the following auxiliary problem for a semi-infinite region $0 \le x < \infty$ subjected to a homogeneous boundary condition of the third

[1] For example, consider the case with the boundary condition of the third kind at $x = 0$ and the second kind at $x = L$. We choose H_1 = finite and $H_2 = 0$; Eqs. 2-24c and 2-24d become

$$K(\beta_m, x) = \sqrt{2} \cdot \frac{\beta_m \cos \beta_m x + H_1 \sin \beta_m x}{[L(\beta_m^2 + H_1^2) + H_1]^{1/2}} \tag{1}$$

where β_m's are the positive roots of

$$\beta \tan \beta L = H_1 \tag{2}$$

The numerator of the kernel $K(\beta_m, x)$ is simplified as follows. Let

$$Z \equiv \beta_m \cos \beta_m x + H_1 \sin \beta_m x \tag{3}$$

In view of relation 2, we have

$$\begin{aligned} Z &= \beta_m \cos \beta_m x + \beta_m \cdot \tan \beta_m L \cdot \sin \beta_m x \\ &= \frac{\beta_m}{\cos \beta_m L} \cdot [\cos \beta_m L \cdot \cos \beta_m x + \sin \beta_m L \cdot \sin \beta_m x] \\ &= \frac{\beta_m}{\cos \beta_m L} \cdot \cos \beta_m (L - x) = \beta_m \sqrt{1 + \tan^2 \beta_m L} \cdot \cos \beta_m (L - x) \\ &= \sqrt{\beta_m^2 + H_1^2} \cdot \cos \beta_m (L - x) \end{aligned} \tag{4}$$

Substituting Eq. 4 into Eq. 1,

$$K(\beta_m, x) = \sqrt{2} \cdot \left[\frac{\beta_m^2 + H_1^2}{L(\beta_m^2 + H_1^2) + H_1}\right]^{1/2} \cdot \cos \beta_m (L - x) \tag{5}$$

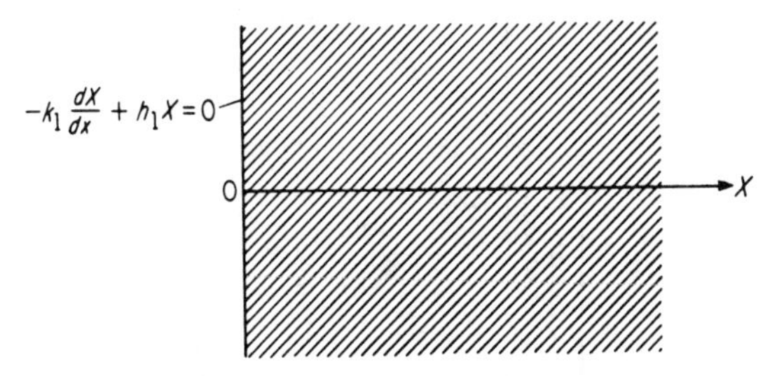

FIG. 2-2. Boundary condition for the auxiliary problem in the region $0 \le x \le \infty$.

kind at $x = 0$. (See Fig. 2-2.)

$$\frac{d^2X(x)}{dx^2} + \beta^2 X(x) = 0 \qquad \text{in } 0 \le x < \infty \tag{2-26a}$$

$$-k_1 \frac{dX(x)}{dx} + h_1 X(x) = 0 \qquad \text{at } x = 0 \tag{2-26b}$$

It can be verified that the function

$$X(\beta, x) = \beta \cos \beta x + H_1 \sin \beta x$$

where

$$H_1 = \frac{h_1}{k_1}$$

satisfies Eq. 2-26. The parameter β is assumed to take all values from 0 to infinity.

An arbitrary function $F(x)$ may be represented in the semi-infinite interval $0 \le x < \infty$ in terms of the $X(\beta, x)$ functions in the form

$$F(x) = \int_{\beta=0}^{\infty} C(\beta) \cdot (\beta \cos \beta x + H_1 \sin \beta x)\, d\beta \qquad 0 \le x \le \infty \tag{2-27}$$

The unknown coefficient $C(\beta)$ has been determined by means of the Laplace transformation by Churchill [3], and the representation 2-27 is given as

$$F(x) = \frac{2}{\pi} \int_{\beta=0}^{\infty} (\beta \cos \beta x + H_1 \sin \beta x) \cdot \left[\int_{x'=0}^{\infty} \frac{\beta \cos \beta x' + H_1 \sin \beta x'}{\beta^2 + H_1^2} F(x')\, dx' \right] d\beta \tag{2-28}$$

Equation 2-28 is rearranged to define the integral transform and the inversion formula of a function $F(x)$ in a semi-infinite interval $0 \le x < \infty$ as [4]

$$\begin{pmatrix}\text{Inversion}\\ \text{formula}\end{pmatrix} \qquad F(x) = \int_{\beta=0}^{\infty} K(\beta, x) \cdot \bar{F}(\beta) \cdot d\beta \tag{2-29a}$$

TABLE 2-1

THE KERNELS AND THE EIGENVALUES FOR USE IN THE INTEGRAL TRANSFORM (FOURIER TRANSFORM) AND THE INVERSION FORMULA FOR A FINITE REGION $0 \le x \le L$ IN THE CARTESIAN COORDINATE SYSTEM*

$$\begin{pmatrix}\text{Integral}\\ \text{transform}\end{pmatrix} \bar{F}(\beta_m) = \int_{x'=0}^{L} K(\beta_m, x') \cdot F(x') \cdot dx'$$

$$\begin{pmatrix}\text{Inversion}\\ \text{formula}\end{pmatrix} F(x) = \sum_{m=1}^{\infty} K(\beta_m, x) \cdot \bar{F}(\beta_m)$$

$-K_1 \frac{dX}{dx} + h_1 X = 0$ (at 0); $K_2 \frac{dX}{dx} + h_2 X = 0$ (at L); x

Boundary Condition† at $x = 0$	Boundary Condition† at $x = L$	Kernel $K(\beta_m, x)$	Eigenvalues β_m's Are Positive Roots of
3rd kind H_1 = finite	3rd kind H_2 = finite	$\sqrt{2} \dfrac{\beta_m \cos \beta_m x + H_1 \sin \beta_m x}{\left[(\beta_m^2 + H_1^2)\left(L + \dfrac{H_2}{\beta_m^2 + H_2^2}\right) + H_1\right]^{1/2}}$	$\tan \beta L = \dfrac{\beta(H_1 + H_2)}{\beta^2 - H_1 H_2}$
3rd kind H_1 = finite	2nd kind $H_2 = 0$ (i.e., $h_2 = 0$)	$\sqrt{2}\left[\dfrac{\beta_m^2 + H_1^2}{L(\beta_m^2 + H_1^2) + H_1}\right]^{1/2} \cdot \cos \beta_m (L - x)$	$\beta \tan \beta L = H_1$
3rd kind H_1 = finite	1st kind $H_2 = \infty$ (i.e., $k_2 = 0$)	$\sqrt{2}\left[\dfrac{\beta_m^2 + H_1^2}{L(\beta_m^2 + H_1^2) + H_1}\right]^{1/2} \cdot \sin \beta_m (L - x)$	$\beta \cot \beta L = -H_1$
2nd kind $H_1 = 0$ (i.e., $h_1 = 0$)	3rd kind H_2 = finite	$\sqrt{2} \cdot \left[\dfrac{\beta_m^2 + H_2^2}{L(\beta_m^2 + H_2^2) + H_2}\right]^{1/2} \cdot \cos \beta_m x$	$\beta \tan \beta L = H_2$
2nd kind $H_1 = 0$ (i.e., $h_1 = 0$)	2nd kind $H_2 = 0$ (i.e., $h_2 = 0$)	$\sqrt{\dfrac{2}{L}} \cdot \cos \beta_m x$††	$\sin \beta L = 0$

2nd kind $H_1 = 0$ (i.e., $h_1 = 0$)	1st kind $H_2 = \infty$ (i.e., $k_2 = 0$)	$\sqrt{\frac{2}{L}} \cdot \cos \beta_m x$	$\cos \beta L = 0$
1st kind $H_1 = \infty$ (i.e., $k_1 = 0$)	3rd kind $H_2 =$ finite	$\sqrt{2} \left[\frac{\beta_m^2 + H_2^2}{L(\beta_m^2 + H_2^2) + H_2} \right]^{1/2} \cdot \sin \beta_m x$	$\beta \cot \beta L = -H_2$
1st kind $H_1 = \infty$ (i.e., $k_1 = 0$)	2nd kind $H_2 = 0$ (i.e., $h_2 = 0$)	$\sqrt{\frac{2}{L}} \cdot \sin \beta_m x$	$\cos \beta L = 0$
1st kind $H_1 = \infty$ (i.e., $k_1 = 0$)	1st kind $H_2 = \infty$ (i.e., $k_2 = 0$)	$\sqrt{\frac{2}{L}} \cdot \sin \beta_m x$	$\sin \beta L = 0$

*(From: M. N. Özışık, ASME paper 67-H-67, 1967).

†$H_1 \equiv h_1/k_1$ and $H_2 \equiv h_2/k_2$.

††For this particular case replace $\sqrt{2/L}$ by $\sqrt{1/L}$ when β is zero.

$$\begin{pmatrix}\text{Integral}\\ \text{transform}\end{pmatrix} \quad \bar{F}(\beta) = \int_{x'=0}^{\infty} K(\beta, x') \cdot F(x') \cdot dx' \tag{2-29b}$$

where

$$K(\beta, x) = \sqrt{\frac{2}{\pi}} \frac{\beta \cos \beta x + H_1 \sin \beta x}{\sqrt{\beta^2 + H_1^2}} \tag{2-29c}$$

and

$$H_1 = \frac{h_1}{k_1}$$

The kernel 2-29c for use in the integral transform and the inversion formula as defined above is for a boundary condition of the third kind at $x = 0$. The kernel $K(\beta, x)$ for boundary condition of the first kind and second kind at $x = 0$ is immediately obtained from Eq. 2-29c by choosing the value of H_1 as infinity and zero respectively.

Table 2-2 summarizes the kernel $K(\beta, x)$ for use in the integral transform and the inversion formula given above for the three different boundary conditions at $x = 0$.

TABLE 2-2

KERNELS FOR USE IN THE INTERGRAL TRANSFORM (FOURIER TRANSFORM) AND THE INVERSION FORMULA FOR A SEMI-INFINITE REGION $0 \le x < \infty$ IN THE CARTESIAN COORDINATE SYSTEM*

$$\begin{pmatrix}\text{Integral}\\ \text{transform}\end{pmatrix} \quad \bar{F}(\beta) = \int_{x'=0}^{\infty} K(\beta, x') \cdot F(x') \cdot dx'$$

$$\begin{pmatrix}\text{Inversion}\\ \text{formula}\end{pmatrix} \quad F(x) = \int_{\beta=0}^{\infty} K(\beta, x) \cdot \bar{F}(\beta) \cdot d\beta$$

$-h_1 \frac{dX}{dx} + h_1 X = 0$

0 x

Boundary Condition at $x = 0$†	Kernel $K(\beta, x)$
Third kind, H_1 = finite	$\sqrt{\frac{2}{\pi}} \frac{\beta \cos \beta x + H_1 \sin \beta x}{\sqrt{\beta^2 + H_1^2}}$
Second kind, $H_1 = 0$ (i.e., $h_1 = 0$)	$\sqrt{\frac{2}{\pi}} \cos \beta x$
First kind, $H_1 = \infty$ (i.e., $k_1 = 0$)	$\sqrt{\frac{2}{\pi}} \sin \beta x$

*(From: M. N. Özışık, ASME paper 67-H-67, 1967)

†$H_1 \equiv h_1/k_1$.

The representation of an arbitrary function $F(x)$ as given in Eq. 2-29 is valid when $F(x)$ and $dF(x)/dx$ are sectionally continuous on each interval in the range $x \ge 0$, provided $\int_0^{\infty} |F(x)|\, dx$ exists, and if $F(x)$ is defined as its mean value at each point of discontinuity [3].

Infinite region, $-\infty < x < \infty$. For a one-dimensional infinite region $-\infty < x < \infty$, the integral transform and the inversion formula of a function $F(x)$ can be derived by considering the following auxiliary equation.

$$\frac{d^2X(x)}{dx^2} + \beta^2 X(x) = 0 \qquad \text{in } -\infty < x < \infty \tag{2-30}$$

The function

$$X(\beta, x) = a(\beta) \cdot \cos \beta x + b(\beta) \cdot \sin \beta x \tag{2-31}$$

satisfies the differential equation 2-30, and the parameter β takes all values from zero to infinity continuously.

An arbitrary function $F(x)$ is represented in the infinite interval $-\infty < x < \infty$, in terms of $X(\beta, x)$ functions in the form

$$F(x) = \int_{\beta=0}^{\infty} [a(\beta) \cdot \cos \beta x + b(\beta) \cdot \sin \beta x]\, d\beta \qquad -\infty < x < \infty \tag{2-32}$$

Equation 2-32 is the well-known Fourier representation of an arbitrary function $F(x)$ in the interval $-\infty < x < \infty$, and the unknown coefficients $a(\beta)$ and $b(\beta)$ are given as [5]

$$a(\beta) = \frac{1}{\pi} \int_{x'=-\infty}^{\infty} F(x') \cdot \cos \beta x' \cdot dx' \tag{2-33}$$

$$b(\beta) = \frac{1}{\pi} \int_{x'=-\infty}^{\infty} F(x') \cdot \sin \beta x' \cdot dx' \tag{2-34}$$

We substitute these coefficients in Eq. 2-32 and make use of the following relation.

$$\cos \beta x \cdot \cos \beta x' + \sin \beta x \cdot \sin \beta x' = \cos \beta(x - x') \tag{2-35}$$

Then representation 2-32 becomes

$$F(x) = \frac{1}{\pi} \int_{x'=-\infty}^{\infty} \int_{\beta=0}^{\infty} F(x') \cdot dx' \cos \beta(x - x') \cdot d\beta \tag{2-36}$$

An equivalent formula for Eq. 2-36 is given as [5][2]

$$F(x) = \frac{1}{2\pi} \int_{\beta=-\infty}^{\infty} e^{-i\beta x} \left[\int_{x'=-\infty}^{\infty} e^{i\beta x'} \cdot F(x') \cdot dx' \right] \cdot d\beta \tag{2-37}$$

By rearranging Eq. 2-37 we define the integral transform and the inversion formula with respect to the space variable x of function $F(x)$ in the infinite interval $-\infty < x < \infty$ in the cartesian coordinate system as

$$\begin{pmatrix}\text{Inversion}\\ \text{formula}\end{pmatrix} \quad F(x) = \frac{1}{2\pi} \int_{\beta=-\infty}^{\infty} e^{-i\beta x} \cdot \bar{F}(\beta) \cdot d\beta \tag{2-38a}$$

[2]Equation 1.1.6, Ref. 5, p. 3.

$$\begin{pmatrix}\text{Integral}\\ \text{transform}\end{pmatrix} \quad \bar{F}(\beta) = \int_{x'=-\infty}^{\infty} e^{i\beta x'} \cdot F(x') \cdot dx' \tag{2-38b}$$

The representative of an arbitrary function $F(x)$ in the infinite interval $-\infty < x < \infty$ as given by Eq. 2-38 is valid if $F(x)$ satisfies Dirichlet's conditions and that $\int_{-\infty}^{\infty} |F(x)|\, dx$ exists.

2-2. ONE-DIMENSIONAL HOMOGENEOUS BOUNDARY-VALUE PROBLEMS OF HEAT CONDUCTION IN FINITE REGIONS (SOLUTION WITH SEPARATION OF VARIABLES)

In this section we consider the solution of homogeneous boundary-value problems of heat conduction for a finite region $0 \le x \le L$ subject to the homogeneous boundary condition of the third kind at the boundary surfaces, by the separation of variables.

Consider a slab $0 \le x \le L$ which is initially at temperature $F(x)$. For times $t > 0$ there is heat dissipation by convection from its bounding surfaces at $x = 0$ and $x = L$ into a surrounding at zero temperature (i.e., homogeneous boundary condition of the third kind). Figure 2-3

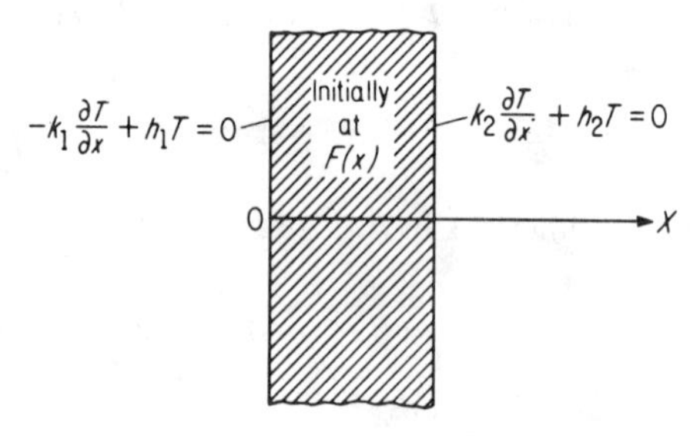

FIG. 2-3. Slab with homogeneous boundary conditions of the third kind.

shows the geometry and the boundary conditions. The boundary-value problem of heat conduction is given as

$$\frac{\partial^2 T}{\partial x^2} = \frac{1}{\alpha}\frac{\partial T}{\partial t} \quad \text{in } 0 \le x \le L,\ t > 0 \tag{2-39a}$$

$$-k_1 \frac{\partial T}{\partial x} + h_1 T = 0 \quad \text{at } x = 0,\ t > 0 \tag{2-39b}$$

$$k_2 \frac{\partial T}{\partial x} + h_2 T = 0 \quad \text{at } x = L,\ t > 0 \tag{2-39c}$$

$$T = F(x) \quad \text{in } 0 \le x \le L,\ t = 0 \tag{2-39d}$$

where $T \equiv T(x, t)$.

Assuming separation of the variables in the form

$$T(x,t) = X(x)\cdot\Gamma(t) \tag{2-40}$$

The separation equation for the time variable $\Gamma(t)$ is

$$\frac{d\Gamma(t)}{dt} + \alpha\beta^2\Gamma(t) = 0 \tag{2-41}$$

and the space variable $X(x)$ satisfies the following homogeneous system.

$$\frac{d^2X(x)}{dx^2} + \beta^2 X(x) = 0 \qquad \text{in } 0 \le x \le L \tag{2-42a}$$

$$-k_1\frac{dX(x)}{dx} + h_1 X(x) = 0 \qquad \text{at } x = 0 \tag{2-42b}$$

$$k_2\frac{dX(x)}{dx} + h_2 X(x) = 0 \qquad \text{at } x = L \tag{2-42c}$$

Solution for the time variable $\Gamma(t)$ is in the form

$$\Gamma(t){:}e^{-\alpha\beta^2 t} \tag{2-43}$$

If $X(\beta_m, x)$ is the eigenfunction and β_m's are the eigenvalues of the eigenvalue problem 2-42, the general solution of the homogeneous boundary-value problem of heat conduction, Eq. 2-39, is given in the form

$$T(x,t) = \sum_{m=1}^{\infty} c_m \cdot X(\beta_m, x)\cdot e^{-\alpha\beta_m^2 t} \tag{2-44}$$

This solution should satisfy the initial condition of the problem, i.e., Eq. 2-39d; for $t = 0$ we have

$$F(x) = \sum_{m=1}^{\infty} c_m \cdot X(\beta_m, x) \qquad 0 \le x \le L \tag{2-45}$$

Equation 2-45 is representation of an arbitrary function $F(x)$ in the finite interval $0 \le x \le L$ in an infinite series of the eigenfunctions of the eigenvalue problem 2-42. This problem was treated in the previous section and the unknown coefficient c_m is given by Eq. 2-4 as

$$c_m = \frac{\displaystyle\int_0^L F(x)\cdot X(\beta_m, x)\cdot dx}{N} \tag{2-46}$$

where the norm is

$$N = \int_0^L X^2(\beta_m, x)\cdot dx$$

Substituting Eq. 2-46 into Eq. 2-44, and defining kernel $K(\beta_m, x)$ as

$$K(\beta_m, x) = \frac{X(\beta_m, x)}{\sqrt{N}} \tag{2-47}$$

The solution of the homogeneous boundary-value problem of heat conduction (2-39) becomes

$$T(x,t) = \sum_{m=1}^{\infty} e^{-\alpha\beta_m^2 t} \cdot K(\beta_m, x) \cdot \int_{x'=0}^{L} K(\beta_m, x') \cdot F(x') \cdot dx' \tag{2-48}$$

where the summation is taken over all eigenvalues β_m.

The kernel $K(\beta_m, x)$ and the eigenvalues β_m for boundary condition of the third kind at both ends are obtained from Table 2-1 as

$$K(\beta_m, x) = \sqrt{2}\, \frac{\beta_m \cos \beta_m x + H_1 \sin \beta_m x}{\left[(\beta_m^2 + H_1^2)\left(L + \dfrac{H_2}{\beta_m^2 + H_2^2} \right) + H_1 \right]^{1/2}} \tag{2-49a}$$

and the eigenvalues β_m's are the positive roots of the transcendental equation

$$\tan \beta L = \frac{\beta(H_1 + H_2)}{\beta^2 - H_1 \cdot H_2} \tag{2-49b}$$

The homogeneous boundary-value problem of heat conduction, Eq. 2-39, considered above includes nine combinations of boundary conditions. The kernels $K(\beta_m, x)$ and the eigenvalues β_m for use in the solution 2-48 for each of these combinations are obtained from Table 2-1. The special case of homogeneous boundary condition of the second kind at both ends (i.e., both boundary surfaces are insulated) will include a term

$$\frac{1}{L} \int_0^L F(x') \cdot dx' \tag{2-50}$$

in the solution. The physical significance of this term is that, after the temperature transients have passed, the temperature in the region attains the equilibrium temperature, Eq. 2-50, which is the mean of the initial temperature distribution over the region $0 \le x \le L$.

We now examine the physical significance of the eigenvalues given by the transcendental equation 2-49b. For convenience we write this equation in the form

$$Z = \cot \xi \tag{2-51a}$$

$$Z = \frac{1}{B_1 + B_2}\left(\xi - \frac{B_1 B_2}{\xi} \right) \tag{2-51b}$$

where

$$B \equiv \frac{hL}{k} = H \cdot L = \text{Biot number}$$

$$\xi = \beta L$$

Assuming further equal heat-transfer coefficients for both surfaces, we have

$$Z = \cot \xi \tag{2-52a}$$

$$Z = \frac{1}{2}\left(\frac{\xi}{B} - \frac{B}{\xi}\right) \tag{2-52b}$$

where $\xi = \beta L$.

Equation 2-52a is a cotangent curve and Eq. 2-52b is a hyperbola. Figure 2-4 illustrates a plot of these two curves; the intersection of these curves corresponds to

$$\xi_m = \beta_m L$$

hence an estimate is made of the numerical equivalents of the eigenvalues β_m corresponding to the transcendental equation 2-52. Only the positive

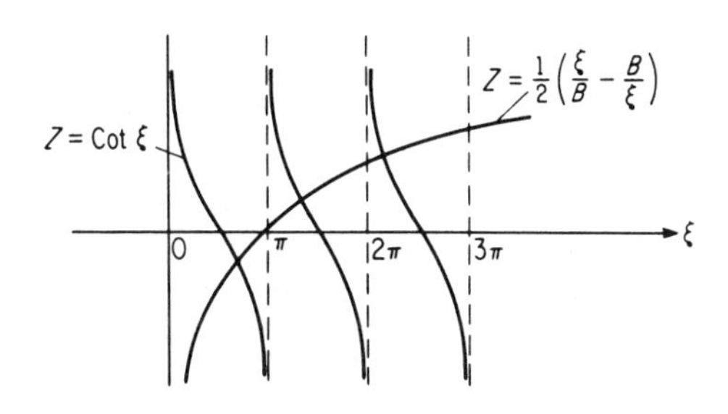

FIG. 2-4. Geometrical determination of the roots of the transcendental equation 2-52.

roots are to be considered, since negative roots are equal in their absolute magnitude to positive roots. Roots are easily determined with greater accuracy using a digital computer.

The roots of the following transcendental equations are presented in Appendix I.

$$\begin{aligned} \tan \xi &= c \\ \cot \xi &= -c \end{aligned} \tag{2-53}$$

In the following examples we examine some special cases of the solution 2-48.

EXAMPLE 2-1. A slab $0 \leq x \leq L$ is initially at temperature $F(x)$. For times $t > 0$ heat is dissipated by convection from both surfaces into a surrounding at zero temperature. We assume that heat-transfer coefficients h are equal on both sides. This problem is similar to the general problem Eq. 2-39 considered above, except the boundary conditions are

$$-k\frac{\partial T}{\partial x} + hT = 0 \quad \text{at} \quad x = 0,\ t > 0 \tag{2-55a}$$

$$k\frac{\partial T}{\partial x} + hT = 0 \quad \text{at} \quad x = L,\ t > 0. \tag{2-55b}$$

Since $h_1 = h_2 \equiv h$, we have

$$H_1 = H_2 \equiv H$$

Then the kernel $K(\beta_m, x)$ becomes

$$K(\beta_m, x) = \sqrt{2}\,\frac{\beta_m \cos\beta_m x + H \sin\beta_m x}{[L(\beta_m^2 + H^2) + 2H]^{1/2}} \tag{2-56}$$

and eigenvalues β_m are the positive roots of the transcendental equation

$$\tan\beta L = \frac{2\beta H}{\beta^2 - H^2}$$

where

$$H = \frac{h}{k} \tag{2-57}$$

Then the solution of the above problem is given by Eq. 2-48 and the kernel $K(\beta_m, x)$ and eigenvalues β_m are taken as given by Eqs. 2-56 and 2-57 respectively.

EXAMPLE 2-2. A slab $0 \le x \le L$ is initially at temperature $F(x)$. For times $t > 0$ the boundary at $x = 0$ is insulated and that at $x = L$ dissipates heat by convection into a surrounding at zero temperature.

This is a special case of the homogeneous boundary-value problem of heat conduction, Eq. 2-39, with boundary conditions given as

$$\frac{\partial T}{\partial x} = 0 \quad \text{at} \quad x = 0,\ t > 0 \tag{2-58a}$$

$$k_2\frac{\partial T}{\partial x} + h_2 T = 0 \quad \text{at} \quad x = L,\ t > 0 \tag{2-58b}$$

For this problem $H_1 = 0$ and $H_2 = \frac{h_2}{k_2} =$ finite. The kernel $K(\beta_m, x)$ and the eigenvalues β_m for a boundary condition of the second kind at $x = 0$ and of the third kind at $x = L$ are immediately obtained from Table 2-1. Substituting the corresponding kernel in Eq. 2-48, the solution of the present problem becomes

$$T(x, t) = 2\sum_{m=1}^{\infty} e^{-\alpha\beta_m^2 t}\,\frac{\beta_m^2 + H_2^2}{L(\beta_m^2 + H_2^2) + H_2}\cdot\cos\beta_m x\cdot\int_{x'=0}^{L} F(x')\cdot\cos\beta_m x'\cdot dx' \tag{2-59}$$

where the summation is taken over all positive roots of the transcendental equation

$$\beta\tan\beta L = H_2$$

or

$$\beta L\cdot\tan\beta L = H_2 L \tag{2-60}$$

The roots of the transcendental equation 2-60 are given in Appendix I.

For the special case of constant initial temperature,

$$F(x) = T_0 = \text{const.}$$

Integration in Eq. 2-59 is performed,

$$T(x, t) = 2T_0\sum_{m=1}^{\infty} e^{-\alpha\beta_m^2 t}\cdot\frac{\beta_m^2 + H_2^2}{L(\beta_m^2 + H_2^2) + H_2}\cdot\frac{\sin\beta_m L}{\beta_m}\cdot\cos\beta_m x \tag{2-61a}$$

By making use of the transcendental equation 2-60, solution 2-61a is written in the form

$$T(x,t) = 2T_0 \sum_{m=1}^{\infty} e^{-\alpha\beta_m^2 t} \cdot \frac{H_2}{L(\beta_m^2 + H_2^2) + H_2} \cdot \frac{\cos \beta_m x}{\cos \beta_m L} \tag{2-61b}$$

EXAMPLE 2-3. A slab $0 \le x \le L$ is initially at temperature $F(x)$. For times $t > 0$ the boundaries at $x = 0$ and $x = L$ are kept insulated.

This is again a special case of problem 2-39 and the boundary conditions are

$$\frac{\partial T}{\partial x} = 0 \quad \text{at} \quad x = 0,\ t > 0 \tag{2-62a}$$

$$\frac{\partial T}{\partial x} = 0 \quad \text{at} \quad x = L,\ t > 0 \tag{2-62b}$$

For this problem $H_1 = H_2 = 0$. From Table 2-1 the kernel $K(\beta_m, x)$ and the eigenvalues β_m for boundary condition of the second kind at both boundaries are immediately obtained. When they are substituted in the solution 2-48, the solution of the present problem becomes

$$T(x,t) = \frac{1}{L} \int_0^L F(x') \cdot dx' + \frac{2}{L} \sum_{m=1}^{\infty} e^{-\alpha\beta_m^2 t} \cdot \cos \beta_m x \cdot \int_{x'=0}^{L} F(x') \cdot \cos \beta_m x' \cdot dx' \tag{2-63}$$

where β_m's are

$$\beta_m = \frac{m\pi}{L} \qquad m = 1, 2, 3, \ldots,$$

It is to be noted that the first term on the right-hand side of Eq. 2-63 is the temperature in the solid after temperature transients have passed. This term is the mean of the initial temperature over the region $0 \le x \le L$.

2-3. ONE-DIMENSIONAL NONHOMOGENEOUS BOUNDARY-VALUE PROBLEMS OF HEAT CONDUCTION IN FINITE REGIONS (SOLUTION WITH INTEGRAL TRANSFORM)

In this section we consider the application of the integral-transform technique to the solution of a one-dimensional boundary-value problem of heat conduction in a finite region $0 \le x \le L$ with heat generation within the solid and with generalized nonhomogeneous boundary conditions. It is to be noted that the homogeneous problem considered in the previous section by the method of the separation of variables could easily be solved with the integral-transform technique described in the present section.

Consider a slab $0 \le x \le L$ which is initially at temperature $F(x)$. For times $t > 0$ heat is generated within the solid at a rate of $g(x, t)$

Btu/hr ft³, while heat is dissipated by convection from the boundary surfaces $x = 0$ and $x = L$ into a surrounding temperature which varies with time. For generality we assume that heat-transfer coefficients are not the same for both surfaces. Figure 2-5 shows the geometry and the boundary conditions.

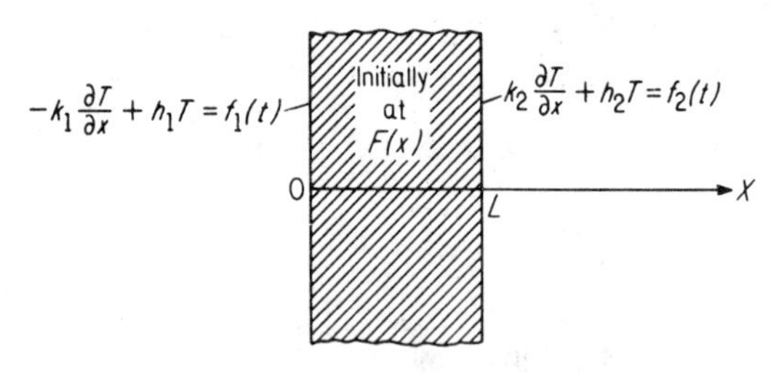

FIG. 2-5. Slab with non-homogeneous boundary conditions of the third kind.

The boundary-value problem of heat conduction is given as

$$\frac{\partial^2 T}{\partial x^2} + \frac{g(x,t)}{k} = \frac{1}{\alpha}\frac{\partial T}{\partial t} \quad \text{in } 0 \le x \le L,\ t > 0 \tag{2-64a}$$

$$-k_1 \frac{\partial T}{\partial x} + h_1 T = f_1(t) \quad \text{at } x = 0,\ t > 0 \tag{2-64b}$$

$$k_2 \frac{\partial T}{\partial x} + h_2 T = f_2(t) \quad \text{at } x = L,\ t > 0 \tag{2-64c}$$

$$T = F(x) \quad \text{in } 0 \le x \le L,\ t = 0 \tag{2-64d}$$

where $T \equiv T(x, t)$.

We define the integral transform and the inversion formula of temperature function $T(x, t)$ with respect to the space variable x, in the range $0 \le x \le L$ as

$$\begin{pmatrix}\text{Integral}\\ \text{transform}\end{pmatrix} \quad \bar{T}(\beta_m, t) = \int_{x'=0}^{L} K(\beta_m, x') \cdot T(x', t) \cdot dx' \tag{2-65a}$$

$$\begin{pmatrix}\text{Inversion}\\ \text{formula}\end{pmatrix} \quad T(x, t) = \sum_{m=1}^{\infty} K(\beta_m, x) \cdot \bar{T}(\beta_m, t) \tag{2-65b}$$

where the kernel $K(\beta_m, x)$ is the normalized eigenfunction of the following eigenvalue problem

$$\frac{d^2 X(x)}{dx^2} + \beta^2 X(x) = 0 \quad \text{in} \quad 0 \le x \le L \tag{2-66a}$$

$$-k_1 \frac{dX(x)}{dx} + h_1 X(x) = 0 \quad \text{at} \quad x = 0 \tag{2-66b}$$

$$k_2 \frac{dX(x)}{dx} + h_2 X(x) = 0 \quad \text{at} \quad x = L \tag{2-66c}$$

It is to be noted that the kernels $K(\beta_m, x)$ and the eigenvalues β_m are tabulated in Table 2-1 for nine different combinations of boundary conditions at the surfaces $x = 0$ and $x = L$.

By applying the transform 2-65a we take the integral transform of the differential equation 2-64a,

$$\int_0^L K(\beta_m, x) \cdot \frac{\partial^2 T(x,t)}{\partial x^2} \cdot dx + \frac{1}{k} \int_0^L K(\beta_m, x) \cdot g(x,t) \cdot dx$$

$$= \frac{1}{\alpha} \int_0^L K(\beta_m, x) \cdot \frac{\partial T(x,t)}{\partial t} \cdot dx \tag{2-67}$$

which is written in the form

$$\int_0^L K(\beta_m, x) \cdot \frac{\partial^2 T(x,t)}{\partial x^2}\, dx + \frac{1}{k}\, \bar{g}(\beta_m, t) = \frac{1}{\alpha} \frac{d\bar{T}(\beta_m, t)}{dt} \tag{2-68}$$

where quantities with bars refer to the integral transform according to transform, Eq. 2-65a.

The first term on the left-hand side of Eq. 2-68 is evaluated by making use of Green's theorem as explained in the previous chapter (or immediately obtained from the general relation 1-64 derived in the previous chapter). The result is

$$\int_0^L K(\beta_m, x) \cdot \frac{\partial^2 T(x,t)}{\partial x^2} \cdot dx = -\beta_m^2 \cdot \bar{T}(\beta_m, t)$$

$$+ \left[\left. \frac{K(\beta_m, x)}{k_1} \right|_{x=0} \cdot f_1(t) + \left. \frac{K(\beta_m, x)}{k_2} \right|_{x=L} \cdot f_2(t) \right] \tag{2-69}$$

Substituting Eq. 2-69 into Eq. 2-68

$$\frac{d\bar{T}(\beta_m, t)}{dt} + \alpha \beta_m^2 \cdot \bar{T}(\beta_m, t) = A(\beta_m, t) \tag{2-70a}$$

where

$$A(\beta_m, t) = \frac{\alpha}{k}\, \bar{g}(\beta_m, t)$$

$$+ \alpha \left[\left. \frac{K(\beta_m, x)}{k_1} \right|_{x=0} \cdot f_1(t) + \left. \frac{K(\beta_m, x)}{k_2} \right|_{x=L} \cdot f_2(t) \right] \tag{2-70b}$$

Thus we removed from the system 2-64 the second partial derivative with respect to the space variable x and reduced it to a first-order ordinary differential equation with respect to the time variable t, for the integral transform of temperature $\bar{T}(\beta_m, t)$.

The initial condition for the ordinary differential Eq. 2-70 is obtained by taking the integral transform of the initial condition 2-64d by applying the transform, Eq. 2-65a. That is,

$$\bar{T}(\beta_m, t)\Big|_{t=0} = \int_{x'=0}^{L} K(\beta_m, x') \cdot F(x') \cdot dx' \equiv \bar{F}(\beta_m) \tag{2-71}$$

Solution of the ordinary differential equation 2-70 subject to the transformed initial condition 2-71 is

$$\bar{T}(\beta_m, t) = e^{-\alpha\beta_m^2 t}\left[\bar{F}(\beta_m) + \int_{t'=0}^{t} e^{\alpha\beta_m^2 t'} \cdot A(\beta_m, t') \cdot dt'\right] \tag{2-72}$$

Substituting the integral transform, Eq. 2-72, into the inversion formula 2-65b, we obtain the general solution of the boundary-value problem of heat conduction, Eq. 2-64, as

$$T(x, t) = \sum_{m=1}^{\infty} e^{-\alpha\beta_m^2 t} \cdot K(\beta_m, x) \cdot \left[\bar{F}(\beta_m) + \int_{t'=0}^{t} e^{\alpha\beta_m^2 t'} \cdot A(\beta_m, t') \cdot dt'\right] \tag{2-73a}$$

where

$$A(\beta_m, t') = \frac{\alpha}{k}\,\bar{g}(\beta_m, t') + \alpha\left[\frac{K(\beta_m, x)}{k_1}\bigg|_{x=0} \cdot f_1(t') + \frac{K(\beta_m, x)}{k_2}\bigg|_{x=L} \cdot f_2(t')\right] \tag{2-73b}$$

$$\bar{F}(\beta_m) = \int_{x'=0}^{L} K(\beta_m, x') \cdot F(x') \cdot dx' \tag{2-73c}$$

$$\bar{g}(\beta_m, t') = \int_{x'=0}^{L} K(\beta_m, x') \cdot g(x', t') \cdot dx' \tag{2-73d}$$

Table 2-1 gives the kernel $K(\beta_m, x)$ and the eigenvalues β_m for use in the solution 2-73 for the nine combinations of boundary conditions. In applying the solution 2-73 for various combination of boundary conditions shown in Table 2-1 special attention should be paid to the following case:

When one or both boundary conditions are of the first kind (i.e., prescribed temperature at the boundary) the parameters k_1 or k_2 or both are zero, but these parameters are in the denominator of the terms in $A(\beta_m, t')$ in Eq. 2-73b. To avoid difficulty in such cases the following changes should be made in Eq. 2-73b:

$$\text{When } k_1 = 0, \quad \text{replace} \quad \frac{K(\beta_m, x)}{k_1}\bigg|_{x=0} \quad \text{by} \quad \frac{1}{h_1}\frac{dK(\beta_m, x)}{dx}\bigg|_{x=0} \tag{2-74a}$$

$$\text{When } k_2 = 0, \quad \text{replace} \quad \left.\frac{K(\beta_m, x)}{k_2}\right|_{x=L} \quad \text{by} \quad -\frac{1}{h_2}\left.\frac{dK(\beta_m, x)}{dx}\right|_{x=L} \tag{2-74b}$$

The validity of the relations in Eq. 2-74 is apparent from the boundary conditions (2-66b and 2-66c) of the eigenvalue problem. Then solution 2-73 is applicable for the boundary conditions of the first kind and the appropriate kernels and corresponding eigenvalues are obtainable from Table 2-1.

Application of the general solution 2-73 to several special cases is illustrated below with examples.

EXAMPLE 2-4. A slab $0 \le x \le L$ is initially at temperature $F(x)$. For times $t > 0$ heat is generated within the slab at a rate of $g(x, t)$ Btu/hr ft^3, while the boundary at $x = 0$ is kept insulated and the boundary at $x = L$ is at a temperature which varies with time. The boundary-value problem of heat conduction for this particular case is given as

$$\frac{\partial^2 T}{\partial x^2} + \frac{g(x,t)}{k} = \frac{1}{\alpha}\frac{\partial T}{\partial t} \quad \text{in} \quad 0 \le x \le L,\ t > 0 \tag{2-75a}$$

$$\frac{\partial T}{\partial x} = 0 \quad \text{at} \quad x = 0,\ t > 0 \tag{2-75b}$$

$$T = \phi(t) \quad \text{at} \quad x = L,\ t > 0 \tag{2-75c}$$

$$T = F(x) \quad \text{in} \quad 0 \le x \le L,\ t > 0 \tag{2-75d}$$

where $T \equiv T(x, t)$.

For the present problem the boundary condition at $x = 0$ is of the second kind and that at $x = L$ is of the first kind; therefore, the kernel $K(\beta_m, x)$ and the eigenvalues β_m as obtained from Table 2-1 are

$$K(\beta_m, x) = \sqrt{\frac{2}{L}}\cos\beta_m x \tag{2-76a}$$

and β_m's are the positive roots of

$$\cos\beta L = 0 \tag{2-76b}$$

Since the boundary condition at $x = L$ in problem 2-75 is of the first kind, in the general solution 2-73 the term $\left.\frac{K(\beta_m, x)}{k_2}\right|_{x=L}$ should be replaced by $-\frac{1}{h_2}\left.\frac{dK(\beta_m, x)}{dx}\right|_{x=L}$. Furthermore, by comparing the boundary conditions for the above problem 2-75 with that for the general problem 2-64 we notice that

$$k_1 = 1 \qquad h_1 = 0 \qquad f_1(t) = 0 \tag{2-77a}$$

$$k_2 = 0 \qquad h_2 = 1 \qquad f_2(t) = \phi(t) \tag{2-77b}$$

Substituting Eqs. 2-76 and 2-77 in the general solution 2-73 and replacing

$$\left.\frac{K(\beta_m, x)}{k_2}\right|_{x=L} \quad \text{by} \quad -\frac{1}{h_2}\left.\frac{dK(\beta_m, x)}{dx}\right|_{x=L}$$

we obtain the solution of the boundary-value problem 2-75 as

$$T(x,t) = \frac{2}{L}\sum_{m=0}^{\infty} e^{-\alpha\beta_m^2 t}\cdot\cos\beta_m x\cdot\left\{\int_0^L F(x')\cdot\cos\beta_m x'\cdot dx' + \int_{t'=0}^{t} e^{\alpha\beta_m^2 t'}\cdot\left[\frac{\alpha}{k}\int_0^L g(x',t')\cdot\cos\beta_m x'\cdot dx' + (-1)^m\cdot\alpha\cdot\beta_m\cdot\phi(t')\right]\cdot dt'\right\} \tag{2-78}$$

where

$$\beta_m = \frac{(2m+1)\pi}{2L} \quad \text{with} \quad m = 0, 1, 2, 3, \ldots,$$

As a special case of solution 2-78 we consider zero initial temperature and no heat generation within the solid. Taking

$$F(x) = g(x,t) = 0$$

Eq. 2-78 reduces to

$$T(x,t) = \frac{2\alpha}{L}\sum_{m=0}^{\infty} e^{-\alpha\beta_m^2 t}\cdot(-1)^m\cdot\beta_m\cdot\cos\beta_m x\cdot\int_{t'=0}^{t} e^{\alpha\beta_m^2 t'}\cdot\phi(t')\cdot dt' \tag{2-79}$$

where

$$\beta_m = \frac{(2m+1)\pi}{2L} \quad \text{with} \quad m = 0, 1, 2, 3, \ldots,$$

EXAMPLE 2-5. A slab $0 \le x \le L$ is initially at zero temperature. A plane-surface heat source of strength $g_s(t)$ Btu/hr ft^2 (i.e., strength includes heat released from both surfaces of the surface heat source) is situated at $x = b$ within the solid as shown in Fig. 2-6. For times $t > 0$ the heat source is releasing heat

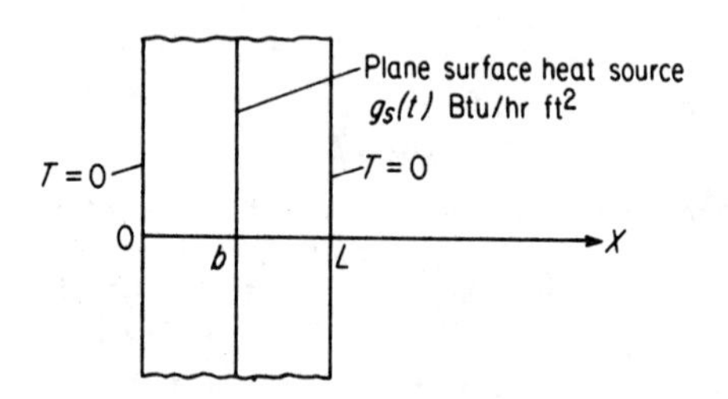

FIG. 2-6. Slab with a plane-surface heat source at $x = b$.

continuously while the boundary surfaces at $x = 0$ and $x = L$ are kept at zero temperature.

In this problem the heat source is a plane-surface heat source of strength $g_s(t)$ Btu/hr ft^2. We relate this surface heat source to the volume heat source

$g(x, t)$ Btu/hr ft^3 by using the Dirac delta function[3] as

$$g(x, t) = g_s(t) \cdot \delta(x - b) \tag{2-80}$$

where $\delta(x - b) = 0$ everywhere $x \neq b$.

Then the boundary-value problem of heat conduction is given as

$$\begin{aligned} \frac{\partial^2 T}{\partial x^2} + \frac{g_s(t)}{k} \cdot \delta(x - b) &= \frac{1}{\alpha}\frac{\partial T}{\partial t} \quad \text{in} \quad 0 \leq x \leq L,\ t > 0 \\ T &= 0 \quad \text{at} \quad x = 0,\ t > 0 \\ T &= 0 \quad \text{at} \quad x = L,\ t > 0 \\ T &= 0 \quad \text{in} \quad 0 \leq x \leq L,\ t = 0 \end{aligned} \tag{2-81}$$

where $T \equiv T(x, t)$.

For the present problem the boundary conditions are both of the first kind; from Table 2-1 the kernel $K(\beta_m, x)$ and the eigenvalues β_m are given as

$$K(\beta_m, x) = \sqrt{\frac{2}{L}} \sin \beta_m x \tag{2-82a}$$

and β_m's are the positive roots of

$$\sin \beta L = 0 \tag{2-82b}$$

Furthermore, by comparing the present boundary-value problem 2-81 with the general problem 2-64 we find that

$$\begin{aligned} k_1 &= 0 \quad h_1 = 1 \quad f_1(t) = 0 \\ k_2 &= 0 \quad h_2 = 1 \quad f_2(t) = 0 \\ F(x) &= 0 \\ g(x, t) &= g_s(t) \cdot \delta(x - b) \end{aligned} \tag{2-83}$$

Substituting Eqs. 2-82 and 2-83 in the general solution 2-73, we obtain the solution of problem 2-81 as

$$\begin{aligned} T(x, t) = \frac{2}{L} \sum_{m=1}^{\infty} & e^{-\alpha \beta_m^2 t} \cdot \sin \beta_m x \\ & \cdot \frac{\alpha}{k} \int_{t'=0}^{t} e^{\alpha \beta_m^2 t'} \cdot g_s(t') \cdot dt' \int_{x'=0}^{L} \delta(x' - b) \cdot \sin \beta_m x' \cdot dx' \end{aligned} \tag{2-84}$$

[3] The Dirac delta function, which is usually called the *delta function*, has the following properties:

$$\delta(x - b) = 0 \quad \text{everywhere} \quad x \neq b$$

$$\int_{-\infty}^{\infty} \delta(x) \cdot dx = 1$$

$$\int_{-\infty}^{\infty} F(x) \cdot \delta(x - b) \cdot dx = F(b)$$

$$f(x) \cdot \delta(x - b) = f(b) \cdot \delta(x - b)$$

The last integration is easily performed, i.e.,

$$\int_{x'=0}^{L} \delta(x' - b) \cdot \sin \beta_m x' \cdot dx' = \sin \beta_m b$$

and Eq. 2-84 becomes

$$T(x,t) = \frac{2\alpha}{Lk} \sum_{m=1}^{\infty} e^{-\alpha\beta_m^2 t} \cdot \sin \beta_m x \cdot \sin \beta_m b \cdot \int_{t'=0}^{t} e^{\alpha\beta_m^2 t'} \cdot g_s(t') \cdot dt' \tag{2-85}$$

where

$$\beta_m = \frac{m\pi}{L} \quad \text{with} \quad m = 1, 2, 3, \ldots$$

We consider now the following special cases of the solution 2-85.

(a) The strength of the plane-surface heat source is constant, i.e., $g_s(t) = g_s =$ constant. In this case integration with respect to the time variable is performed in Eq. 2-85, and the solution becomes

$$T(x,t) = \frac{2g_s}{Lk} \sum_{m=1}^{\infty} \frac{1 - e^{-\alpha\beta_m^2 t}}{\beta_m^2} \cdot \sin \beta_m b \cdot \sin \beta_m x \tag{2-86}$$

where

$$\beta_m = \frac{m\pi}{L} \quad \text{with} \quad m = 1, 2, 3, \ldots,$$

(b) The heat source is an *Instantaneous* plane-surface heat source of strength g_{si} Btu/ft^2 that releases its heat spontaneously at time $t = 0$. An instantaneous plane-surface heat source g_{si} Btu/ft^2 is related to a plane-surface heat source of continuous strength $g_s(t)$ Btu/hr ft^2 by means of the delta function as

$$g_s(t) = g_{si} \cdot \delta(t - 0) \tag{2-87}$$

Substituting Eq. 2-87 into the solution 2-85 and performing integration with respect to the time variable, i.e.,

$$\int_{t'=0}^{t} g_{si} \cdot e^{\alpha\beta_m^2 t'} \cdot \delta(t - 0) \cdot dt' = g_{si}$$

the solution becomes

$$T(x,t) = \frac{2}{L} \frac{\alpha g_{si}}{k} \sum_{m=1}^{\infty} e^{-\alpha\beta_m^2 t} \cdot \sin \beta_m b \cdot \sin \beta_m x \tag{2-88}$$

Sometimes it is convenient to define the strength of instantaneous plane-surface heat source in °F/ft units in the form

$$S_{si} = \frac{\alpha g_{si}}{k} \ {}^\circ\mathrm{F} \cdot \mathrm{ft} \tag{2-89}$$

Using the definition in Eq. 2-89, solution 2-88 becomes

$$T(x,t) = \frac{2}{L} \cdot S_{si} \cdot \sum_{m=1}^{\infty} e^{-\alpha \beta_m^2 t} \cdot \sin \beta_m b \cdot \sin \beta_m x \tag{2-90}$$

where

$$\beta_m = \frac{m\pi}{L} \quad \text{with} \quad m = 1, 2, 3, \ldots$$

We examine the physical significance of the term

$$\frac{2}{L} \sum_{m=1}^{\infty} e^{-\alpha b_m^2 t} \cdot \sin \beta_m b \cdot \sin \beta_m x \tag{2-91}$$

in Eq. 2-90. This term represents the temperature at x at time t due to an instantaneous plane-surface heat source of strength $S_{si} = 1$ °F · ft situated at $x = b$ and releasing its heat spontaneously at time $t = 0$, inside a slab $0 \le x \le L$, which is initially at zero temperature and for times $t > 0$, boundary surfaces of which are kept at zero temperature. The function given by Eq. 2-91 is called *Green's function* for a slab with zero surface temperature. To reiterate, the form of Green's function for the slab problem considered above depends on the combination of the homogeneous boundary conditions chosen; the Green's function 2-91 is for homogeneous boundary condition of the first kind at the boundaries. Use of Green's function in the solution of heat conduction problems will be discussed later in this book.

EXAMPLE 2-6. A slab $0 \le x \le L$ is initially at zero temperature. An instantaneous plane-surface heat source of strength g_{si} Btu/ft^2 is situated at $x = b$ within the slab. Heat is released by the source spontaneously at time $t = 0$. For times $t > 0$ heat is dissipated by convection from the boundary surfaces at $x = 0$ and $x = L$ into a surrounding at zero temperature. Heat-transfer coefficients at the two boundary surfaces of the slab are different.

First we relate the instantaneous plane-surface heat source g_{si} Btu/ft^2 to the volume heat source $g(x, t)$ Btu/hr ft^3 by means of delta functions as

$$g(x,t) = g_{si} \cdot \delta(x - b) \cdot \delta(t - 0) \tag{2-92}$$

Then the boundary-value problem of heat conduction for this particular problem is given as

$$\begin{aligned} \frac{\partial^2 T}{\partial x^2} + \frac{g_{si}}{k}\, \delta(x - b) \cdot \delta(t - 0) &= \frac{1}{\alpha} \frac{\partial T}{\partial t} \quad \text{in} \quad 0 \le x \le L,\ t > 0 \\ -k_1 \frac{\partial T}{\partial x} + h_1 T &= 0 \quad \text{at} \quad x = 0,\ t > 0 \\ k_2 \frac{\partial T}{\partial x} + h_2 T &= 0 \quad \text{at} \quad x = L,\ t > 0 \\ T &= 0 \quad \text{in} \quad 0 \le x \le L,\ t = 0 \end{aligned} \tag{2-93}$$

where $T \equiv T(x, t)$.

Since both boundary conditions are of the third kind the kernel $K(\beta_m, x)$ and the eigenvalues β_m for use in the general solution 2-73 are immediately obtained from Table 2-1 as

$$K(\beta_m, x) = \sqrt{2}\,\frac{\beta_m \cos \beta_m x + H_1 \sin \beta_m x}{\left[(\beta_m^2 + H_1^2)\left(L + \dfrac{H_2}{\beta_m^2 + H_2^2}\right) + H_1\right]^{1/2}} \tag{2-94a}$$

where β_m's are the positive roots of the transcendental equation

$$\tan \beta L = \frac{\beta(H_1 + H_2)}{\beta^2 - H_1 \cdot H_2} \tag{2-94b}$$

and

$$H_1 = \frac{h_1}{k_1} \qquad H_2 = \frac{h_2}{k_2}$$

By comparing the present boundary-value problem of heat conduction, Eq. 2-93, with the general problem 2-64 of slab we notice that

$$f_1(t) = f_2(t) = F(x) = 0 \tag{2-95a}$$

$$g(x, t) = g_{si} \cdot \delta(x - b) \cdot \delta(t - 0) \tag{2-95b}$$

Substituting Eq. 2-94 and 2-95 in the general solution 2-73 and performing the integrations, the solution of the present problem becomes

$$T(x, t) = 2\,\frac{\alpha g_{si}}{k} \sum_{m=1}^{\infty} e^{-\alpha \beta_m^2 t} \cdot \frac{(\beta_m \cos \beta_m x + H_1 \sin \beta_m x)(\beta_m \cos \beta_m b + H_1 \sin \beta_m b)}{(\beta_m^2 + H_1^2)\left(L + \dfrac{H_2}{\beta_m^2 + H_2^2}\right) + H_1} \tag{2-96}$$

where the summation is taken over all positive roots of the transcendental equation 2-94b.

Defining the strength of the heat source in °F · ft units in the form

$$S_{si} = \frac{\alpha g_{si}}{k} \ \text{°F} \cdot \text{ft} \tag{2-89}$$

Equation 2-96 is written in the form

$$T(x, t) = S_{si} \sum_{m=1}^{\infty} e^{-\alpha \beta_m^2 t} \cdot K(\beta_m, x) \cdot K(\beta_m, b) \tag{2-97}$$

where $K(\beta_m, x)$ is as defined by Eq. 2-94a.

In Eq. 2-97 the term

$$\sum_{m=1}^{\infty} e^{-\alpha \beta_m^2 t} \cdot K(\beta_m, x) \cdot K(\beta_m, b) \tag{2-98}$$

represents the temperature at x at time t due to an instantaneous plane-surface heat source of strength $S_{si} = 1$ °F · ft, situated at $x = b$ and releasing its heat spontaneously at time $t = 0$, in a slab $0 \le x \le L$, which is initially at zero temperature and for times $t > 0$ releases heat by convection from its boundary surfaces at $x = 0$ and $x = L$ into a surrounding at zero temperature.

2-4. ONE-DIMENSIONAL HOMOGENEOUS BOUNDARY-VALUE PROBLEMS OF HEAT CONDUCTION IN A SEMI-INFINITE REGION (SOLUTION WITH SEPARATION OF VARIABLES)

Consider a one-dimensional semi-infinite region $0 \leq x < \infty$, which is initially at temperature $F(x)$. For times $t > 0$ the solid dissipates heat by convection at the boundary at $x = 0$ into a surrounding at zero temperature (i.e., homogeneous boundary condition of the third kind). Figure 2-7 shows the geometry and the boundary condition under con-

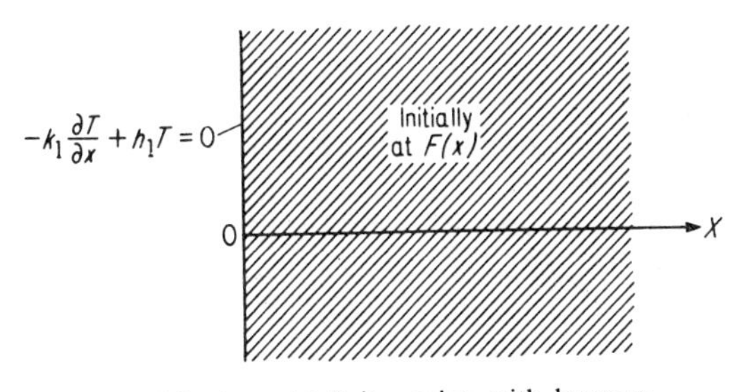

FIG. 2-7. A semi-infinite region with homogeneous boundary condition of the third kind.

sideration. The homogeneous boundary-value problem of heat conduction is given as

$$\frac{\partial^2 T}{\partial x^2} = \frac{1}{\alpha}\frac{\partial T}{\partial t} \quad \text{in} \quad 0 \leq x \leq \infty,\ t > 0 \tag{2-99a}$$

$$-k_1 \frac{\partial T}{\partial x} + h_1 T = 0 \quad \text{at} \quad x = 0, \quad t > 0 \tag{2-99b}$$

$$T = F(x) \quad \text{in} \quad 0 \leq x \leq \infty,\ t = 0 \tag{2-99c}$$

Assuming the separation of the time and space variables, i.e.,

$$T(x, t) = X(x) \cdot \Gamma(t) \tag{2-100}$$

the solution for the time variable is in the form

$$\Gamma(t) : e^{-\alpha\beta^2 t} \tag{2-101}$$

and the space variable $X(x)$ satisfies the following homogeneous system:

$$\frac{d^2 X(x)}{dx^2} + \beta^2 X(x) = 0 \quad \text{in} \quad 0 \leq x \leq \infty \tag{2-102a}$$

$$-k_1 \frac{dX(x)}{dx} + h_1 X(x) = 0 \quad \text{at} \quad x = 0 \tag{2-102b}$$

It can be verified that this homogeneous system is satisfied by the function

$$X(x, \beta) = \beta \cos \beta x + H_1 \sin \beta x \tag{2-103}$$

where $H_1 = h_1/k_1$

β = assumes all values from zero to infinity continuously

Then the general solution of Eq. 2-99 is given in the form

$$T(x, t) = \int_{\beta=0}^{\infty} c(\beta) \cdot e^{-\alpha\beta^2 t} \cdot (\beta \cos \beta x + H_1 \sin \beta x)\, d\beta \tag{2-104}$$

Since this solution should satisfy the initial condition 2-99c of the boundary-value problem of heat conduction, for $t = 0$ we have

$$F(x) = \int_{\beta=0}^{\infty} c(\beta) \cdot (\beta \cos \beta x + H_1 \sin \beta x)\, d\beta \tag{2-105}$$

Equation 2-105 is representation of an arbitrary function $F(x)$ in a semi-infinite interval $0 \le x < \infty$ in terms of trigonometric functions. This representation was treated previously in this chapter and is given as (see Eq. 2-28)

$$F(x) = \int_{\beta=0}^{\infty} K(\beta, x) \int_{x'=0}^{\infty} K(\beta, x') \cdot F(x') \cdot dx'\, d\beta \tag{2-106}$$

where the kernel $K(\beta, x)$ is

$$K(\beta, x) = \sqrt{\frac{2}{\pi}} \, \frac{\beta \cos \beta x + H_1 \sin \beta x}{\sqrt{\beta^2 + H_1^2}} \tag{2-107}$$

$$H_1 \equiv \frac{h_1}{k_1}$$

Therefore,

$$c(\beta) = \frac{2}{\pi} \int_0^{\infty} \frac{\beta \cos \beta x' + H_1 \sin \beta x'}{\beta^2 + H_1^2} F(x')\, dx'$$

and the solution of the homogeneous boundary-value problem of heat conduction, Eq. 2-99, becomes

$$T(x, t) = \int_{\beta=0}^{\infty} e^{-\alpha\beta^2 t} \cdot K(\beta, x) \cdot d\beta \int_{x'=0}^{\infty} K(\beta, x') \cdot F(x') \cdot dx' \tag{2-108}$$

The kernel $K(\beta, x)$ is given by Eq. 2-107. The kernels for use in Eq. 2-108 can also be obtained from Table 2-2.

We now examine some special cases of solution 2-108.

EXAMPLE 2-7. A semi-infinite region $0 \le x < \infty$ is initially at temperature $F(x)$; for time $t > 0$ the boundary surface at $x = 0$ is kept at zero temperature.

For a boundary condition of the first kind at $x = 0$, the kernel for use in the solution 2-108 is obtained from Table 2-2 as

$$K(\beta, x) = \sqrt{\frac{2}{\pi}} \sin \beta x$$

Then the solution of this problem becomes

$$T(x,t) = \frac{2}{\pi} \int_{\beta=0}^{\infty} e^{-\alpha\beta^2 t} \cdot \sin \beta x \cdot d\beta \int_{x'=0}^{\infty} F(x') \cdot \sin \beta x' \cdot dx' \quad \text{(2-109a)}$$

provided that $F(x)$ is well behaved, we change the order of integration

$$T(x,t) = \frac{2}{\pi} \int_{x'=0}^{\infty} F(x') \cdot dx' \int_{\beta=0}^{\infty} e^{-\alpha\beta^2 t} \cdot \sin \beta x \cdot \sin \beta x' d\beta \quad \text{(2-109b)}$$

The integration with respect to β is evaluated by making use of the following relations:

$$2 \sin \beta x \cdot \sin \beta x' = \cos \beta(x - x') - \cos \beta(x + x') \quad \text{(2-110a)}$$

$$\int_{\beta=0}^{\infty} e^{-\alpha\beta^2 t} \cdot \cos \beta(x - x') d\beta = \sqrt{\frac{\pi}{4\alpha t}} \cdot e^{-\frac{(x-x')^2}{4\alpha t}} \quad \text{(2-110b)}$$

$$\int_{\beta=0}^{\infty} e^{-\alpha\beta^2 t} \cdot \cos \beta(x + x') d\beta = \sqrt{\frac{\pi}{4\alpha t}} \cdot e^{-\frac{(x+x')^2}{4\alpha t}} \quad \text{(2-110c)}$$

and the solution 2-109b becomes

$$T(x,t) = \frac{1}{(4\pi\alpha t)^{1/2}} \int_{x'=0}^{\infty} F(x') \cdot \left[e^{-\frac{(x-x')^2}{4\alpha t}} - e^{-\frac{(x+x')^2}{4\alpha t}} \right] dx' \quad \text{(2-111)}$$

For a constant initial temperature in the solid, we take

$$F(x) = T_0 = \text{const.}$$

Equation 2-111 becomes

$$\frac{T(x,t)}{T_0} = \frac{1}{(4\pi\alpha t)^{1/2}} \left[\int_{x'=0}^{\infty} e^{-\frac{(x-x')^2}{4\alpha t}} \cdot dx' - \int_{x'=0}^{\infty} e^{-\frac{(x+x')^2}{4\alpha t}} \cdot dx' \right] \quad \text{(2-112)}$$

Introducing the following new variables,

$$-\eta = \frac{x - x'}{\sqrt{4\alpha t}} \qquad dx' = \sqrt{4\alpha t} \cdot d\eta \text{ for the first integral} \quad \text{(2-113a)}$$

$$\eta = \frac{x + x'}{\sqrt{4\alpha t}} \qquad dx' = \sqrt{4\alpha t} \cdot d\eta \text{ for the second integral} \quad \text{(2-113b)}$$

Eq. 2-112 becomes

$$\frac{T(x,t)}{T_0} = \frac{1}{\sqrt{\pi}} \left[\int_{-\frac{x}{\sqrt{4\alpha t}}}^{\infty} e^{-\eta^2} d\eta - \int_{\frac{x}{\sqrt{4\alpha t}}}^{\infty} e^{-\eta^2} d\eta \right] \quad \text{(2-114)}$$

Since $e^{-\eta^2}$ is symmetrical about $\eta = 0$, Eq. 2-114 is written in the form

$$\frac{T(x,t)}{T_0} = \frac{2}{\sqrt{\pi}} \int_0^{x/\sqrt{4\alpha t}} e^{-\eta^2} \cdot d\eta \quad \text{(2-115)}$$

The right-hand side of Eq. 2-115 is called the *error function*[4] of argument $x/\sqrt{4\alpha t}$ and expressed in the form

$$\frac{T(x,t)}{T_0} = \operatorname{erf}\left(\frac{x}{\sqrt{4\alpha t}}\right) \tag{2-116}$$

Values of error functions are tabulated in Appendix II.

2-5. ONE-DIMENSIONAL NONHOMOGENEOUS BOUNDARY-VALUE PROBLEMS OF HEAT CONDUCTION IN A SEMI-INFINITE REGION (SOLUTION WITH INTEGRAL TRANSFORM)

In this section we consider the solution of nonhomogeneous boundary-value problem of heat conduction in a semi-infinite region using the integral-transform technique.

Consider a semi-infinite region $0 \le x < \infty$ which is initially at temperature $F(x)$. For times $t > 0$ heat is generated within the region at a rate of $g(x,t)$ Btu/hr ft^3, and heat is dissipated by convection from the boundary surface at $x = 0$ into a surrounding temperature which varies with time. Figure 2-8 shows the geometry and the boundary conditions.

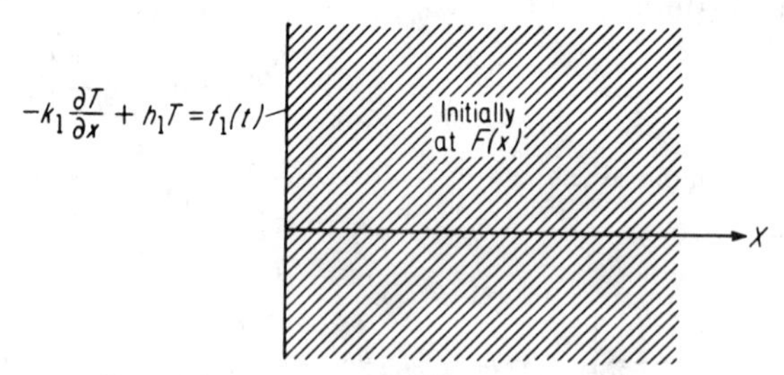

FIG. 2-8. A semi-infinite region with nonhomogeneous boundary conditions of the third kind.

[4] Error function of argument x is defined as

$$\operatorname{erf}(x) = \frac{2}{\sqrt{\pi}} \int_0^x e^{-\eta^2}\, d\eta$$

The error function has the following properties: $\operatorname{erf}(-x) = -\operatorname{erf}(x)$; $\operatorname{erf}(\infty) = 1$. The first and second derivatives of error function are given as

$$\frac{d}{dx}\operatorname{erf}(x) = \frac{2}{\sqrt{\pi}} e^{-x^2} \qquad \frac{d^2}{dx^2}\operatorname{erf} x = -\frac{4}{\sqrt{\pi}}\, x \cdot e^{-x^2}$$

Complimentary error function, erfc, is defined as

$$\operatorname{erfc}(x) = 1 - \operatorname{erf}(x) = \frac{2}{\sqrt{\pi}} \int_x^\infty e^{-\eta^2}\, d\eta$$

The boundary-value problem of heat conduction is given as

$$\frac{\partial^2 T}{\partial x^2} + \frac{g(x,t)}{k} = \frac{1}{\alpha}\frac{\partial T}{\partial t} \quad \text{in} \quad 0 \le x < \infty,\ t > 0 \tag{2-117a}$$

$$-k_1 \frac{\partial T}{\partial x} + h_1 T = f_1(t) \quad \text{at} \quad x = 0,\ t > 0 \tag{2-117b}$$

$$T = F(x) \quad \text{in} \quad 0 \le x < \infty,\ t = 0 \tag{2-117c}$$

where $T \equiv T(x,t)$.

We define the integral transform and the inversion formula for the temperature function $T(x,t)$ with respect to the space variable x, in the semi-infinite region $0 \le x < \infty$ as

$$\begin{pmatrix}\text{Integral}\\ \text{transform}\end{pmatrix} \quad \bar{T}(\beta,t) = \int_{x'=0}^{\infty} K(\beta,x')\cdot T(x',t)\cdot dx' \tag{2-118a}$$

$$\begin{pmatrix}\text{Inversion}\\ \text{formula}\end{pmatrix} \quad T(x,t) = \int_{\beta=0}^{\infty} K(\beta,x)\cdot \bar{T}(\beta,t)\cdot d\beta \tag{2-118b}$$

where the kernel $K(\beta,x)$ is the normalized solution of the following auxiliary problem:

$$\frac{d^2X(x)}{dx^2} + \beta^2 X(x) = 0 \quad \text{in} \quad 0 \le x < \infty \tag{2-119a}$$

$$-k_1 \frac{dX(x)}{dx} + h_1 X(x) = 0 \quad \text{at} \quad x = 0 \tag{2-119b}$$

It is to be noted that the kernel $K(\beta,x)$, for a boundary condition of the third kind at $x = 0$, is given in Table 2-2 as

$$K(\beta,x) = \sqrt{\frac{2}{\pi}}\,\frac{\beta\cos\beta x + H_1\sin\beta x}{\sqrt{\beta^2 + H_1^2}} \tag{2-120}$$

where $H_1 = h_1/k_1$ and β's assume all values from zero to infinity continuously. By applying the transform 2-118a, we take the integral transform of the differential Eq. 2-117a.

$$\int_{x=0}^{\infty} K(\beta,x)\frac{\partial^2 T}{\partial x^2}\cdot dx + \frac{1}{k}\bar{g}(\beta,t) = \frac{1}{\alpha}\frac{d\bar{T}(\beta,t)}{dt} \tag{2-121}$$

where the quantities with bars refer to the integral transform according to transform 2-118a. The first integral on the left-hand side of Eq. 2-121 is evaluated by integrating it by parts twice,

$$\begin{aligned}\int_0^{\infty} K(\beta,x)\cdot\frac{\partial^2 T}{\partial x^2}\,dx &= \left[K\frac{\partial T}{\partial x}\right]_0^{\infty} - \int_0^{\infty}\frac{dK}{dx}\cdot\frac{\partial T}{\partial x}\cdot dx \\ &= \left[K\frac{\partial T}{\partial x} - T\frac{dK}{dx}\right]_0^{\infty} + \int_0^{\infty} T\cdot\frac{d^2K}{dx^2}\cdot dx\end{aligned} \tag{2-122}$$

where $K \equiv K(\beta,x)$.

The terms on the right-hand side of Eq. 2-122 are now evaluated. Assuming that the temperature and its derivative with respect to x vanishes at the upper limit, i.e., $x = \infty$, the first term on the right-hand side becomes

$$\left[K\frac{\partial T}{\partial x} - T\frac{dK}{dx}\right]_0^\infty = \frac{K(\beta,x)}{k_1}\bigg|_{x=0}\cdot f_1(t) \tag{2-123a}$$

The value of the term in the bracket at the lower limit $x = 0$ is evaluated by making use of the boundary condition 2-117b of the heat-conduction problem and Eq. 2-119b of the auxiliary problem.

The integral on the right-hand side of Eq. 2-122 is evaluated by making use of the Eqs. 2-119a and 2-118a.

$$\int_0^\infty T\frac{d^2K}{dx^2}\cdot dx = -\beta^2\int_0^\infty T\cdot K\cdot dx = -\beta^2\cdot\overline{T}(\beta,t) \tag{2-123b}$$

Substituting Eqs. 2-123a and 2-123b into Eq. 2-122, we obtain the integral transformation of the second partial derivative of temperature $\partial^2 T/\partial x^2$ in the semi-infinite range $0 \le x < \infty$ subject to a boundary condition of the third kind at $x = 0$ as

$$\int_0^\infty K(\beta,x)\frac{\partial^2 T}{\partial x^2}\,dx = -\beta^2\overline{T}(\beta,t) + \frac{K(\beta,x)}{k_1}\bigg|_{x=0}\cdot f_1(t) \tag{2-124}$$

Substituting Eq. 2-124 into Eq. 2-121 we obtain

$$\frac{d\overline{T}(\beta,t)}{dt} + \alpha\beta^2\overline{T}(\beta,t) = A(\beta,t) \tag{2-125a}$$

where

$$A(\beta,t) = \frac{\alpha}{k}\,\overline{g}(\beta,t) + \alpha\cdot\frac{K(\beta,x)}{k_1}\bigg|_{x=0}\cdot f_1(t) \tag{2-125b}$$

Thus we removed from the differential equation of heat conduction (2-117a) the second partial derivative with respect to the x variable and reduced it to an ordinary differential equation with respect to the time variable t for the integral transform of temperature $\overline{T}(\beta_m, t)$. The initial condition for the ordinary differential equation 2-125 is obtained by taking the integral transform of the initial condition 2-117c by applying the transform 2-118a. That is,

$$\overline{T}(\beta,t)\,|_{t=0} = \int_0^\infty K(\beta,x')\cdot F(x')\cdot dx' \equiv \overline{F}(\beta) \tag{2-126}$$

Solution of the ordinary differential equation 2-125 subject to the initial condition 2-126 is

$$\overline{T}(\beta,t) = e^{-\alpha\beta^2 t}\left[\overline{F}(\beta) + \int_{t'=0}^{t} e^{\alpha\beta^2 t'}\cdot A(\beta,t')\cdot dt'\right] \tag{2-127}$$

Substituting the integral transform of temperature, Eq. 2-127, into the inversion formula 2-118b, the solution of the nonhomogeneous boundary-value problem of heat conduction, Eq. 2-117, becomes

$$T(x,t) = \int_{\beta=0}^{\infty} e^{-\alpha\beta^2 t} \cdot K(\beta,x) \cdot \left[\bar{F}(\beta) + \int_{t'=0}^{t} e^{\alpha\beta^2 t'} \cdot A(\beta,t') \cdot dt'\right] d\beta \tag{2-128a}$$

where

$$A(\beta,t') = \frac{\alpha}{k}\,\bar{g}(\beta,t') + \alpha \left.\frac{K(\beta,x)}{k_1}\right|_{x=0} \cdot f_1(t') \tag{2-128b}$$

$$\bar{F}(\beta) = \int_{x'=0}^{\infty} K(\beta,x') \cdot F(x') \cdot dx' \tag{2-128c}$$

$$\bar{g}(\beta,t') = \int_{x'=0}^{\infty} K(\beta,x') \cdot g(x',t') \cdot dx' \tag{2-128d}$$

It is to be noted that the special case of this solution for no heat generation (i.e., $g(x,t) = 0$) and homogeneous boundary condition (i.e., $f_1(t) = 0$) is the same as the solution 2-108 for the homogeneous boundary-value problem of heat conduction in a semi-infinite region.

Table 2-2 gives the kernels $K(\beta,x)$ for use in the solution 2-128 for three different boundary conditions at $x = 0$. In applying the solution 2-128 for the boundary condition of the first kind, i.e., $k_1 = 0$, the following change should be made in Eq. 2-128b.

$$\text{When } k_1 = 0, \text{ replace } \left.\frac{K(\beta,x)}{k_1}\right|_{x=0} \text{ by } \left.\frac{1}{h_1}\frac{dK(\beta,x)}{dx}\right|_{x=0} \tag{2-129}$$

In the following examples we illustrate the application of solution 2-128 to special cases.

EXAMPLE 2-8. A semi-infinite solid $0 \le x < \infty$ is initially at temperature $F(x)$. For times $t > 0$ heat is generated within the solid at a rate of $g(x,t)$ Btu/hr ft^3 while the boundary surface at $x = 0$ is kept insulated (i.e., no heat flow).

The boundary-value problem of heat conduction is

$$\frac{\partial^2 T}{\partial x^2} + \frac{g(x,t)}{k} = \frac{1}{\alpha}\frac{\partial T}{\partial t} \quad \text{in } 0 \le x < \infty,\ t > 0 \tag{2-130a}$$

$$\frac{\partial T}{\partial x} = 0 \quad \text{at } x = 0,\ t > 0 \tag{2-130b}$$

$$T = F(x) \quad \text{in } 0 \le x < \infty,\ t = 0 \tag{2-130c}$$

Since the boundary condition at $x = 0$ is of the second kind, the kernel $K(\beta,x)$ from Table 2-2 is

$$K(\beta,x) = \sqrt{\frac{2}{\pi}} \cos \beta x \tag{2-131}$$

Comparing the boundary conditions of the present problem 2-130 with that of the general problem 2-117, we find that

$$h_1 = 0 \qquad f_1(t) = 0 \tag{2-132}$$

Substituting Eqs. 2-131 and 2-132 in the general solution 2-128, we obtain the solution for the present problem as

$$T(x,t) = \frac{2}{\pi} \int_{\beta=0}^{\infty} e^{-\alpha\beta^2 t} \cdot \cos\beta x \cdot \left[\int_{x'=0}^{\infty} F(x') \cdot \cos\beta x' \cdot dx' + \int_{t'=0}^{t} e^{\alpha\beta^2 t'} dt' \cdot \int_{x'=0}^{\infty} \frac{\alpha}{k} g(x',t') \cdot \cos\beta x' \cdot dx' \right] d\beta \tag{2-133}$$

Changing the orders of integrations and rearranging the terms,

$$T(x,t) = \frac{2}{\pi} \int_{x'=0}^{\infty} F(x') \cdot dx' \int_{\beta=0}^{\infty} e^{-\alpha\beta^2 t} \cdot \cos\beta x \cdot \cos\beta x' \cdot d\beta + \frac{2}{\pi}\frac{\alpha}{k} \int_{t'=0}^{t} dt' \int_{x'=0}^{\infty} g(x',t')\, dx' \cdot \int_{\beta=0}^{\infty} e^{-\alpha\beta^2 (t-t')} \cdot \cos\beta x \cdot \cos\beta x' \cdot d\beta \tag{2-134}$$

Integrations with respect to β is performed by making use of the following relation:

$$\frac{2}{\pi} \int_{\beta=0}^{\infty} e^{-\alpha\beta^2 t} \cdot \cos\beta x \cdot \cos\beta x' \, d\beta = \frac{1}{(4\pi\alpha t)^{1/2}} \left[e^{-\frac{(x-x')^2}{4\alpha t}} + e^{-\frac{(x+x')^2}{4\alpha t}} \right] \tag{2-135}$$

Then Eq. 2-134 becomes

$$T(x,t) = \frac{1}{(4\pi\alpha t)^{1/2}} \int_{x'=0}^{\infty} F(x') \cdot \left[e^{-\frac{(x-x')^2}{4\alpha t}} + e^{-\frac{(x+x')^2}{4\alpha t}} \right] dx' + \frac{\alpha}{k} \int_{t'=0}^{t} \frac{dt'}{[4\pi\alpha(t-t')]^{1/2}} \cdot \int_{x'=0}^{\infty} g(x',t') \cdot \left[e^{-\frac{(x-x')^2}{4\alpha(t-t')}} + e^{-\frac{(x+x')^2}{4\alpha(t-t')}} \right] dx' \tag{2-136}$$

We now consider some special cases of solution 2-136.

(a) Solid is initially at zero temperature, and the heat source is a plane-surface heat source of strength $g_s(t)$ Btu/hr ft^2 which is situated at $x = b$. Taking

$$F(x) = 0 \qquad g(x,t) = g_s(t) \cdot \delta(x - b) \tag{2-137}$$

Substituting Eq. 2-137 into Eq. 2-136, and performing integration with respect to space variable, we obtain

$$T(x,t) = \frac{\alpha}{k}\int_{t'=0}^{t} \frac{g_s(t')}{[4\pi\alpha(t-t')]^{1/2}}\left[e^{-\frac{(x-b)^2}{4\alpha(t-t')}} + e^{-\frac{(x+b)^2}{4\alpha(t-t')}}\right]dt' \tag{2-138}$$

(b) Solid is initially at zero temperature, and the heat source is an instantaneous plane-surface heat source of strength g_{si} Btu/ft^2 which is situated at $x = b$ and releasing its heat spontaneously at time $t = \tau$. In this case we have

$$F(x) = 0 \qquad g(x,t) = g_{si}\cdot\delta(x-b)\cdot\delta(t-\tau) \tag{2-139}$$

Substituting Eq. 2-139 in solution 2-136 and performing integrations with respect to the space and time variables, we obtain

$$T(x,t) = \frac{\alpha g_{si}}{k}\frac{1}{[4\pi\alpha(t-\tau)]^{1/2}}\left[e^{-\frac{(x-b)^2}{4\alpha(t-\tau)}} + e^{-\frac{(x+b)^2}{4\alpha(t-\tau)}}\right] \quad \text{for } t > \tau \tag{2-140}$$

Defining the heat source as

$$S_{si} \equiv \frac{\alpha g_{si}}{k} \quad {}^\circ\mathrm{F}\cdot\mathrm{ft}$$

Then in Eq. 2-140 the term

$$\frac{1}{[4\pi\alpha(t-\tau)]^{1/2}}\left[e^{-\frac{(x-b)^2}{4\alpha(t-\tau)}} + e^{-\frac{(x+b)^2}{4\alpha(t-\tau)}}\right] \quad \text{for} \quad t > \tau \tag{2-141}$$

represents temperature at x at time t due to an instantaneous plane-surface heat source of strength $S_{si} = 1°\mathrm{F}\cdot\mathrm{ft}$, situated at $x = b$ and releasing its heat spontaneously at time $t = \tau$ in a semi-infinite region $x \geq 0$ which is initially at zero temperature and for times $t > 0$ its boundary at $x = 0$ is kept insulated. The function 2-141 is called *Green's Function* for a semi-infinite region $x \geq 0$ with insulated boundary at $x = 0$.

Example 2-9. A semi-infinite solid $0 \leq x < \infty$ is initially at temperature $F(x)$. For times $t > 0$ heat is generated within the solid at a rate of $g(x,t)$ Btu/hr ft^3 while the boundary surface at $x = 0$ is kept at zero temperature.

The boundary-value problem of heat conduction is

$$\frac{\partial^2 T}{\partial x^2} + \frac{g(x,t)}{k} = \frac{1}{\alpha}\frac{\partial T}{\partial t} \quad \text{in} \quad 0 \leq x < \infty,\ t > 0 \tag{2-142a}$$

$$T = 0 \quad \text{at} \quad x = 0,\ t > 0 \tag{2-142b}$$

$$T = F(x) \quad \text{in} \quad 0 \leq x < \infty,\ t = 0 \tag{2-142c}$$

Since the boundary condition at $x = 0$ is of the first kind, the kernel $K(\beta, x)$ from Table 2-2 is

$$K(\beta,x) = \sqrt{\frac{2}{\pi}}\sin\beta x \tag{2-143}$$

Comparing the boundary conditions of this problem with that of the general problem 2-117 for a semi-infinite region, we find that

$$k_1 = 0 \qquad f_1(t) = 0 \tag{2-144}$$

Substituting Eqs. 2-143 and 2-144 into the general solution 2-128 and changing the order of integration, we obtain

$$\begin{aligned} T(x,t) &= \frac{2}{\pi}\int_{x'=0}^{\infty} F(x')\cdot dx' \int_{\beta=0}^{\infty} e^{-\alpha\beta^2 t}\cdot \sin\beta x\cdot \sin\beta x'\cdot d\beta \\ &\quad + \frac{2}{\pi}\frac{\alpha}{k}\int_{t'=0}^{t} dt' \int_{x'=0}^{\infty} g(x',t')\cdot dx' \\ &\quad \cdot \int_{\beta=0}^{\infty} e^{-\alpha\beta^2(t-t')}\cdot \sin\beta x\cdot\sin\beta x'\cdot d\beta \end{aligned} \tag{2-145}$$

Integrations with respect to β are performed by making use of the following relation:

$$\begin{aligned} &\frac{2}{\pi}\int_{\beta=0}^{\infty} e^{-\alpha\beta^2 t}\cdot\sin\beta x\cdot\sin\beta x'\cdot d\beta \\ &\qquad = \frac{1}{(4\pi\alpha t)^{1/2}}\left[e^{-\frac{(x-x')^2}{4\alpha t}} - e^{-\frac{(x+x')^2}{4\alpha t}}\right] \end{aligned} \tag{2-146}$$

Then Eq. 2-145 becomes

$$\begin{aligned} T(x,t) &= \frac{1}{(4\pi\alpha t)^{1/2}}\int_{x'=0}^{\infty} F(x')\cdot\left[e^{-\frac{(x-x')^2}{4\alpha t}} - e^{-\frac{(x+x')^2}{4\alpha t}}\right]dx' \\ &\quad + \frac{\alpha}{k}\int_{t'=0}^{t}\frac{dt'}{[4\pi\alpha(t-t')]^{1/2}} \\ &\quad \cdot\int_{x'=0}^{\infty} g(x',t')\cdot\left[e^{-\frac{(x-x')^2}{4\alpha(t-t')}} - e^{-\frac{(x+x')^2}{4\alpha(t-t')}}\right]dx' \end{aligned} \tag{2-147}$$

We consider the following special case of the solution 2-147.

(a) Solid is initially at zero temperature and the heat source is an instantaneous plane-surface heat source of strength g_{si} Btu/ft^2 which is situated at $x = b$ and releases its heat spontaneously at time $t = \tau$.

Taking

$$F(x) = 0 \qquad g(x,t) = g_{si}\cdot\delta(x-b)\cdot\delta(t-\tau) \tag{2-148}$$

and substituting Eq. 2-148 into Eq. 2-147, and performing integrations with respect to the space and time variables, we obtain

$$T(x,t) = \frac{\alpha g_{si}}{k}\frac{1}{[4\pi\alpha(t-\tau)]^{1/2}}\left[e^{-\frac{(x-b)^2}{4\alpha(t-\tau)}} - e^{-\frac{(x+b)^2}{4\alpha(t-\tau)}}\right] \quad \text{for } t>\tau \tag{2-149}$$

It is to be noted that in this solution the term

$$\frac{1}{[4\pi\alpha(t-\tau)]^{1/2}}\left[e^{-\frac{(x-b)^2}{4\alpha(t-\tau)}} - e^{-\frac{(x+b)^2}{4\alpha(t-\tau)}}\right] \tag{2-150}$$

is called *Green's function* for the semi-infinite region $x > 0$ with zero temperature at the boundary surface $x = 0$.

2-6. ONE-DIMENSIONAL HOMOGENEOUS BOUNDARY-VALUE PROBLEM OF HEAT CONDUCTION IN AN INFINITE REGION (SOLUTION WITH SEPARATION OF VARIABLES)

Consider a one-dimensional infinite region which is initially at temperature $F(x)$. The homogeneous boundary-value problem of heat conduction is given as

$$\frac{\partial^2 T}{\partial x^2} = \frac{1}{\alpha}\frac{\partial T}{\partial t} \quad \text{in} \quad -\infty < x < \infty,\ t > 0 \tag{2-151a}$$

$$T = F(x) \quad \text{in} \quad -\infty < x < \infty,\ t = 0 \tag{2-151b}$$

Assuming the separation of the space and time variables, the separated solutions are

$$\begin{aligned} &\Gamma(t): e^{-\alpha\beta^2 t} \\ &X(x): a \cos \beta x + b \sin \beta x \end{aligned} \tag{2-152}$$

where β assumes all values from 0 to infinity continuously.

Then the solution of the problem 2-151 is expressed in the form

$$T(x,t) = \int_{\beta=0}^{\infty} e^{-\alpha\beta^2 t}[a(\beta)\cdot\cos\beta x + b(\beta)\cdot\sin\beta x]\,d\beta \tag{2-153}$$

which should satisfy the initial condition 2-151b—that is,

$$F(x) = \int_{\beta=0}^{\infty} [a(\beta)\cdot\cos\beta x + b(\beta)\cdot\sin\beta x]\,d\beta \tag{2-154}$$

Equation 2-154 is the representation of an arbitrary function $F(x)$ in an infinite interval $-\infty < x < \infty$ in terms of trigonometric function. Such an expansion was considered previously in this chapter and is given by Eq. 2-36 as

$$F(x) = \frac{1}{\pi}\int_{\beta=0}^{\infty} d\beta \int_{x'=-\infty}^{\infty} F(x')\cdot\cos\beta(x - x')\cdot dx' \tag{2-155}$$

In view of Eqs. 2-155 and 2-154, solution 2-153 becomes

$$T(x,t) = \frac{1}{\pi}\int_{\beta=0}^{\infty} e^{-\alpha\beta^2 t}\cdot d\beta \int_{x'=-\infty}^{\infty} F(x')\cdot\cos\beta(x - x')\cdot dx' \tag{2-156}$$

In view of the integral 2-110b, i.e.,

$$\int_{\beta=0}^{\infty} e^{-\alpha\beta^2 t}\cdot\cos\beta(x - x')\cdot d\beta = \sqrt{\frac{\pi}{4\alpha t}}\cdot e^{-\frac{(x-x')^2}{4\alpha t}} \tag{2-110b}$$

Equation 2-156 becomes

$$T(x,t) = \frac{1}{[4\pi\alpha t]^{1/2}}\int_{x'=-\infty}^{\infty} F(x')\cdot e^{-\frac{(x-x')^2}{4\alpha t}}\cdot dx' \tag{2-157}$$

which is a more general form than that given by Eq. 2-156.

In the following example we consider the application of Eq. 2-157 to determining the temperature distribution in an infinite solid for a particular given distribution of the initial temperature.

EXAMPLE 2-10. In a one-dimensional infinite solid, $-\infty < x < \infty$, the region $a < x < b$ is initially at a constant temperature T_0, and everywhere outside this region it is at zero temperature. That is,

$$\begin{aligned} T &= T_0 \quad a < x < b \\ T &= 0 \quad \text{everywhere outside this region} \end{aligned} \tag{2-158}$$

Substituting this condition in Eq. 2-157, we have

$$T(x, t) = \frac{T_0}{[4\pi\alpha t]^{1/2}} \int_a^b e^{-\frac{(x-x')^2}{4\alpha t}} \cdot dx' \tag{2-159}$$

This result is put into a more convenient form by defining a new independent variable

$$\eta = -\frac{x - x'}{\sqrt{4\alpha t}} \quad \therefore \; dx' = \sqrt{4\alpha t}\, d\eta \tag{2-160}$$

Then Eq. 2-159 becomes

$$T(x, t) = \frac{T_0}{\sqrt{\pi}} \int_{\frac{a-x}{\sqrt{4\alpha t}}}^{\frac{b-x}{\sqrt{4\alpha t}}} e^{-\eta^2} d\eta$$

$$= \frac{T_0}{2}\left[\frac{2}{\sqrt{\pi}} \int_0^{\frac{b-x}{\sqrt{4\alpha t}}} e^{-\eta^2} d\eta - \frac{2}{\sqrt{\pi}} \int_0^{\frac{a-x}{\sqrt{4\alpha t}}} e^{-\eta^2} d\eta\right] \tag{2-161}$$

which is written in the form

$$\frac{T(x, t)}{T_0} = \frac{1}{2}\left[\operatorname{erf}\left(\frac{b - x}{\sqrt{4\alpha t}}\right) - \operatorname{erf}\left(\frac{a - x}{\sqrt{4\alpha t}}\right)\right] \tag{2-162}$$

2-7. ONE-DIMENSIONAL NONHOMOGENEOUS BOUNDARY-VALUE PROBLEM OF HEAT CONDUCTION IN AN INFINITE REGION (SOLUTION WITH INTEGRAL TRANSFORM)

We now examine the use of the integral-transform technique in the solution of nonhomogeneous boundary-value problems of heat conduction in a one-dimensional infinite region, which is initially at a temperature $F(x)$ and for times $t > 0$ there is heat generation within the solid at a rate of $g(x, t)$ Btu/hr ft^3.

The boundary-value problem of heat conduction is

$$\frac{\partial^2 T}{\partial x^2} + \frac{g(x, t)}{k} = \frac{1}{\alpha}\frac{\partial T}{\partial t} \quad \text{in} \quad -\infty < x < \infty,\; t > 0 \tag{2-163a}$$

$$T = F(x) \quad \text{in} \quad -\infty < x < \infty,\; t = 0 \tag{2-163b}$$

We define the integral transform and the inversion formula for the temperature function $T(x, t)$ with respect to the x variable $(-\infty < x < \infty)$ as (see Eq. 2-38):

$$\begin{pmatrix}\text{Integral}\\ \text{transform}\end{pmatrix} \quad \bar{T}(\beta, t) = \int_{x'=-\infty}^{\infty} e^{i\beta x'} \cdot T(x', t) \cdot dx' \tag{2-164a}$$

$$\begin{pmatrix}\text{Inversion}\\ \text{formula}\end{pmatrix} \quad T(x, t) = \frac{1}{2\pi} \int_{\beta=-\infty}^{\infty} e^{-i\beta x} \cdot \bar{T}(\beta, t) \cdot d\beta \tag{2-164b}$$

We take the integral transform of the differential Eq. 2-163a by applying the transform, Eq. 2-164a.

$$\int_{-\infty}^{\infty} e^{i\beta x} \cdot \frac{\partial^2 T}{\partial x^2} \cdot dx + \frac{1}{k} \bar{g}(\beta, t) = \frac{1}{\alpha} \frac{d\bar{T}(\beta, t)}{dt} \tag{2-165}$$

where quantities with bar refer to integral transform according to transform 2-164a.

The first integral on the left-hand side of Eq. 2-165 is evaluated by integrating it by parts twice:[6]

$$\int_{-\infty}^{\infty} \frac{\partial^2 T}{\partial x^2} \cdot e^{i\beta x} \cdot dx = \left[\frac{\partial T}{\partial x} \cdot e^{i\beta x} - i\beta T e^{i\beta x}\right]_{-\infty}^{\infty} - \beta^2 \int_{-\infty}^{\infty} T \cdot e^{i\beta x} \cdot dx$$

$$= -\beta^2 \int_{-\infty}^{\infty} T \cdot e^{i\beta x} \cdot dx \equiv -\beta^2 \cdot \bar{T}(\beta, t) \tag{2-166}$$

In obtaining the result in Eq. 2-166 it is assumed that T and $\partial T/\partial x$ both become zero as $x \to \pm\infty$.

Substituting Eq. 2-166 into Eq. 2-165 we obtain

$$\frac{d\bar{T}(\beta, t)}{dt} + \alpha\beta^2 \bar{T}(\beta, t) = \frac{\alpha}{k} \bar{g}(\beta, t) \tag{2-167}$$

where

$$\bar{g}(\beta, t) = \int_{x'=-\infty}^{\infty} e^{i\beta x'} \cdot g(x', t) \cdot dx'$$

The integral transform of the initial condition, Eq. 2-163b, by applying the integral transform 2-164a, is

$$\bar{T}(\beta, t)\Big|_{t=0} = \int_{x'=-\infty}^{\infty} e^{i\beta x'} \cdot F(x') dx' \equiv \bar{F}(\beta) \tag{2-168}$$

The solution of Eq. 2-167 subject to the initial condition, (2-168), gives

$$\bar{T}(\beta, t) = e^{-\alpha\beta^2 t}\left[\bar{F}(\beta) + \frac{\alpha}{k} \int_{t'=0}^{t} \bar{g}(\beta, t') \cdot e^{\alpha\beta^2 t'} \cdot dt'\right] \tag{2-169}$$

Substituting the integral transform of temperature Eq. 2-169 into the inversion Eq. 2-164b, the solution of the heat-conduction problem becomes

$$T(x,t) = \frac{1}{2\pi}\int_{\beta=-\infty}^{\infty} e^{-\alpha\beta^2 t - i\beta x}\left[\bar{F}(\beta) + \frac{\alpha}{k}\int_{t'=0}^{t} e^{\alpha\beta^2 t'}\cdot\bar{g}(\beta,t')\cdot dt'\right]d\beta \tag{2-170}$$

where

$$\bar{F}(\beta) = \int_{x'=-\infty}^{\infty} e^{i\beta x'}\cdot F(x')\cdot dx' \tag{2-171}$$

$$\bar{g}(\beta,t') = \int_{x'=-\infty}^{\infty} e^{i\beta x'}\cdot g(x',t')\cdot dx' \tag{2-172}$$

Changing the order of integration and rearranging the terms, Eq. 2-170 becomes

$$\begin{aligned} T(x,t) = {} & \frac{1}{2\pi}\int_{x'=-\infty}^{\infty} F(x')\cdot dx' \int_{\beta=-\infty}^{\infty} e^{-\alpha\beta^2 t - i\beta(x-x')}\cdot d\beta \\ & + \frac{1}{2\pi}\frac{\alpha}{k}\int_{t'=0}^{t} dt' \int_{x'=-\infty}^{\infty} g(x',t')\cdot dx' \\ & \cdot\int_{\beta=-\infty}^{\infty} e^{-\alpha\beta^2(t-t') - i\beta(x-x')}\cdot d\beta \end{aligned} \tag{2-173}$$

In this equation integrations with respect to β are evaluated by making use of the following relation.

$$\frac{1}{2\pi}\int_{\beta=-\infty}^{\infty} e^{-\alpha\beta^2 t - i\beta x}\, d\beta = \frac{1}{(4\pi\alpha t)^{1/2}}\cdot e^{-\frac{x^2}{4\alpha t}} \tag{2-174}$$

Then Eq. 2-173 becomes

$$\begin{aligned} T(x,t) = {} & \frac{1}{(4\pi\alpha t)^{1/2}}\int_{x'=-\infty}^{\infty} F(x')\cdot e^{-\frac{(x-x')^2}{4\alpha t}}\cdot dx' \\ & + \frac{\alpha}{k}\int_{t'=0}^{t}\frac{dt'}{[4\pi\alpha(t-t')]^{1/2}}\int_{x'=-\infty}^{\infty} g(x',t')\cdot e^{-\frac{(x-x')^2}{4\alpha(t-t')}}\cdot dx' \end{aligned} \tag{2-175}$$

When there is no heat generation within the region (i.e., $g(x,t) = 0$), Eq. 2-175 reduces to the solution of the homogeneous boundary-value problem of heat conduction in a one-dimensional infinite region given by Eq. 2-157.

In the following examples we consider the application of the general solution, Eq. 2-175, to some special cases.

EXAMPLE 2-11. A one-dimensional infinite region $-\infty < x < \infty$ is initially at zero temperature. An instantaneous plane-surface heat source of strength g_{si} Btu/ft^2, which is situated at $x = a$, releases its heat spontaneously at time $t = \tau$.

The solution of this problem is immediately obtained from Eq. 2-175, taking

$$F(x) = 0 \qquad g(x,t) = g_{si} \cdot \delta(x - a) \cdot \delta(t - \tau) \tag{2-176}$$

After performing the integrations we obtain

$$T(x,t) = \frac{\alpha g_{si}}{k} \cdot \frac{e^{-\frac{(x-a)^2}{4\alpha(t-\tau)}}}{[4\pi\alpha(t - \tau)]^{1/2}} \quad \text{for } t > \tau \tag{2-177}$$

Defining

$$S_{si} = \frac{\alpha g_{si}}{k} \ {}^\circ\text{F} \cdot \text{ft} \tag{2-178}$$

as the strength of the instantaneous plane-surface heat source in °F · ft, the term

$$\frac{1}{[4\pi\alpha(t - \tau)]^{1/2}} \cdot e^{-\frac{(x-a)^2}{4\alpha(t-\tau)}} \tag{2-179}$$

represents temperature at x at time t due to an instantaneous plane-surface heat source of strength $S_{si} = 1$°F · ft situated at $x = a$ and releasing its heat spontaneously at time $t = \tau$ in an infinite region $-\infty < x < \infty$ which is initially at zero temperature. The function 2-179 is called *Green's* function for one-dimensional infinite region in the cartesian coordinate system.

EXAMPLE 2-12. A one-dimensional infinite solid $-\infty < x < \infty$ is initially at zero temperature. A plane-surface heat source of strength $g_s(t)$ Btu/hr ft^2, which is situated at $x = a$, releases its heat continuously for times $t > 0$.

The solution of this problem is obtained from Eq. 2-175 by taking

$$F(x) = 0 \qquad g(x,t) = g_s(t) \cdot \delta(x - a) \tag{2-180}$$

Substituting Eq. 2-180 into Eq. 2-175 and performing the integration with respect to space variable, we obtain

$$T(x,t) = \frac{\alpha}{k} \int_{t'=0}^{t} \frac{g_s(t')}{[4\pi\alpha(t - t')]^{1/2}} \cdot e^{-\frac{(x-a)^2}{4\alpha(t-t')}} \cdot dt' \tag{2-181}$$

For a special case of constant heat generation, i.e.,

$$g_s(t) = g_s = \text{const.} \tag{2-182}$$

Equation 2-181 can be integrated. In this case, defining a new variable η as

$$\eta = \frac{x - a}{[4\alpha(t - t')]^{1/2}} \quad \therefore dt' = \frac{1}{\eta^3} \frac{(x - a)^2}{2\alpha} d\eta \tag{2-183}$$

Equation 2-181 becomes

$$T(x,t) = g_s \frac{x - a}{2k\sqrt{\pi}} \int_{\frac{x-a}{\sqrt{4\alpha t}}}^{\infty} \frac{e^{-\eta^2}}{\eta^2} d\eta \tag{2-184}$$

Integrating Eq. 2-184 by parts, we obtain

$$T(x,t) = \frac{\alpha g_s}{k}\left[\left(\frac{t}{\alpha\pi}\right)^{1/2}\cdot e^{-\frac{(x-a)^2}{4\alpha t}} - \frac{|x-a|}{2\alpha}\operatorname{erfc}\frac{|x-a|}{\sqrt{4\alpha t}}\right] \tag{2-185}$$

2-8. TWO- AND THREE-DIMENSIONAL NONHOMOGENEOUS BOUNDARY-VALUE PROBLEMS OF HEAT CONDUCTION (SOLUTION BY INTEGRAL TRANSFORM)

In this section we examine the solution of the nonhomogeneous boundary-value problems of heat conduction involving more than one space variable by means of the integral-transform technique. A straightforward method is the repeated application of one-dimensional integral transform with respect to the space variables; the partial derivatives with respect to the space variables are removed, the resulting ordinary differential equation is solved, and the multiple transform of temperature successively inverted.

An essentially equivalent method is the use of multiple integral transform and the corresponding inversion formula. By applying the multiple transform the partial derivatives with respect to the space variables are removed in one step, the resulting ordinary differential equation is solved and the transform of temperature is inverted.

Both approaches will be used to illustrate their application.

A Finite Rectangle. Consider a finite rectangle $0 \le x \le a$, $0 \le y \le b$ which is initially at temperature $F(x, y)$. For times $t > 0$ there is heat generation within the solid at a rate of $g(x, y, t)$ Btu/hr ft^3, while heat is dissipated by convection from the boundary surfaces into a surrounding temperature which varies both with time and position along the boundaries. Figure 2-9 shows the geometry and the boundary conditions under consideration.

The boundary-value problem of heat conduction is given as

$$\frac{\partial^2 T}{\partial x^2} + \frac{\partial^2 T}{\partial y^2} + \frac{g(x,y,t)}{k} = \frac{1}{\alpha}\frac{\partial T}{\partial t} \quad \text{in } 0 \le x \le a, 0 \le y \le b,\ t > 0 \tag{2-186}$$

$$-k_1\frac{\partial T}{\partial x} + h_1 T = f_1(y,t) \qquad \text{at } x = 0, \qquad t > 0 \tag{2-187}$$

$$k_2\frac{\partial T}{\partial x} + h_2 T = f_2(y,t) \qquad \text{at } x = a, \qquad t > 0 \tag{2-188}$$

$$-k_3\frac{\partial T}{\partial y} + h_3 T = f_3(x,t) \qquad \text{at } y = 0, \qquad t > 0 \tag{2-189}$$

$$k_4\frac{\partial T}{\partial y} + h_4 T = f_4(x,t) \qquad \text{at } y = b, \qquad t > 0 \tag{2-190}$$

$$T = F(x,y) \qquad \text{in } 0 \le x \le a, 0 \le y \le b, t = 0 \tag{2-191}$$

where $T \equiv T(x, y, t)$.

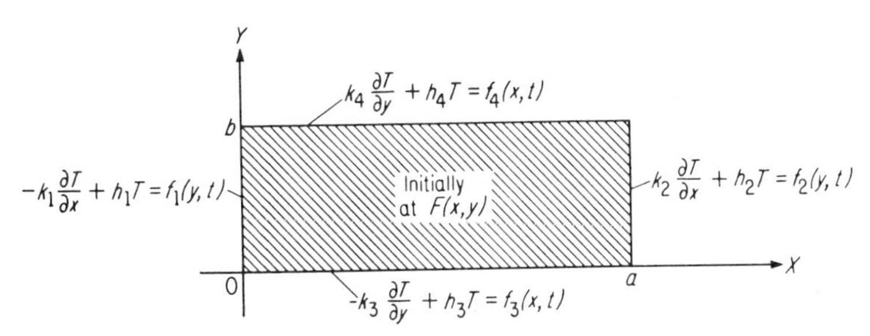

FIG. 2-9. A finite rectangle with nonhomogeneous boundary conditions of the third kind.

We shall solve this problem by the application of multiple-integral transform and the corresponding multiple-inversion formula. Since there are two space variables involved and the range of these space variables is finite, we define a "double-integral transform" and the corresponding "double-inversion formula" with respect to the space variables x, y, of the temperature function $T(x, y, t)$ in the finite range $0 \le x \le a$, $0 \le y \le b$ as

$$\begin{pmatrix}\text{Double-integral}\\ \text{transform}\end{pmatrix} \quad \bar{T}(\beta_m, \nu_n, t) = \int_{x'=0}^{a} \int_{y'=0}^{b} K(\beta_m, x') \cdot K(\nu_n, y') \cdot T(x', y', t) \cdot dx' dy' \tag{2-192a}$$

$$\begin{pmatrix}\text{Double-inversion}\\ \text{formula}\end{pmatrix} \quad T(x, y, t) = \sum_{m=1}^{\infty} \sum_{n=1}^{\infty} K(\beta_m, x) \cdot K(\nu_n, y) \cdot \bar{T}(\beta_m, \nu_n, t) \tag{2-192b}$$

where the summation is taken over all eigenvalues β_m and ν_n.

It is to be noted that the double-integral transform and the double-inversion formula as defined above are merely the repeated application of one-dimensional finite integral transform and the inversion formula for the x and y space variables in the regions $0 \le x \le a$ and $0 \le y \le b$ respectively. Therefore the kernels $K(\beta_m, x)$ and $K(\nu_n, y)$ in Eq. 2-192 are the same as those given in Table 2-1.

We now take the integral transform of the differential equation 2-186 by applying the double transform 2-192a,

$$\int_{x=0}^{a} \int_{y=0}^{b} K(\beta_m, x) \cdot K(\nu_n, y) \cdot \left(\frac{\partial^2 T}{\partial x^2} + \frac{\partial^2 T}{\partial y^2}\right) dx \cdot dy + \frac{1}{k} \bar{g}(\beta_m, \nu_n, t) = \frac{1}{\alpha} \frac{d\bar{T}(\beta_m, \nu_n, t)}{dt} \tag{2-193}$$

where the bars refer to the double transform according to the integral transform 2-192a.

Equation 2-193 is similar in form to that given by Eq. 1-60 in Chapter 1. The first integral on the left-hand side of Eq. 2-193 is the integral transform of the Laplacian which is evaluated by means of Green's theorem as discussed in Sec. 1-8 and Eq. 2-193 becomes (See Eq. 1-65)

$$\frac{d\bar{T}(\beta_m, \nu_n, t)}{dt} + \alpha(\beta_m^2 + \nu_n^2)\bar{T}(\beta_m, \nu_n, t) = A(\beta_m, \nu_n, t) \tag{2-194}$$

where bars denote double transform according to the integral transform 2-192a; and $A(\beta_m, \nu_n, t)$ is defined below. This equation is subject to the transformed initial condition:

$$\bar{T}(\beta_m, \nu_n, t)\big|_{t=0} = \bar{F}(\beta_m, \nu_n) \tag{2-195}$$

The solution of Eq. 2-194 subject to the initial condition 2-195 is straightforward and gives the double transform of temperature $\bar{T}(\beta_m, \nu_n, t)$. Substituting the resulting double transform of temperature into the inversion formula 2-192b, we obtain the solution of the boundary-value problem of heat conduction as

$$T(x, y, t) = \sum_{m=1}^{\infty}\sum_{n=1}^{\infty} e^{-\alpha(\beta_m^2+\nu_n^2)t} \cdot K(\beta_m, x) \cdot K(\nu_n, y)\left[\bar{F}(\beta_m, \nu_n) + \int_{t'=0}^{t} e^{\alpha(\beta_m^2+\nu_n^2)t'} \cdot A(\beta_m, \nu_n, t') \cdot dt'\right] \cdots \tag{2-196}$$

where

$$\begin{aligned} A(\beta_m, \nu_n, t') = {} & \frac{\alpha}{k}\bar{g}(\beta_m, \nu_n, t') \\ & + \alpha\left\{\frac{K(\beta_m, x)}{k_1}\bigg|_{x=0} \cdot \int_{y'=0}^{b} K(\nu_n, y') \cdot f_1(y', t') \cdot dy'\right. \\ & + \frac{K(\beta_m, x)}{k_2}\bigg|_{x=a} \cdot \int_{y'=0}^{b} K(\nu_n, y') \cdot f_2(y', t') \cdot dy' \\ & + \frac{K(\nu_n, y)}{k_3}\bigg|_{y=0} \cdot \int_{x'=0}^{a} K(\beta_m, x') \cdot f_3(x', t') \cdot dx' \\ & \left. + \frac{K(\nu_n, y)}{k_4}\bigg|_{y=b} \cdot \int_{x'=0}^{a} K(\beta_m, x') \cdot f_4(x', t') \cdot dx'\right\} \cdots \end{aligned} \tag{2-197}$$

$$\bar{g}(\beta_m, \nu_n, t') = \int_{x'=0}^{a}\int_{y'=0}^{b} K(\beta_m, x') \cdot K(\nu_n, y') \cdot g(x', y', t') \cdot dx'dy' \cdots \tag{2-198}$$

$$\bar{F}(\beta_m, \nu_n) = \int_{x'=0}^{a} \int_{y'=0}^{b} K(\beta_m, x') \cdot K(\nu_n, y') \cdot F(x', y') \cdot dx'dy' \cdots \tag{2-199}$$

where the summation is taken over all eigenvalues β_m and ν_n.

The kernels $K(\beta_m, x)$, $K(\nu_n, y)$ and the corresponding eigenvalues β_m, ν_n are obtained from Table 2-1 for various combinations of boundary conditions.

In applying the above solution to the cases with boundary condition of the first kind we set $k_i = 0$. To avoid difficulty associated with k_i being zero, the following changes should be made in $A(\beta_m, \nu_n, t')$, i.e., Eq. 2-197.

$$\text{When} \quad k_1 = 0, \quad \text{replace} \quad \left.\frac{K(\beta_m, x)}{k_1}\right|_{x=0} \quad \text{by} \quad \frac{1}{h_1} \left.\frac{dK(\beta_m, x)}{dx}\right|_{x=0} \tag{2-200a}$$

$$\text{When} \quad k_2 = 0, \quad \text{replace} \quad \left.\frac{K(\beta_m, x)}{k_2}\right|_{x=a} \quad \text{by} \quad -\frac{1}{h_2} \left.\frac{dK(\beta_m, x)}{dx}\right|_{x=a} \tag{2-200b}$$

$$\text{When} \quad k_3 = 0, \quad \text{replace} \quad \left.\frac{K(\nu_n, y)}{k_3}\right|_{y=0} \quad \text{by} \quad \frac{1}{h_3} \left.\frac{dK(\nu_n, y)}{dy}\right|_{y=0} \tag{2-200c}$$

$$\text{When} \quad k_4 = 0, \quad \text{replace} \quad \left.\frac{K(\nu_n, y)}{k_4}\right|_{y=b} \quad \text{by} -\frac{1}{h_4} \left.\frac{dK(\nu_n, y)}{dy}\right|_{y=b} \tag{2-200d}$$

We examine some special cases of the solution 2-196.

EXAMPLE 2-13. A finite rectangle $0 \le x \le a$, $0 \le y \le b$ is initially at temperature $F(x, y)$. For times $t > 0$ heat is generated within the solid at a rate of $g(x, y, t)$ Btu/hr ft^3, while the boundary at $y = 0$ is kept at temperature $\phi(x, t)$ and the remaining boundaries are kept at zero temperature. (See Fig. 2-10.)

The boundary and the initial conditions for this problem are given as

$$\left.\begin{aligned} T &= 0 && \text{at} \quad x = 0 \\ T &= 0 && \text{at} \quad x = a \\ T &= \phi(x, t) && \text{at} \quad y = 0 \\ T &= 0 && \text{at} \quad y = b \end{aligned}\right\} t > 0 \tag{2-201}$$

$$T = F(x, y) \quad \text{in} \quad 0 \le x \le a, 0 \le y \le b, t = 0$$

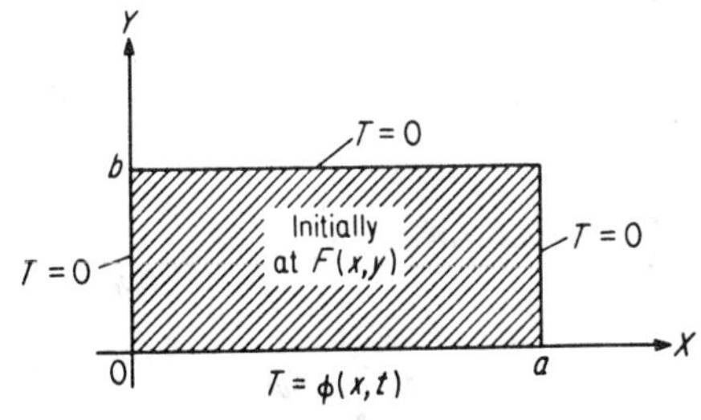

FIG. 2-10. Boundary conditions for a rectangle.

Since the x variable in the range $0 \le x \le a$ and the boundary conditions at $x = 0$ and $x = a$ are both of the first kind, the kernel $K(\beta_m, x)$ and the eigenvalues β_m are immediately obtained from Table 2-1 as

$$K(\beta_m, x) = \sqrt{\frac{2}{a}} \sin \beta_m x \tag{2-202}$$

where

$$\beta_m = \frac{m\pi}{a} \qquad m = 1,2,3,\ldots,$$

Since the y variable is in the range $0 \le y \le b$, and the boundary conditions at $y = 0$ and $y = b$ are of the first kind, the kernel $K(\nu_n, y)$, the eigenvalues ν_n are obtained from Table 2-1 as

$$K(\nu_n, y) = \sqrt{\frac{2}{b}} \sin \nu_n y \tag{2-203}$$

where

$$\nu_n = \frac{n\pi}{b} \qquad n = 1,2,3,\ldots$$

By comparing the boundary-condition functions for the present problem with those for the more general problem 2-186 through 2-191 we note that

$$f_1(y,t) = f_2(y,t) = f_4(x,t) = 0,\ k_1 = k_2 = k_3 = k_4 = 0$$
$$\text{and } f_3(x,t) = \phi(x,t) \text{ for } h_3 = 1 \tag{2-204}$$

Since $k_3 = 0$, the following change should be made in $A(\beta_m, \nu_n, t')$,

$$\text{replace } \left.\frac{K(\nu_n, y)}{k_3}\right|_{y=0} \text{ by } \left.\frac{1}{h_3}\frac{dK(\nu_n, y)}{dy}\right|_{y=0} \tag{2-205}$$

Then the solution of the present problem becomes

$$\begin{aligned}
T(x,y,t) = {} & \frac{4}{ab} \sum_{m=1}^{\infty} \sum_{n=1}^{\infty} e^{-\alpha(\beta_m^2+\nu_n^2)t} \cdot \sin \beta_m x \\
& \cdot \sin \nu_n y \int_{x'=0}^{a} \int_{y'=0}^{b} F(x',y') \cdot \sin \beta_m x' \cdot \sin \nu_n y' \cdot dx'dy' \\
& + \frac{4}{ab} \sum_{m=1}^{\infty} \sum_{n=1}^{\infty} e^{-\alpha(\beta_m^2+\nu_n^2)t} \cdot \sin \beta_m x \cdot \sin \nu_n y \int_{t'=0}^{t} e^{\alpha(\beta_m^2+\nu_n^2)t'}\, dt' \\
& \cdot \int_{x'=0}^{a} \int_{y'=0}^{b} \frac{\alpha}{k} \sin \beta_m x' \cdot \sin \nu_n y' \cdot g(x',y',t') \cdot dx'dy' \\
& + \frac{4}{ab} \sum_{m=1}^{\infty} \sum_{n=1}^{\infty} e^{-\alpha(\beta_m^2+\nu_n^2)t} \cdot \sin \beta_m x \cdot \sin \nu_n y \int_{t'=0}^{t} e^{\alpha(\beta_m^2+\nu_n^2)t'} \\
& \cdot dt' \int_{x'=0}^{a} \alpha \cdot \phi(x',t') \cdot \nu_n \cdot \sin \beta_m x' \cdot dx'
\end{aligned} \tag{2-206}$$

where

$$\beta_m = \frac{m\pi}{a} \qquad m = 1, 2, 3, \ldots$$

$$\nu_n = \frac{n\pi}{b} \qquad n = 1, 2, 3, \ldots$$

Summation is taken over all eigenvalues.

It is to be noted that the above solution includes three different terms. The first term on the right-hand side is for the effects of the initial condition in the solid, the second term for the effects of heat generation and the third term for the effects of prescribed temperature distribution at the boundary $y = 0$.

EXAMPLE 2-14. We now examine some special cases of solution 2-206.

(a) If we assume further that the solid is initially at zero temperature (i.e., $F(x, y) = 0$), and that there is no heat generation within the solid (i.e., $g(x, y) = 0$) and the boundary-condition function at $y = 0$ is independent of time but a function of x only [i.e., $\phi(x, t) = \phi(x)$], Eq. 2-206 simplifies to

$$T(x, y, t) = \frac{4}{ab} \sum_{m=1}^{\infty} \sum_{n=1}^{\infty} \frac{\nu_n}{\beta_m^2 + \nu_n^2} \sin \beta_m x$$

$$\cdot \sin \nu_n y \left(1 - e^{-\alpha(\beta_m^2 + \nu_n^2)t}\right) \int_{x'=0}^{a} \phi(x') \cdot \sin \beta_m x' \cdot dx' \qquad (2\text{-}207)$$

The steady-state temperature distribution is obtainable from Eq. 2-207 by letting $t \to \infty$. Then

$$T(x, y, \infty) = \frac{4}{ab} \sum_{m=1}^{\infty} \sum_{n=1}^{\infty} \frac{\nu_n}{\beta_m^2 + \nu_n^2} \sin \beta_m x \cdot \sin \nu_n y \int_{x'=0}^{a} \phi(x') \cdot \sin \beta_m x' \cdot dx' \qquad (2\text{-}208)$$

If we make use of the following relation (see Eq. 2-281),

$$\frac{2}{b} \sum_{n=1}^{\infty} \frac{\nu_n}{\beta_m^2 + \nu_n^2} \sin \nu_n y = \frac{\sinh \beta_m (b - y)}{\sinh \beta_m b}$$

double summation of Eq. 2-208 is reduced to a single summation:

$$T(x, y, \infty) = \frac{2}{a} \sum_{m=1}^{\infty} \frac{\sinh \beta_m (b - y)}{\sinh \beta_m b} \cdot \sin \beta_m x \int_{x'=0}^{a} \phi(x') \cdot \sin \beta_m x' \cdot dx' \qquad (2\text{-}209)$$

EXAMPLE 2-15. A finite rectangle $0 \le x \le a$, $0 \le y \le b$ is initially at zero temperature. An instantaneous line heat source of strength g_{Li} Btu/ft is situated at $x = x_1$, $y = y_1$ inside the rectangle perpendicular to x-y plane. The heat source releases its heat spontaneously at time $t = \tau$, while the boundaries of the region are kept at zero temperature.

In this problem the instantaneous line heat source g_{Li} Btu/ft is related to the volume heat source $g(x, y, t)$ Btu/hr · ft^3 by

$$g(x, y, t) = g_{\text{Li}}\,\delta(x - x_1)\cdot\delta(y - y_1)\cdot\delta(t - \tau) \tag{2-210}$$

Since the boundary conditions are all of the first kind the kernels $K(\beta_m, x)$ and $K(\nu_n, y)$ are obtained from Table 2-1 as

$$\begin{aligned} K(\beta_m, x) &= \sqrt{\frac{2}{a}}\sin\beta_m x \qquad \beta_m = \frac{m\pi}{a} \qquad m = 1, 2, 3, \ldots \\ K(\nu_n, y) &= \sqrt{\frac{2}{b}}\sin\nu_n y \qquad \nu_n = \frac{n\pi}{b} \qquad n = 1, 2, 3, \ldots \end{aligned} \tag{2-211}$$

For the present problem we have

$$F(x, y) = f_1(y, t) = f_2(y, t) = f_3(x, t) = f_4(x, t) = 0 \tag{2-212}$$

Substituting Eqs. 2-210, 2-211, and 2-212 into the general solution 2-196, we obtain

$$\begin{aligned} T(x, y, t) = \frac{4}{ab}\frac{g_{\text{Li}}\alpha}{k}\sum_{m=1}^{\infty}\sum_{n=1}^{\infty} e^{-\alpha(\beta_m^2+\nu_n^2)t}\cdot\sin\beta_m x \\ \cdot\sin\nu_n y\int_{t'=0}^{t} e^{\alpha(\beta_m^2+\nu_n^2)t'}\cdot\delta(t' - \tau)\,dt' \\ \cdot\int_{x'=0}^{a}\int_{y'=0}^{b}\delta(x' - x_1)\cdot\delta(y' - y_1)\cdot\sin\beta_m x'\cdot\sin\nu_n y'\cdot dx'dy' \end{aligned} \tag{2-213}$$

Performing the integrations involving delta functions,

$$\begin{aligned} T(x, y, t) = \frac{4}{ab}\frac{g_{\text{Li}}\alpha}{k}\sum_{m=1}^{\infty}\sum_{n=1}^{\infty} e^{-\alpha\lambda_{mn}^2(t-\tau)}\cdot\sin\beta_m x_1 \\ \cdot\sin\nu_n y_1\cdot\sin\beta_m x\cdot\sin\nu_n y \quad \text{for } t > \tau \end{aligned} \tag{2-214}$$

where

$$\begin{aligned} \lambda_{mn}^2 &= \beta_m^2 + \nu_n^2 \\ \beta_m &= \frac{m\pi}{a} \qquad m = 1, 2, 3, \ldots \\ \nu_n &= \frac{n\pi}{b} \qquad n = 1, 2, 3, \ldots \end{aligned}$$

Defining

$$S_{\text{Li}} \equiv \frac{\alpha g_{\text{Li}}}{k} \quad \text{in } {}^\circ\text{F}\cdot\text{ft}^2 \tag{2-215}$$

as the strength of the instantaneous line heat source in °F · ft^2, then in Eq. 2-214 the term

$$\frac{4}{ab}\sum_{m=1}^{\infty}\sum_{n=1}^{\infty} e^{-\alpha\lambda_{mn}^2(t-\tau)}\cdot\sin\beta_m x_1\cdot\sin\nu_n y_1\cdot\sin\beta_m x\cdot\sin\nu_n y \tag{2-216}$$

represents temperature at (x, y) at time t, due to an instantaneous line heat source of strength $S_{Li} = 1$ °F · ft², situated at (x_1, y_1) and releasing it spontaneously at time $t = \tau$, inside a finite rectangle $0 \leq x \leq a$, $0 \leq y \leq b$, which is initially at zero temperature and for times $t > 0$ the boundaries of which are kept at zero temperature. The function 2-216 is called *Green's* function for a finite rectangle $0 \leq x \leq a$, $0 \leq y \leq b$, for zero surface temperature.

Semi-Infinite Rectangular Strip. Consider a semi-infinite rectangular strip $0 \leq y \leq b$, $0 \leq x < \infty$ which is initially at temperature $F(x, y)$. For times $t > 0$ heat is generated within the solid at a rate of $g(x, y, t)$ Btu/hr ft³, while there is convection from the boundaries into a surrounding temperature of which varies both with time and position along the boundary. For generality we assume that the heat-transfer coefficients are different at each boundary surface. Figure 2-11 shows the geometry and the boundary conditions under consideration.

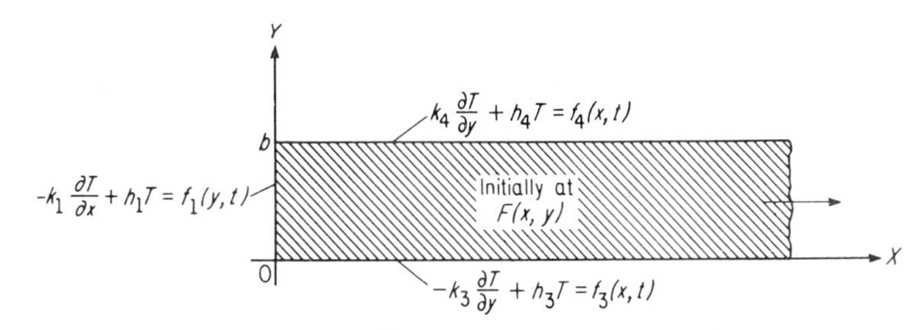

FIG. 2-11. Boundary conditions for a semi-infinite rectangular strip.

The boundary-value problem of heat conduction is given as

$$\frac{\partial^2 T}{\partial x^2} + \frac{\partial^2 T}{\partial y^2} + \frac{g(x, y, t)}{k} = \frac{1}{\alpha} \frac{\partial T}{\partial t} \quad \text{in} \quad 0 \leq x < \infty, 0 \leq y \leq b, t > 0$$

$$-k_1 \frac{\partial T}{\partial x} + h_1 T = f_1(y, t) \quad \text{at} \quad x = 0, t > 0$$

$$-k_3 \frac{\partial T}{\partial y} + h_3 T = f_3(x, t) \quad \text{at} \quad y = 0, t > 0$$

$$k_4 \frac{\partial T}{\partial y} + h_4 T = f_4(x, t) \quad \text{at} \quad y = b, t > 0$$

$$T = F(x, y) \quad \text{in} \quad 0 \leq x < \infty, t = 0 \tag{2-217}$$

where $T \equiv T(x, y, t)$.

We shall solve this problem by the successive application of the one-dimensional integral transform.

Since the range of space variable y is finite, $0 \leq y \leq b$, we define the integral transform and the inversion formula with respect to the y variable of the temperature function $T(x,y,t)$ in the finite range $0 \le y \le b$ as

$$\begin{pmatrix}\text{Integral}\\ \text{transform}\end{pmatrix} \quad \bar{T}(x,\nu_n,t) = \int_{y'=0}^{b} K(\nu_n,y')\cdot T(x',y',t)\,dy' \tag{2-218a}$$

$$\begin{pmatrix}\text{Inversion}\\ \text{formula}\end{pmatrix} \quad T(x,y,t) = \sum_{n=1}^{\infty} K(\nu_n,y)\cdot \bar{T}(x,\nu_n,y) \tag{2-218b}$$

We take integral transform of system 2-217 with respect to the y variable by applying the transform 2-218a.

$$\frac{\partial^2 \bar{T}(x,\nu_n,t)}{\partial x^2} + \left\{-\nu_n^2\cdot \bar{T}(x,\nu_n,t) + \left.\frac{K(\nu_n,y)}{k_3}\right|_{y=0}\cdot \bar{f}_3(x,t) + \left.\frac{K(\nu_n,y)}{k_4}\right|_{y=b}\cdot \bar{f}_4(x,t)\right\} + \frac{\bar{g}(x,\nu_n,t)}{k} = \frac{1}{\alpha}\frac{\partial \bar{T}(x,\nu_n,t)}{\partial t} \quad \text{in} \quad 0 \le x < \infty,\ t > 0$$

$$-k_1\frac{\partial \bar{T}(x,\nu_n,t)}{\partial x} + h_1\bar{T}(x,\nu_n,t) = \bar{f}_1(\nu_n,t) \quad \text{at} \quad x = 0,\ t > 0,$$

$$\bar{T}(x,\nu_n,t) = \bar{F}(x,\nu_n) \quad \text{in} \quad 0 \le x < \infty,\ t = 0 \tag{2-219}$$

Thus we removed from the system 2-217 the second partial derivative with respect to the y variable.

We now define the integral transform and the inversion formula of the temperature function $\bar{T}(x,\nu_n,t)$ with respect to the x variable in the semi-infinite range $0 \le x < \infty$ as (see Eq. 2-29):

$$\begin{pmatrix}\text{Integral}\\ \text{transform}\end{pmatrix} \quad \bar{\bar{T}}(\beta,\nu_n,t) = \int_{x'=0}^{\infty} K(\beta,x')\cdot \bar{T}(x',\nu_n,t)\cdot dx' \tag{2-220a}$$

$$\begin{pmatrix}\text{Inversion}\\ \text{formula}\end{pmatrix} \quad \bar{T}(x,\nu_n,t) = \int_{\beta=0}^{\infty} K(\beta,x)\cdot \bar{\bar{T}}(\beta,\nu_n,t)\cdot d\beta \tag{2-220b}$$

We take the integral transform of system 2-219 by applying the transform 2-220a.

$$\left\{-\beta^2\cdot \bar{\bar{T}}(\beta,\nu_n,t) + \left.\frac{K(\beta,x)}{k_1}\right|_{x=0}\cdot \bar{f}_1(\nu_n,t)\right\} + \left\{-\nu_n^2\cdot \bar{\bar{T}}(\beta,\nu_n,t) + \left.\frac{K(\nu_n,y)}{k_3}\right|_{y=0}\cdot \bar{\bar{f}}_3(\beta,t) + \left.\frac{K(\nu_n,y)}{k_4}\right|_{y=b}\cdot \bar{\bar{f}}_4(\beta,t)\right\} + \frac{\bar{\bar{g}}(\beta,\nu_n,t)}{k} = \frac{1}{\alpha}\frac{d\bar{\bar{T}}(\beta,\nu_n,t)}{dt} \tag{2-221a}$$

$$\bar{\bar{T}}(\beta, \nu_n, t) = \bar{\bar{F}}(\beta, \nu_n) \quad \text{for } t = 0 \tag{2-221b}$$

Equation 2-221 is rearranged as

$$\frac{d\bar{\bar{T}}(\beta, \nu_n, t)}{dt} + \alpha(\beta^2 + \nu_n^2) \cdot \bar{\bar{T}}(\beta, \nu_n, t) = A(\beta, \nu_n, t) \tag{2-222a}$$

$$\bar{\bar{T}}(\beta, \nu_n, t) = \bar{\bar{F}}(\beta, \nu_n) \quad \text{for } t = 0 \tag{2-222b}$$

where

$$A(\beta, \nu_n, t) \equiv \frac{\alpha}{k} \bar{\bar{g}}(\beta, \nu_n, t) + \alpha \left\{ \left. \frac{K(\beta, x)}{k_1} \right|_{x=0} \cdot \bar{f}_1(\nu_n, t) + \left. \frac{K(\nu_n, y)}{k_3} \right|_{y=0} \cdot \bar{f}_3(\beta, t) + \left. \frac{K(\nu_n, y)}{k_4} \right|_{y=b} \cdot \bar{f}_4(\beta, t) \right\}$$

Thus we removed from system 2-217 the partial derivatives with respect to the space variables and reduced it to a first-order ordinary differential equation for the double transform of temperature $\bar{\bar{T}}(\beta, \nu_n, t)$ with respect to the time variable. The solution of differential equation 2-222a subject to the initial condition 2-222b is straightforward. Inverting this solution twice, first with the inversion formula 2-220b and then with the inversion formula 2-218b, we obtain the solution of system 2-217 as

$$T(x, y, t) = \sum_{n=1}^{\infty} K(\nu_n, y) \cdot \int_{\beta=0}^{\infty} K(\beta, x) \cdot e^{-\alpha(\beta^2 + \nu_n^2) \cdot t} \cdot \left[\bar{\bar{F}}(\beta, \nu_n) + \int_{t'=0}^{t} e^{\alpha(\beta^2 + \nu_n^2)t'} \cdot A(\beta, \nu_n, t') \cdot dt' \right] d\beta \tag{2-223}$$

where

$$A(\beta, \nu_n, t') \equiv \frac{\alpha}{k} \bar{\bar{g}}(\beta, \nu_n, t') + \alpha \left\{ \left. \frac{K(\beta, x)}{k_1} \right|_{x=0} \cdot \bar{f}_1(\nu_n, t') + \left. \frac{K(\nu_n, y)}{k_3} \right|_{y=0} \cdot \bar{f}_3(\beta, t') + \left. \frac{K(\nu_n, y)}{k_4} \right|_{y=b} \cdot \bar{f}_4(\beta, t') \right\}$$

$$\bar{\bar{F}}(\beta, \nu_n) = \int_{x'=0}^{\infty} K(\beta, x') \cdot dx' \int_{y'=0}^{b} K(\nu_n, y') \cdot F(x', y') \cdot dy'$$

$$\bar{\bar{g}}(\beta, \nu_n, t') = \int_{x'=0}^{\infty} K(\beta, x') \cdot dx' \int_{y'=0}^{b} K(\nu_n, y') \cdot g(x', y', t') \cdot dy'$$

$$\bar{f}(\nu_n, t') = \int_{y'=0}^{b} K(\nu_n, y') \cdot f(y', t') \cdot dy'$$

$$\bar{f}(\beta, t') = \int_{x'=0}^{\infty} K(\beta, x') \cdot f(x', t') \cdot dx'$$

where the summation is taken over all eigenvalues ν_n.

The kernel $K(\nu_n, y)$ and the eigenvalues ν_n for the finite region $0 \le y \le b$ are obtained from Table 2-1, and the kernel $K(\beta, x)$ for the semi-infinite region $0 \le x \le \infty$ is obtained from Table 2-2.

In applying the general solution 2-223 for a boundary condition of the first kind (i.e., when $k_i = 0$), to avoid difficulty, the following changes should be made in $A(\beta, \nu_n, t)$.

When $k_1 = 0$, replace $\left.\frac{K(\beta,x)}{k_1}\right|_{x=0}$ by $\left.\frac{1}{h_1}\frac{dK(\beta,x)}{dx}\right|_{x=0}$

When $k_3 = 0$, replace $\left.\frac{K(\nu_n,y)}{k_3}\right|_{y=0}$ by $\left.\frac{1}{h_3}\frac{dK(\nu_n,y)}{dy}\right|_{y=0}$

When $k_4 = 0$, replace $\left.\frac{K(\nu_n,y)}{k_4}\right|_{y=b}$ by $\left.-\frac{1}{h_4}\frac{dK(\nu_n,y)}{dy}\right|_{y=b}$

We illustrate the application of the general solution 2-223 to some special cases with the following examples.

EXAMPLE 2-16. A semi-infinite rectangular strip, $0 \le x < \infty$, $0 \le y \le b$ is initially at temperature $F(x, y)$, and for times $t > 0$ all boundaries are kept at zero temperature. We assume no heat generation within the solid.

Since both boundary conditions at $y = 0$ and $y = b$ are of the first kind, the kernel $K(\nu_n, y)$ and the eigenvalues ν_n from Table 2-1 are

$$K(\nu_n, y) = \sqrt{\frac{2}{b}} \sin \nu_n y \tag{2-224a}$$

where

$$\nu_n = \frac{n\pi}{b} \qquad n = 1, 2, 3, \ldots$$

For a boundary condition of the first kind at $x = 0$, the kernel $K(\beta, x)$ from Table 2-2 is

$$K(\beta, x) = \sqrt{\frac{2}{\pi}} \sin \beta x \tag{2-224b}$$

For the present problem we have

$$g(x, y, t) = f_1(y, t) = f_3(x, t) = f_4(x, t) = 0 \tag{2-224c}$$

Substituting Eqs. 2-224a, 2-224b, 2-224c into the general solution 2-223, we obtain

$$T(x, y, t) = \frac{4}{\pi b} \sum_{n=1}^{\infty} \int_{\beta=0}^{\infty} e^{-\alpha(\beta^2+\nu_n^2)t} \cdot \sin \beta x$$

$$\cdot \sin \nu_n y \int_{x'=0}^{\infty} \int_{y'=0}^{b} F(x', y') \cdot \sin \beta x' \cdot \sin \nu_n y' dx' dy' d\beta \tag{2-224}$$

where

$$\nu_n = \frac{n\pi}{b}, \quad n = 1, 2, 3, \ldots$$

EXAMPLE 2-17. A semi-infinite rectangular strip $0 \le x < \infty$, $0 \le y \le b$ is initially at zero temperature. For times $t > 0$ there is heat generation within the solid at a rate of $g(x, y, t)$ Btu/hr/ft^3, while the boundary at $y = b$ is kept at temperature $\phi(x, t)$ and the remaining boundaries are kept at zero temperature.

The boundary and initial conditions for this problem are

$$
\begin{aligned}
T &= 0 & &\text{at} \quad x = 0, & & t > 0, \\
T &= 0 & &\text{at} \quad y = 0 & & t > 0, \\
T &= \phi(x, t) & &\text{at} \quad y = b, & & t > 0, \\
T &= 0 & &\text{in} \quad 0 \le x < \infty, 0 \le y \le b, & & t = 0.
\end{aligned}
\tag{2-225}
$$

Since all the boundary conditions are of the first kind, the kernels $K(\beta, x)$ and $K(\nu_n, y)$ from Tables 2-2 and 2-1 respectively are

$$K(\beta, x) = \sqrt{\frac{2}{\pi}} \sin \beta x \tag{2-226a}$$

$$K(\nu_n, y) = \sqrt{\frac{2}{b}} \sin \nu_n y \tag{2-226b}$$

where

$$\nu_n = \frac{n\pi}{b} \qquad n = 1, 2, 3, \ldots \tag{2-226c}$$

By comparing the boundary and initial conditions for the present problem with that for the general problem as given by Eq. 2-217, we obtain the following relations:

$$
\begin{aligned}
F(x, y) &= f_1(y, t) = f_3(x, t) = 0 \\
k_1 &= k_3 = k_4 = 0 \\
f_4(x, t) &= \phi(x, t) \\
h_1 &= h_3 = h_4 = 1
\end{aligned}
\tag{2-227}
$$

Substituting Eqs. 2-226 and 2-227 into the solution 2-223, and replacing the term

$$\left.\frac{K(\nu_n, y)}{k_4}\right|_{y=0} \quad \text{by} \quad -\left.\frac{dK(\nu_n, y)}{dy}\right|_{y=b}$$

we obtain

$$
\begin{aligned}
T(x, y, t) = \frac{4}{\pi b} \sum_{n=1}^{\infty} \sin \nu_n y \int_{\beta=b}^{\infty} \sin \beta x \cdot e^{-\alpha(\beta^2+\nu_n^2)t} \cdot \int_{t'=0}^{t} e^{\alpha(\beta^2+\nu_n^2)t'} \\
\cdot \left\{ \frac{\alpha}{k} \int_{x'=0}^{\infty} \sin \beta x' \cdot dx' \int_{y'=0}^{b} \sin \nu_n y' \cdot g(x', y', t') \cdot dy' \right. \\
\left. - \alpha \cdot \nu_n \cdot (-1)^n \int_{x'=0}^{\infty} \phi(x', t') \cdot \sin \beta x' \cdot dx' \right\} dt' \cdot d\beta
\end{aligned}
\tag{2-228}
$$

where

$$\nu_n = \frac{n\pi}{b} \qquad n = 1, 2, 3, \ldots$$

In this relation the term $(-1)^n \cdot \nu_n$ results from

$$\left.\frac{d(\sin \nu_n y)}{dy}\right|_{y=b} = (-1)^n \cdot \nu_n$$

We now examine a special case of the solution 2-228.

(a) There is no heat generation within the solid and that the boundary at $y = b$ is kept at a constant temperature T_0.

For this special case we take

$$g(x, y, t) = 0 \quad \text{and} \quad \phi(x, t) = T_0 = \text{const.}$$

and the solution 2-228 reduces to

$$T(x, y, t) = \frac{4T_0}{\pi b} \sum_{n=1}^{\infty} \sin \nu_n y \int_{\beta=0}^{\infty} \sin \beta x \cdot \frac{1 - e^{-\alpha(\beta^2+\nu_n^2)t}}{\beta^2 + \nu_n^2} (-1)^{n+1} \cdot \frac{\nu_n}{\beta} \cdot d\beta \tag{2-229}$$

where

$$\nu_n = \frac{n\pi}{b} \qquad n = 1, 2, 3, \ldots$$

In obtaining Eq. 2-229 we formally wrote [2, p. 188]

$$\int_{x'=0}^{\infty} \sin \beta x' \cdot dx' = \frac{1}{\beta} \tag{2-230a}$$

and performed the integral

$$\int_{t'=0}^{t} e^{\alpha(\beta^2+\nu_n^2)t'} \cdot dt' = \frac{e^{\alpha(\beta^2+\nu_n^2)t} - 1}{\alpha(\beta^2 + \nu_n^2)} \tag{2-230b}$$

Three-Dimensional Infinite Medium. We consider now the heat conduction problem in an infinite medium $-\infty < x < \infty$, $-\infty < y < \infty$, $-\infty < z < \infty$, which is initially at temperature $F(x, y, z)$, and for times $t > 0$ heat is generated within the solid at a rate of $g(x, y, z, t)$ Btu/hr ft^3.

The boundary-value problem of heat conduction is given as

$$\frac{\partial^2 T}{\partial x^2} + \frac{\partial^2 T}{\partial y^2} + \frac{\partial^2 T}{\partial z^2} + \frac{g(x, y, z, t)}{k} = \frac{1}{\alpha}\frac{\partial T}{\partial t} \quad \text{for} \quad t > 0 \tag{2-231a}$$

$$T = F(x, y, z) \quad \text{for} \quad t = 0 \tag{2-231b}$$

in the region $-\infty < x < \infty$, $-\infty < y < \infty$, $-\infty < z < \infty$, and $T \equiv T(x, y, z, t)$.

We define the triple-integral transform and the corresponding triple-inversion formula for the space variables x, y, z of the temperature function $T(x, y, z, t)$ in the infinite region as:

$$\begin{pmatrix}\text{Triple-integral}\\ \text{transform}\end{pmatrix} \bar{\bar{\bar{T}}}(\beta, \nu, \eta, t)$$

$$= \int_{-\infty}^{\infty}\int_{-\infty}^{\infty}\int_{-\infty}^{\infty} e^{i(\beta x' + \nu y' + \eta z')} \cdot T(x', y', z', t) \cdot dx' dy' dz' \tag{2-232a}$$

$$\left(\begin{matrix}\text{Triple-inversion}\\ \text{formula}\end{matrix}\right) \quad T(x, y, z, t) = \frac{1}{(2\pi)^3}$$

$$\int_{-\infty}^{\infty}\int_{-\infty}^{\infty}\int_{-\infty}^{\infty} e^{-(\beta x+\nu y+\eta z)i} \cdot \bar{T}(\beta, \nu, \eta, t)\cdot d\beta \cdot d\nu \cdot d\eta \tag{2-232b}$$

The triple-integral transform and the inversion formula as defined above is merely the repeated application of the one-dimensional integral transform and the inversion formula for an infinite region as defined by Eq. 2-38 for the x, y, z space variables.

We take the integral transform of system 2-231 by applying the triple transform 2-232a,

$$\frac{d\bar{T}(\beta, \nu, \eta, t)}{dt} + \alpha(\beta^2 + \nu^2 + \eta^2)\cdot \bar{T}(\beta, \nu, \eta, t) = \frac{\alpha}{k}\,\bar{g}(\beta, \nu, \eta, t) \tag{2-233a}$$

$$\bar{T}(\beta, \nu, \eta, t)\big|_{t=0} = \bar{F}(\beta, \nu, \eta) \tag{2-233b}$$

The solution of this ordinary differential equation is

$$\bar{T}(\beta, \nu, \eta, t) = e^{-\alpha(\beta^2+\nu^2+\eta^2)t}\cdot\left[\bar{F}(\beta, \nu, \eta) + \frac{\alpha}{k}\cdot\int_{t'=0}^{t}\bar{g}(\beta, \nu, \eta, t')\cdot e^{\alpha(\beta^2+\nu^2+\eta^2)t'}\cdot dt'\right] \tag{2-234}$$

Substituting this solution into the triple-inversion formula 2-232b we obtain the solution of problem 2-231 as

$$T(x, y, z, t) = \frac{1}{(2\pi)^3}\int_{-\infty}^{\infty}\int_{-\infty}^{\infty}\int_{-\infty}^{\infty} e^{-\alpha(\beta^2+\nu^2+\eta^2)t - i(\beta x + \nu y + \eta z)} \cdot\left[\bar{F}(\beta, \nu, \eta) + \frac{\alpha}{k}\int_{t'=0}^{t}\bar{g}(\beta, \nu, \eta, t')\cdot e^{\alpha(\beta^2+\nu^2+\eta^2)\cdot t'}\cdot dt'\right]\cdot d\beta\, d\nu\, d\eta \tag{2-235}$$

where

$$\bar{F}(\beta, \nu, \eta) = \int_{-\infty}^{\infty}\int_{-\infty}^{\infty}\int_{-\infty}^{\infty} F(x', y', z')\cdot e^{i(\beta x'+\nu y'+\eta z')}\cdot dx'\cdot dy' dz'$$

$$\bar{g}(\beta, \nu, \eta, t') = \int_{-\infty}^{\infty}\int_{-\infty}^{\infty}\int_{-\infty}^{\infty} g(x', y', z', t')\cdot e^{i(\beta x'+\nu y'+\eta z')}\cdot dx' dy' dz'$$

In Eq. 2-235 integrations with respect to the dummy variables β, ν, η can be performed by making use of the relations

$$\frac{1}{2\pi}\int_{-\infty}^{\infty} e^{-\alpha\beta^2 t - i\beta(x-x')} d\beta = \frac{1}{(4\pi\alpha t)^{1/2}}\, e^{-\frac{(x-x')^2}{4\alpha t}} \tag{2-236}$$

and similar relations for (ν, y) and (η, z) combinations.

Then Eq. 2-235 becomes

$$T(x,y,z,t) = \frac{1}{(4\pi\alpha t)^{3/2}} \int_{-\infty}^{\infty}\int_{-\infty}^{\infty}\int_{-\infty}^{\infty} F(x',y',z') \cdot e^{-\frac{(x-x')^2+(y-y')^2+(z-z')^2}{4\alpha t}}$$

$$\cdot dx'\,dy'\,dz' + \frac{\alpha}{k}\int_{t'=0}^{t} \frac{dt'}{[4\pi\alpha(t-t')]^{3/2}} \int_{-\infty}^{\infty}\int_{-\infty}^{\infty}\int_{-\infty}^{\infty}$$

$$\cdot g(x',y',z',t') e^{-\frac{(x-x')^2+(y-y')^2+(z-z')^2}{4\alpha(t-t')}} \cdot dx'\,dy'\,dz' \qquad (2\text{-}237)$$

In this relation the first term is for the effects of the initial temperature distribution $F(x, y, z)$, and the second term is for the effects of heat generation $g(x, y, z, t)$ within the solid.

We now consider one special case of solution 2-237.

EXAMPLE 2-18. A three-dimensional infinite medium is initially at zero temperature. A point heat source of strength $g_p(t)$ Btu/hr is situated at (x_1, y_1, z_1) and releases its heat continuously for times $t > 0$.

In this problem the point heat source $g_p(t)$ Btu/hr is related to volume heat source $g(x, y, z, t)$ Btu/hr ft³ by

$$g(x,y,z,t) = g_p(t) \cdot \delta(x - x_1) \cdot \delta(y - y_1) \cdot \delta(z - z_1) \qquad (2\text{-}238)$$

and its initial condition is

$$F(x,y,z) = 0 \qquad (2\text{-}239)$$

Substituting Eqs. 2-238 and 2-239 in the general solution 2-237 and performing integrations involving delta functions we obtain

$$T(x,y,z,t) = \frac{\alpha}{k}\int_{t'=0}^{t} \frac{g_p(t')}{[4\pi\alpha(t-t')]^{3/2}} \cdot e^{-\frac{(x-x_1)^2+(y-y_1)^2+(z-z_1)^2}{4\alpha(t-t')}} \cdot dt' \qquad (2\text{-}240)$$

We now consider a special case of solution 2-240.

(a) The point heat source is an instantaneous point heat source of strength g_{pi} Btu, which releases its heat spontaneously at time $t = \tau$.

Instantaneous point heat source g_{pi} Btu is related to time-dependent continuous-point heat source $g_p(t)$ Btu/hr by

$$g_p(t) = g_{pi} \cdot \delta(t - \tau) \qquad (2\text{-}241)$$

substituting Eq. 2-241 into Eq. 2-240 and performing the integration

$$T(x,y,z,t) = \frac{\alpha g_{pi}}{k} \frac{1}{[4\pi\alpha(t-\tau)]^{3/2}} \cdot e^{-\frac{(x-x_1)^2+(y-y_1)^2+(z-z_1)^2}{4\alpha(t-\tau)}} \quad \text{for } t > \tau \qquad (2\text{-}242)$$

defining the strength of instantaneous point heat source in °F · ft³ units as

$$S_{pi} = \frac{\alpha g_{pi}}{k} \ {}^\circ\text{F} \cdot \text{ft}^3$$

Then in Eq. 2-242 the term

$$\frac{1}{[4\pi\alpha(t-\tau)]^{3/2}} \cdot e^{-\frac{(x-x_1)^2+(y-y_1)^2+(z-z_1)^2}{4\alpha(t-\tau)}} \quad \text{for } t > \tau \qquad (2\text{-}243)$$

represents the temperature at (x, y, z) at time t due to an instantaneous point heat source of strength $S_{pi}|$ °F · ft³ situated at the point (x_1, y_1, z_1) and releasing its heat spontaneously at time $t = \tau$ in an infinite medium which is initially at zero temperature. The function given by Eq. 2-243 is called *Green's function* for a three-dimensional infinite medium.

Three-Dimensional Finite Rectangular Parallelepiped. A rectangular parallelepiped $0 \le x \le a$, $0 \le y \le b$, $0 \le z \le c$ is initially at temperature $F(x, y, z)$. For times $t > 0$ heat is generated within the solid at a rate of $g(x, y, z, t)$ Btu/hr ft³ while the boundary surfaces are kept at zero temperature.

The boundary-value problem of heat conduction is given as

$$\frac{\partial^2 T}{\partial x^2} + \frac{\partial^2 T}{\partial y^2} + \frac{\partial^2 T}{\partial z^2} + \frac{g(x, y, z, t)}{k} = \frac{1}{\alpha}\frac{\partial T}{\partial t}$$
$$\text{in} \quad 0 \le x \le a,\ 0 \le y \le b,\ 0 \le z \le c, \quad \text{for } t > 0$$
$$T = 0$$
$$\text{at} \quad \text{all boundary surfaces} \qquad \text{for } t > 0 \qquad \text{(2-244)}$$
$$T = F(x, y, z)$$
$$\text{in} \quad 0 \le x \le a,\ 0 \le y \le b,\ 0 \le z \le c, \quad \text{for } t = 0$$

where $T \equiv T(x, y, z, t)$.

We define the triple-integral transform and the corresponding inversion formula for the x, y, z variables of the temperature function $T(x, y, z, t)$ as

$$\begin{pmatrix}\text{Triple-integral}\\ \text{transform}\end{pmatrix} \quad \bar{T}(\beta_m, \nu_n, \eta_p, t) = \int_{x'=0}^{a}\int_{y'=0}^{b}\int_{z'=0}^{c} K(\beta_m, x')$$
$$\cdot K(\nu_n, y') \cdot K(\eta_n, z') \cdot T(x', y', z', t) \cdot dx'\,dy'\,dz' \qquad \text{(2-245a)}$$

$$\begin{pmatrix}\text{Triple-inversion}\\ \text{formula}\end{pmatrix} \quad T(x, y, z, t) = \sum_{m=1}^{\infty}\sum_{n=1}^{\infty}\sum_{p=1}^{\infty} K(\beta_m, x)$$
$$\cdot K(\nu_n, y) \cdot K(\eta_p, z) \cdot \bar{T}(\beta_m, \nu_n, \eta_p, t) \qquad \text{(2-245b)}$$

The triple-integral transform and the triple-inversion formula as defined above are the repeated application of the one-dimensional finite transform and the inversion Eq. 2-7 for the x, y, z variables in the range $0 \le x \le a$, $0 \le y \le b$, $0 \le z \le c$. The kernels and the eigenvalues are given in Table 2-1; since the boundary conditions are of the first kind at all boundaries they are given as

$$K(\beta_m, x) = \sqrt{\frac{2}{a}} \sin \beta_m x, \quad \text{where} \quad \beta_m = \frac{m\pi}{a}, \quad m = 1, 2, 3, \ldots \qquad \text{(2-246a)}$$

$$K(\nu_n, y) = \sqrt{\frac{2}{b}} \sin \nu_n y, \quad \text{where} \quad \nu_n = \frac{n\pi}{b}, \quad n = 1, 2, 3, \ldots \qquad \text{(2-246b)}$$

$$K(\eta_p, z) = \sqrt{\frac{2}{c}} \sin \eta_p z, \quad \text{where} \quad \eta_p = \frac{p\pi}{c}, \quad p = 1, 2, 3, \ldots \qquad \text{(2-246c)}$$

We take the integral transform of system 2-244 by applying the triple transform, Eq. 2-245a,

$$\frac{d\bar{T}(\beta_m, \nu_n, \eta_p, t)}{dt} + \alpha(\beta_m^2 + \nu_n^2 + \eta_p^2) \cdot \bar{T}(\beta_m, \nu_n, \eta_p, t) = \frac{\alpha}{k}\,\bar{g}(\beta_m, \nu_n, \eta_p, t) \tag{2-247a}$$

$$\bar{T}(\beta_m, \nu_n, \eta_p, t)\,\big|_{t=0} = \bar{F}(\beta_m, \nu_n, \eta_p) \tag{2-247b}$$

where quantities with bars refer to the integral transform according to the triple-integral transform 2-245a. It is to be noted that Eq. 2-247 does not involve any boundary-condition functions because the boundaries are at zero temperature.

The solution of Eq. 2-247 is

$$\bar{T}(\beta_m, \nu_n, \eta_p, t) = e^{-\alpha(\beta_m^2 + \nu_n^2 + \eta_p^2)t}\left[\bar{F}(\beta_m, \tau_n, \eta_p) + \frac{\alpha}{k}\int_{t'=0}^{t} \bar{g}(\beta_m, \nu_n, \eta_p, t') \cdot e^{\alpha(\beta_m^2 + \nu_n^2 + \eta_p^2)t'} \cdot dt'\right] \tag{2-248}$$

Inverting this result by means of the inversion Eq. 2-45b, the solution of the boundary-value problem of heat conduction becomes

$$T(x, y, z, t) = \sum_{m=1}^{\infty}\sum_{n=1}^{\infty}\sum_{p=1}^{\infty} K(\beta_m, x) \cdot K(\nu_n, y) \cdot K(\eta_p, z) \cdot e^{-\alpha(\beta_m^2 + \nu_n^2 + \eta_p^2)t} \cdot \left[\bar{F}(\beta_m, \nu_n, \eta_p) + \frac{\alpha}{k}\int_{t'=0}^{t} \bar{g}(\beta_m, \nu_n, \eta_p, t') \cdot e^{\alpha(\beta_m^2 + \nu_n^2 + \eta_p^2)t'} \cdot dt'\right] \tag{2-249}$$

where

$$\bar{F}(\beta_m, \nu_n, \eta_p) = \int_{x'=0}^{a}\int_{y'=0}^{b}\int_{z'=0}^{c} K(\beta_m, x') \cdot K(\nu_n, y') \cdot K(\eta_p, z') \cdot F(x', y', z', t') \cdot dx' \cdot dy' \cdot dz'$$

$$\bar{g}(\beta_m, \nu_n, \eta_p, t') = \int_{x'=0}^{a}\int_{y'=0}^{b}\int_{z'=0}^{c} K(\beta_m, x') \cdot K(\nu_n, y') \cdot K(\eta_p, z') \cdot g(x', y', z', t') \cdot dx' \cdot dy'dz'$$

$$K(\beta_m, x) = \sqrt{\frac{2}{a}}\sin \beta_m x \qquad \beta_m = \frac{m\pi}{a} \qquad m = 1, 2, 3, \ldots$$

$$K(\nu_n, y) = \sqrt{\frac{2}{b}}\sin \nu_n y \qquad \nu_n = \frac{n\pi}{b} \qquad n = 1, 2, 3, \ldots$$

$$K(\eta_p, z) = \sqrt{\frac{2}{c}}\sin \eta_p z \qquad \eta_p = \frac{p\pi}{c} \qquad p = 1, 2, 3, \ldots$$

In the following example we consider one special case of solution 2-249.

EXAMPLE 2-19. A rectangular parallelepiped $0 \leq x \leq a, 0 \leq y \leq b, 0 \leq z \leq c$ is initially at zero temperature, and an instantaneous point heat source of strength g_{pi} Btu, which is situated at the point (x_1, y_1, z_1) inside the solid, releases its heat spontaneously at the time $t = \tau$.

For this particular case we have

$$F(x, y, z) = 0 \tag{2-250}$$

and the heat source is given as

$$g(x', y', z', t') = g_{pi} \cdot \delta(x' - x_1) \cdot \delta(y' - y_1) \cdot \delta(z' - z_1) \cdot \delta(t' - \tau) \tag{2-251}$$

Substituting Eqs. 2-250 and 2-251 into the solution 2-249, and performing the integrations involving delta functions we obtain

$$T(x, y, z, t) = \frac{8}{abc} \frac{\alpha g_{pi}}{k} \sum_{m=1}^{\infty} \sum_{n=1}^{\infty} \sum_{p=1}^{\infty} e^{-\alpha(\beta_m^2 + \nu_n^2 + \eta_p^2)(t-\tau)} \cdot \sin \beta_m x_1 \cdot \sin \nu_n y_1 \cdot \sin \eta_p z_1 \cdot \sin \beta_m x \cdot \sin \nu_n y \cdot \sin \eta_p z \quad \text{for } t > \tau \tag{2-252}$$

where

$$\beta_m = \frac{m\pi}{a} \qquad \nu_n = \frac{n\pi}{b} \qquad \eta_p = \frac{p\pi}{c}$$

If we define the strength of the instantaneous point heat source in °F · ft³ units in the form

$$S_{pi} = \frac{\alpha g_{pi}}{k}, \quad \text{in °F} \cdot \text{ft}^3 \tag{2-253}$$

Then in the solution 2-252 the function

$$\frac{8}{abc} \sum_{m=1}^{\infty} \sum_{n=1}^{\infty} \sum_{p=1}^{\infty} e^{-\alpha(\beta_m^2 + \nu_n^2 + \eta_p^2)(t-\tau)} \cdot \sin \beta_m x_1 \cdot \sin \nu_n y_1 \cdot \sin \eta_p z_1 \cdot \sin \beta_m x \cdot \sin \nu_n y \cdot \sin \eta_p z \tag{2-254}$$

represents temperature at (x, y, z) at time t due to an instantaneous point heat source of strength $S_{pi} = 1$ °F · ft³ situated at (x_1, y_1, z_1) and releasing its heat spontaneously at time $t = \tau$ in a rectangular parallelepiped $0 \leq x \leq a, 0 \leq y \leq b, 0 \leq z \leq c$, which is initially at zero temperature and for times $t > 0$ the boundary surfaces of which are kept at zero temperature. The function 2-254 is called *Green's function* for a rectangular parallelepiped for zero surface temperature.

2-9. STEADY-STATE PROBLEMS

The steady-state heat conduction with heat generation within the solid satisfies a *Poisson's* equation subject to prescribed boundary conditions on the surfaces, and is given in the form

$$\nabla^2 T(x, y, z) + \frac{g(x, y, z)}{k} = 0 \quad \text{in region } R \tag{2-255a}$$

$$k_i \frac{\partial T(x,y,z)}{\partial n_i} + h_i T(x,y,z) = f_i \quad \text{on the boundary surface } i \qquad (2\text{-}255\text{b})$$

where $\partial/\partial n_i$ = differentiation along the outward-drawn normal to the boundary surface i

f_i = the boundary condition function which may be a function of position along the boundary surface.

We assume further that all h_i's do not vanish simultaneously.

Poisson's equation 2-255 can be transformed into Laplace's equation by means of a change of dependent variable, as described below:

A new dependent variable $\theta(x,y,z)$ is defined as

$$\theta(x,y,z) = T(x,y,z) - P(x,y,z) \qquad (2\text{-}256)$$

where the function $P(x,y,z)$ is so chosen that when Eq. 2-256 is substituted into system 2-255, Poisson's equation reduces to Laplace's equation; that is, the system 2-255 becomes

$$\nabla^2\theta(x,y,z) = 0 \qquad (2\text{-}257)$$

subject to boundary conditions in terms of the new variable. Table 2-3[7] shows $P(x,y,z)$ functions for use in Eq. 2-256 to transform Poisson's equation into Laplace's equation in the cartesian coordinate system for

$$\frac{g}{k} = \text{const.}$$

When the heat generation term $g(x,y,z)$ is a function of position, similar results may be worked out depending on the functional form of $g(x,y,z)$.

TABLE 2-3
FUNCTION $P(x,y,z)$

Poisson's equation $\nabla^2 T(x,y,z) = -\dfrac{g_0}{k}$ = constant is transformed to Laplace's equation $\nabla^2\theta(x,y,z) = 0$ by means of a change of variable in the form $T(x,y,z) = \theta(x,y,z) + P(x,y,z)$.

T is a function of	$P(x,y,z)$
x	$-\dfrac{g_0 x^2}{2k}$
y	$-\dfrac{g_0 y^2}{2k}$
z	$-\dfrac{g_0 z^2}{2k}$
x, y	$-\dfrac{g_0 x^2}{2k}$ or $-\dfrac{g_0 y^2}{2k}$
x, y, z	$-\dfrac{g_0 x^2}{2k}$ or $-\dfrac{g_0 y^2}{2k}$ or $-\dfrac{g_0 z^2}{2k}$

When the resulting Laplace's equation involves more than one nonhomogeneous boundary condition, the principle of *superposition* is applied; that is, the problem is reduced to a number of Laplace equations each with one nonhomogeneous boundary condition. The resulting simpler equations are solved either by means of the *separation of variables* or by applying *integral-transform* technique.

We now examine the principle of superposition, the method of separation of variables and the integral-transform technique as applied to the solution of Laplace's equation.

Superposition. By the principle of superposition the Laplace equation involving more than one nonhomogeneous boundary condition is separated into a set of Laplace's equations each with one nonhomogeneous boundary condition. Consider the following steady-state heat-conduction problem:

$$\nabla^2 T = 0 \quad \text{in region } R$$
$$k_i \frac{\partial T}{\partial n_i} + h_i T = f_i \quad \text{on boundary } S_i \tag{2-258}$$

where $i = 1, 2, 3, \ldots, s$ the number of continuous bounding surfaces of the solid

$\partial/\partial n_i =$ differentiation along the outward-drawn normal to the boundary surface-i

The boundary-value problem of heat conduction is split up into simpler problems by putting

$$T = \sum_{j=1}^{s} T_j \tag{2-259}$$

where T_j's are the solution of the following system of Laplace equations each with one nonhomogeneous boundary condition.

$$\nabla^2 T_j = 0$$
$$k_i \frac{\partial T_j}{\partial n_i} + h_i T_j = \delta_{ij} f_i \tag{2-260}$$

where $i = 1, 2, 3, \ldots, s =$ number of continuous bounding surfaces of the solid

$j = 1, 2, 3, \ldots, s$

$\delta_{ij} =$ Kroneker delta

To illustrate the superposition principle described above we consider the steady-state heat-conduction problem for a finite rectangle $0 \le x \le a$, $0 \le y \le b$. We assume that the boundary surfaces are kept at different temperatures which are functions of position along the boundary. The boundary-value problem of heat conduction is given as

$$\frac{\partial^2 T}{\partial x^2} + \frac{\partial^2 T}{\partial y^2} = 0 \quad \text{in} \quad 0 \le x \le a,\ 0 \le y \le b$$

$$\begin{aligned} T &= f_1(y) \quad \text{at} \quad x = 0 \\ T &= f_2(y) \quad \text{at} \quad x = a \\ T &= f_3(x) \quad \text{at} \quad y = 0 \\ T &= f_4(x) \quad \text{at} \quad y = b \end{aligned} \tag{2-261}$$

where $T \equiv T(x, y)$.

By the principle of superposition we separate system 2-261 into four simple problems by putting

$$T(x, y) = T_1(x, y) + T_2(x, y) + T_3(x, y) + T_4(x, y)$$

where $T_1(x, y)$ function satisfies

$$\begin{aligned} \frac{\partial^2 T_1}{\partial x^2} + \frac{\partial^2 T_1}{\partial y^2} &= 0 \quad \text{in} \quad 0 \le x \le a,\ 0 \le y \le b \\ T_1 &= f_1(y) \quad \text{at} \quad x = 0 \\ T_1 &= 0 \quad \text{at} \quad x = a \\ T_1 &= 0 \quad \text{at} \quad y = 0 \\ T_1 &= 0 \quad \text{at} \quad y = b \end{aligned} \tag{2-262}$$

and so forth.

Figure 2-12 illustrates the splitting up of the system 2-261 into four simpler problems. The solution of each of these problems is easily obtained either by the separation of variables or by the integral-transform technique.

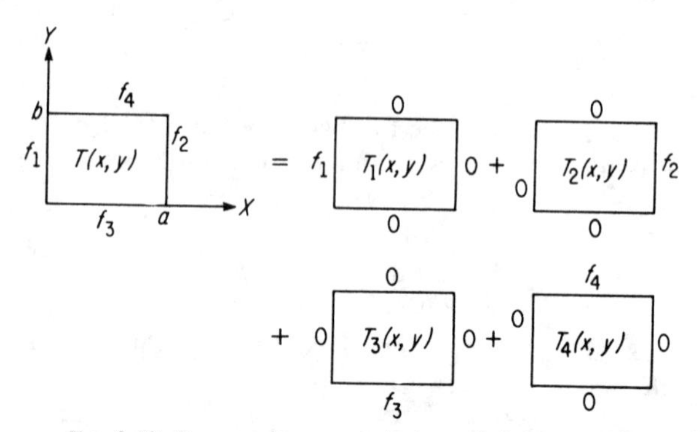

FIG. 2-12. Superposition as applied to a finite rectangle.

Separation of Variables. The separation of variables is a straightforward method in the solution of Laplace's equation subject to one nonhomogeneous boundary condition. We illustrate the method by the following example.

Semi-Infinite Rectangular Strip. Consider a semi-infinite rectangular strip $0 \le x < \infty$, $0 \le y \le b$ with boundary surface at $x = 0$ is kept at

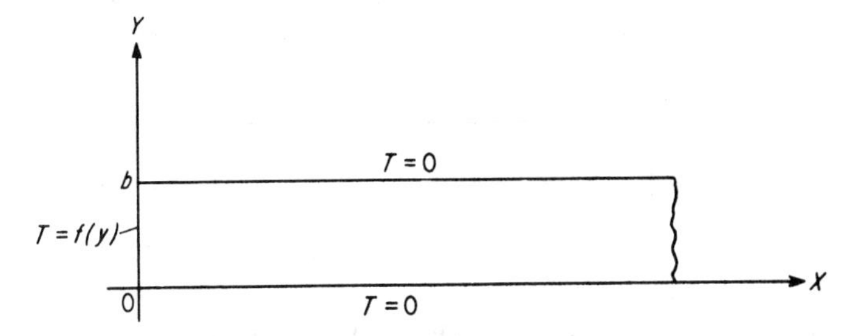

FIG. 2-13. Boundary conditions for a semi-infinite rectangular strip.

temperature $f(y)$ and the remaining boundaries kept at zero temperature. Figure 2-13 shows the boundary conditions. The boundary-value problem of heat conduction is

$$\frac{\partial^2 T}{\partial x^2} + \frac{\partial^2 T}{\partial y^2} = 0 \quad \text{in} \quad 0 \le x \le \infty, 0 \le y \le b$$

$$T = f(y) \quad \text{at} \quad x = 0 \tag{2-263}$$

$$T = 0 \quad \text{at} \quad y = 0$$

$$T = 0 \quad \text{at} \quad y = b$$

We assume separation in the form

$$T(x, y) = X(x) \cdot Y(y)$$

The separated equations are:

$$\frac{d^2 X}{dx^2} - \beta^2 X = 0 \tag{2-264a}$$

$$\frac{d^2 Y}{dy^2} + \beta^2 Y = 0 \tag{2-264b}$$

The sign of the separation constant is so chosen that it yields solution for the y-separation in trigonometric functions.

A particular solution for the $Y(y)$ is in the form

$$Y(y){:}\sin \beta_n y \tag{2-265}$$

where

$$\beta_n = \frac{n\pi}{b}$$

$\cos \beta y$ is excluded from the solution because of the boundary condition at $y = 0$.

A particular solution for the $X(x)$ is in the form

$$X(x){:}e^{-\beta_n x} \tag{2-266}$$

The function $e^{\beta x}$ is excluded from the solution because temperature should remain finite as x becomes infinite.

Complete solution of the problem is in the form

$$T(x, y) = \sum_{n=1}^{\infty} c_n \cdot e^{-\beta_n x} \cdot \sin \beta_n y \qquad (2\text{-}267)$$

This solution should satisfy the boundary condition at $x = 0$, that is,

$$f(y) = \sum_{n=1}^{\infty} c_n \cdot \sin \beta_n y \qquad 0 \leq y \leq b \qquad (2\text{-}268)$$

The unknown coefficients c_n are immediately determined from the orthogonality of the trigonometric functions, and given as

$$c_n = \frac{2}{b} \int_0^b f(y') \cdot \sin \beta_n y' \cdot dy' \qquad (2\text{-}269)$$

The solution of the problem becomes:

$$T(x, y) = \frac{2}{b} \sum_{n=1}^{\infty} e^{-\beta_n x} \cdot \sin \beta_n y \int_{y'=0}^{b} f(y') \cdot \sin \beta_n y' \cdot dy' \qquad (2\text{-}270)$$

$$\beta_n = \frac{n\pi}{b} \qquad n = 1, 2, 3, \ldots$$

Integral-Transform Technique. By applying the integral-transform technique we remove from the differential equation of heat conduction the partial derivatives with respect to the space variables. The procedure is similar to that described previously for the transient problems except for the present problem there is no time-dependence in the differential equation. Therefore, if all the partial derivatives are removed by applying the integral transform the resulting equation is an algebraic equation for the integral transform of temperature which is then inverted by means of the inversion formula. Generally all the partial derivatives are not removed, but the partial differential equation is reduced to an ordinary differential equation which is solved with the standard methods, and the resulting transform of temperature is inverted by means of the inversion formula. The advantage of the latter method is that the resulting solution, for example, for finite regions will have one infinite series less than that in the solution obtained by the former method. We illustrate this with the following example.

A Finite Rectangle. Consider a finite rectangle $0 \leq x \leq a$, $0 \leq y \leq b$ with boundary at $y = 0$ is kept at temperature $f(x)$ while the others are kept at zero temperature. Figure 2-14 shows the geometry and the boundary conditions. For no heat generation within the solid the steady-state temperature satisfies the following system:

$$\frac{\partial^2 T}{\partial x^2} + \frac{\partial^2 T}{\partial y^2} = 0 \quad \text{in} \quad 0 \leq x \leq a, 0 \leq y \leq b$$

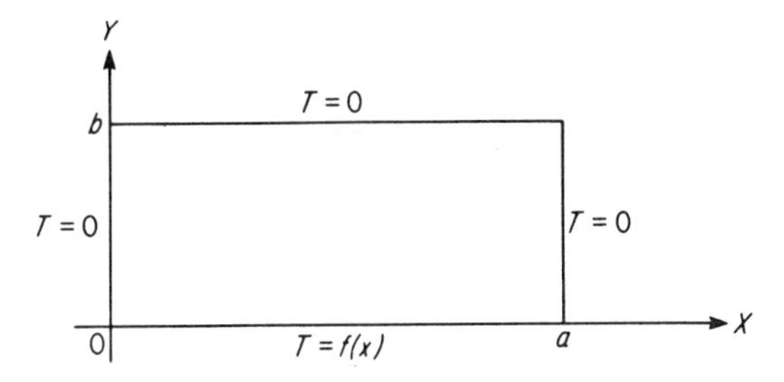

FIG. 2-14. Boundary conditions for a finite rectangle.

$$
\begin{aligned}
T &= 0 \quad \text{at} \quad x = 0 \\
T &= 0 \quad \text{at} \quad x = a \\
T &= f(x) \quad \text{at} \quad y = 0 \\
T &= 0 \quad \text{at} \quad y = b
\end{aligned} \tag{2-271}
$$

where $T \equiv T(x, y)$. We solve this problem by two different methods.

(a) We remove only the partial derivative with respect to the x variable from the differential equation of heat conduction by the application of integral-transform, solve the resulting ordinary differential equation for the transform of temperature, and invert the result.

Since the range of x is finite, i.e., $0 \le x \le a$, we define the integral transform and the inversion formula with respect to the space variable x of the temperature function $T(x, y)$ as

$$\begin{pmatrix}\text{Integral}\\ \text{transform}\end{pmatrix} \quad \bar{T}(\beta_m, y) = \int_{x'=0}^{a} K(\beta_m, x') \cdot T(x', y) \cdot dx' \tag{2-272a}$$

$$\begin{pmatrix}\text{Inversion}\\ \text{formula}\end{pmatrix} \quad T(x, y) = \sum_{m=1}^{\infty} K(\beta_m, x) \cdot \bar{T}(\beta_m, y) \tag{2-272b}$$

The kernels $K(\beta_m, x)$ and the eigenvalues β_m are obtained from Table 2-1; for the boundary condition of the first kind at both boundaries they are given as

$$K(\beta_m, x) = \sqrt{\frac{2}{a}} \sin \beta_m x \tag{2-273}$$

where

$$\beta_m = \frac{m\pi}{a} \qquad m = 1, 2, 3, \ldots$$

We take the integral transform of system 2-271 by applying the transform, Eq. 2-272a,

$$\frac{d^2\bar{T}(\beta_m, y)}{dy^2} - \beta_m^2 \bar{T}(\beta_m, y) = 0 \quad \text{in} \quad 0 \le y \le b$$
$$\bar{T}(\beta_m, y) = \bar{f}(\beta_m) \quad \text{at} \quad y = 0 \tag{2-274}$$
$$\bar{T}(\beta_m, y) = 0 \quad \text{at} \quad y = b$$

Equation 2-274 is an ordinary differential equation the solution of which is

$$\bar{T}(\beta_m, y) = \bar{f}(\beta_m) \cdot \frac{\sinh \beta_m (b - y)}{\sinh \beta_m b} \tag{2-275}$$

When this result is inverted by means of inversion formula 2-272b, the solution of the Prob. 2-271 becomes

$$T(x, y) = \frac{2}{a} \sum_{m=1}^{\infty} \sin \beta_m x \cdot \frac{\sinh \beta_m (b - y)}{\sinh \beta_m b} \cdot \int_{x'=0}^{a} f(x') \cdot \sin \beta_m x' \cdot dx' \tag{2-276}$$

where

$$\beta_m = \frac{m\pi}{a} \qquad m = 1, 2, 3, \ldots,$$

(b) We solve the same problem by removing both partial derivatives with respect to the x, y variables from the differential equation by the application of integral-transform.

Since the range of space variables is $0 \le x \le a$, $0 \le y \le b$, we define the double-integral transform and the corresponding inversion formula with respect to the space variables x and y of the temperature function $T(x, y)$ as

$$\begin{pmatrix}\text{Double-integral}\\ \text{transform}\end{pmatrix} \bar{T}(\beta_m, \nu_n) = \int_{x'=0}^{a} \int_{y'=0}^{b} K(\beta_m, x') \cdot K(\nu_n, y') \cdot T(x', y') \cdot dx' dy' \tag{2-277a}$$

$$\begin{pmatrix}\text{Double-inversion}\\ \text{formula}\end{pmatrix} T(x, y) = \sum_{m=1}^{\infty} \sum_{n=1}^{\infty} K(\beta_m, x) \cdot K(\nu_n, y) \cdot \bar{T}(\beta_m, \nu_n) \tag{2-277b}$$

The kernels $K(\beta_m, x)$, $K(\nu_n, y)$ and the eigenvalues β_m, ν_n are obtained from Table 2-1; for boundary condition of the first kind at both boundaries they are given as

$$K(\beta_m, x) = \sqrt{\frac{2}{a}} \sin \beta_m x, \qquad \beta_m = \frac{m\pi}{a} \quad m = 1, 2, 3, \ldots \tag{2-278a}$$

$$K(\nu_n, y) = \sqrt{\frac{2}{b}} \sin \nu_n y, \qquad \nu_n = \frac{n\pi}{b} \quad n = 1, 2, 3, \ldots \tag{2-278b}$$

We take the integral transform of system 2-271 by applying the double-integral transform Eq. 2-277a. (See Eq. 1-64 for the integral transform of Laplacian for finite regions.) The result is

$$-(\beta_m^2 + \nu_n^2)\cdot \bar{T}(\beta_m, \nu_n) + \left.\frac{dK(\nu_n, y)}{dy}\right|_{y=0} \cdot \int_{x'=0}^{a} K(\beta_m, x')\cdot f(x')\cdot dx' = 0 \qquad \text{(2-279a)}$$

or in the form

$$\bar{T}(\beta_m, \nu_n) = \frac{1}{\beta_m^2 + \nu_n^2} \left.\frac{dK(\nu_n, y)}{dy}\right|_{y=0} \cdot \int_{x'=0}^{a} K(\beta_m, x')\cdot f(x')\cdot dx' \qquad \text{(2-279b)}$$

Inverting this result by means of the inversion formula 2-277b, and substituting the kernels, the solution of system 2-271 becomes

$$T(x, y) = \frac{4}{ab} \sum_{m=1}^{\infty} \sum_{n=1}^{\infty} \frac{\nu_n}{\beta_m^2 + \nu_n^2} \sin \beta_m x \cdot \sin \nu_n y \int_{x'=0}^{a} f(x')\cdot \sin \beta_m x' \cdot dx' \qquad \text{(2-280)}$$

where

$$\beta_m = \frac{m\pi}{a}, \qquad \nu_n = \frac{n\pi}{b}$$

It is to be noted that the present solution includes one more infinite series than the solution 2-276. Comparing the solutions 2-276 and 2-280 we obtain the following relation for the infinite series.

$$\frac{2}{b} \sum_{n=1}^{\infty} \frac{\nu_n}{\beta_m^2 + \nu_n^2} \sin \nu_n y = \frac{\sinh \beta_m (b - y)}{\sinh \beta_m b} \qquad \text{(2-281)}$$

Finite Rectangle with Heat Generation. Consider a finite rectangle $0 \le x \le a$, $0 \le y \le b$ with heat generation at a constant rate of g_0 Btu/hr ft^3 within the solid. The boundaries at $x = 0$ and $y = 0$ are insulated, and boundaries at $x = a$ and $y = b$ dissipate heat by convection into a surrounding at zero temperature.

The steady-state temperature satisfies the following system:

$$\frac{\partial^2 T}{\partial x^2} + \frac{\partial^2 T}{\partial y^2} + \frac{g_0}{k} = 0 \quad \text{in} \quad 0 \le x \le a, 0 \le y \le b$$

$$\frac{\partial T}{\partial x} = 0 \quad \text{at} \quad x = 0$$

$$k\frac{\partial T}{\partial x} + hT = 0 \quad \text{at} \quad x = a \qquad \text{(2-282)}$$

$$\frac{\partial T}{\partial y} = 0 \quad \text{at} \quad y = 0$$

$$k\frac{\partial T}{\partial y} + hT = 0 \quad \text{at} \quad y = b$$

Poisson's equation with constant heat generation is reduced to Laplace's equation by defining a new dependent variable θ as

$$\theta(x, y) = T(x, y) + \frac{x^2}{2} \cdot \frac{g_0}{k} - A \tag{2-283}$$

The arbitrary constant A is introduced for convenience and will be defined later.

Substituting Eq. 2-283 into Eq. 2-282, we obtain a Laplace equation with two nonhomogeneous boundary conditions.

$$\begin{aligned}
\frac{\partial^2 \theta}{\partial x^2} + \frac{\partial^2 \theta}{\partial y^2} &= 0 \quad \text{in} \quad 0 \le x \le a,\ 0 \le y \le b \\
\frac{\partial \theta}{\partial x} &= 0 \quad \text{at} \quad x = 0 \\
k\frac{\partial \theta}{\partial x} + h\theta = h\left(\frac{a^2 g_0}{2k} + \frac{a g_0}{h} - A\right) & \quad \text{at} \quad x = a \\
\frac{\partial \theta}{\partial y} &= 0 \quad \text{at} \quad y = 0 \\
k\frac{\partial \theta}{\partial y} + h\theta = h\left(\frac{x^2 g_0}{2k} - A\right) & \quad \text{at} \quad y = b
\end{aligned} \tag{2-284}$$

By choosing

$$A = \frac{a^2 g_0}{2k} + \frac{a g_0}{h} \tag{2-285}$$

we reduce system 2-284 to Laplace's equation with one nonhomogeneous boundary condition:

$$\begin{aligned}
\frac{\partial^2 \theta}{\partial x^2} + \frac{\partial^2 \theta}{\partial y^2} &= 0 \quad \text{in} \quad 0 \le x \le a,\ 0 \le y \le b \\
\frac{\partial \theta}{\partial x} &= 0 \quad \text{at} \quad x = 0 \\
k\frac{\partial \theta}{\partial x} + h\theta &= 0 \quad \text{at} \quad x = a \\
\frac{\partial \theta}{\partial y} &= 0 \quad \text{at} \quad y = 0 \\
k\frac{\partial \theta}{\partial y} + h\theta = h\left(\frac{x^2 g_0}{2k} - \frac{a^2 g_0}{k} - \frac{a g_0}{h}\right) &\equiv f(x) \quad \text{at} \quad y = b
\end{aligned} \tag{2-286}$$

System 2-286 can be solved by any one of the methods discussed previously; hence the solution of system 2-282 is considered as complete.

2-10. STEADY-STATE PROBLEM INVOLVING PERIODIC BOUNDARY CONDITION

Steady-state solution of heat-conduction problems subjected to a boundary condition which is a periodic function of time is of interest in many engineering applications. Consider a region which is initially at temperature $F(\bar{r})$ and for times $t > 0$ one of its continuous bounding surfaces is subjected to a boundary condition which is a periodic function of time while the remaining boundary surfaces are subjected to homogeneous boundary conditions. We assume no heat generation within the region. The boundary-value problem of heat conduction is given as

$$\nabla^2 T = \frac{1}{\alpha}\frac{\partial T}{\partial t} \quad \text{in region } R, \text{ for } t > 0 \tag{2-287a}$$

$$k_j \frac{\partial T}{\partial n_j} + h_j T = \delta_{0j}\cdot f_j \cdot \cos \omega t \quad \text{on boundary surface } j,\ t > 0 \tag{2-287b}$$

$$T = F(\bar{r}) \quad \text{in region } R, \text{ for } t = 0 \tag{2-287c}$$

where δ_{0j} = Kroneker delta
j = 0, 1, 2, ..., number of continuous bounding surfaces of the region
$\partial/\partial n_j$ = differentiation along the outward-drawn normal at the boundary surface j

To obtain the solution of system 2-287 for large times we consider the following auxiliary problem:

$$\nabla^2 \tilde{T} = \frac{1}{\alpha}\frac{\partial \tilde{T}}{\partial t} \quad \text{in region } R, \text{ for } t > 0 \tag{2-288a}$$

$$k_j \frac{\partial \tilde{T}}{\partial n_j} + h_j \tilde{T} = \delta_{0j}\cdot f_j \cdot \sin \omega t \quad \text{on boundary surface } j,\ t > 0 \tag{2-288b}$$

$$\tilde{T} = F(\bar{r}) \quad \text{in region } R, \text{ for } t = 0 \tag{2-288c}$$

It is to be noted that the auxiliary Prob. 2-288 is similar to the original problem 2-287 except the periodic boundary-condition function has a shift by $\pi/2$ (i.e., $\sin \omega t$ instead of $\cos \omega t$).

We define a complex variable T_c as

$$T_c = T + i\tilde{T} \tag{2-289}$$

where $i = \sqrt{-1}$.

By multiplying the auxiliary Eq. 2-288 by i and adding it to Eq. 2-287 it can be shown that the complex variable T_c satisfies the following system:

$$\nabla^2 T_c = \frac{1}{\alpha}\frac{\partial T_c}{\partial t} \quad \text{in the region, } t > 0 \tag{2-290a}$$

$$k_j \frac{\partial T_c}{\partial n_j} + h_j T_c = \delta_{0j} f_j e^{i\omega t} \quad \text{on boundary surface } j,\ t > 0 \tag{2-290b}$$

$$T_c = (1 + i)F(\bar{r}) \quad \text{in the region, } t = 0 \tag{2-290c}$$

where $e^{i\omega t} = \cos \omega t + i \sin \omega t$.

We are interested in the solution of system 2-290 only for large times. Solution of Eq. 2-290 for large times is taken in the form

$$T_c = \psi(\bar{r})\cdot e^{i\omega t} \tag{2-292}$$

which need not satisfy the initial condition 2-290c because it is valid only for the large times. Substituting Eq. 2-292 into Eqs. 2-290a and 2-290b yields the following system for the space variable $\psi(\bar{r})$.

$$\nabla^2\psi(\bar{r}) - \frac{i\omega}{\alpha}\psi(\bar{r}) = 0 \quad \text{in region } R \tag{2-293a}$$

$$k_j\frac{\partial\psi(\bar{r})}{\partial n_j} + h_j\psi(\bar{r}) = \delta_{0j}f_j \quad \text{on boundary } j \tag{2-293b}$$

Boundary-value Prob. 2-293 is easily solved with any one of the conventional methods. Knowing $\psi(\bar{r})$, the function T_c for large times is given by Eq. 2-292. Since the complex function T_c is related to T and $\tilde{T}$ by Eq. 2-289, the real part of T_c function yields the temperature T, i.e., the solution of system 2-287 for large times.

We illustrate the application of the above technique with the following example.

EXAMPLE 2-20. Consider a semi-infinite region $0 \le x < \infty$, initially at zero temperature and for times $t > 0$ the boundary surface at $x = 0$ is subjected to a periodic heat flux in the form $g_0 \cos \omega t$. The temperature distribution in the solid for large times will be determined.

The boundary-value problem of heat conduction is given as

$$\frac{\partial^2 T}{\partial x^2} = \frac{1}{\alpha}\frac{\partial T}{\partial t} \quad \text{in} \quad 0 \le x < \infty,\ t > 0 \tag{2-294a}$$

$$-k\frac{\partial T}{\partial x} = g_0 \cos\omega t \quad \text{at} \quad x = 0,\ t > 0 \tag{2-294b}$$

$$T = 0 \quad \text{in} \quad 0 \le x < \infty,\ t = 0 \tag{2-294c}$$

We consider the following auxiliary problem,

$$\frac{\partial^2 \tilde{T}}{\partial x^2} = \frac{1}{\alpha}\frac{\partial \tilde{T}}{\partial t} \quad \text{in} \quad 0 \le x < \infty,\ t > 0 \tag{2-295a}$$

$$-k\frac{\partial \tilde{T}}{\partial x} = g_0 \sin\omega t \quad \text{at} \quad x = 0,\ t > 0 \tag{2-295b}$$

$$\tilde{T} = 0 \quad \text{in} \quad 0 \le x < \infty,\ t = 0 \tag{2-295c}$$

and define the complex function T_c as

$$T_c = T + i\tilde{T}$$

The complex function T_c satisfies the following system:

$$\frac{\partial^2 T_c}{\partial x^2} = \frac{1}{\alpha}\frac{\partial T_c}{\partial t} \quad \text{in the region, } t > 0 \tag{2-296a}$$

$$-k\frac{\partial T_c}{\partial x} = g_0 \cdot e^{i\omega t} \quad \text{at } x = 0,\ t > 0 \tag{2-296b}$$

$$T_c = 0 \quad \text{in the region, } t = 0 \tag{2-296c}$$

For large times the solution of Eq. 2-296 is taken in the form

$$T_c = \psi(x) \cdot e^{i\omega t} \tag{2-297}$$

where the space variable function $\psi(x)$ satisfies the following system:

$$\frac{d^2\psi(x)}{dx^2} - i\frac{\omega}{\alpha}\psi(x) = 0 \quad \text{in} \quad 0 \le x < \infty \tag{2-298a}$$

$$-k\frac{d\psi(x)}{dx} = g_0 \quad \text{at} \quad x = 0 \tag{2-298b}$$

$$\text{and } \psi(x) \text{ is finite} \quad \text{at} \quad x \to \infty \tag{2-298c}$$

The solution of Eq. 2-298a that satisfies the boundary condition Eq. 2-298c is in the form

$$\psi(x) = c \cdot e^{-\sqrt{i\frac{\omega}{\alpha}}x} \tag{2-299}$$

The unknown coefficient c is determined from the boundary condition at $x = 0$, Eq. 2-298b; then the solution for $\psi(x)$ becomes

$$\psi(x) = \frac{1}{\sqrt{i}}\frac{g_0}{k}\sqrt{\frac{\alpha}{\omega}} \cdot e^{-\sqrt{i\frac{\omega}{\alpha}}x} \tag{2-300}$$

Substituting $\psi(x)$ into Eq. 2-297, the solution for the complex function T_c becomes

$$T_c = \frac{g_0}{k} \cdot \sqrt{\frac{\alpha}{\omega}} \cdot \frac{1}{\sqrt{i}} e^{i\omega t - \sqrt{i}\sqrt{\frac{\omega}{\alpha}}x} \tag{2-301}$$

Noting that

$$\sqrt{i} = \frac{1+i}{\sqrt{2}} \quad \text{and} \quad \frac{1}{\sqrt{i}} = \frac{1-i}{\sqrt{2}} \tag{2-302}$$

Equation 2-301 is written in the form

$$T_c = \frac{g_0}{2k}\sqrt{\frac{2\alpha}{\omega}} \cdot e^{-\sqrt{\frac{\omega}{2\alpha}}x} \cdot (1-i) \cdot e^{i\left(\omega t - \sqrt{\frac{\omega}{2\alpha}}x\right)} \tag{2-303}$$

or in the form

$$T_c = \frac{g_0}{2k}\sqrt{\frac{2\alpha}{\omega}} \cdot e^{-\sqrt{\frac{\omega}{2\alpha}}x} \cdot (1-i)\left[\cos\left(\omega t - \sqrt{\frac{\omega}{2\alpha}}x\right) + i\sin\left(\omega t - \sqrt{\frac{\omega}{2\alpha}}x\right)\right] \tag{2-304}$$

Since the solution of heat conduction problem 2-294 for large times is given by the *real part* of the complex function 2-304 we obtain

$$T = \frac{g_0}{2k}\sqrt{\frac{2\alpha}{\omega}} \cdot e^{-\sqrt{\frac{\omega}{2\alpha}} \cdot x} \left[\cos\left(\omega t - \sqrt{\frac{\omega}{2\alpha}}x\right) + \sin\left(\omega t - \sqrt{\frac{\omega}{2\alpha}}x\right)\right] \tag{2-305}$$

By making use of the following trigonometric relation

$$\cos\left(A - \frac{\pi}{4}\right) = \cos A \cdot \cos\frac{\pi}{4} + \sin A \cdot \sin\frac{\pi}{4} = \frac{\sqrt{2}}{2}(\cos A + \sin A)$$

Equation 2-305 is written in the form

$$T = \frac{g_0}{\sqrt{2}k}\sqrt{\frac{2\alpha}{\omega}}\cdot e^{-\sqrt{\frac{\omega}{2\alpha}}\cdot x}\cdot \cos\left(\omega t - \sqrt{\frac{\omega}{2\alpha}}\cdot x - \frac{\pi}{4}\right) \tag{2-306}$$

The amplitude of steady temperature, Eq. 2-306, is given by the factor

$$\frac{g_0}{\sqrt{2}k}\sqrt{\frac{2\alpha}{\omega}}\cdot e^{-\sqrt{\frac{\omega}{2\alpha}}\cdot x}$$

The larger is the ω or the smaller is the thermal diffusivity α, the smaller is the amplitude of oscillations at any given depth in the solid. For a given value of ω and α, the amplitude of temperature oscillations is smaller at greater depths in the solid.

The time at which a maximum or minimum of temperature will occur at any given point in the solid is obtained from Eq. 2-306 as

$$\left(\omega t - x\sqrt{\frac{\omega}{2\alpha}} - \frac{\pi}{4}\right) = n\pi \tag{2-307}$$

where n = *even* gives maximum
n = *odd* gives minimum

2-11. TRANSIENT TEMPERATURE CHARTS

Temperature–time charts are useful for rapid estimation of temperature history in solids. Charts have been prepared by several investigators for simple geometries for a variety of boundary conditions. Heisler [8] has developed temperature–time charts for one-dimensional heat conduction in slabs, long cylinders, and spheres subjected to convection at the boundary surfaces. In the case of plate problem, a region of uniform thickness $2L$ was considered to be initially at uniform temperature T_i and for times $t > 0$ subjected to convection from both boundary surfaces into a surrounding at constant temperature T_e with equal heat-transfer coefficients at both surfaces. Choosing the origin of the x-axis at the center-plane of the plate, the boundary-value problem of heat conduction is given as

$$\frac{\partial^2 T}{\partial x^2} = \frac{1}{\alpha}\frac{\partial T}{\partial t} \quad \text{in} \quad 0 \le x \le L,\ t > 0 \tag{2-308a}$$

$$\frac{\partial T}{\partial x} = 0 \quad \text{at} \quad x = 0,\ t > 0 \tag{2-308b}$$

$$-k\frac{\partial T}{\partial x} = h(T - T_e) \quad \text{at} \quad x = L,\ t > 0 \tag{2-308c}$$

$$T = T_i \quad \text{in} \quad 0 \le x \le L,\ t = 0 \tag{2-308d}$$

The solution of this problem is straightforward or is immediately obtained from Eq. 2-61.

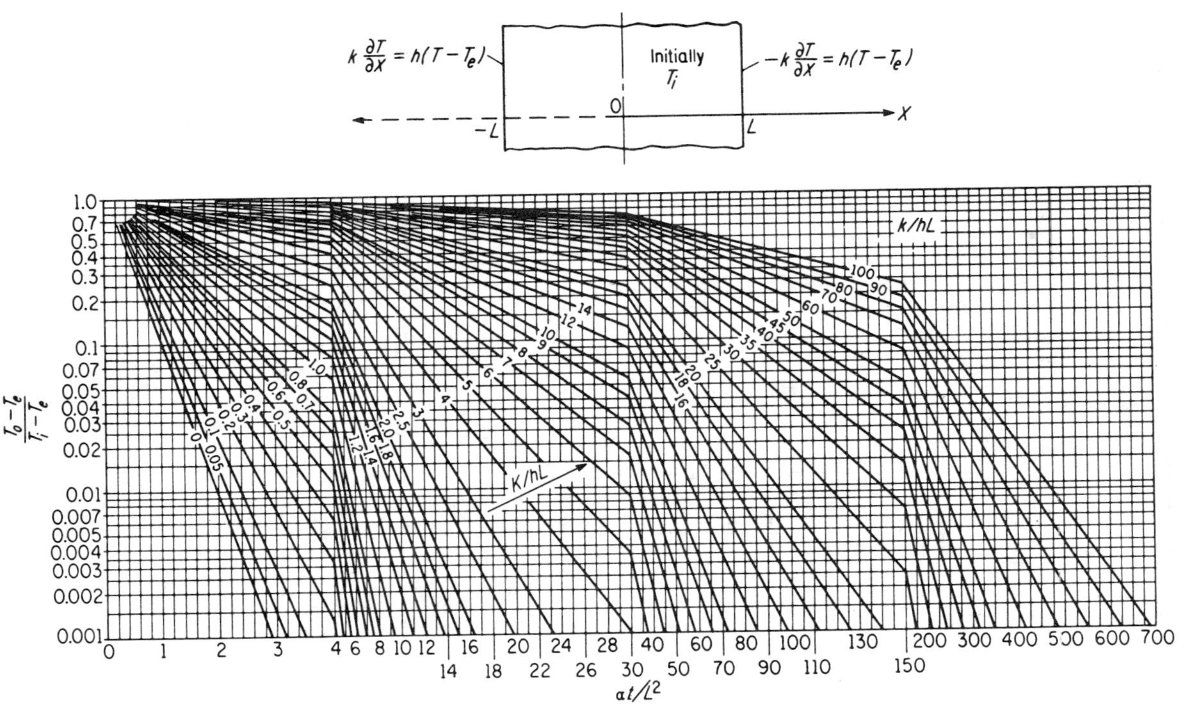

FIG. 2-15. Transient temperature T_0 at the center-plane of a slab of thickness $2L$ subjected to convection at both boundary surfaces. (From Heisler, Ref. 8.)

Figure 2-15 shows the Heisler chart for the dimensionless center-plane temperature, $(T_0 - T_e)/(T_i - T_e)$, i.e., temperature at $x = 0$, as a function of the dimensionless time $\alpha t/L^2$ for several different values of the heat-transfer parameter k/hL. Figure 2-16 is a position-correction

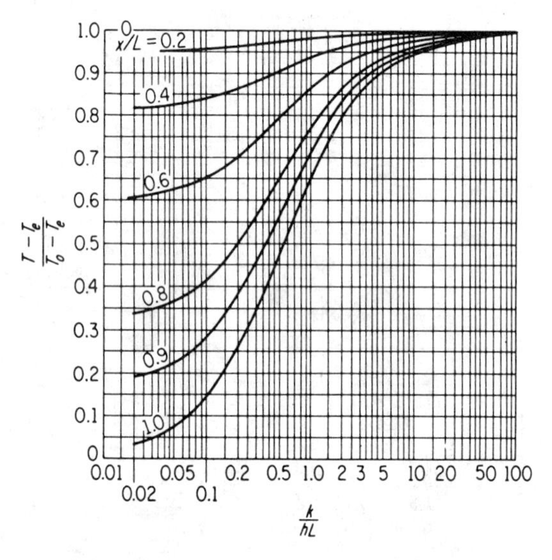

FIG. 2-16. Position-correction chart for use with Fig. 2-15. (From Heisler, Ref. 8.)

factor chart for use in conjunction with the chart 2-15 in order to determine the temperature history in the slab at other positions. Temperature at any position x/L for any given value of k/hL and $\alpha t/L^2$ can be determined by multiplying the value of $(T_0 - T_e)/(T_i - T_e)$ from Fig. 2-15 by the value of $(T - T_e)/(T_0 - T_e)$ at the corresponding value of x/L and k/hL from Fig. 2-16.

Similar charts can easily be prepared for other boundary conditions. For example, Fig. 2-17 shows the solution of the above slab problem when the boundary at $x = 0$ is insulated (i.e., $\partial T/\partial x = 0$) and that $x = L$ is kept at a constant temperature T_b. In this case the dimensionless temperature $(T - T_b)/(T_i - T_b)$ is plotted as a function of $\alpha t/L^2$ at several different locations, i.e., x/L, in the plate.

Schneider [10] prepared temperature–time chart for a semi-infinite solid $(0 \leq x < \infty)$, initially at uniform temperature T_i and for times $t > 0$ subjected to convection from the boundary surface at $x = 0$ into

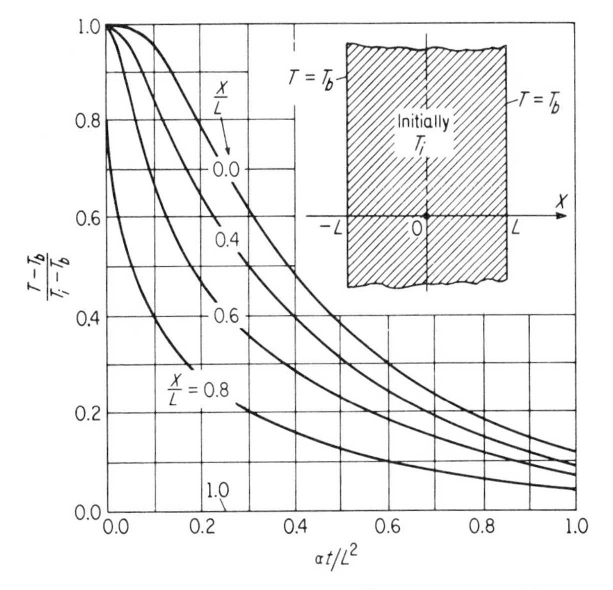

FIG. 2-17. Transient temperature distribution in a slab subjected to constant and equal temperature T_b at both boundary surfaces.

an environment at constant temperature T_e. The boundary-value problem of heat conduction is given as

$$\frac{\partial^2 T}{\partial x^2} = \frac{1}{\alpha}\frac{\partial T}{\partial t} \quad \text{in} \quad 0 \le x < \infty,\ t > 0 \tag{2-309a}$$

$$k\frac{\partial T}{\partial x} = h(T - T_e) \quad \text{at} \quad x = 0,\ t > 0 \tag{2-309b}$$

$$T = T_i \quad \text{in} \quad 0 \le x < \infty,\ t = 0 \tag{2-309c}$$

Solution of this problem is obtained in the form

$$\frac{T - T_i}{T_e - T_i} = \text{erfc}\left(\frac{x}{\sqrt{4\alpha t}}\right) - \left\{\exp\left[\frac{h\sqrt{\alpha t}}{k}\left(\frac{x}{\sqrt{\alpha t}} + \frac{h\sqrt{\alpha t}}{k}\right)\right]\right\} \cdot \text{erfc}\left[\frac{x}{\sqrt{4\alpha t}} + \frac{h\sqrt{\alpha t}}{k}\right] \tag{2-310}$$

Figure 2-18 shows a plot of dimensionless temperature $(T - T_i)/(T_e - T_i)$ as evaluated from Eq. 2-310 as a function of the dimensionless distance $x/\sqrt{4\alpha t}$ for several different values of the dimensionless time $h\sqrt{\alpha t}/k$.

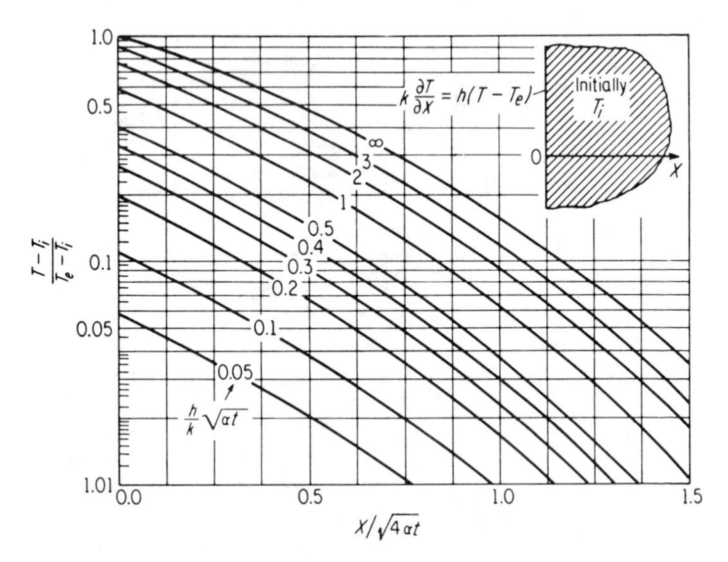

FIG. 2-18. Transient temperature distribution in the semi-infinite solid subjected to convection at the boundary surface $x = 0$. (From Schneider, Ref. 9.)

2-12. SUMMARY OF INTEGRAL TRANSFORM OF LAPLACIAN IN THE CARTESIAN COORDINATE SYSTEM

The partial derivatives with respect to the space variables can be removed from a differential equation by the application of integral transform. The type of integral transform and the inversion formula for use in any given problem depend on the range of space variables and the type of the boundary conditions. In this section we present the integral transform of the Laplacian associated with the heat-conduction problem in the cartesian coordinate system for various combinations of finite, semi-infinite and infinite regions. For generality boundary condition of the third kind is assumed at all boundaries, i.e.,

$$k_i \frac{\partial T}{\partial n_i} + h_i T = f_i \quad \text{for boundary surface } i \tag{2-311}$$

where $\partial/\partial n_i$ = differentiation along the outward-drawn normal to the boundary surface i. Results for other boundary conditions are obtainable by setting k_i equal to zero (i.e., prescribed temperature) or h_i equal to zero (i.e., prescribed heat flux).

Table 2-4 summarizes the integral transform of the Laplacian in the cartesian coordinate system for one, two, and three-dimensional finite, semi-infinite, and infinite regions with boundary condition of the third kind at the boundary surfaces. The kernels referred to in this table can be obtained from Tables 2-1 and 2-2.

TABLE 2-4

INTEGRAL TRANSFORM (I.E., FOURIER TRANSFORM) OF LAPLACIAN IN THE CARTESIAN COORDINATE SYSTEM*

Region	Definition of Integral Transform*	Definition of Inversion Formula*	Form of $\nabla^2 T$	Integral Transform of $\nabla^2 T$†
Finite Region 1. $0 \le x \le a$	$\bar{T} = \int_0^a K(\beta_m, x') \cdot T \cdot dx'$	$T = \sum_{m=1}^{\infty} K(\beta_m, x) \cdot \bar{T}$	$\frac{\partial^2 T}{\partial x^2}$	$-\beta_m^2 \bar{T} + \sum_{i=1}^{2} \frac{K_i}{k_i} f_i$
2. $0 \le x \le a$ $0 \le y \le b$	$\bar{T} = \int_0^b \int_0^a Z \cdot T \cdot dx' dy'$ $Z \equiv K(\beta_m, x') \cdot K(\nu_n, y')$	$T = \sum_{m=1}^{\infty} \sum_{n=1}^{\infty} Z \cdot \bar{T}$ $Z \equiv K(\beta_m, x) \cdot K(\nu_n, y)$	$\frac{\partial^2 T}{\partial x^2} + \frac{\partial^2 T}{\partial y^2}$	$-(\beta_m^2 + \nu_n^2) \bar{T} + \sum_{i=1}^{4} \int_{\text{Boundary path-}i} \frac{Z_i}{k_i} f_i dl_i$
3. $0 \le x \le a$ $0 \le y \le b$ $0 \le z \le c$	$\bar{T} = \int_0^c \int_0^b \int_0^a Z T dx' dy' dz'$ $Z \equiv K(\beta_m, x') \cdot K(\nu_n, y') \cdot K(\eta_p, z')$	$T = \sum_{m=1}^{\infty} \sum_{n=1}^{\infty} \sum_{p=1}^{\infty} Z \cdot \bar{T}$ $Z \equiv K(\beta_m, x) \cdot K(\nu_n, y) \cdot K(\eta_p, z)$	$\frac{\partial^2 T}{\partial x^2} + \frac{\partial^2 T}{\partial y^2} + \frac{\partial^2 T}{\partial z^2}$	$-(\beta_m^2 + \nu_n^2 + \eta_p^2) \bar{T} + \sum_{i=1}^{6} \iint_{\text{Boundary surface } i} \frac{Z_i}{k_i} f_i ds_i$
Semi-infinite Region 4. $0 \le x < \infty$	$\bar{T} = \int_0^{\infty} K(\beta, x') \cdot T \cdot dx'$	$T = \int_0^{\infty} K(\beta, x) \cdot \bar{T}, d\beta$	$\frac{\partial^2 T}{\partial x^2}$	$-\beta^2 \bar{T} + \left[\frac{K_i}{k_i} \cdot f_i\right]_{x=0}$
5. $0 \le x < \infty$ $0 \le y < \infty$	$\bar{T} = \int_0^{\infty} \int_0^{\infty} Z \cdot T \cdot dx' dy'$ $Z \equiv K(\beta, x') \cdot K(\nu, y')$	$T = \int_0^{\infty} \int_0^{\infty} Z \cdot \bar{T} d\beta \cdot d\nu$ $Z = K(\beta, x) \cdot K(\nu, y)$	$\frac{\partial^2 T}{\partial x^2} + \frac{\partial^2 T}{\partial y^2}$	$-(\beta^2 + \nu^2) \bar{T} + \sum_{i=1}^{2} \int_{\text{Boundary path } i} \frac{Z_i}{k_i} f_i dl_i$

*(From: M. N. Özısık, ASME paper 67-H-67, 1967)

TABLE 2-4 (Continued)

Region	Definition of Integral Transform*	Definition of Inversion Formula*	Form of $\nabla^2 T$	Integral Transform of $\nabla^2 T$†
6. $0 \le x < \infty$ $0 \le y < \infty$ $0 \le z < \infty$	$\bar{T} = \int_0^\infty \int_0^\infty \int_0^\infty \cdot ZT dx' dy' dz'$ $Z \equiv K(\beta, x') \cdot K(\nu, y') \cdot K(\eta, z')$	$T = \int_0^\infty \int_0^\infty \cdot \int_0^\infty Z\bar{T} d\beta\, d\nu\, d\eta$ $Z \equiv K(\beta, x) \cdot K(\nu, y) \cdot K(\eta, z)$	$\frac{\partial^2 T}{\partial x^2} + \frac{\partial^2 T}{\partial y^2} + \frac{\partial^2 T}{\partial z^2}$	$-(\beta^2 + \nu^2 + \eta^2)\bar{T} + \sum_{i=1}^{3} \iint_{\text{Boundary surface } i} \frac{Z_i}{k_i} f_i ds_i$
Infinite Region 7. $-\infty < x < \infty$	$\bar{T} = \int_{-\infty}^{\infty} e^{i\beta x'} \cdot T \cdot dx'$	$T = \frac{1}{2\pi} \int_{-\infty}^{\infty} e^{-i\beta x} \bar{T} d\beta$	$\frac{\partial^2 T}{\partial x^2}$	$-\beta^2 \cdot \bar{T}$
8. $-\infty < x < \infty$ $-\infty < y < \infty$	$\bar{T} = \int_{-\infty}^{\infty} \int_{-\infty}^{\infty} e^{i(\beta x' + \nu y')} \cdot T \cdot dx' dy'$	$T = \frac{1}{(2\pi)^2} \cdot \int_{-\infty}^{\infty} \int_{-\infty}^{\infty} e^{-i(\beta x + \nu y)} \cdot \bar{T} d\beta d\nu$	$\frac{\partial^2 T}{\partial x^2} + \frac{\partial^2 T}{\partial y^2}$	$-(\beta^2 + \nu^2) \cdot \bar{T}$
9. $-\infty < x < \infty$ $-\infty < y < \infty$ $-\infty < z < \infty$	$\bar{T} = \int_{-\infty}^{\infty} \int_{-\infty}^{\infty} \int_{-\infty}^{\infty} e^{i(\beta x' + \nu y' + \eta z')} \cdot T \cdot dx' dy' dz'$	$T = \frac{1}{(2\pi)^3} \cdot \int_{-\infty}^{\infty} \int_{-\infty}^{\infty} \int_{-\infty}^{\infty} e^{-i(\beta x + \nu y + \eta z)} \cdot \bar{T} \cdot d\beta d\nu d\eta$	$\frac{\partial^2 T}{\partial x^2} + \frac{\partial^2 T}{\partial y^2} + \frac{\partial^2 T}{\partial z^2}$	$-(\beta^2 + \nu^2 + \eta^2)\bar{T}$
Mixed Region 10. $0 \le y \le b$ $0 \le x < \infty$	$\bar{T} = \int_{x'=0}^{\infty} \int_{y'=0}^{b} Z \cdot T \cdot dx' \cdot dy'$ $Z \equiv K(\beta, x') \cdot K(\nu_n, y')$	$T = \sum_{n=1}^{\infty} \int_{\beta=0}^{\infty} Z \cdot \bar{T} \cdot d\beta$ $Z \equiv K(\beta, x) \cdot K(\nu_n, y')$	$\frac{\partial^2 T}{\partial x^2} + \frac{\partial^2 T}{\partial y^2}$	$-(\beta^2 + \nu_n^2)\bar{T} + \sum_{i=1}^{2} \int_{\text{Boundary path } i} \frac{Z_i}{k_i} f_i dl_i$

11. $0 \le y \le b$ $-\infty < x < \infty$	$\bar{T} = \int_{x'=-\infty}^{\infty} \int_{y'=0}^{b} Z \cdot T \cdot dx' \cdot dy'$ $Z = e^{i\beta x'} \cdot K(\nu_n, y')$	$T = \frac{1}{2\pi} \sum_{n=1}^{\infty} \int_{\beta=-\infty}^{\infty} Z \cdot \bar{T} \cdot d\beta$ $Z = e^{-i\beta x} \cdot K(\nu_n, y)$	$\frac{\partial^2 T}{\partial x^2} + \frac{\partial^2 T}{\partial y^2}$	$-(\beta^2 + \nu_n^2)\,\bar{T} + \sum_{i=1}^{2} \int_{\text{Boundary path } i} \frac{Z_i}{k_i} f_i\, dl_i$
12. $0 \le y < \infty$ $-\infty < x < \infty$	$\bar{T} = \int_{x'=-\infty}^{\infty} \int_{y'=0}^{\infty} Z \cdot T \cdot dx' \cdot dy'$ $Z = e^{i\beta x'} \cdot K(\nu, y')$	$T = \frac{1}{2\pi} \int_{\nu=0}^{\infty} \int_{\beta=-\infty}^{\infty} Z \cdot \bar{T} \cdot d\beta \cdot d\nu$ $Z = e^{-i\beta x} \cdot K(\nu, y)$	$\frac{\partial^2 T}{\partial x^2} + \frac{\partial^2 T}{\partial y^2}$	$-(\beta^2 + \nu^2)\,\bar{T} + \int_{\text{Boundary path } i} \frac{Z_i}{k_i} f_i\, dl_i$
13. $0 \le z \le c$ $0 \le y \le b$ $0 \le x < \infty$	$\bar{T} = \int_{z'=0}^{c} \int_{y'=0}^{b} \int_{x'=0}^{\infty} Z \cdot T \cdot dx' \cdot dy' \cdot dz'$ $Z \equiv K(\beta, x') \cdot K(\nu_n, y') \cdot K(\eta_p, z')$	$T = \sum_{p=1}^{\infty} \sum_{n=1}^{\infty} \int_{\beta=0}^{\infty} Z \cdot \bar{T} \cdot d\beta$ $Z \equiv K(\beta, x) \cdot K(\nu_n, y) \cdot K(\eta_p, z)$	$\frac{\partial^2 T}{\partial x^2} + \frac{\partial^2 T}{\partial y^2} + \frac{\partial^2 T}{\partial z^2}$	$-(\beta^2 + \nu_n^2 + \eta_p^2)\,\bar{T} + \sum_{i=1}^{5} \iint_{\text{Boundary surface } i} \frac{Z_i}{k_i} f_i\, ds_i$
14. $0 \le z \le c$ $0 \le y \le b$ $-\infty < x < \infty$	$\bar{T} = \int_{z'=0}^{c} \int_{y'=0}^{b} \int_{x'=-\infty}^{\infty} Z \cdot T \cdot dx' \cdot dy' \cdot dz'$ $Z \equiv e^{i\beta x'} \cdot K(\nu_n, y') \cdot K(\eta_p, z')$	$\bar{T} = \frac{1}{2\pi} \cdot \sum_{p=1}^{\infty} \sum_{n=1}^{\infty} \int_{\beta=0}^{\infty} Z \cdot \bar{T} \cdot d\beta$ $Z \equiv e^{-i\beta x} \cdot K(\nu_n, y) \cdot K(\eta_p, z)$	$\frac{\partial^2 T}{\partial x^2} + \frac{\partial^2 T}{\partial y^2} + \frac{\partial^2 T}{\partial z^2}$	$-(\beta^2 + \nu_n^2 + \eta_p^2)\,\bar{T} + \sum_{i=1}^{4} \iint_{\text{Boundary surface } i} \frac{Z_i}{k_i} f_i\, ds_i$

TABLE 2-4 (Continued)

Region	Definition of Integral Transform*	Definition of Inversion Formula*	Form of $\nabla^2 T$	Integral Transform of $\nabla^2 T$†
15. $0 \le z \le c$ $0 \le y < \infty$ $0 \le x < \infty$	$\bar{T} = \int_{z'=0}^{c}\int_{y'=0}^{\infty}\int_{x'=0}^{\infty} Z \cdot T \cdot dx' \cdot dy' \cdot dz'$ $Z \equiv K(\beta, x') \cdot K(\nu, y') \cdot K(\eta_p, z')$	$T = \sum_{p=1}^{\infty}\int_{\nu=0}^{\infty}\int_{\beta=0}^{\infty} Z \cdot \bar{T} \cdot d\beta \cdot d\nu$ $Z \equiv K(\beta, x) \cdot K(\nu, y) \cdot K(\eta_p, z)$	$\frac{\partial^2 T}{\partial x^2} + \frac{\partial^2 T}{\partial y^2} + \frac{\partial^2 T}{\partial z^2}$	$-(\beta^2 + \nu^2 + \eta_p^2)\bar{T} + \sum_{i=1}^{4} \iint_{\text{Boundary surface } i} \frac{Z_i}{k_i} f_i \, ds_i$
16. $0 \le z \le c$ $0 \le y < \infty$ $-\infty < x < \infty$	$\bar{T} = \int_{z'=0}^{c}\int_{y'=0}^{\infty}\int_{x'=-\infty}^{\infty} Z \cdot T \cdot dx' \cdot dy' \cdot dz'$ $Z \equiv e^{i\beta x'} \cdot K(\nu, y') \cdot K(\eta_p, z')$	$T = \frac{1}{2\pi}\sum_{p=1}^{\infty}\int_{\nu=0}^{\infty}\int_{\beta=-\infty}^{\infty} Z \cdot \bar{T} \cdot d\beta \cdot d\nu$ $Z \equiv e^{-\beta x} \cdot K(\nu, y) \cdot K(\eta_p, z)$	$\frac{\partial^2 T}{\partial x^2} + \frac{\partial^2 T}{\partial y^2} + \frac{\partial^2 T}{\partial z^2}$	$-(\beta^2 + \nu^2 + \eta_p^2)\bar{T} + \sum_{i=1}^{3} \iint_{\text{Boundary surface } i} \frac{Z_i}{k_i} f_i \, ds_i$
17. $0 \le z \le c$ $-\infty < y < \infty$ $-\infty < x < \infty$	$\bar{T} = \int_{z'=0}^{c}\int_{y'=-\infty}^{\infty}\int_{x'=-\infty}^{\infty} Z \cdot T \cdot dx' \cdot dy' \cdot dz'$ $Z \equiv e^{i\beta x'} \cdot e^{i\nu y'} \cdot K(\eta_p, z')$	$T = \frac{1}{(2\pi)^2}\sum_{p=1}^{\infty}\int_{\nu=0}^{\infty}\int_{\beta=-\infty}^{\infty} Z \cdot \bar{T} \cdot d\beta \cdot d\nu$ $Z \equiv e^{-i\beta x} \cdot e^{-i\nu y} \cdot K(\eta_p, z)$	$\frac{\partial^2 T}{\partial x^2} + \frac{\partial^2 T}{\partial y^2} + \frac{\partial^2 T}{\partial z^2}$	$-(\beta^2 + \nu^2 + \eta_p^2)\bar{T} + \sum_{i=1}^{2} \iint_{\text{Boundary surface } i} \frac{Z_i}{k_i} f_i \, ds_i$

*Kernels are given in Tables 2-1 and 2-2.

†For boundary condition of the first kind at the boundary i, replace $\frac{K_i}{k_i}$ by $-\frac{1}{h_i}\frac{dK_i}{dn_i}$, where $\frac{d}{dn_i}$ denotes differentiation along the outward-drawn normal to the boundary i.

REFERENCES

1. H. S. Carslaw and J. C. Jaeger, *Conduction of Heat in Solids*, at the Clarendon Press, Oxford, 1959.

2. Ian N. Sneddon, *Fourier Transform*, McGraw-Hill Book Company, New York, 1951, pp. 70–74.

3. Ruel V. Churchill, *Operational Mathematics*, 2d ed., McGraw-Hill Book Company, New York, 1958, pp. 228–229.

4. Ian N. Sneddon, *Fourier Transform*, McGraw-Hill Book Company, New York, 1951, p. 173.

5. E. C. Titchmarsh, *Fourier Integrals*, 2d ed., Clarendon Press, Oxford, 1948, p. 1.

6. E. C. Titchmarsh, *Fourier Integrals*, 2d ed., Clarendon Press, Oxford, 1948, p. 281.

7. Parry Moon and D. E. Spencer, *Field Theory for Engineers*, D. Van Nostrand Company, Inc., Princeton, N. J., 1961, p. 409.

8. M. P. Heisler, "Transient Charts for Induction and Constant-Temperature Heating," *Trans. ASME*, Vol. 69 (April 1947), pp. 227–236.

9. P. J. Schneider, *Conduction Heat Transfer*, Addison-Wesley Publishing Company, Inc., Reading, Mass., 1955, p. 266.

PROBLEMS

1. Determine a relation for one-dimensional, transient temperature distribution in a slab $0 \leq x \leq L$ which is initially at temperature $F(x)$ and for times $t > 0$ the boundary surface at $x = 0$ is kept at zero temperature and that of $x = L$ dissipates heat by convection into a medium at zero temperature.

2. A slab $0 \leq x \leq L$ is initially at zero temperature. A plane-surface heat source of strength $g_s(t)$ Btu/hr ft^2 is situated within the solid at $x = x_1$. For times $t > 0$ the plane heat source releases its heat continuously while the boundary surface at $x = 0$ is kept at zero temperature and that at $x = L$ dissipates heat by convection into a medium at zero temperature. Determine a relation for the temperature distribution in the slab for times $t > 0$. Evaluate this relation for the special case of instantaneous plane-surface heat source of strength g_{si} Btu/ft^2 that releases its heat spontaneously at time $t = 0$. Discuss the physical significance of the latter solution when the strength of the instantaneous plane-surface heat source is

$$\frac{\alpha g_{si}}{k} = 1°\text{F} \cdot \text{ft}$$

3. A semi-infinite solid $0 \leq x \leq \infty$ is initially at zero temperature. For times $t > 0$ the boundary surface at $x = 0$ is kept at constant temperature T_0, and temperature of the solid at a distance $x = L$ from the boundary surface is recorded. Explain how you would determine the thermal diffusivity of the solid from the temperature measurement taken. If the thermal diffusivity of the solid were 0.5 ft^2/hr how long would it take for the temperature to reach $1/2\ T_0$ at a depth of 0.5 ft from the boundary surface?

4. A rectangular region $0 \leq x \leq a$, $0 \leq y \leq b$ is initially at zero temperature. For times $t > 0$ the boundary surface at $x = 0$ is kept at temperature

$f(y)$ while the remaining boundaries are kept at zero temperature. Determine a relation for temperature distribution in the solid for times $t > 0$.

5. Find a relation for temperature distribution in a region $0 \leq x < \infty$, $-\infty < y < \infty$ which is initially at zero temperature for times $t > 0$ the boundary surface $x = 0$ kept at temperature $f(y, t)$.

6. An instantaneous-point heat source of strength g_{pi} Btu is situated at x_1, y_1, z_1 inside a solid rectangular parallelepiped $0 \leq x \leq a$, $0 \leq y \leq b$, $0 \leq z \leq c$ which is initially at zero temperature. The point heat source releases its heat spontaneously at time $t = 0$ while for times $t > 0$ the boundary surfaces of the solid are kept at zero temperature. Determine a relation for the temperature distribution in the solid for times $t > 0$.

7. Find a relation for steady-state temperature distribution in a rectangular parallelepiped $0 \leq x \leq a$, $0 \leq y \leq b$, $0 \leq z \leq c$ having its boundary surface at $z = c$ kept at temperature $f(x, y)$ and the remaining boundary surfaces at zero temperature.

8. Find a relation for temperature distribution in a region $0 \leq x \leq \infty$, $-\infty < y < \infty$, $-\infty < z < \infty$, which is initially at zero temperature and for times $t > 0$ the boundary surface at $x = 0$ is kept at temperature $f(y, z)$.

9. An instantaneous line heat source of strength g_{Li} Btu/ft is situated at (x_1, y_1) perpendicular to the x–y plane in a region $0 \leq x < \infty$, $0 \leq y \leq b$ which is initially at zero temperature. The heat source releases its heat spontaneously at time $t = 0$ while the boundaries of the solid are kept at zero temperature for times $t > 0$. Find a relation for temperature distribution in the solid.

10. A semi-infinite region $0 \leq x < \infty$ is subjected to a periodic temperature $T_0 \cos \omega t$ at the boundary surface $x = 0$. Determine a relation for the steady-state temperature distribution in the solid. Explain how you would determine the thermal diffusivity of the solid from the measurements of amplitude of temperature.

11. A rectangular region $-a \leq x \leq a$, $-b \leq y \leq b$, has its center at the origin and the sides parallel to the x and y axis. Heat is generated within the solid at a constant rate of g Btu/hr ft^3 while the boundaries are kept at zero temperature. Determine a relation for steady-state temperature distribution in the region.

3

Heat Conduction in the Cylindrical Coordinate System

In this chapter we examine the solution of boundary-value problems of heat conduction in the circular cylindrical coordinate system. Homogeneous boundary-value problems of heat conduction will be solved by the method of separation of variables, and the nonhomogeneous problem with the integral-transform technique. Since the solutions in the cylindrical coordinate system involve Bessel functions, a brief discussion of these will be given.

3-1. SEPARATION OF HOMOGENEOUS DIFFERENTIAL EQUATION OF HEAT CONDUCTION

Consider the three-dimensional, homogeneous differential equation of heat conduction in the cylindrical coordinate system,

$$\frac{\partial^2 T}{\partial r^2} + \frac{1}{r}\frac{\partial T}{\partial r} + \frac{1}{r^2}\frac{\partial^2 T}{\partial \phi^2} + \frac{\partial^2 T}{\partial z^2} = \frac{1}{\alpha}\frac{\partial T}{\partial t} \tag{3-1}$$

where $T \equiv T(r, \phi, z, t)$.
Assume a separation of variables in the form

$$T(r, \phi, z, t) = R(r) \cdot \Phi(\phi) \cdot Z(z) \cdot \Gamma(t) \tag{3-2}$$

Then differential equation 3-1 becomes

$$\frac{1}{R}\left[\frac{d^2R}{dr^2} + \frac{1}{r}\frac{dR}{dr}\right] + \frac{1}{r^2}\frac{1}{\Phi}\frac{d^2\Phi}{d\phi^2} + \frac{1}{Z}\frac{d^2Z}{dz^2} = \frac{1}{\alpha}\frac{1}{\Gamma}\frac{d\Gamma}{dt} \tag{3-3}$$

Since the functions R, Φ, Z, and Γ are independent of each other, the only way Eq. 3-3 is satisfied if each of the groups equals to a constant; that is

$$\frac{1}{\alpha\Gamma}\frac{d\Gamma}{dt} = -\lambda^2 \quad \text{or} \quad \frac{d\Gamma}{dt} + \alpha\lambda^2\Gamma = 0 \tag{3-4a}$$

$$\frac{1}{\Phi}\frac{d^2\Phi}{d\phi^2} = -\nu^2 \quad \text{or} \quad \frac{d^2\Phi}{d\phi^2} + \nu^2\Phi = 0 \tag{3-4b}$$

$$\frac{1}{Z}\frac{d^2Z}{dz^2} = -\eta^2 \quad \text{or} \quad \frac{d^2Z}{dz^2} + \eta^2 Z = 0 \tag{3-4c}$$

where λ, ν, and η are the separation parameters and are constants. Then R function satisfies

$$\left(\frac{d^2R}{dr^2} + \frac{1}{r}\frac{dR}{dr}\right) + \left(\beta^2 - \frac{\nu^2}{r^2}\right)R = 0 \tag{3-4d}$$

where $\beta^2 = \lambda^2 - \eta^2$.

Equations 3-4a, 3-4b, and 3-4c have particular solutions in the form

$$\Gamma : e^{-\alpha\lambda^2 t} \qquad \Phi(\phi) : \begin{matrix}\sin \nu\phi \\ \cos \nu\phi\end{matrix} \qquad Z(z) : \begin{matrix}\sin \eta z \\ \cos \eta z\end{matrix}$$

The differential equation 3-4d is called *Bessel's differential equation* of order ν and the particular solution of which is in the form

$$R(r) : \begin{matrix} J_\nu(\beta r) \\ Y_\nu(\beta r)\end{matrix}$$

where $\beta^2 + \eta^2 = \lambda^2$. The functions $J_\nu(\beta r)$ and $Y_\nu(\beta r)$ are called *Bessel functions* of order ν of the first and second kind respectively.

We now examine the separation of Eq. 3-1 for special cases.

(1) Temperature has no ϕ-dependence. Equation 3-1 becomes

$$\frac{\partial^2 T}{\partial r^2} + \frac{1}{r}\frac{\partial T}{\partial r} + \frac{\partial^2 T}{\partial z^2} = \frac{1}{\alpha}\frac{\partial T}{\partial t}$$

and the separation equations and their particular solutions may be given in the form:

$$\frac{d\Gamma}{dt} + \alpha\lambda^2\Gamma = 0 \qquad \Gamma : e^{-\alpha\lambda^2 t} \tag{3-5a}$$

$$\frac{d^2R}{dr^2} + \frac{1}{r}\frac{dR}{dr} + \beta^2 R = 0 \qquad R : \begin{matrix} J_0(\beta r) \\ Y_0(\beta r)\end{matrix} \tag{3-5b}$$

$$\frac{d^2Z}{dz^2} + \eta^2 Z = 0 \qquad Z : \begin{matrix}\sin \eta z \\ \cos \eta z\end{matrix} \tag{3-5c}$$

The solution for the R variable is a zero-order Bessel function, because with no ϕ dependence we have $\nu = 0$.

(2) Temperature has no z-dependence. Equation 3-1 becomes

$$\frac{\partial^2 T}{\partial r^2} + \frac{1}{r}\frac{\partial T}{\partial r} + \frac{1}{r^2}\frac{\partial^2 T}{\partial \phi^2} = \frac{1}{\alpha}\frac{\partial T}{\partial t}$$

and the separation equations and their particular solutions may be given as

$$\frac{d\Gamma}{dt} + \alpha\beta^2\Gamma = 0 \qquad \Gamma : e^{-\alpha\beta^2 t} \tag{3-6a}$$

$$\frac{d^2\Phi}{d\Phi^2} + \nu^2\Phi = 0 \qquad \Phi : \begin{matrix}\sin \nu\phi \\ \cos \nu\phi\end{matrix} \tag{3-6b}$$

$$\frac{d^2R}{dr^2} + \frac{1}{r}\frac{dR}{dr} + \left(\beta^2 - \frac{\nu^2}{r^2}\right)R = 0 \qquad R: \begin{matrix} J_\nu(\beta r) \\ Y_\nu(\beta r) \end{matrix} \tag{3-6c}$$

(3) Temperature has no time-dependence. Equation 3-1 becomes

$$\frac{\partial^2 T}{\partial r^2} + \frac{1}{r}\frac{\partial T}{\partial r} + \frac{1}{r^2}\frac{\partial^2 T}{\partial \phi^2} + \frac{\partial^2 T}{\partial z^2} = 0$$

and the separation equation and their solutions may be given as

$$\frac{d^2\Phi}{d\phi^2} + \nu^2\Phi = 0 \qquad \Phi: \begin{matrix} \sin \nu\phi \\ \cos \nu\phi \end{matrix} \tag{3-7a}$$

$$\frac{d^2Z}{dz^2} + \eta^2 Z = 0 \qquad Z: \begin{matrix} \sin \eta z \\ \cos \eta z \end{matrix} \tag{3-7b}$$

$$\frac{d^2R}{dr^2} + \frac{1}{r}\frac{dR}{dr} - \left(\eta^2 + \frac{\nu^2}{r^2}\right)R = 0 \qquad R: \begin{matrix} J_\nu(i\eta r) \\ Y_\nu(i\eta r) \end{matrix} \quad \text{or} \quad \begin{matrix} I_\nu(\eta r) \\ K_\nu(\eta r) \end{matrix} \tag{3-7c}$$

For this case the separation equation for the R variable is called *Bessel's modified differential equation* of order ν, and it is obtainable from Bessel's differential equation if in Bessel's differential equation r is replaced by ir. The solution of Bessel's modified differential equation may be given in the imaginary argument $i\eta r$ in Bessel functions, i.e., $J_\nu(i\eta r)$ and $Y_\nu(i\eta r)$; or instead of using the imaginary argument the functions of real argument $I_\nu(\eta r)$ and $K_\nu(\eta r)$, called *modified Bessel functions of order* ν of the first and second kind respectively are used.[1]

(4) Temperature is a function of r and ϕ only. In this case we have

$$\frac{\partial^2 T}{\partial r^2} + \frac{1}{r}\frac{\partial T}{\partial r} + \frac{1}{r^2}\frac{\partial^2 T}{\partial \phi^2} = 0$$

the separation equations and their solutions may be given as

$$\frac{d^2\Phi}{d\phi^2} + \nu^2\Phi = 0 \qquad \Phi: \begin{matrix} \sin \nu\phi \\ \cos \nu\phi \end{matrix} \tag{3-8a}$$

$$\frac{d^2R}{dr^2} + \frac{1}{r}\frac{dR}{dr} - \frac{\nu^2}{r^2}R = 0 \qquad \begin{matrix} R = c_1 r^\nu + c_2 r^{-\nu}, & \nu \neq 0 \\ R = c_3 + c_4 \ln r, & \nu = 0 \end{matrix} \tag{3-8b}$$

In this particular case the separation equation for the R variable is Euler's homogeneous (i.e., equidimensional) differential equation because the separation parameter β is zero.

[1]The relation between Bessel functions of real and imaginery arguments for $\nu = n$ as real integer is given as

$$I_n(x) = i^{-n} J_n(ix)$$

$$K_n(x) = i^{n+1} \frac{\pi}{2} [J_n(ix) + i\, Y_n(ix)]$$

3-2. BESSEL FUNCTIONS

We have seen that the separation equation for the R variable, for certain cases, is a Bessel's differential equation or Bessel's modified differential equation. In this section we examine briefly the important properties of Bessel functions that will be needed in the solution of heat-conduction problem in the cylindrical coordinate system. For detailed treatment of Bessel functions the reader should refer to standard texts on Bessel functions[1, 2].

Bessel Functions. Consider Bessel's differential equation of order ν, as obtained from Eq. 3-4d by defining a new independent variable $x = \beta r$,

$$\frac{d^2R}{dx^2} + \frac{1}{x}\frac{dR}{dx} + \left(1 - \frac{\nu^2}{x^2}\right)R = 0 \tag{3-9}$$

The general solution of this equation is in the form

$$R(x) = c_1 J_\nu(x) + c_2 Y_\nu(x) \tag{3-10}$$

where $J_\nu(x)$ and $Y_\nu(x)$ are Bessel functions of order ν of the first and second kind respectively and are linearly independent solutions for all values of ν (i.e., integral, nonintegral, real, imaginary or complex).

For nonintegral values of ν solution 3-10 may be written in the form

$$R(x) = c_1 J_\nu(x) + c_2 J_{-\nu}(x) \tag{3-11}$$

and solutions $J_\nu(x)$ and $J_{-\nu}(x)$ are two linearly independent solutions so long as ν is nonintegral. When ν is integral (i.e., $\nu = n = 0, 1, 2, 3, \ldots$) solutions $J_\nu(x)$ and $J_{-\nu}(x)$ are related to each other in the form

$$J_{-n}(x) = (-1)^n J_n(x)$$

then solutions $J_{-n}(x)$ and $J_n(x)$ are no longer independent and the general solution 3-10 should be used. For instance, for $\nu = 1/3$ (i.e., nonintegral) either Eq. 3-10 or Eq. 3-11 is used as the solution of Bessel's differential equation 3-9. But when, say, $\nu = n = 1$ (i.e., integral) solution 3-11 is no longer valid and solution 3-10 should be used.

Figure 3-1 shows a plot of zero and first-order Bessel functions of the first and second kind. The $Y(x)$ functions become infinite as x goes to zero, and both functions have oscillatory behavior like the trigonometric functions for $x > 0$. The $J_0(x)$ function is an even function and behaves like $\cos x$, whereas $J_1(x)$ is an odd function and behaves like $\sin x$.

The numerical values and some properties of the Bessel functions of the first and second kind are given in Appendices III and IV.

Modified Bessel Functions. Consider Bessel's modified differential equation of order ν, which is obtained from Eq. 3-7c by defining a new independent variable, $x = \eta r$.

$$\frac{d^2R}{dx^2} + \frac{1}{x}\frac{dR}{dx} - \left(1 + \frac{\nu^2}{x^2}\right)R = 0 \tag{3-12}$$

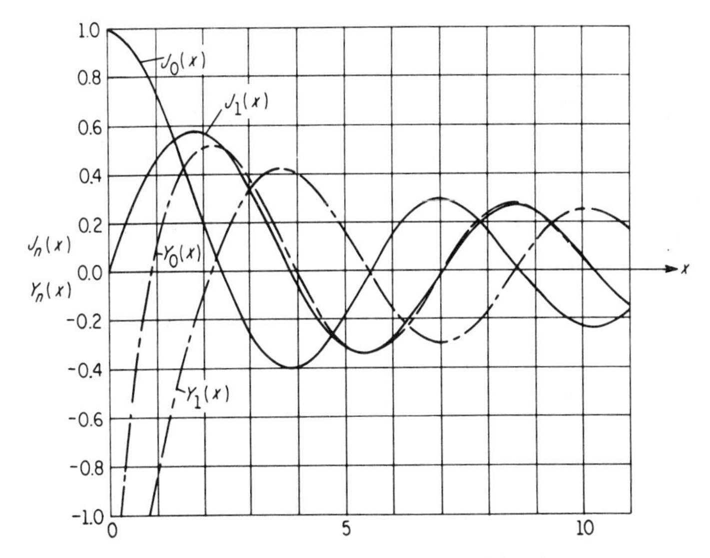

FIG. 3-1. $J_0(x)$, $Y_0(x)$ and $J_1(x)$, $Y_1(x)$ functions.

The solution of this equation is in the form

$$R = c_1 I_\nu(x) + c_2 K_\nu(x) \tag{3-13}$$

where $I_\nu(x)$ and $K_\nu(x)$ are modified Bessel functions of order ν of the first and second kind respectively. The functions $I_\nu(x)$ and $K_\nu(x)$ are two linearly independent solutions of Eq. 3-12 and are valid for all values of ν. When ν is nonintegral the solution of Eq. 3-12 may be given in the form

$$R = c_1 I_\nu(x) + c_2 I_{-\nu}(x) \tag{3-14}$$

and functions $I_\nu(x)$ and $I_{-\nu}(x)$ are two independent solutions so long as ν remains nonintegral. When ν is an integer they are no longer independent but related to each other in the form

$$I_{-n}(x) = I_n(x) \text{ for } n = 0,1,2,3,\ldots$$

Therefore solution of Eq. 3-12 in the form of Eq. 3-13 is always valid, but solution in the form of Eq. 3-14 is valid only when ν is nonintegral.

Figure 3-2 shows a plot of zero and first-order modified Bessel functions of the first and second kind. It is to be noted that $K(x)$ functions become infinite as x goes to zero, whereas $I(x)$ functions become infinite as x goes to infinity. The numerical values and some properties of modified Bessel functions of the first and second kind are given in Appendices III and IV.

Generalized Bessel Solution. Sometimes a given differential equation, after suitable transformation of the independent variable, yields a solution which is a linear combination of Bessel functions. A very convenient way

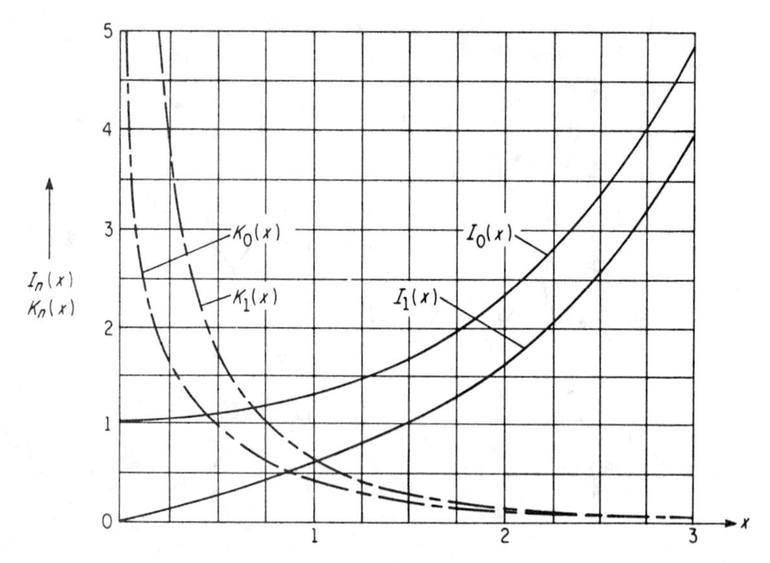

FIG. 3-2. $I_0(x)$, $K_0(x)$ and $I_1(x)$, $K_1(x)$ functions.

of finding out whether a given differential equation possesses a solution in terms of Bessel functions is to compare it with the *generalized Bessel equation* as developed by Douglass[3] in the form

$$\frac{d^2R}{dx^2} + \left[\frac{1-2m}{x} - 2\alpha\right] \cdot \frac{dR}{dx}$$

$$+ \left[p^2a^2x^{2p-2} + \alpha^2 + \frac{\alpha(2m-1)}{x} + \frac{m^2 - p^2\nu^2}{x^2}\right]R = 0 \qquad \text{(3-15a)}$$

and the corresponding solution of which is

$$R = x^m \cdot e^{\alpha x}[c_1 J_\nu(ax^p) + c_2 Y_\nu(ax^p)] \qquad \text{(3-15b)}$$

where c_1 and c_2 are arbitrary constants and the other terms are as defined by the differential equation 3-15a.

For example, by comparing the differential equation

$$\frac{d^2R}{dx^2} + \frac{1}{x}\frac{dR}{dx} - \frac{\beta}{x}R = 0 \qquad \text{(3-16a)}$$

with the generalized Bessel equation 3-15a we find

$$\alpha = 0 \qquad m = 0 \qquad p = 1/2 \qquad p^2\nu^2 = -\beta \qquad a = 2i\sqrt{\beta} \qquad \nu = 0$$

Hence, the solution of differential equation 3-16a is in the form

$$R = c_1 J_0(2i\sqrt{\beta x}) + c_2 Y_0(2i\sqrt{\beta x})$$

or

$$R = c_1 I_0(2\sqrt{\beta x}) + c_2 K_0(2\sqrt{\beta x}) \qquad \text{(3-16b)}$$

which involves Bessel functions.

3-3. THE INTEGRAL TRANSFORM (HANKEL TRANSFORM) AND THE INVERSION FORMULA

The partial derivatives with respect to the space variable r can be removed from the differential equation of heat conduction by applying an integral transform, which is generally called *Hankel Transform*; such transforms and the inversion formulas are derived from the expansion of an arbitrary function over a given interval in an infinite series of Bessel functions. In this section we examine the construction of Hankel transform and the corresponding inversion formula for finite, semi-finite and infinite regions in the cylindrical coordinate system for different types of boundary conditions.

Finite Region. Consider a region between two concentric cylinders of radii $r = a$ and $r = b$. The integral transform (*i.e. Hankel Transform of order* ν) and the inversion formula of function $F(r)$ in the region $a \leq r \leq b$ is given in the form

$$\begin{pmatrix}\text{Inversion}\\ \text{formula}\end{pmatrix} F(r) = \sum_{m=1}^{\infty} K_\nu(\beta_m, r) \cdot \bar{F}(\beta_m) \tag{3-17a}$$

$$\begin{pmatrix}\text{Integral}\\ \text{transform}\end{pmatrix} \bar{F}(\beta_m) = \int_{r'=a}^{b} r' \cdot K_\nu(\beta_m, r') \cdot F(r') \cdot dr' \tag{3-17b}$$

where summation is taken over all eigenvalues β_m. It will be shown that the relationship given by Eq. 3-17 is equivalent to representation of an arbitrary function $F(r)$ in the interval $a \leq r \leq b$ in terms of Bessel functions.

Consider the following auxiliary problem

$$\frac{d}{dr}\left(r\frac{dR}{dr}\right) + \left(\beta^2 r - \frac{\nu^2}{r}\right)R = 0 \quad \text{in} \quad a \leq r \leq b \tag{3-18a}$$

$$-k_1\frac{dR}{dr} + h_1 R = 0 \qquad \text{at} \quad r = a \tag{3-18b}$$

$$k_2\frac{dR}{dr} + h_2 R = 0 \qquad \text{at} \quad r = b \tag{3-18c}$$

Equation 3-18a is Bessel's differential equation of order ν. Since Eq. 3-18 is a special case of Sturm-Liouville system, the eigenfunctions $R_\nu(\beta_m, r)$ constitute an orthogonal set in the interval $a \leq r \leq b$ with respect to the weighting function $w = r$. That is,

$$\int_a^b r' \cdot R_\nu(\beta_m, r') \cdot R_\nu(\beta_n, r')\, dr' = \begin{cases} 0 & \text{for } m \neq n \\ \text{number} & \text{for } m = n \end{cases}$$

We now consider the expansion of an arbitrary function $F(r)$ in the interval $a \leq r \leq b$ in terms of the eigenfunctions $R_\nu(\beta_m, r)$ in the form

$$F(r) = \sum_{m=1}^{\infty} c_m R_\nu(\beta_m, r) \tag{3-19}$$

where the summation is taken over all eigenvalues β_m.

By making use of the orthogonality of eigenfunctions $R_\nu(\beta_m, r)$, the unknown coefficients c_m are immediately determined from Eq. 3-19 as

$$c_m = \frac{1}{N} \int_a^b r' \cdot R_\nu(\beta_m, r') \cdot F(r') \cdot dr' \tag{3-20a}$$

where the norm N is

$$N = \int_a^b r' \cdot R_\nu^2(\beta_m, r') \cdot dr' \tag{3-20b}$$

Then, the expansion Eq. 3-19 becomes

$$F(r) = \sum_{m=1}^{\infty} \frac{R_\nu(\beta_m, r)}{\sqrt{N}} \int_a^b r' \cdot \frac{R_\nu(\beta_m, r')}{\sqrt{N}} \cdot F(r') \cdot dr' \tag{3-21}$$

Expansion Eq. 3-21 is valid if function $F(r)$ satisfies Dirichlet's conditions in the interval $a \leq r \leq b$, and at each point in the interval the function $F(r)$ is continuous.

Defining for convenience the *normalized eigenfunctions* as

$$K_\nu(\beta_m, r) \equiv \frac{R_\nu(\beta_m, r)}{\sqrt{N}} \tag{3-22}$$

we rewrite expansion Eq. 3-21 as the *inversion formula* for the *integral transform* of function $F(r)$ in the finite interval $a \leq r \leq b$ in the form

$$\begin{pmatrix}\text{Inversion}\\ \text{formula}\end{pmatrix} \qquad F(r) = \sum_{m=1}^{\infty} K_\nu(\beta_m, r) \cdot \bar{F}(\beta_m) \tag{3-23a}$$

$$\begin{pmatrix}\text{Integral}\\ \text{transform}\end{pmatrix} \qquad \bar{F}(\beta_m) = \int_{r'=a}^{b} r' \cdot K_\nu(\beta_m, r') \cdot F(r') \cdot dr' \tag{3-23b}$$

The kernels $K_\nu(\beta_m, r)$ and eigenvalues β_m depend on the range of the space variable (i.e., $0 \leq r \leq b$ or $a \leq r \leq b$) and the type of boundary conditions. We examine below the evaluation of the kernels and the eigenvalues.

Region $0 \leq r \leq b$ (i.e., Solid Cylinder). The integral transform (*i.e. Hankel Transform of order ν*) and the inversion formula for a finite region $0 \leq r \leq b$ are defined as

$$\begin{pmatrix}\text{Integral}\\ \text{transform}\end{pmatrix} \qquad \bar{F}(\beta_m) = \int_{r'=0}^{b} r' \cdot K_\nu(\beta_m, r') \cdot F(r') \cdot dr' \tag{3-24a}$$

$$\begin{pmatrix}\text{Inversion}\\ \text{formula}\end{pmatrix} \qquad F(r) = \sum_{m=1}^{\infty} K_\nu(\beta_m, r) \cdot \bar{F}(\beta_m) \tag{3-24b}$$

where

$$K_\nu(\beta_m, r) = \frac{R_\nu(\beta_m, r)}{\sqrt{N}}$$

The eigenfunctions $K_\nu(\beta_m, r)$ satisfy the following eigenvalue problem:

$$\frac{d}{dr}\left(r\frac{dR}{dr}\right) + \left(\beta^2 r - \frac{\nu^2}{r}\right)R = 0 \quad \text{in} \quad 0 \le r \le b \tag{3-25a}$$

$$k_2\frac{dR}{dr} + h_2 R = 0 \qquad \text{at} \quad r = b \tag{3-25b}$$

Since the region includes the origin (i.e., $r = 0$) the $Y_\nu(\beta_m, r)$ function is excluded from the solution and the eigenfunction is in the form

$$R_\nu(\beta_m, r) = J_\nu(\beta_m r) \tag{3-26}$$

the eigenvalues β_m are determined by making use of the boundary condition Eq. 3-25b, and are the positive roots of the transcendental equation

$$k_2\beta J_\nu'(\beta b) + h_2 J_\nu(\beta b) = 0 \tag{3-27}$$

where

$$J_\nu'(\beta b) \equiv \left[\frac{dJ_\nu(r)}{dr}\right]_{r=\beta_m b} \tag{3-28}$$

The norm N is evaluated as [Ref. 2, p. 110]

$$N = \int_0^b r \cdot J_\nu^2(\beta_m r) \cdot dr = \left[\frac{r^2}{2}\left\{J_\nu'^2(\beta_m r) + \left(1 - \frac{\nu^2}{\beta_m^2 r^2}\right) \cdot J_\nu^2(\beta_m r)\right\}\right]_0^b$$
$$= \frac{b^2}{2}\left[J_\nu'^2(\beta_m b) + \left(1 - \frac{\nu^2}{\beta_m^2 b^2}\right) \cdot J_\nu^2(\beta_m b)\right] \tag{3-29}$$

In view of relationship 3-27 the norm N is written in a more convenient form as

$$N = \frac{b^2}{2}\left[\frac{H_2^2}{\beta_m^2} + \left(1 - \frac{\nu^2}{\beta_m^2 b^2}\right)\right] \cdot J_\nu^2(\beta_m b) \tag{3-29a}$$

Then the kernel $K_\nu(\beta_m, r)$ and the eigenvalues β_m for use in the transform 3-24 for a finite region $0 \le r \le b$ subjected to a boundary condition of the third kind at $r = b$ are given as

$$K_\nu(\beta_m, r) \equiv \frac{R_\nu(\beta_m, r)}{\sqrt{N}} = \frac{\sqrt{2}}{b} \frac{1}{\left[\frac{H_2^2}{\beta_m^2} + \left(1 - \frac{\nu^2}{\beta_m^2 b^2}\right)\right]^{1/2}} \frac{J_\nu(\beta_m r)}{J_\nu(\beta_m b)} \tag{3-30a}$$

The eigenvalues β_m are the positive roots of the transcendental equation

$$\beta J'_\nu(\beta b) + H_2 \cdot J_\nu(\beta b) = 0 \tag{3-30b}$$

where

$$H_2 = \frac{h_2}{k_2}$$

and

$$\nu + \tfrac{1}{2} \geq 0$$

Special cases of Eq. 3-30 for boundary conditions of the second kind and the first kind at $r = b$ are evaluated below.

Boundary Condition at $r = b$ is of the Second Kind. For this special case setting $h_2 = 0$, Eq. 3-27 becomes

$$J'_\nu(\beta b) = 0 \tag{3-31}$$

and Eq. 3-29 reduces to

$$N = \frac{b^2}{2}\left(1 - \frac{\nu^2}{\beta_m^2 b^2}\right) \cdot J_\nu^2(\beta_m b) \tag{3-32}$$

Hence the kernel $K_\nu(\beta_m, r)$ and the eigenvalues β_m for use in the transform Eq. 3-24 are

$$K_\nu(\beta_m, r) = \frac{\sqrt{2}}{b} \frac{1}{\left[1 - \dfrac{\nu^2}{\beta_m^2 b^2}\right]^{1/2}} \cdot \frac{J_\nu(\beta_m r)}{J_\nu(\beta_m b)} \tag{3-33a}$$

and the eigenvalues β_m are the positive roots of the transcendental equation

$$J'_\nu(\beta b) = 0 \tag{3-33b}$$

Boundary Condition at $r = b$ is of the First Kind. For this special case setting $k_2 = 0$, Eq. 3-27 becomes

$$J_\nu(\beta b) = 0 \tag{3-34}$$

In view of this relation Eq. 3-29 reduces to

$$N = \frac{b^2}{2} J'^2_\nu(\beta_m b) \tag{3-35}$$

Then the kernel $K_\nu(\beta_m, r)$ and the eigenvalues β_m for use in the transform Eq. 3-24 are

$$K_\nu(\beta_m, r) = \frac{\sqrt{2}}{b} \frac{J_\nu(\beta_m r)}{J'_\nu(\beta_m b)} \tag{3-36a}$$

and the eigenvalues β_m are the positive roots of the transcendental equation

$$J_\nu(\beta b) = 0 \tag{3-36b}$$

Table 3-1 summarizes the kernels $K_\nu(\beta_m, r)$ and the eigenvalues β_m for a finite region $0 \leq r \leq b$ in the cylindrical coordinate system for the boundary conditions of the first, the second and the third kinds.

TABLE 3-1.
KERNELS $K(\beta_m, r)$ FOR USE IN THE INTEGRAL TRANSFORM (HANKEL TRANSFORM) AND THE INVERSION FORMULA FOR A FINITE REGION $0 \le r \le b$ IN THE CYLINDRICAL COORDINATE SYSTEM.

(Integral transform) $$\bar{F}(\beta_m) = \int_{r'=0}^{b} r' \cdot K_\nu(\beta_m, r') \cdot F(r') \cdot dr'$$

(Inversion formula) $$F(r) = \sum_{m=1}^{\infty} K_\nu(\beta_m, r) \cdot \bar{F}(\beta_m)$$

b; $k_2 \dfrac{dR}{dr} + h_2 R = 0$

Boundary Condition* at $r = b$	The Kernel $K_\nu(\beta_m, r)$	Eigenvalues β_m are positive roots of
Third kind: H_2 = finite (i.e., k_2, h_2 finite)	$\dfrac{\sqrt{2}}{b} \dfrac{1}{\left[\dfrac{H_2^2}{\beta_m^2} + \left(1 - \dfrac{\nu^2}{\beta_m^2 b^2}\right)\right]^{1/2}} \dfrac{J_\nu(\beta_m r)}{J_\nu(\beta_m b)}$	$\beta J'_\nu(\beta b) + H_2 J_\nu(\beta b) = 0$
Second kind: $H_2 = 0$ (i.e., $h_2 = 0$)	$\dfrac{\sqrt{2}}{b} \dfrac{1}{\left[1 - \dfrac{\nu^2}{\beta_m^2 b^2}\right]^{1/2}} \cdot \dfrac{J_\nu(\beta_m r)}{J_\nu(\beta_m b)}$	$J'_\nu(\beta b) = 0$†
First kind: $H_2 = \infty$ (i.e., $k_2 = 0$)	$\dfrac{\sqrt{2}}{b} \dfrac{J_\nu(\beta_m r)}{J'_\nu(\beta_m b)}$	$J_\nu(\beta b) = 0$

*$H_2 \equiv h_2/k_2$.

†For this particular case $\beta = 0$ is also a root; when evaluating the kernel for this case with $\nu = 0$, choose $\nu = 0$ prior to choosing $\beta = 0$.

Region $a \le r \le b$ (Hollow Cylinder). The integral transform and the inversion formula are defined by Eq. 3-23—that is,

(Inversion formula) $$F(r) = \sum_{m=1}^{\infty} K_\nu(\beta_m, r) \cdot \bar{F}(\beta_m) \tag{3-37a}$$

(Integral transform) $$\bar{F}(\beta_m) = \int_{r'=a}^{b} r' \cdot K_\nu(\beta_m, r') \cdot F(r') \cdot dr' \tag{3-37b}$$

where

$$K_\nu(\beta_m, r) = \frac{R_\nu(\beta_m, r)}{\sqrt{N}}$$

The eigenfunctions $R_\nu(\beta_m, r)$ satisfy the eigenvalue Prob. 3-18 and include Bessel functions of order ν of the first and second kind—that is,

$$R_\nu(\beta_m, r) : \begin{matrix} J_\nu(\beta_m r) \\ Y_\nu(\beta_m r) \end{matrix} \tag{3-38}$$

For convenience, choosing the eigenfunctions $R_\nu(\beta_m, r)$ in the form

$$R_\nu(\beta_m, r) = \frac{J_\nu(\beta_m r)}{k_2\beta_m J_\nu'(\beta_m b) + h_2 J_\nu(\beta_m b)} - \frac{Y_\nu(\beta_m r)}{k_2\beta_m Y_\nu'(\beta_m b) + h_2 Y_\nu(\beta_m b)} \tag{3-39}$$

then in system 3-18 the boundary condition at $r = b$ is satisfied. The solution 3-39 will satisfy the boundary condition at $r = a$ if we choose

$$\frac{-\beta_m k_1 J_\nu'(\beta_m a) + h_1 J_\nu(\beta_m a)}{\beta_m k_2 J_\nu'(\beta_m b) + h_2 J_\nu(\beta_m b)} - \frac{-\beta_m k_1 Y_\nu'(\beta_m a) + h_1 Y_\nu(\beta_m a)}{\beta_m k_2 Y_\nu'(\beta_m b) + h_2 Y_\nu(\beta_m b)} = 0 \tag{3-40}$$

The positive roots of the transcendental Eq. 3-40 gives the eigenvalues β_m.

The norm N is given as (Ref. 2, p. 110)

$$N = \int_a^b r \cdot R_\nu^2(\beta_m, r)\cdot dr = \left[\frac{r^2}{2}\left\{R_\nu'^2(\beta_m, r) + \left(1 - \frac{\nu^2}{\beta_m^2 r^2}\right)\cdot R_\nu^2(\beta_m, r)\right\}\right]_a^b$$

$$= \frac{b^2}{2}\left[R_\nu'^2(\beta_m, b) + \left(1 - \frac{\nu^2}{\beta_m^2 b^2}\right)\cdot R_\nu^2(\beta_m, b)\right]$$

$$- \frac{a^2}{2}\left[R_\nu'^2(\beta_m a) + \left(1 - \frac{\nu^2}{\beta_m^2 a^2}\right)\cdot R_\nu^2(\beta_m, a)\right] \tag{3-41}$$

which can be put into a more convenient form by making use of the fact that the eigenfunctions $R_\nu(\beta_m, r)$ should satisfy the boundary conditions 3-18b and 3-18c—that is,

$$-k_1\beta_m R_\nu'(\beta_m, a) + h_1 R_\nu(\beta_m, a) = 0 \tag{3-42a}$$

$$k_2\beta_m R_\nu'(\beta_m, b) + h_2 R_\nu(\beta_m, b) = 0 \tag{3-42b}$$

In view of the relation Eq. 3-42 the derivatives $R_\nu'(\beta_m, a)$ and $R_\nu'(\beta_m, b)$ are removed from Eq. 3-41, and the norm N becomes

$$N = \frac{b^2}{2}\left[\frac{h_2^2}{k_2^2\beta_m^2} + \left(1 - \frac{\nu^2}{\beta_m^2 b^2}\right)\right]R_\nu^2(\beta_m, b)$$

$$- \frac{a^2}{2}\left[\frac{h_1^2}{k_1^2\beta_m^2} + \left(1 - \frac{\nu^2}{\beta_m^2 a^2}\right)\right]R_\nu^2(\beta_m, a) \tag{3-43a}$$

where

$$R_\nu(\beta_m, r) = \frac{J_\nu(\beta_m r)}{k_2\beta_m J_\nu'(\beta_m b) + h_2 J_\nu(\beta_m b)} - \frac{Y_\nu(\beta_m r)}{k_2\beta_m Y_\nu'(\beta_m b) + h_2 Y_\nu(\beta_m b)} \tag{3-43b}$$

and β_m's are the positive roots of the transcendental equation

$$\frac{-\beta k_1 J_\nu'(\beta a) + h_1 J_\nu(\beta a)}{\beta k_2 J_\nu'(\beta b) + h_2 J_\nu(\beta b)} - \frac{-\beta k_1 Y_\nu'(\beta a) + h_1 Y_\nu(\beta a)}{\beta k_2 Y_\nu'(\beta b) + h_2 Y_\nu(\beta b)} = 0 \quad \text{(3-43c)}$$

and the kernel $K_\nu(\beta_m, r)$ is defined as

$$K_\nu(\beta_m, r) = \frac{R_\nu(\beta_m, r)}{\sqrt{N}} \quad \text{(3-43d)}$$

The results given by Eq. 3-43 are for a finite region $a \le r \le b$ subjected to boundary conditions of the third kind at both boundary surfaces. The kernels and the eigenvalues for the remaining eight combinations are obtained by setting h_1, h_2, k_1, k_2 as zero or finite. (h_1, k_1 not simultaneously zero; h_2, k_2 not simultaneously zero.)

Table 3-2 summarizes the eigenfunctions $K_\nu(\beta_m, r)$, the norm N and the eigenvalues β_m for the nine combinations of boundary conditions at $r = a$ and $r = b$.

Infinite Region $0 \le r < \infty$. When the region extends from $r = 0$ to infinity in the cylindrical coordinate system, the integral transform (i.e. Hankel Transform) and the inversion formula of a function $F(r)$ in the region $0 \le r < \infty$ are defined as[4]:

$$\begin{pmatrix}\text{Inversion}\\ \text{formula}\end{pmatrix} \qquad F(r) = \int_{\beta=0}^{\infty} \beta \cdot J_\nu(\beta r) \cdot \bar{F}(\beta) \cdot d\beta \quad \text{(3-44a)}$$

$$\begin{pmatrix}\text{Integral}\\ \text{transform}\end{pmatrix} \qquad \bar{F}(\beta) = \int_{r'=0}^{\infty} r' \cdot J_\nu(\beta r') \cdot F(r') \cdot dr' \quad \text{(3-44b)}$$

It is to be noted that the integral transform and inversion formula as defined by Eq. 3-44 are essentially equivalent to representation of an arbitrary function $F(r)$ in an infinite interval $0 \le r < \infty$ in terms of the solution of the following differential equation

$$\frac{d}{dr}\left(r\frac{dR}{dr}\right) + \left(\beta^2 r - \frac{\nu^2}{r}\right)R = 0 \quad \text{in} \quad 0 \le r \le \infty$$

3-4. ONE-DIMENSIONAL HOMOGENEOUS BOUNDARY-VALUE PROBLEM OF HEAT CONDUCTION. (SOLUTION WITH SEPARATION OF VARIABLES)

In this section we examine the solution of homogeneous boundary-value problem of heat conduction involving the space variable r by the separation of variables for a long solid cylinder and a hollow cylinder.

Solid cylinder $0 \le r \le b$. Consider a long solid cylinder of radius $r = b$ which is initially at temperature $F(r)$. For times $t > 0$ heat is dissipated by convection from the boundary surface at $r = b$ into a surrounding at zero temperature. (See Fig. 3-3.) The boundary-value

TABLE 3-2.
THE KERNELS FOR USE IN THE INTEGRAL TRANSFORM (HANKEL TRANSFORM) AND INVERSION FORMULA FOR A FINITE REGION $a \leq r \leq b$ IN THE CYLINDRICAL-COORDINATE SYSTEM

(Integral transform) $$\bar{F}(\beta_m) = \int_{r'=a}^{b} r' \cdot K_\nu(\beta_m, r) \cdot F(r') \cdot dr'$$

(Inversion formula) $$F(r) = \sum_{m=1}^{\infty} K_\nu(\beta_m, r) \cdot \bar{F}(\beta_m)$$

where

$$K_\nu(\beta_m, r) \equiv \frac{R_\nu(\beta_m, r)}{\sqrt{N}}$$

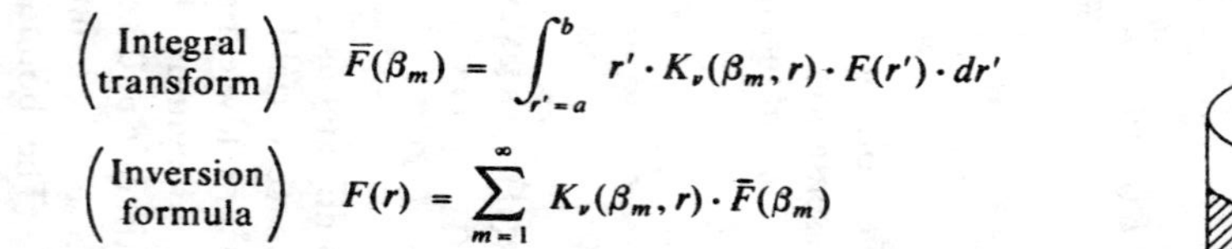

Boundary Condition at $r = a$	Boundary Condition at $r = b$	$R_\nu(\beta_m, r)$ and N	β_m's are the positive roots of*
Third kind: k_1, h_1 are finite	Third kind: k_2, h_2 are finite	$R_\nu(\beta_m, r) = \dfrac{J_\nu(\beta_m r)}{k_2 \beta_m J'_\nu(\beta_m b) + h_2 J_\nu(\beta_m b)} - \dfrac{Y_\nu(\beta_m r)}{k_2 \beta_m Y'_\nu(\beta_m b) + h_2 Y_\nu(\beta_m b)}$ $N = \dfrac{b^2}{2}\left[\dfrac{h_2^2}{k_2^2 \beta_m^2} + \left(1 - \dfrac{\nu^2}{\beta_m^2 b^2}\right)\right] R_\nu^2(\beta_m, b) - \dfrac{a^2}{2}\left[\dfrac{h_1^2}{k_1^2 \beta_m^2} + \left(1 - \dfrac{\nu^2}{\beta_m^2 a^2}\right)\right] R_\nu^2(\beta_m, a)$	$\dfrac{-\beta k_1 J'_\nu(\beta a) + h_1 J_\nu(\beta a)}{\beta k_2 J'_\nu(\beta b) + h_2 J_\nu(\beta b)} - \dfrac{-\beta k_1 Y'_\nu(\beta a) + h_1 Y_\nu(\beta a)}{\beta k_2 Y'_\nu(\beta b) + h_2 Y_\nu(\beta b)} = 0$

Third kind: k_1, h_1 are finite	Second kind: $h_2 = 0$, $k_2 = 1$, $R'_\nu(\beta_m, b) = 0$	$R_\nu(\beta_m, r) = \dfrac{J_\nu(\beta_m r)}{\beta_m J'_\nu(\beta_m b)} - \dfrac{Y_\nu(\beta_m r)}{\beta_m Y'_\nu(\beta_m b)}$ $N = \dfrac{b^2}{2}\left(1 - \dfrac{\nu^2}{\beta_m^2 b^2}\right) R_\nu^2(\beta_m, b) - \dfrac{a^2}{2}\left[\dfrac{h_1^2}{k_1^2 \beta_m^2} + \left(1 - \dfrac{\nu^2}{\beta_m^2 a^2}\right)\right] R_\nu^2(\beta_m, a)$	$\dfrac{-\beta k_1 J'_\nu(\beta a) + h_1 J_\nu(\beta a)}{J'_\nu(\beta b)} - \dfrac{-\beta k_1 Y'(\beta a) + h_1 Y_\nu(\beta a)}{Y'_\nu(\beta b)} = 0$
Third kind: k_1, h_1 are finite	First kind: $k_2 = 0$, $h_2 = 1$, $R_\nu(\beta_m, b) = 0$	$R_\nu(\beta_m, r) = \dfrac{J_\nu(\beta_m r)}{J_\nu(\beta_m b)} - \dfrac{Y_\nu(\beta_m r)}{Y_\nu(\beta_m b)}$ $N = \dfrac{b^2}{2} R_\nu'^2(\beta_m, b) - \dfrac{a^2}{2}\left[\dfrac{h_1^2}{k_1^2 \beta_m^2} + \left(1 - \dfrac{\nu^2}{\beta_m^2 a^2}\right)\right] R_\nu^2(\beta_m, a)$	$\dfrac{-\beta k_1 J'_\nu(\beta a) + h_1 J_\nu(\beta a)}{J_\nu(\beta b)} - \dfrac{-\beta k_1 Y'_\nu(\beta a) + h_1 Y_\nu(\beta a)}{Y_\nu(\beta b)} = 0$
Second kind: $h_1 = 0$, $k_1 = 1$, $R_\nu(\beta, a) = 0$	Third kind: h_2, k_2 are finite	$R_\nu(\beta_m, r) = \dfrac{J_\nu(\beta_m r)}{k_2 \beta_m J'_\nu(\beta_m b) + h_2 J_\nu(\beta_m b)} - \dfrac{Y_\nu(\beta_m r)}{k_2 \beta_m Y'_\nu(\beta_m b) + h_2 Y_\nu(\beta_m b)}$ $N = \dfrac{b^2}{2}\left[\dfrac{h_2^2}{k_2^2 \beta_m^2} + \left(1 - \dfrac{\nu^2}{\beta_m^2 b^2}\right)\right] R_\nu^2(\beta_m, b) - \dfrac{a^2}{2}\left(1 - \dfrac{\nu^2}{\beta_m^2 a^2}\right) R_\nu^2(\beta_m, a)$	$\dfrac{J'_\nu(\beta a)}{\beta k_2 J'_\nu(\beta b) + h_2 J_\nu(\beta b)} - \dfrac{Y'_\nu(\beta a)}{\beta k_2 Y'_\nu(\beta b) + h_2 Y_\nu(\beta b)} = 0$

TABLE 3-2 (Continued)

Second kind: $h_1 = 0$, $k_1 = 1$, $R'_\nu(\beta, a) = 0$	Second kind: $h_2 = 0$, $k_2 = 1$, $R'_\nu(\beta, b) = 0$	$R_\nu(\beta_m, r) = \dfrac{J_\nu(\beta_m r)}{\beta_m J'_\nu(\beta_m b)} - \dfrac{Y_\nu(\beta_m r)}{\beta_m Y'_\nu(\beta_m b)}$ $N = \dfrac{b^2}{2}\left(1 - \dfrac{\nu^2}{\beta_m^2 b^2}\right) R_\nu^2(\beta_m, b) - \dfrac{a^2}{2}\left(1 - \dfrac{\nu^2}{\beta_m^2 a^2}\right) R_\nu^2(\beta_m, a)$	$\dfrac{J'_\nu(\beta a)}{J'_\nu(\beta b)} - \dfrac{Y'_\nu(\beta a)}{Y'_\nu(\beta b)} = 0$
Second kind: $h_1 = 0$, $k_1 = 1$, $R'_\nu(\beta, a) = 0$	First kind: $h_2 = 1$, $k_2 = 0$, $R_\nu(\beta, b) = 0$	$R_\nu(\beta_m, r) = \dfrac{J_\nu(\beta_m r)}{J_\nu(\beta_m b)} - \dfrac{Y_\nu(\beta_m r)}{Y_\nu(\beta_m b)}$ $N = \dfrac{b^2}{2} R'^2_\nu(\beta_m, b) - \dfrac{a^2}{2}\left(1 - \dfrac{\nu^2}{\beta_m^2 a^2}\right) R_\nu^2(\beta_m, a)$	$\dfrac{J'_\nu(\beta a)}{J_\nu(\beta b)} - \dfrac{Y'_\nu(\beta a)}{Y_\nu(\beta b)} = 0$
First kind: $h_1 = 1$, $k_1 = 0$, $R_\nu(\beta, a) = 0$	Third kind: h_2, k_2 are finite	$R_\nu(\beta_m, r) = \dfrac{J_\nu(\beta_m r)}{k_2 \beta_m J'_\nu(\beta_m b) + h_2 J_\nu(\beta_m b)} - \dfrac{Y_\nu(\beta_m r)}{k_2 \beta_m Y'_\nu(\beta_m b) + h_2 Y_\nu(\beta_m b)}$ $N = \dfrac{b^2}{2}\left[\dfrac{h_2^2}{k_2^2 \beta_m^2} + \left(1 - \dfrac{\nu^2}{\beta_m^2 b^2}\right)\right] R_\nu^2(\beta_m, b) - \dfrac{a^2}{2} R'^2_\nu(\beta_m, a)$	$\dfrac{J_\nu(\beta a)}{\beta k_2 J'_\nu(\beta b) + h_2 J_\nu(\beta b)} - \dfrac{Y_\nu(\beta a)}{\beta k_2 Y'_\nu(\beta b) + h_2 Y_\nu(\beta b)} = 0$

First kind: $h_1 = 1$, $k_1 = 0$, $R_\nu(\beta, a) = 0$	Second kind: $h_2 = 0$, $k_2 = 1$, $R'_\nu(\beta, b) = 0$	$R_\nu(\beta_m, r) = \frac{J_\nu(\beta_m r)}{\beta_m J'_\nu(\beta_m b)} - \frac{Y_\nu(\beta_m r)}{\beta_m Y'_\nu(\beta_m b)}$ $N = \frac{b^2}{2}\left(1 - \frac{\nu^2}{\beta_m^2 b^2}\right) R_\nu^2(\beta_m, b) - \frac{a^2}{2} R'^2(\beta_m, a)$	$\frac{J_\nu(\beta a)}{J'_\nu(\beta b)} - \frac{Y_\nu(\beta a)}{Y'_\nu(\beta b)} = 0$
First kind: $h_1 = 1$, $k_1 = 0$, $R_\nu(\beta, a) = 0$	First kind: $h_2 = 1$, $k_2 = 0$, $R_\nu(\beta, b) = 0$	$R_\nu(\beta_m, r) = \frac{J_\nu(\beta_m r)}{J_\nu(\beta_m b)} - \frac{Y_\nu(\beta_m r)}{Y_\nu(\beta_m b)}$ $N = \frac{b^2}{2} R'^2_\nu(\beta_m, b) - \frac{a^2}{2} R'^2_\nu(\beta_m, a)$	$\frac{J_\nu(\beta a)}{J_\nu(\beta b)} - \frac{Y_\nu(\beta a)}{Y_\nu(\beta b)} = 0$

*The roots of the transcendental equation may be determined using a numerical method, i.e., L. S. Allan, and C. W. Tittle "Some Functions in the Theory of Neutron Logging,"—*Journal of the Graduate Research Center*, Southern Methodist University, Vol. 33, No. 1, (1964), pp. 33–54. Bessel functions may be evaluated by means of a computer subroutine, i.e., "Generation of Bessel Functions, C3–Utex Bessel," by Control Data Corporation (February 1963).

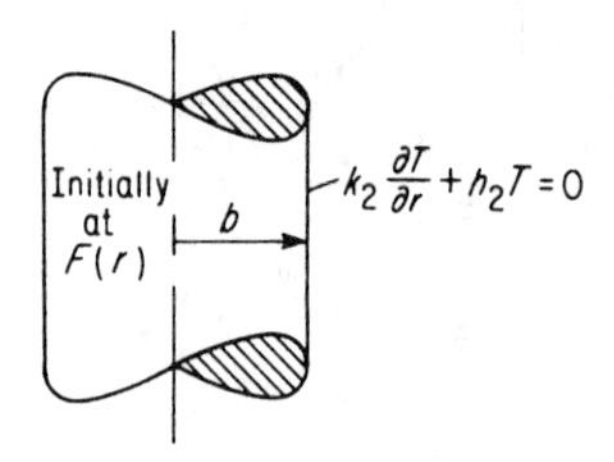

FIG. 3-3. Homogeneous boundary condition of the third kind at the boundary $r = b$ for a solid cylinder.

problem of heat conduction is given as

$$\frac{\partial^2 T}{\partial r^2} + \frac{1}{r}\frac{\partial T}{\partial r} = \frac{1}{\alpha}\frac{\partial T}{\partial t} \quad \text{in} \quad 0 \le r \le b, t > 0 \tag{3-45a}$$

$$k_2 \frac{\partial T}{\partial r} + h_2 T = 0 \quad \text{at} \quad r = b, \quad t > 0 \tag{3-45b}$$

$$T = F(r) \quad \text{in} \quad 0 \le r \le b, \ t = 0 \tag{3-45c}$$

where $T \equiv T(r, t)$.

Assume separation in the form

$$T(r, t) = \Gamma(t) \cdot R(r)$$

Solution for the separated time variable is in the form

$$\Gamma(t) : e^{-\alpha\beta^2 t} \tag{3-46}$$

The space variable $R(r)$ satisfies the following eigenvalue problem:

$$\frac{d^2R}{dr^2} + \frac{1}{r}\frac{dR}{dr} + \beta^2 R = 0 \quad \text{in} \quad 0 \le r \le b \tag{3-47a}$$

$$k_2 \frac{dR}{dr} + h_2 R = 0 \quad \text{at} \quad r = b \tag{3-47b}$$

Equation 3-47a is Bessel's differential equation of order zero. Let $R_0(\beta_m, r)$ be the eigenfunctions and β_m the eigenvalues of Eq. 3-47, then the solution of the heat-conduction problem 3-45 is in the form

$$T(r, t) = \sum_{m=1}^{\infty} c_m e^{-\alpha\beta_m^2 t} \cdot R_0(\beta_m, r) \tag{3-48}$$

This solution should satisfy the initial condition for the problem; hence for $t = 0$ it becomes

$$F(r) = \sum_{m=1}^{\infty} c_m R_0(\beta_m, r) \tag{3-49}$$

The unknown coefficients c_m are determined by making use of the orthogonality of the eigenfunctions $R_0(\beta_m, r)$ in the interval $0 \leq r \leq b$, and given as

$$c_m = \frac{1}{N} \int_{r'=0}^{b} r' \cdot R_0(\beta_m, r') \cdot F(r') \cdot dr' \tag{3-50}$$

The complete solution becomes

$$T(r, t) = \sum_{m=1}^{\infty} e^{-\alpha\beta_m^2 t} \cdot K_0(\beta_m, r) \int_{r'=0}^{b} r' \cdot K_0(\beta_m, r') \cdot F(r') \cdot dr' \tag{3-51}$$

where

$$K_0(\beta_m, r) = \frac{R_0(\beta_m, r)}{\sqrt{N}}$$

summation is taken over all eigenvalues β_m. The kernel $K_0(\beta_m, r)$ and the eigenvalues β_m are obtainable from Table 3-1 by setting $\nu = 0$; for the case with boundary condition of the third kind at $r = b$ they are given as

$$K_0(\beta_m, r) = \frac{\sqrt{2}}{b} \frac{\beta_m}{\sqrt{H_2^2 + \beta_m^2}} \frac{J_0(\beta_m r)}{J_0(\beta_m b)} \tag{3-52}$$

and β_m's are the positive roots of the transcendental equation

$$\beta J_0'(\beta b) + H_2 J_0(\beta b) = 0 \tag{3-53a}$$

or

$$\beta J_1(\beta b) = H_2 J_0(\beta b) \tag{3-53b}$$

where

$$H_2 = \frac{h_2}{k_2}$$

The roots of the transcendental Equation 3-53b are given in Appendix III.

Substituting the kernel 3-52 into Eq. 3-51 the complete solution of the heat-conduction problem becomes

$$T(r, t) = \frac{2}{b^2} \sum_{m=1}^{\infty} e^{-\alpha\beta_m^2 t} \frac{\beta_m^2}{H_2^2 + \beta_m^2} \frac{J_0(\beta_m r)}{J_0^2(\beta_m b)} \int_{r'=0}^{b} r' \cdot J_0(\beta_m r') \cdot F(r') \cdot dr' \tag{3-54}$$

where the summation is taken over all positive roots of the transcendental equation 3-53.

For constant initial temperature, i.e.,

$$F(r) = T_0 = \text{const.}$$

the integration in Eq. 3-54 is performed, i.e.,

$$\int_{r'=0}^{b} r' \cdot J_0(\beta_m r') \cdot dr' = \left[\frac{r'}{\beta_m} J_1(\beta_m r')\right]_0^b = \frac{b}{\beta_m} J_1(\beta_m b)$$

and solution 3-54 becomes

$$T(r, t) = \frac{2T_0}{b} \sum_{m=1}^{\infty} e^{-\alpha\beta_m^2 t} \cdot \frac{\beta_m}{H_2^2 + \beta_m^2} \frac{J_1(\beta_m b)}{J_0^2(\beta_m b)} \cdot J_0(\beta_m r) \tag{3-55}$$

In view of the relation 3-53b, Eq. 3-55 is written in the form

$$T(r, t) = \frac{2T_0}{b} \sum_{m=1}^{\infty} e^{-\alpha\beta_m^2 t} \cdot \frac{H_2}{H_2^2 + \beta_m^2} \frac{J_0(\beta_m r)}{J_0(\beta_m b)} \tag{3-56}$$

EXAMPLE 3-1. A long solid cylinder of radius b is initially at temperature $F(r)$. For times $t > 0$ the boundary at $r = b$ is kept at zero temperature. (See Fig. 3-4a.)

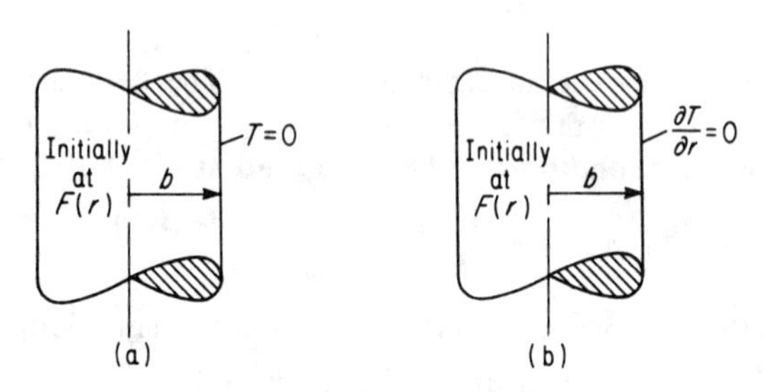

FIG. 3-4. Boundary conditions for a solid cylinder.

The solution of this problem is of the same form as that given by Eq. 3-51, except the kernel $K_0(\beta_m, r)$ and the eigenvalues β_m are obtained from Table 3-1, by setting $\nu = 0$ for the case of boundary condition of the first kind at $r = b$. Then the kernel $K_0(\beta_m, r)$ is given as

$$K_0(\beta_m, r) = \frac{\sqrt{2}}{b} \frac{J_0(\beta_m r)}{J_0'(\beta_m b)} = \frac{\sqrt{2}}{b} \frac{J_0(\beta_m r)}{-J_1(\beta_m b)} \tag{3-57}$$

and β_m's are the positive roots of the transcendental equation

$$J_0(\beta b) = 0 \tag{3-58}$$

Substituting the kernel 3-57 into Eq. 3-51, we obtain the solution of the present problem as

$$T(r, t) = \frac{2}{b^2} \sum_{m=1}^{\infty} e^{-\alpha\beta_m^2 t} \cdot \frac{J_0(\beta_m r)}{J_1^2(\beta_m b)} \int_{r'=0}^{b} r' \cdot J_0(\beta_m r') \cdot F(r') \cdot dr' \tag{3-59}$$

where the summation is taken over all positive roots of the transcendental Eq. 3-58.

EXAMPLE 3-2. A long solid cylinder of radius b is initially at temperature $F(r)$. For times $t > 0$ the boundary surface at $r = b$ is kept insulated. (See Fig. 3-4b.)

The kernel $K_0(\beta_m, r)$ and eigenvalues β_m are obtained from Table 3-1 by setting $\nu = 0$; for a boundary condition of the second kind they are given as

$$K_0(\beta_m, r) = \frac{\sqrt{2}}{b} \frac{J_0(\beta_m r)}{J_0(\beta_m b)} \tag{3-60}$$

and β_m's are the positive roots of the transcendental equation

$$J_0'(\beta b) = 0 \quad \text{or} \quad J_1(\beta b) = 0 \tag{3-61}$$

Substituting Eq. 3-60 into Eq. 3-51, wc obtain

$$T(r, t) = \frac{2}{b^2} \int_{r'=0}^{b} r' F(r') \cdot dr' + \frac{2}{b^2} \sum_{m=1}^{\infty} e^{-\alpha \beta_m^2 t} \cdot \frac{J_0(\beta_m r)}{J_0^2(\beta_m b)} \int_{r'=0}^{b} r' \cdot J_0(\beta_m r') \cdot F(r') \cdot dr' \tag{3-62}$$

where β_m's are the positive roots of the transcendental equation 3-61 and the term

$$\frac{2}{b^2} \int_{r'=0}^{b} r' \cdot F(r') \cdot dr'$$

is the average of the initial temperature distribution over the cross section. Since the region is insulated and there are no heat losses or heat additions to the solid, after the temperature transients have passed the temperature of the region will be represented with the first term in Eq. 3-62.

Hollow Cylinder of Radius $a \leq r \leq b$. Consider a long hollow cylinder of radius $a \leq r \leq b$ which is initially at temperature $F(r)$. For times $t > 0$ heat is dissipated by convection from the boundary surfaces at $r = a$ and $r = b$ into a surrounding at zero temperature (See Fig. 3-5).

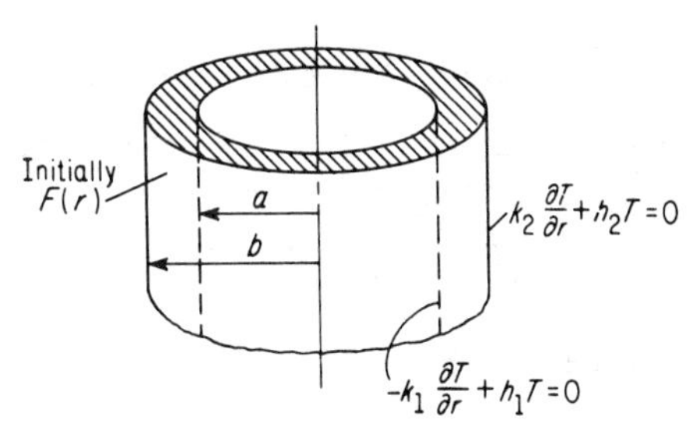

FIG. 3-5. Hollow cylinder subject to homogeneous boundary condition of the third kind at $r = a$ and $r = b$.

The boundary-value problem of heat conduction is given as

$$\frac{\partial^2 T}{\partial r^2} + \frac{1}{r} \frac{\partial T}{\partial r} = \frac{1}{\alpha} \frac{\partial T}{\partial t} \quad \text{in} \quad a \leq r \leq b, t > 0 \tag{3-63a}$$

$$-k_1 \frac{\partial T}{\partial r} + h_1 T = 0 \quad \text{at} \quad r = a, \quad t > 0 \tag{3-63b}$$

$$k_2 \frac{\partial T}{\partial r} + h_2 T = 0 \qquad \text{at} \quad r = b, \qquad t > 0 \tag{3-63c}$$

$$T = F(r) \qquad \text{in} \quad a \leq r \leq b, t = 0 \tag{3-63d}$$

Assuming separation of variables, the solution for the separated time variable is in the form

$$\Gamma(t) : e^{-\alpha\beta^2 t} \tag{3-64}$$

and the separated space variable R satisfies the following eigenvalue problem:

$$\frac{d^2R}{dr^2} + \frac{1}{r}\frac{dR}{dr} + \beta^2 R = 0 \quad \text{in} \quad a \leq r \leq b \tag{3-65a}$$

$$-k_1 \frac{dR}{dr} + h_1 R = 0 \qquad \text{at} \quad r = a \tag{3-65b}$$

$$k_2 \frac{dR}{dr} + h_2 R = 0 \qquad \text{at} \quad r = b \tag{3-65c}$$

Let $R_0(\beta_m, r)$ be the eigenfunctions and β_m the eigenvalues of the eigenvalue Prob. 3-65. Then the solution of the problem 3-63 is given in the form

$$T(r,t) = \sum_{m=1}^{\infty} c_m \cdot e^{-\alpha\beta_m^2 t} \cdot R_0(\beta_m, r) \tag{3-66}$$

This solution should satisfy the initial condition for the problem, and for $t = 0$ it becomes

$$F(r) = \sum_{m=1}^{\infty} c_m \cdot R_0(\beta_m, r) \quad a \leq r \leq b \tag{3-67}$$

The unknown coefficients c_m are immediately determined from the orthogonality of $R_0(\beta_m, r)$ functions in the interval $a \leq r \leq b$, and given as

$$c_m = \frac{1}{N} \int_{r'=a}^{b} r' \cdot R_0(\beta_m, r') \cdot F(r') \cdot dr' \tag{3-68}$$

The solution of system 3-63 becomes

$$T(r,t) = \sum_{m=1}^{\infty} e^{-\alpha\beta_m^2 t} \cdot K_0(\beta_m, r) \cdot \int_{r'=a}^{b} r' \cdot K_0(\beta_m, r') \cdot F(r') \cdot dr' \tag{3-69a}$$

where the kernel $K_0(\beta_m, r)$ is

$$K_0(\beta_m, r) = \frac{R_0(\beta_m, r)}{\sqrt{N}} \tag{3-69b}$$

The eigenfunctions $R_0(\beta_m, r)$ and the norm N for use in Eq. 3-69b are obtained from Table 3-2 for nine different combinations of boundary conditions at the surfaces $r = a$ and $r = b$ by setting $\nu = 0$. For the boundary condition of the third kind at both boundaries they are given as

$$R_0(\beta_m, r) = \frac{J_0(\beta_m r)}{k_2\beta_m J_0'(\beta_m b) + h_2 J_0(\beta_m b)} - \frac{Y_0(\beta_m r)}{k_2\beta_m Y_0'(\beta_m b) + h_2 Y_0(\beta_m b)} \tag{3-70a}$$

$$N = \frac{b^2}{2}\left[1 + \frac{h_2^2}{k_2^2\beta_m^2}\right] R_0^2(\beta_m, b) - \frac{a^2}{2}\left[1 + \frac{h_1^2}{k_1^2\beta_m^2}\right] R_0^2(\beta_m, a) \tag{3-70b}$$

and β_m's are the positive roots of the transcendental equation

$$\frac{-\beta k_1 J_0'(\beta a) + h_1 J_0(\beta a)}{\beta k_2 J_0'(\beta b) + h_2 J_0(\beta b)} - \frac{-\beta k_1 Y_0'(\beta a) + h_1 Y_0(\beta a)}{\beta k_2 Y_0'(\beta b) + h_2 Y_0(\beta b)} = 0 \tag{3-70c}$$

We illustrate the application of solution 3-69 to other combinations of homogeneous boundary conditions with the following examples.

EXAMPLE 3-3. A long hollow cylinder of radius $a \le r \le b$ is initially at temperature $F(r)$. For times $t > 0$ the boundaries at $r = a$ and $r = b$ are kept at zero temperature.

For this special case the boundary conditions at both boundaries are of the first kind. The eigenfunctions $R_0(\beta_m, r)$, the norm N and the eigenvalues β_m for use in the solution 3-69 are immediately obtained from Table 3-2 (the last case) by setting $\nu = 0$. The results are

$$R_0(\beta_m, r) = \frac{J_0(\beta_m r)}{J_0(\beta_m b)} - \frac{Y_0(\beta_m r)}{Y_0(\beta_m b)} \tag{3-71a}$$

$$N = \frac{b^2}{2} R_0'^2(\beta_m, b) - \frac{a^2}{2} R_0'^2(\beta_m, a) \tag{3-71b}$$

and β_m's are the positive roots of the transcendental equation

$$\frac{J_0(\beta a)}{J_0(\beta b)} - \frac{Y_0(\beta a)}{Y_0(\beta b)} = 0 \tag{3-72}$$

These results are simplified further by making use of the following relation for Bessel functions [5]

$$\begin{vmatrix} J_\nu(r) & Y_\nu(r) \\ J_\nu'(r) & Y_\nu'(r) \end{vmatrix} = \frac{2}{\pi r} \tag{3-73}$$

For zero-order Bessel functions of argument βr, Eq. 3-73 is written in the form

$$\begin{vmatrix} J_0(\beta r) & Y_0(\beta r) \\ J_0'(\beta r) & Y_0'(\beta r) \end{vmatrix} = \frac{2}{\pi \beta r} \tag{3-74}$$

For convenience we write Eq. 3-72 in the form

$$\frac{J_0(\beta a)}{J_0(\beta b)} = \frac{Y_0(\beta a)}{Y_0(\beta b)} \equiv S \tag{3-75}$$

We evaluate the functions $R_0'(\beta_m, b)$ and $R_0'(\beta_m, a)$ by making use of Eqs. 3-74 and 3-75 as follows.

$$R_0'(\beta, b) = \frac{J_0'(\beta b)}{J_0(\beta b)} - \frac{Y_0'(\beta b)}{Y_0(\beta b)} = -\frac{1}{J_0(\beta b)\cdot Y_0(\beta b)}\cdot\frac{2}{\pi\beta b} \tag{3-76}$$

Similarly, we have

$$\begin{aligned} R_0'(\beta, a) &= \frac{J_0'(\beta a)}{J_0(\beta b)} - \frac{Y_0'(\beta a)}{Y_0(\beta b)} \\ &= \frac{1}{J_0(\beta b)\cdot Y_0(\beta b)}[J_0'(\beta a)\cdot Y_0(\beta b) - Y_0'(\beta a)\cdot J_0(\beta b)] \\ &= \frac{1}{J_0(\beta b)\cdot Y_0(\beta b)}\cdot\frac{1}{S}\cdot[J_0'(\beta a)\cdot Y_0(\beta a) - Y_0'(\beta a)\cdot J_0(\beta a)] \\ &= -\frac{1}{J_0(\beta b)\cdot Y_0(\beta b)}\cdot\frac{1}{S}\cdot\frac{2}{\pi\beta a} \end{aligned} \tag{3-77}$$

Substituting Eqs. 3-76 and 3-77 into Eq. 3-71,

$$N = \frac{2}{\pi^2}\frac{1}{\beta_m^2\cdot J_0^2(\beta_m b)\cdot Y_0^2(\beta_m b)}\left[1 - \frac{J_0^2(\beta_m b)}{J_0^2(\beta_m a)}\right] \tag{3-78}$$

Then the kernel $K_0(\beta_m, r)$ becomes

$$K_0(\beta_m, r) = \frac{R_0(\beta_m, r)}{\sqrt{N}} = \frac{\pi}{\sqrt{2}}\frac{\beta_m J_0(\beta_m b)\cdot Y_0(\beta_m b)}{\left[1 - \frac{J_0^2(\beta_m b)}{J_0^2(\beta_m a)}\right]^{1/2}}\left[\frac{J_0(\beta_m r)}{J_0(\beta_m b)} - \frac{Y_0(\beta_m r)}{Y_0(\beta_m b)}\right] \tag{3-79}$$

Substituting the kernel $K_0(\beta_m, r)$ from Eq. 3-79 into Eq. 3-69a the solution of the present problem is given as

$$T(r, t) = \sum_{m=1}^{\infty} e^{-\alpha\beta_m^2 t}\cdot K_0(\beta_m, r)\cdot\int_{r'=a}^{b} r'\cdot K_0(\beta_m, r')\cdot F(r')\cdot dr' \tag{3-80}$$

where

$$K_0(\beta_m, r) = \frac{\pi}{\sqrt{2}}\frac{\beta_m J_0(\beta_m b)\cdot Y_0(\beta_m b)}{\left[1 - \frac{J_0^2(\beta_m b)}{J_0^2(\beta_m a)}\right]^{1/2}}\left[\frac{J_0(\beta_m r)}{J_0(\beta_m b)} - \frac{Y_0(\beta_m r)}{Y_0(\beta_m b)}\right] \tag{3-81}$$

and β_m's are the positive roots of the transcendental

$$\frac{J_0(\beta a)}{J_0(\beta b)} - \frac{Y_0(\beta a)}{Y_0(\beta b)} = 0 \tag{3-82}$$

The summation is taken over all eigenvalues β_m.

3-5. ONE-DIMENSIONAL NONHOMOGENEOUS BOUNDARY-VALUE PROBLEMS OF HEAT CONDUCTION (SOLUTION WITH INTEGRAL TRANSFORM)

We now examine the solution of one-dimensional nonhomogeneous boundary-value problems of heat conduction involving the space variable r by means of integral-transform technique. The problems involving a solid cylinder ($0 \leq r \leq b$), a hollow cylinder ($a \leq r \leq b$), and an infinite region ($0 \leq r < \infty$) will be examined separately.

Solid Cylinder of Radius b. Consider a long cylinder of radius $r = b$ which is initially at temperature $F(r)$. For time $t > 0$ heat is generated within the solid at a rate of $g(r, t)$ Btu/hr ft³ and heat is dissipated by convection from the boundary surface at $r = b$ into a medium whose temperature varies with time. Figure 3-6 shows the geometry and

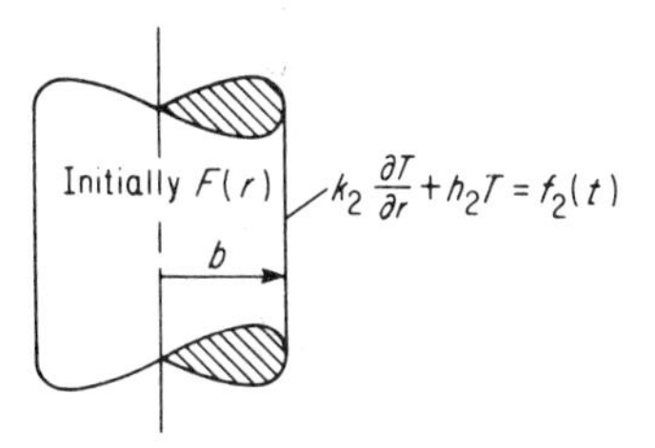

FIG. 3-6. Solid cylinder with nonhomogeneous boundary condition of the third kind at $r = b$.

the boundary condition under consideration. The nonhomogeneous boundary-value problem of heat conduction is given as

$$\frac{\partial^2 T}{\partial r^2} + \frac{1}{r}\frac{\partial T}{\partial r} + \frac{g(r,t)}{k} = \frac{1}{\alpha}\frac{\partial T}{\partial t} \quad \text{in} \quad 0 \leq r \leq b, t > 0 \tag{3-83a}$$

$$k_2 \frac{\partial T}{\partial r} + h_2 T = f_2(t) \quad \text{at} \quad r = b, \quad t > 0 \tag{3-83b}$$

$$T = F(r) \quad \text{in} \quad 0 \leq r \leq b, t = 0 \tag{3-83c}$$

We define the integral transform and inversion formula of temperature function $T(r, t)$ with respect to the r variable in the range $0 \leq r \leq b$ as

$$\begin{pmatrix}\text{Integral}\\ \text{transform}\end{pmatrix} \quad \bar{T}(\beta_m, t) = \int_{r'=0}^{b} r' \cdot K_0(\beta_m, r') \cdot T(r', t) \cdot dr' \tag{3-84a}$$

$$\begin{pmatrix}\text{Inversion}\\ \text{formula}\end{pmatrix} \quad T(r, t) = \sum_{m=1}^{\infty} K_0(\beta_m, r) \cdot \bar{T}(\beta_m, t) \tag{3-84b}$$

where the kernel $K_0(\beta_m, r)$ is the normalized eigenfunction of the following auxiliary eigenvalue problem.

$$\frac{d^2R}{dr^2} + \frac{1}{r}\frac{dR}{dr} + \beta^2 R = 0 \quad \text{in} \quad 0 \leq r \leq b \tag{3-85a}$$

$$k_2 \frac{dR}{dr} + h_2 R = 0 \quad \text{at} \quad r = b \tag{3-85b}$$

and can be obtained from Table 3-1 by setting $\nu = 0$.

By applying integral transform 3-84a we take integral transform of the differential Eq. 3-83a:

$$\int_0^b r\cdot K_0(\beta_m, r)\cdot \nabla^2 T(r,t)\cdot dr + \frac{1}{k}\bar{g}(\beta_m, t) = \frac{1}{\alpha}\frac{d\bar{T}(\beta_m, t)}{dt} \tag{3-86}$$

where

$$\nabla^2 \equiv \frac{\partial^2}{\partial r^2} + \frac{1}{r}\frac{\partial}{\partial r}$$

The first integral on the right-hand side of Eq. 3-86 can be evaluated by means of the general relationship given in Chapter 1 (i.e., see Eq. 1-64).[2]

$$\int_0^b r\cdot K_0(\beta_m, r)\cdot \nabla^2 T(r,t)\cdot dr = -\beta_m^2\cdot \bar{T}(\beta_m, t) + b\left.\frac{K_0(\beta_m, r)}{k_2}\right|_{r=b}\cdot f_2(t) \tag{3-87}$$

Substituting Eq. 3-87 into Eq. 3-86,

$$\frac{d\bar{T}(\beta_m, t)}{dt} + \alpha\beta_m^2\bar{T}(\beta_m, t) = A(\beta_m, t) \tag{3-88a}$$

where

$$A(\beta_m, t) = \frac{\alpha}{k}\bar{g}(\beta_m, t) + \alpha\cdot b\cdot\left.\frac{K_0(\beta_m, r)}{k_2}\right|_{r=b}\cdot f_2(t) \tag{3-88b}$$

Taking the integral transform of the initial condition 3-83c by applying the transform 3-84a we obtain

$$\bar{T}(\beta_m, t)\,|_{t=0} = \bar{F}(\beta_m) \tag{3-89}$$

The solution of Eq. 3-88 subject to the initial condition 3-89 is straightforward and gives the integral transform of temperature $\bar{T}(\beta_m, t)$. Substituting the resulting integral transform into the inversion formula 3-84b we obtain the solution of the nonhomogeneous boundary-value problem of heat conduction, Eq. 3-83, as

$$T(r,t) = \sum_{m=1}^{\infty} e^{-\alpha\beta_m^2 t}\cdot K_0(\beta_m, r)\cdot\left[\bar{F}(\beta_m) + \int_{t'=0}^{t} e^{\alpha\beta_m^2 t'}\cdot A(\beta_m, t')\cdot dt'\right] \tag{3-90a}$$

where

$$A(\beta_m, t') = \frac{\alpha}{k}\bar{g}(\beta_m, t') + \alpha\cdot b\cdot\left.\frac{K_0(\beta_m, r)}{k_2}\right|_{r=b}\cdot f_2(t') \tag{3-90b}$$

$$\bar{g}(\beta_m, t') = \int_{r'=0}^{b} r'\cdot K_0(\beta_m, r')\cdot g(r', t')\cdot dr' \tag{3-90c}$$

$$\bar{F}(\beta_m) = \int_{r'=0}^{b} r'\cdot K_0(\beta_m, r')\cdot F(r')\cdot dr' \tag{3-90d}$$

[2] The same result could be obtained by integrating it by parts twice and by making use of Bessel's differential equation of order zero, Eq. 3-85a and the boundary conditions 3-83b and 3-85b.

The kernel $K_0(\beta_m, r)$ and the eigenvalues β_m are obtained from Table 3-1 by setting $\nu = 0$. For the boundary condition of the third kind at $r = b$ the kernel is given as

$$K_0(\beta_m, r) = \frac{\sqrt{2}}{b} \frac{\beta_m}{[H_2^2 + \beta_m^2]^{1/2}} \cdot \frac{J_0(\beta_m r)}{J_0(\beta_m b)} \tag{3-91}$$

The eigenvalues β_m are the positive roots of the transcendental equation

$$\frac{\beta J_1(\beta b)}{J_0(\beta b)} = H_2 \tag{3-92}$$

$$H_2 = \frac{h_2}{k_2} \tag{3-93}$$

When boundary conditions at $r = b$ is of the first kind k_2 is set equal to zero. To avoid difficulty associated with k_2 being zero at the denominator, the following change should be made in Eq. 3-90b:

$$\text{When } k_2 = 0, \quad \text{replace } \left.\frac{K_0(\beta_m, r)}{k_2}\right|_{r=b} \text{ by } -\frac{1}{h_2} \left.\frac{dK_0(\beta_m, r)}{dr}\right|_{r=b} \tag{3-94}$$

We now examine the application of solution 3-90 to some special cases.

EXAMPLE 3-4. A long solid cylinder of radius $r = b$ is initially at zero temperature. For times $t > 0$ heat is generated within the solid at a rate of $g(r, t)$ Btu/hr ft^3 while the boundary at $r = b$ is kept at zero temperature.

The boundary-value problem of heat conduction is given as

$$\frac{\partial^2 T}{\partial r^2} + \frac{1}{r}\frac{\partial T}{\partial r} + \frac{g(r, t)}{k} = \frac{1}{\alpha}\frac{\partial T}{\partial t} \quad \text{in} \quad 0 \le r \le b, t > 0 \tag{3-95a}$$

$$T = 0 \quad \text{at} \quad r = b, \quad t > 0 \tag{3-95b}$$

$$T = 0 \quad \text{in} \quad 0 \le r \le b, t = 0 \tag{3-95c}$$

This problem is a special case of the above problem, with

$$F(r) = 0, \qquad f(t) = 0 \tag{3-96}$$

The kernel $K_0(\beta_m, r)$ and the eigenvalues β_m for the boundary condition of the first kind at $r = b$ are obtained from Table 3-1 by setting $\nu = 0$.

$$K_0(\beta_m, r) = \frac{\sqrt{2}}{b} \frac{J_0(\beta_m r)}{J_0'(\beta_m b)} = -\frac{\sqrt{2}}{b} \frac{J_0(\beta_m r)}{J_1(\beta_m b)} \tag{3-97a}$$

the eigenvalues β_m's are the positive roots of the transcendental equation

$$J_0(\beta b) = 0 \tag{3-97b}$$

Substituting Eqs. 3-96 and 3-97 into solution 3-90 we obtain

$$T(r, t) = \frac{2}{b^2} \frac{\alpha}{k} \sum_{m=1}^{\infty} e^{-\alpha\beta_m^2 t} \cdot \frac{J_0(\beta_m r)}{J_1^2(\beta_m b)} \cdot \int_{t'=0}^{t} e^{\alpha\beta_m^2 t'} dt' \int_{r'=0}^{b} r' \cdot J_0(\beta_m r') \cdot g(r', t') \cdot dr' \tag{3-98}$$

where summation is taken over all positive roots of Eq. 3-97b.

We now consider some special cases of solution 3-98.

(a) Heat source is uniform throughout the region but varies with time. Taking

$$g(r', t') = g(t') \text{ Btu/hr ft}^3$$

integration with respect to the space variable is performed and Eq. 3-98 becomes

$$T(r, t) = \frac{2}{b} \frac{\alpha}{k} \sum_{m=1}^{\infty} e^{-\alpha\beta_m^2 t} \cdot \frac{J_0(\beta_m r)}{\beta_m J_1(\beta_m b)} \int_{t'=0}^{t} e^{\alpha\beta_m^2 t'} \cdot g(t') \cdot dt' \tag{3-99}$$

Summation is taken over all positive roots of the transcendental Eq. 3-97b.

(b) Heat source is of constant strength. Taking

$$g(r', t') = g_0 = \text{const., Btu/hr ft}^3$$

integrations with respect to both time variable t' and space variable r' are performed and Eq. 3-98 becomes

$$T(r, t) = \frac{2}{b} \frac{g_0}{k} \sum_{m=1}^{\infty} \frac{J_0(\beta_m r)}{J_1(\beta_m b)} \frac{1 - e^{-\alpha\beta_m^2 t}}{\beta_m^3} \tag{3-100}$$

Summation is taken over all positive roots of Eq. 3-97b.

(c) The heat source is an instantaneous volume-heat source of uniform strength $g_i(r')$ Btu/ft^3 and releases its heat spontaneously at time $t = 0$. In this case we relate the instantaneous volume-heat source $g_i(r')$ Btu/ft^3 to the continuous-volume heat source $g(r, t)$ Btu/hr ft^3 by

$$g(r', t') = g_i(r') \cdot \delta(t' - 0) \tag{3-101}$$

Substituting Eq. 3-101 into Eq. 3-98, and performing the integration with respect to the t' variable we obtain

$$T(r, t) = \frac{2}{b^2} \sum_{m=1}^{\infty} e^{-\alpha\beta_m^2 t} \cdot \frac{J_0(\beta_m r)}{J_1^2(\beta_m b)} \int_{r'=0}^{b} r' \cdot J_0(\beta_m r') \cdot \frac{\alpha g_i(r')}{k} \cdot dr' \tag{3-102}$$

In this equation the quantity $\dfrac{\alpha g_i(r')}{k}$ has the unit of temperature. Denoting

$$F(r') = \frac{\alpha g_i(r')}{k}$$

as the temperature distribution in the region at time $t = 0$, then solution 3-102 is exactly the same as the solution of homogeneous boundary-value problem of heat conduction for a solid cylinder $0 \le r \le b$ which is initially at temperature $F(r)$ and with zero temperature at the boundaries (i.e., Eq. 3-59).

(d) The heat source is a line-heat source of strength $g_L(t)$ Btu/hr ft situated at the center of the cylinder along the z-axis.

In the cylindrical coordinate system a line-heat source $g_L(t)$ Btu/hr ft is related to the volume heat source $g(r', t')$ Btu/hr ft^3, by[3]

$$g(r', t') = \frac{g_L(t')}{2\pi r'} \cdot \delta(r' - 0) \tag{3-103}$$

[3]It is to be noted that r' in the denominator of Eq. 3-103 will cancel out when Eq. 3-103 is substituted into Eq. 3-98.

Substituting Eq. 3-103 into Eq. 3-98, we obtain

$$T(r, t) = \frac{1}{\pi b^2} \frac{\alpha}{k} \sum_{m=1}^{\infty} e^{-\alpha\beta_m^2 t} \cdot \frac{J_0(\beta_m r)}{J_1^2(\beta_m b)} \int_{t'=0}^{t} e^{\alpha\beta_m^2 t'} \cdot g_L(t') \cdot dt' \quad (3\text{-}104)$$

(e) The heat source is an instantaneous line heat source of strength g_{Li} Btu/ft, situated at the center of the cylinder along the z-axis, and releases its heat spontaneously at time $t = \tau$.

Instantaneous line heat source g_{Li} Btu/ft is related to a continuous line-heat source $g_L(t')$ Btu/hr ft by

$$g_L(t') = g_{Li}\,\delta(t' - \tau) \quad (3\text{-}105)$$

Substituting Eq. 3-105 into Eq. 3-104, we obtain

$$T(r, t) = \frac{1}{\pi b^2} \cdot \frac{\alpha g_{Li}}{k} \sum_{m=1}^{\infty} e^{-\alpha\beta_m^2 (t-\tau)} \cdot \frac{J_0(\beta_m r)}{J_1^2(\beta_m b)} \quad \text{for } t > \tau \quad (3\text{-}106)$$

Defining for convenience

$$S_{Li} \equiv \frac{\alpha g_{Li}}{k}\ {}^\circ\text{F} \cdot \text{ft}^2$$

Eq. 3-106 becomes

$$T(r, t) = \frac{S_{Li}}{\pi b^2} \sum_{m=1}^{\infty} e^{-\alpha\beta_m^2 (t-\tau)} \cdot \frac{J_0(\beta_m r)}{J_1^2(\beta_m b)} \quad \text{for } t > \tau \quad (3\text{-}107)$$

EXAMPLE 3-5. A long solid cylinder of radius $r = b$ is initially at zero temperature. A cylindrical surface-heat source of radius $r = r_1$ $(0 < r_1 < b)$ and of strength $g_{\text{cyl}}(t)$ Btu/hr ft of linear length of the cylinder, is situated concentrically inside the cylinder as shown in Fig. 3-7. For times $t > 0$ the heat source

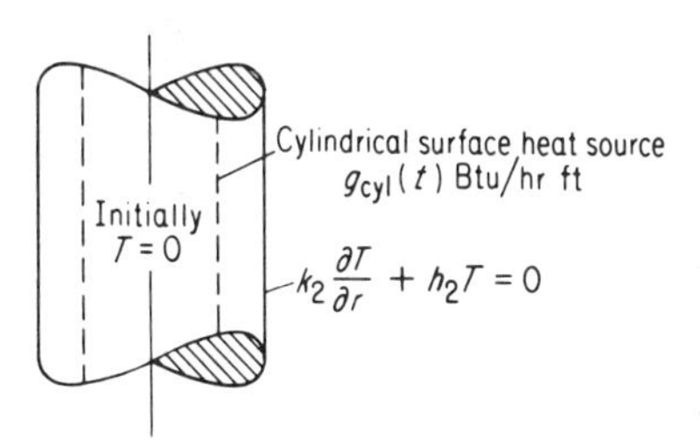

FIG. 3-7. A solid cylinder with a cylindrical-surface heat source.

releases its heat continuously, while heat is dissipated by convection from the boundary surface at $r = b$ into a medium at zero temperature.

The boundary-value problem of heat conduction is given as

$$\frac{\partial^2 T}{\partial r^2} + \frac{1}{r}\frac{\partial T}{\partial r} + \frac{g_{\text{cyl}}(t) \cdot \delta(r - r_1)}{2\pi r \cdot k} = \frac{1}{\alpha}\frac{\partial T(r, t)}{\partial t} \quad \text{in} \quad 0 \le r \le b,\ t > 0$$

$$k_2 \frac{\partial T}{\partial r} + h_2 T = 0 \quad \text{at} \quad r = b,\ t > 0 \tag{3-108}$$

$$T = 0 \quad \text{in} \quad 0 \le r \le b,\ t = 0$$

This problem is a special case of the problem 3-83. By comparing the system 3-108 with system 3-83 we find that

$$g(r, t) = \frac{g_{\text{cyl}}(t)}{2\pi r} \cdot \delta(r - r_1)$$
$$f_2(t) = F(r) = 0 \tag{3-109}$$

Substituting Eq. 3-109 in the solution 3-90, we obtain

$$T(r, t) = \frac{\alpha}{k 2\pi} \sum_{m=1}^{\infty} e^{-\alpha\beta_m^2 t} \cdot K_0(\beta_m, r) \int_{t'=0}^{t} e^{\alpha\beta_m^2 t'}$$
$$\cdot g_{\text{cyl}}(t') \cdot dt' \int_{r'=0}^{b} K_0(\beta_m, r') \cdot \delta(r' - r_1) \cdot dr' \tag{3-110a}$$

where

$$K_0(\beta_m, r) = \frac{\sqrt{2}}{b} \frac{\beta_m}{[H_2^2 + \beta_m^2]^{1/2}} \cdot \frac{J_0(\beta_m r)}{J_0(\beta_m b)} \tag{3-110b}$$

and β_m's are the positive roots of the transcendental equation

$$\frac{\beta J_1(\beta b)}{J_0(\beta b)} = H_2$$
$$H_2 \equiv \frac{h_2}{k_2} \tag{3-110c}$$

In Eq. 3-110a integration with respect to the space variable r' is performed:

$$T(r, t) = \frac{1}{\pi b^2} \frac{\alpha}{k} \sum_{m=1}^{\infty} e^{-\alpha\beta_m^2 t} \cdot \frac{\beta_m^2}{H_2^2 + \beta_m^2} \frac{J_0(\beta_m r) \cdot J_0(\beta_m r_1)}{J_0^2(\beta_m b)}$$
$$\cdot \int_{t'=0}^{t} e^{\alpha\beta_m^2 t'} \cdot g_{\text{cyl}}(t') \cdot dt' \tag{3-111}$$

where the summation is taken over all positive roots of the transcendental Eq. 3-110c.

We now consider some special cases of solution 3-111.

(a) The cylindrical surface-heat source is of constant strength and releases its heat continuously for times $t > 0$. For this case taking

$$g_{\text{cyl}}(t) \equiv g_{\text{cyl}} = \text{const. (Btu/hr ft)} \tag{3-112}$$

Performing the integration with respect to the time variable t', Eq. 3-111 becomes

$$T(r, t) = \frac{1}{\pi b^2} \frac{g_{\text{cyl}}}{k} \sum_{m=1}^{\infty} \frac{1}{H_2^2 + \beta_m^2} \frac{J_0(\beta_m r) \cdot J_0(\beta_m r_1)}{J_0^2(\beta_m b)} [1 - e^{-\alpha\beta_m^2 t}] \tag{3-113}$$

(b) The cylindrical surface-heat source is an instantaneous surface-heat source of strength $g_{\text{cyl},i}$ Btu/ft that releases its heat spontaneously at time $t = \tau$.

In this case we take

$$g_{\text{cyl}}(t) = g_{\text{cyl},i} \cdot \delta(t - \tau) \tag{3-114}$$

Substituting Eq. 3-114 into Eq. 3-111,

$$T(r, t) = \frac{1}{\pi b^2} \frac{\alpha g_{\text{cyl},i}}{k} \sum_{m=1}^{\infty} e^{-\alpha \beta_m^2 (t-\tau)} \cdot \frac{\beta_m{}^2}{H_2^2 + \beta_m^2} \frac{J_0(\beta_m r) \cdot J_0(\beta_m r_1)}{J_0^2(\beta_m b)} \quad \text{for } t > \tau \tag{3-115}$$

In view of the relation 3-110c, Eq. 3-115 is written in the form

$$T(r, t) = \frac{1}{\pi b^2} \frac{\alpha g_{\text{cyl},i}}{k} \sum_{m=1}^{\infty} e^{-\alpha \beta_m^2 (t-\tau)} \cdot \frac{J_0(\beta_m r) \cdot J_0(\beta_m r_1)}{J_0^2(\beta_m b) + J_1^2(\beta_m b)} \quad \text{for } t > \tau \tag{3-116}$$

If the strength of the instantaneous surface-heat source is expressed as

$$S_{\text{cyl},i} \equiv \frac{\alpha g_{\text{cyl},i}}{k} \quad {}^\circ\text{F} \cdot \text{ft}^2 \text{ per unit length of cylinder} \tag{3-117}$$

Then, in Eq. 3-116 the term

$$\frac{1}{\pi b^2} \sum_{m=1}^{\infty} e^{-\alpha \beta_m^2 (t-\tau)} \cdot \frac{J_0(\beta_m r) \cdot J_0(\beta_m r_1)}{J_0^2(\beta_m b) + J_1^2(\beta_{mb})} \quad \text{for } t > \tau \tag{3-118}$$

represents the temperature at r at time t due to an instantaneous cylindrical surface-heat source of strength $S_{\text{cyl},i} = 1$ °F · ft^2 per unit length, situated at $r = r_1$ and releasing its heat spontaneously at time $t = \tau$ in a long solid cylinder of radius $r = b$, which is initially at zero temperature and for times $t > 0$ heat is dissipated from its boundary surface by convection into a surrounding at zero temperature.

Hollow Cylinder $a \leq r \leq b$. Consider a hollow cylinder of radius $a \leq r \leq b$ which is initially at temperature $F(r)$. For times $t > 0$ heat is generated within the region at a rate of $g(r, t)$ Btu/hr ft^3 and heat is dissipated by convection from the boundary surfaces at $r = a$ and $r = b$ into a surrounding temperature of which varies with time. Figure 3-8 shows the geometry and the boundary conditions under consideration.

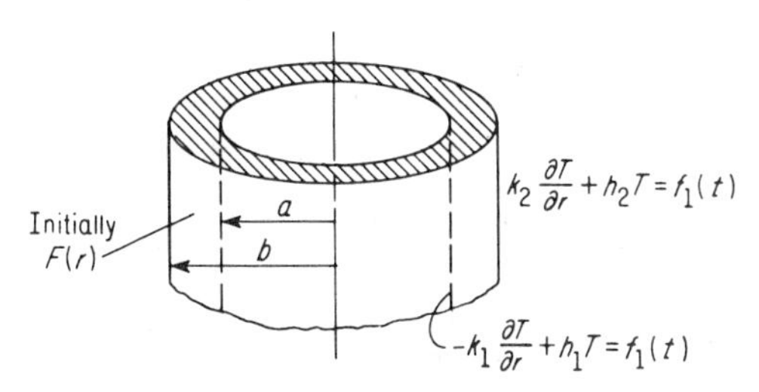

FIG. 3-8. Hollow cylinder with nonhomogeneous boundary conditions of the third kind at $r = a$ and $r = b$.

The boundary-value problem of heat conduction is given as

$$\begin{aligned}
\frac{\partial^2 T}{\partial r^2} + \frac{1}{r}\frac{\partial T}{\partial r} + \frac{g(r,t)}{k} &= \frac{1}{\alpha}\frac{\partial T}{\partial t} \quad \text{in} \quad a \le r \le b,\ t > 0 \\
-k_1 \frac{\partial T}{\partial r} + h_1 T &= f_1(t) \quad \text{at} \quad r = a,\ t > 0 \\
k_2 \frac{\partial T}{\partial r} + h_2 T &= f_2(t) \quad \text{at} \quad r = b,\ t > 0 \\
T &= F(r) \quad \text{in} \quad a \le r \le b,\ t = 0
\end{aligned} \tag{3-119}$$

We define the integral transform and the inversion formula of the temperature function $T(r, t)$ with respect to the space variable r as

$$\begin{pmatrix}\text{Integral}\\ \text{transform}\end{pmatrix} \quad \bar{T}(\beta_m, t) = \int_{r'=a}^{b} r' \cdot K_0(\beta_m, r') \cdot T(r', t) \cdot dr' \tag{3-120a}$$

$$\begin{pmatrix}\text{Inversion}\\ \text{formula}\end{pmatrix} \quad T(r, t) = \sum_{m=1}^{\infty} K_0(\beta_m, r) \cdot \bar{T}(\beta_m, t) \tag{3-120b}$$

where the kernel $K_0(\beta_m, r)$ is the normalized eigenfunction of the following eigenvalue problem.

$$\frac{d^2 R}{dr^2} + \frac{1}{r}\frac{dR}{dr} + \beta^2 R = 0 \quad \text{in} \quad a \le r \le b \tag{3-121a}$$

$$-k_1 \frac{dR}{dr} + h_1 R = 0 \quad \text{at} \quad r = a \tag{3-121b}$$

$$k_2 \frac{dR}{dr} + h_2 R = 0 \quad \text{at} \quad r = b \tag{3-121c}$$

We now take the integral transform of the system 3-119 by applying the transform Eq. 3-120a,[4]

$$\frac{d\bar{T}(\beta_m, t)}{dt} + \alpha\beta_m^2 \bar{T}(\beta_m, t) = A(\beta_m, t) \tag{3-122a}$$

with

$$\bar{T}(\beta_m, t)\big|_{t=0} = \bar{F}(\beta_m) \tag{3-122b}$$

[4] The transform of the differential operator $\dfrac{\partial^2 T}{\partial r^2} + \dfrac{1}{r}\dfrac{\partial T}{\partial r}$ is

$$\int_{r'=a}^{b} r' K_0(\beta_m, r') \cdot \left[\frac{\partial^2 T}{\partial r^2} + \frac{1}{r}\frac{\partial T}{\partial r}\right] dr' = -\beta_m^2 \bar{T}(\beta_m, t) + a \left.\frac{K_0(\beta_m, r)}{k_1}\right|_{r=a} \cdot f_1(t) + b \left.\frac{K_0(\beta_m, r)}{k_2}\right|_{r=b} \cdot f_2(t)$$

where

$$A(\beta_m, t') \equiv \frac{\alpha}{k}\,\bar{g}(\beta_m, t') + \alpha\left[a \cdot \frac{K_0(\beta_m, r)}{k_1}\bigg|_{r=a} \cdot f_1(t') + b \cdot \frac{K_0(\beta_m, r)}{k_2}\bigg|_{r=b} \cdot f_2(t')\right]$$

and quantities with bars refer to integral transform according to transform Eq. 3-120a.

The solution of the ordinary differential Eq. 3-122a subject to the initial condition Eq. 3-122b is straightforward. Substituting the resulting integral transform into the inversion formula 3-120b, we obtain the solution of the boundary-value problem of heat conduction (Eq. 3-119) as

$$T(r, t) = \sum_{m=1}^{\infty} e^{-\alpha\beta_m^2 t} \cdot K_0(\beta_m, r) \cdot \left[\bar{F}(\beta_m) + \int_{t'=0}^{t} e^{\alpha\beta_m^2 t'} \cdot A(\beta_m, t') \cdot dt'\right] \tag{3-123a}$$

where

$$A(\beta_m, t') = \frac{\alpha}{k}\,\bar{g}(\beta_m, t') + \alpha\left[a \cdot \frac{K_0(\beta_m, r)}{k_1}\bigg|_{r=a} \cdot f_1(t') + b\,\frac{K_0(\beta_m, r)}{k_2}\bigg|_{r=b} \cdot f_2(t')\right] \tag{3-123b}$$

$$\bar{g}(\beta_m, t') = \int_{r'=a}^{b} r' \cdot K_0(\beta_m, r') \cdot g(r', t') \cdot dr' \tag{3-123c}$$

$$\bar{F}(\beta_m) = \int_{r'=a}^{b} r' \cdot K_0(\beta_m, r') \cdot F(r') \cdot dr' \tag{3-123d}$$

and the summation is taken over all eigenvalues β_m.

The kernel $K_0(\beta_m, r)$ in Eq. 3-123 is defined as

$$K_0(\beta_m, r) = \frac{R_0(\beta_m, r)}{\sqrt{N}} \tag{3-124}$$

where the eigenfunctions $R_0(\beta_m, r)$, the norm N and the eigenvalues β_m are obtained from Table 3-2 for any combination of the boundary conditions at $r = a$ and $r = b$ by setting $\nu = 0$.

For a boundary condition of the first kind we set k_1 or k_2 or both equal to zero. To avoid difficulty for such cases the following changes should be made in Eq. 3-123b.

$$\text{When } k_1 = 0, \text{ replace } \frac{K_0(\beta_m, r)}{k_1}\bigg|_{r=a} \text{ by } \frac{1}{h_1}\frac{dK_0(\beta_m, r)}{dr}\bigg|_{r=a} \tag{3-125a}$$

$$\text{When } k_2 = 0, \text{ replace } \frac{K_0(\beta_m, r)}{k_2}\bigg|_{r=b} \text{ by } -\frac{1}{h_2}\frac{dK_0(\beta_m, r)}{dr}\bigg|_{r=b} \tag{3-125b}$$

We examine the application of solution 3-123 to some special cases with the following examples.

EXAMPLE 3-6. A long hollow cylinder of radius $a \le r \le b$ is initially at temperature $F(r)$. For times $t > 0$ heat is generated within the solid at a rate of $g(r, t)$ Btu/hr ft^3, while the boundaries at $r = a$ and $r = b$ are kept at temperatures $\phi_1(t)$ and $\phi_2(t)$ respectively.

The boundary-value problem of heat conduction is given as

$$\begin{aligned} \frac{\partial^2 T}{\partial r^2} + \frac{1}{r}\frac{\partial T}{\partial r} + \frac{g(r,t)}{k} &= \frac{1}{\alpha}\frac{\partial T}{\partial t} \quad \text{in } a \le r \le b,\ t > 0 \\ T &= \phi_1(t) \quad \text{at } r = a,\ t > 0 \\ T &= \phi_2(t) \quad \text{at } r = b,\ t > 0 \\ T &= F(r) \quad \text{in } a \le r \le b,\ t = 0 \end{aligned} \tag{3-126}$$

By comparing this problem with the general problem 3-119 for a hollow cylinder we note that

$$\begin{aligned} k_1 &= k_2 = 0 \\ h_1 &= h_2 = 1 \\ f_1(t) &= \phi_1(t) \\ f_2(t) &= \phi_2(t) \end{aligned} \tag{3-127}$$

Since k_1 and k_2 are zero, we make changes in the general solution 3-123 according to Eq. 3-125, then we substitute Eq. 3-127 into the general solution 3-123. The resulting solution is

$$T(r,t) = \sum_{m=1}^{\infty} e^{-\alpha\beta_m^2 t} \cdot K_0(\beta_m, r)\left[\bar{F}(\beta_m) + \int_{t'=0}^{t} e^{\alpha\beta_m^2 t'} \cdot A(\beta_m, t') \cdot dt'\right] \tag{3-128}$$

where

$$\begin{aligned} A(\beta_m, t') &= \frac{\alpha}{k}\,\bar{g}(\beta_m, t') \\ &\quad + \alpha\left[a\left.\frac{dK_0(\beta_m, r)}{dr}\right|_{r=a} \cdot \phi_1(t') - b\left.\frac{dK_0(\beta_m, r)}{dr}\right|_{r=b} \cdot \phi_2(t')\right] \end{aligned}$$

$\bar{g}(\beta_m, t)$ and $\bar{F}(\beta_m)$ are as defined by Eq. 3-123c and d.

The kernel $K_0(\beta_m, r)$ is defined as

$$K_0(\beta_m, r) = \frac{R_0(\beta_m, r)}{\sqrt{N}} \tag{3-129}$$

and the values of $R_0(\beta_m, r)$, N and β_m are obtained from Table 3-2 (the last case) by setting $\nu = 0$. For a boundary condition of the first kind at both boundaries they are given as

$$R_0(\beta_m, r) = \frac{J_0(\beta_m r)}{J_0(\beta_m b)} - \frac{Y_0(\beta_m r)}{Y_0(\beta_m b)} \tag{3-130}$$

$$N = \frac{b^2}{2} R_0'^2(\beta_m, b) - \frac{a^2}{2} R_0'^2(\beta_m, a) \tag{3-131}$$

and β_m's are the positive roots of the transcendental equation

$$\frac{J_0(\beta a)}{J_0(\beta b)} - \frac{Y_0(\beta a)}{Y_0(\beta b)} = 0 \tag{3-132}$$

After evaluating the functions $R_0'(\beta_m, b)$ and $R_0'(\beta_m, a)$ in a similar manner discussed in example 3-3, the kernel $K_0(\beta_m, r)$ for use in the solution 3-128 becomes (See Eq. 3-81.)

$$K_0(\beta_m, r) = \frac{\pi}{\sqrt{2}} \frac{\beta_m \cdot J_0(\beta_m b) \cdot Y_0(\beta_m b)}{\left[1 - \dfrac{J_0^2(\beta_m b)}{J_0^2(\beta_m a)}\right]^{1/2}} \left[\frac{J_0(\beta_m r)}{J_0(\beta_m b)} - \frac{Y_0(\beta_m r)}{Y_0(\beta_m b)}\right] \tag{3-133}$$

and the derivatives are

$$\left.\frac{dK_0(\beta_m, r)}{dr}\right|_{r=b} = -\frac{1}{\sqrt{N}} \frac{1}{J_0(\beta_m b) \cdot Y_0(\beta_m b)} \frac{2}{\pi \beta_m b} \tag{3-134a}$$

$$\left.\frac{dK_0(\beta_m, r)}{dr}\right|_{r=a} = -\frac{1}{\sqrt{N}} \frac{1}{J_0(\beta_m b) \cdot Y_0(\beta_m b)} \cdot \frac{1}{S} \cdot \frac{2}{\pi \beta_m a} \tag{3-134b}$$

where

$$N = \frac{2}{\pi^2} \frac{1}{\beta_m^2 \cdot J_0^2(\beta_m b) \cdot Y_0^2(\beta_m b)} \cdot \left(1 - \frac{J_0^2(\beta_m b)}{J_0^2(\beta_m a)}\right) \tag{3-135a}$$

$$S \equiv \frac{J_0(\beta_m a)}{J_0(\beta_m b)} = \frac{Y_0(\beta_m a)}{Y_0(\beta_m b)} \tag{3-135b}$$

We examine some special cases of the solution 3-128.

(a) Boundaries at $r = a$ and $r = b$ are kept at zero temperature and that there is no heat generation within the solid. For this special case we have

$$\phi_1(t) = \phi_2(t) = g(r, t) = 0 \tag{3-136}$$

Substituting Eq. 3-136 into the solution 3-128, we obtain

$$T(r, t) = \sum_{m=1}^{\infty} e^{-\alpha \beta_m^2 t} \cdot K_0(\beta_m, r) \int_{r'=a}^{b} r' \cdot K_0(\beta_m, r') \cdot F(r') \cdot dr' \tag{3-137}$$

where the kernel $K_0(\beta_m, r)$ is given by Eq. 3-133 and summation is taken over all eigenvalues β_m which are the positive roots of Eq. 3-132. It is to be noted that this special case, Eq. 3-137, is the same as the solution 3-80 of the homogeneous boundary-value problem of heat conduction for a hollow cylinder with initial temperature $F(r)$ and with zero surface temperature.

(b) The hollow cylinder is initially at zero temperature. For times $t > 0$ an instantaneous cylindrical-surface heat source of radius $r = r_1$ (i.e., $a < r_1 < b$) and of strength $g_{\text{cyl},i}$ Btu/ft of linear length of the cylinder, which is situated inside the cylinder coaxially, releases its heat spontaneously at time $t = \tau$, while the boundary surfaces at $r = a$ and $r = b$ are kept at zero temperature.

For this special case we have

$$F(r) = \phi_1(t) = \phi_2(t) = 0 \tag{3-138}$$

and the instantaneous cylindrical-surface heat source is related to the volume heat source by

$$g(r', t') = \frac{g_{\text{cyl},i}}{2\pi r'} \cdot \delta(r' - r_1) \cdot \delta(t' - \tau) \tag{3-139}$$

Substituting Eqs. 3-138 and 3-139 into the solution 3-128 we obtain

$$T(r,t) = \frac{\alpha}{k} \frac{g_{\text{cyl},i}}{2\pi} \sum_{m=1}^{\infty} e^{-\alpha\beta_m^2 t} \cdot K_0(\beta_m, r') \int_{t'=0}^{t} e^{\alpha\beta_m^2 t'} \cdot \delta(t' - \tau) \cdot dt' \int_{r'=a}^{b} K_0(\beta_m, r') \cdot \delta(r' - r_1) \cdot dr' \tag{3-140}$$

Integrations involving delta functions are immediately performed and Eq. 3-140 becomes

$$T(r,t) = \frac{1}{2\pi} \frac{\alpha g_{\text{cyl},i}}{k} \sum_{m=1}^{\infty} e^{-\alpha\beta_m^2 (t-\tau)} \cdot K_0(\beta_m, r) \cdot K_0(\beta_m, r_1) \quad \text{for} \quad t > \tau \tag{3-141}$$

If the strength of the instantaneous cylindrical-surface source $g_{\text{cyl},i}$ Btu/ft is expressed as

$$S_{\text{cyl},i} = \frac{\alpha g_{\text{cyl},i}}{k} \ {}^\circ\text{F} \cdot \text{ft}^2 \tag{3-142}$$

then Eq. 3-141 becomes

$$T(r,t) = \frac{S_{\text{cyl},i}}{2\pi} \sum_{m=1}^{\infty} e^{-\alpha\beta_m^2 (t-\tau)} \cdot K_0(\beta_m, r) \cdot K_0(\beta_m, r_1) \quad \text{for} \quad t > \tau \tag{3-143}$$

where the kernel $K_0(\beta_m, r)$ is given by Eq. 3-133 and the summation is taken over all positive roots of Eq. 3-132.

Infinite Region $0 \le r < \infty$. We now consider heat conduction in an infinite region $0 \le r < \infty$ which is initially at temperature $F(r)$ and for times $t > 0$ with heat generation within the solid at a rate of $g(r,t)$ Btu/hr ft^3. The boundary-value problem of heat conduction is given as

$$\frac{\partial^2 T}{\partial r^2} + \frac{1}{r}\frac{\partial T}{\partial r} + \frac{g(r,t)}{k} = \frac{1}{\alpha}\frac{\partial T}{\partial t} \quad \text{in} \quad 0 \le r < \infty,\ t > 0 \tag{3-144a}$$

$$T = F(r) \quad \text{in} \quad 0 \le r < \infty,\ t = 0 \tag{3-144b}$$

We define the integral transform and the inversion formula of the temperature function $T(r,t)$ in the infinite interval $0 \le r < \infty$ as (see Eq. 3-44 for $\nu = 0$):

$$\begin{pmatrix}\text{Integral}\\ \text{transform}\end{pmatrix} \quad \bar{T}(\beta, t) = \int_{r'=0}^{\infty} r' \cdot J_0(\beta r') \cdot T(r', t) \cdot dr' \tag{3-145a}$$

$$\begin{pmatrix}\text{Inversion}\\ \text{formula}\end{pmatrix} \quad T(r,t) = \int_{\beta=0}^{\infty} \beta \cdot J_0(\beta r) \cdot \bar{T}(\beta, t) \cdot d\beta \tag{3-145b}$$

It is to be noted that the kernel $J_0(\beta r)$ is the solution of the following auxiliary differential equation

$$\frac{d^2R}{dr^2} + \frac{1}{r}\frac{dR}{dr} + \beta^2 R = 0 \quad \text{in} \quad 0 \le r < \infty \tag{3-146}$$

with R remaining finite within the region.

We take the integral transform of system 3-144 by applying the integral transform 3-145a:

$$\int_{r=0}^{\infty} r \cdot J_0(\beta r)\left[\frac{\partial^2 T}{\partial r^2} + \frac{1}{r}\frac{\partial T}{\partial r}\right] dr + \frac{1}{k}\,\bar{g}(\beta, t) = \frac{1}{\alpha}\frac{d\bar{T}(\beta, t)}{dt} \tag{3-147a}$$

$$\bar{T}(\beta, t)\,|_{t=0} = \bar{F}(\beta) \tag{3-147b}$$

The first integral on the left-hand side of Eq. 3-147a is evaluated by integrating it by parts twice and by making use of the auxiliary equation 3-146. If it is assumed further that $r \cdot J_0(\beta r) \cdot \partial T/\partial r$ and $rTJ_0'(\beta r)$ vanish as r becomes zero or infinite, we obtain

$$\int_{r=0}^{\infty} rJ_0(\beta r) \cdot \left[\frac{\partial^2 T}{\partial r^2} + \frac{1}{r}\frac{\partial T}{\partial r}\right] dr = -\beta^2 \bar{T}(\beta, t) \tag{3-148}$$

Substituting Eq. 3-148 into Eq. 3-147a we obtain an ordinary differential equation subject to the transformed initial condition 3-147b. When this ordinary differential equation is solved and the resulting transform of temperature is substituted into the inversion formula 3-145b we obtain

$$T(r, t) = \int_{\beta=0}^{\infty} \beta J_0(\beta r) \cdot e^{-\alpha\beta^2 t}\left[\int_{r'=0}^{\infty} r' J_0(\beta r) \cdot F(r') \cdot dr' \right.$$

$$\left. + \frac{\alpha}{k}\int_{t'=0}^{t} e^{\alpha\beta^2 t'}\, dt' \int_{r'=0}^{\infty} r' J_0(\beta r') \cdot g(r', t') \cdot dr' \right] d\beta \tag{3-149}$$

which is the solution of the boundary-value problem of heat conduction Eq. 3-144. We examine some special cases of this solution.

EXAMPLE 3-7. An infinite medium $0 \le r < \infty$ is initially at temperature $F(r)$, but there is no heat generation within the solid.

Setting in Eq. 3-149,

$$g(r', t') = 0$$

the solution of the present problem becomes

$$T(r, t) = \int_{\beta=0}^{\infty} e^{-\alpha\beta^2 t} \cdot \beta J_0(\beta r) \int_{r'=0}^{\infty} r' J_0(\beta r') \cdot F(r') \cdot dr' d\beta \tag{3-150}$$

This result is simplified further by making use of the integral [6]

$$\int_{\beta=0}^{\infty} e^{-\alpha\beta^2 t} \beta J_0(\beta r) J_0\,(\beta r') d\beta = \frac{1}{2\alpha t} e^{-\frac{r^2 + r'^2}{4\alpha t}} \cdot I_0\left(\frac{rr'}{2\alpha t}\right) \tag{3-151}$$

Equation 3-150 becomes

$$T(r,t) = \frac{1}{2\alpha t}\int_{r'=0}^{\infty} r' e^{-\frac{r^2+r'^2}{4\alpha t}} \cdot I_0\left(\frac{rr'}{2\alpha t}\right) \cdot F(r') \cdot dr' \tag{3-152}$$

EXAMPLE 3-8. An infinite region $0 \leq r < \infty$ is initially at zero temperature. A cylindrical-surface heat source of radius $r = r_1$ and of strength $g_{cyl}(t)$ Btu/hr ft of linear length of cylinder is situated in this region with axis along the z-axis. The heat source releases its heat continuously for times $t > 0$.

We relate the cylindrical surface-heat source $g_{cyl}(t)$ Btu/hr ft to the volume-heat source $g(r, t)$ Btu/hr ft^3 by

$$g(r', t') = \frac{g_{cyl}(t')}{2\pi r'}\delta(r' - r_1) \tag{3-153}$$

Substituting Eq. 3-153 into Eq. 3-149 and taking $F(r') = 0$, we obtain

$$T(r,t) = \frac{1}{2\pi}\frac{\alpha}{k}\int_{\beta=0}^{\infty} \beta J_0(\beta r) \cdot e^{-\alpha\beta^2 t}\int_{t'=0}^{t} e^{\alpha\beta^2 t'} \cdot g_{cyl}(t') \cdot dt'$$

$$\cdot \int_{r'=0}^{\infty} J_0(\beta r') \cdot \delta(r' - r_1) \cdot dr' \cdot d\beta \tag{3-154}$$

Performing the integration with respect to the r' variable and changing the order of integration,

$$T(r,t) = \frac{1}{2\pi}\frac{\alpha}{k}\int_{t'=0}^{t}\int_{\beta=0}^{\infty} e^{\alpha\beta^2 t'} \cdot g_{cyl}(t') \cdot dt' \; e^{-\alpha\beta^2 t} \cdot \beta \cdot J_0(\beta r) \cdot J_0(\beta r_1) \cdot d\beta \tag{3-155}$$

In view of the relation 3-151 the last integral in Eq. 3-155 is evaluated and we obtain

$$T(r,t) = \frac{1}{4\pi k}\int_{t'=0}^{t} \frac{1}{t-t'} e^{-\frac{r^2+r_1^2}{4\alpha(t-t')}} \cdot I_0\left(\frac{rr_1}{2\alpha(t-t')}\right) g_{cyl}(t') \cdot dt' \tag{3-156}$$

We consider the following special case of solution 3-156.

(a) The heat source is an instantaneous cylindrical surface-heat source of strength $g_{cyl,i}$ Btu/ft of linear length of cylinder and releases its heat spontaneously at time $t = 0$.

The instantaneous cylindrical surface-heat source is related to the continuous cylindrical surface-heat source by

$$g_{cyl}(t') = g_{cyl,i} \cdot \delta(t' - 0) \tag{3-157}$$

Substituting Eq. 3-157 into Eq. 3-156 and performing the integration

$$T(r,t) = \frac{g_{cyl,i}}{4\pi k t} \cdot e^{-\frac{r^2+r_1^2}{4\alpha t}} \cdot I_0\left(\frac{rr_1}{2\alpha t}\right) \tag{3-158}$$

Defining

$$S_{cyl,i} \equiv \frac{\alpha g_{cyl,i}}{k} \text{ °F} \cdot \text{ft}^2$$

Eq. 3-158 becomes

$$T(r,t) = \frac{S_{\text{cyl},i}}{4\pi\alpha t} \cdot e^{-\frac{r^2+r_1^2}{4\alpha t}} \cdot I_0\left(\frac{rr_1}{2\alpha t}\right) \tag{3-159}$$

EXAMPLE 3-9. An infinite region $0 \leq r < \infty$ is initially at zero temperature. An instantaneous line-heat source of strength $g_{L,i}$ Btu/ft is situated along the z-axis in the medium and releases its heat spontaneously at time $t = \tau$.

The solution of this problem is immediately obtained from solution 3-149. For this case we have

$$F(r') = 0$$
$$g(r',t') = \frac{g_{L,i}}{2\pi r'} \cdot \delta(r' - 0) \cdot \delta(t' - \tau) \tag{3-160}$$

Substituting Eq. 3-160 into Eq. 3-149,

$$T(r,t) = \frac{1}{2\pi}\frac{\alpha g_{L,i}}{k} \int_{\beta=0}^{\infty} e^{-\alpha\beta^2 t} \cdot \beta \cdot J_0(\beta r) \cdot d\beta \int_{t'=0}^{t} e^{\alpha\beta^2 t'} \cdot \delta(t' - \tau) \cdot dt'$$
$$\cdot \int_{r'=0}^{\infty} J_0(\beta r') \cdot \delta(r' - 0) \cdot dr' \tag{3-161}$$

Performing the integrations with respect to the r' and t' variables,

$$T(r,t) = \frac{1}{2\pi}\frac{\alpha g_{L,i}}{k} \int_{\beta=0}^{\infty} e^{-\alpha\beta^2(t-\tau)} \cdot \beta \cdot J_0(\beta r) \cdot d\beta \quad \text{for } t > \tau \tag{3-162}$$

In view of the integral[7]

$$\int_{\beta=0}^{\infty} e^{-\alpha\beta^2(t-\tau)} \cdot \beta \cdot J_0(\beta r) \cdot d\beta = \frac{1}{2\alpha(t-\tau)} e^{-\frac{r^2}{4\alpha(t-\tau)}} \tag{3-163}$$

Eq. 3-162 becomes

$$T(r,t) = \frac{g_{L,i}}{4\pi k(t-\tau)} \cdot e^{-\frac{r^2}{4\alpha(t-\tau)}} \quad \text{for } t > \tau \tag{3-164}$$

3-6. HOMOGENEOUS BOUNDARY-VALUE PROBLEMS OF HEAT CONDUCTION INVOLVING MORE THAN ONE SPACE VARIABLE (SOLUTION WITH SEPARATION OF VARIABLES)

In this section we examine the solution of homogeneous boundary-value problems of heat conduction involving more than one space variable by the method of separation of variables. Such problems include so many combinations of space variables, geometry and boundary conditions that it is not possible to consider them all in the limited space available. Only a few representative cases will be examined in order to illustrate the method.

Long Solid Cylinder $0 \le r \le b, -\pi \le \phi \le \pi$. Consider a long solid cylinder of radius $r = b$ which is initially at temperature $F(r, \phi)$. For times $t > 0$ heat is dissipated by convection from the boundary surface at $r = b$ into a medium at zero temperature. Figure 3-9 shows the geometry and the boundary conditions.

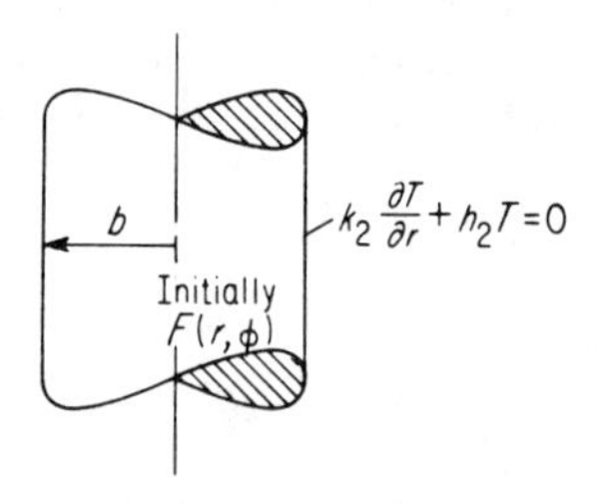

FIG. 3-9. Boundary conditions for a solid cylinder.

The homogeneous boundary-value problem of heat conduction is given as

$$\frac{\partial^2 T}{\partial r^2} + \frac{1}{r}\frac{\partial T}{\partial r} + \frac{1}{r^2}\frac{\partial^2 T}{\partial \phi^2} = \frac{1}{\alpha}\frac{\partial T}{\partial t} \quad \text{in} \quad 0 \le r \le b, -\pi \le \phi \le \pi, \text{ for } t > 0 \tag{3-165a}$$

$$k_2 \frac{\partial T}{\partial r} + h_2 T = 0 \quad \text{at} \quad r = b, \text{ for } t > 0 \tag{3-165b}$$

$$T = F(r, \phi) \quad \text{in} \quad 0 \le r \le b, -\pi \le \phi \le \pi, \text{ for } t = 0 \tag{3-165c}$$

where $T \equiv T(r, \phi, t)$.

We assume separation of variables in the form

$$T(r, \phi, t) = \Gamma(t) \cdot R(r) \cdot \Phi(\phi) \tag{3-166}$$

Then the separated equations and their particular solutions are

$$\frac{d\Gamma}{dt} + \alpha\beta^2\Gamma = 0 \qquad \Gamma : e^{-\alpha\beta^2 t} \tag{3-167a}$$

$$\frac{d^2R}{dr^2} + \frac{1}{r}\frac{dR}{dr} + \left(\beta^2 - \frac{\nu^2}{r^2}\right)R = 0 \quad R : J_\nu(\beta r) \quad 0 \le r \le b \tag{3-167b}$$

$$\frac{d^2\Phi}{d\phi^2} + \nu^2\Phi = 0 \qquad \Phi : \begin{matrix} \sin \nu\phi \\ \cos \nu\phi \end{matrix} \quad -\pi \le \phi \le \pi \tag{3-167c}$$

The $Y_\nu(\beta r)$ function was excluded from the solution since the region includes $r = 0$. In this problem the eigenvalues β and ν constitute a discrete set of points and they are determined by the following considerations.

The temperature function should be cyclic with a period of 2π with respect to the ϕ variable and this condition is satisfied if

$$\nu = 0, 1, 2, 3, \ldots \tag{3-168}$$

The solution for the R separation should satisfy the boundary condition at $r = b$, that is,

$$\beta J_\nu'(\beta b) + H_2 J_\nu(\beta b) = 0 \tag{3-169}$$

β_m's are the positive roots of the transcendental equation 3-169. Then the complete solution of the problem 3-165 is given in the form

$$T(r, \phi, t) = \sum_{m=1}^{\infty} \sum_{\nu=0}^{\infty} e^{-\alpha\beta_m^2 t}[A_{m\nu} \cos \nu\phi + B_{m\nu} \sin \nu\phi] \cdot J_\nu(\beta_m r) \tag{3-170}$$

This solution should satisfy the initial condition 3-165c, hence, for $t = 0$ it becomes

$$F(r, \phi) = \sum_{m=1}^{\infty} \sum_{\nu=0}^{\infty} [A_{m\nu} \cos \nu\phi + B_{m\nu} \sin \nu\phi] \cdot J\nu(\beta_m r) \tag{3-171}$$

where the double summation is taken over all eigenvalues β_m and ν. The unknown coefficients $A_{m\nu}$ and $B_{m\nu}$ are determined by making use of the orthogonality of eigenfunctions.

$B_{m\nu}$ is determined by operating successively both sides of Eq. 3-171 by

$$\int_{-\pi}^{\pi} \sin \nu\phi \, d\phi \quad \text{and} \quad \int_{r'=0}^{b} r' \cdot J_\nu(\beta_m r') \cdot dr'$$

which gives

$$B_{m\nu} = \frac{1}{N_\phi N_r} \int_{\phi'=-\pi}^{\pi} \int_{r'=0}^{b} r' J_\nu(\beta_m r') \sin \nu\phi' \cdot F(r', \phi') \cdot dr' \cdot d\phi' \tag{3-172a}$$

where N_ϕ and N_r are the norms for the ϕ and r separations respectively and given as (See Eq. 3-29a for evaluation of N_r.)

$$N_r = \int_{r'=0}^{b} r' J_\nu^2(\beta_m r') \cdot dr' = \frac{b^2}{2}\left[\frac{H_2^2}{\beta_m^2} + \left(1 - \frac{\nu^2}{\beta_m^2 b^2}\right)\right] \cdot J_\nu^2(\beta_m b) \tag{3-172b}$$

$$N_\phi = \int_{\phi'=-\pi}^{\pi} \sin^2 \nu\phi' \cdot d\phi' = \pi \tag{3-172c}$$

The unknown coefficients $A_{m\nu}$ are determined by operating both sides of Eq. 3-171 successively by

$$\int_{-\pi}^{\pi} \cos \nu\phi \cdot d\nu \quad \text{and} \quad \int_{r'=0}^{b} r' \cdot J_\nu(\beta_m r') \cdot dr'$$

which gives

$$A_{m\nu} = \frac{1}{N_\phi N_r} \int_{\phi'=-\pi}^{\pi} \int_{r'=0}^{b} r' J_\nu(\beta_m r') \cdot \cos \nu\phi' \cdot F(r', \phi') \cdot dr' d\phi' \tag{3-173a}$$

where the norms N_r and N_ϕ are

$$N_r = \frac{b^2}{2}\left[\frac{H_2^2}{\beta_m^2} + \left(1 - \frac{\nu^2}{\beta_m^2 b^2}\right)\right] \cdot J_\nu^2(\beta_m b) \tag{3-173b}$$

$$N_\phi = \int_{\phi'=-\pi}^{\pi} \cos^2 \nu\phi' \cdot d\phi = \begin{cases} \pi & \text{for} \quad \nu = 1, 2, 3, \ldots \\ 2\pi & \text{for} \quad \nu = 0 \end{cases} \tag{3-173c}$$

Substituting the coefficients $B_{m\nu}$ and $A_{m\nu}$ as given above into the solution 3-170 we obtain

$$T(r, \phi, t) = \frac{2}{\pi} \sum_{m=1}^{\infty} \sum_{\nu=0}^{\infty} e^{-\alpha\beta_m^2 t} \frac{\beta_m^2}{b^2(H_2^2 + \beta_m^2) - \nu^2} \frac{J_\nu(\beta r)}{J_\nu^2(\beta b)}$$

$$\cdot \int_{\phi'=-\pi}^{\pi} \int_{r'=0}^{b} r' \cdot J_\nu(\beta_m r') \cdot \cos \nu(\phi - \phi') \cdot F(r', \phi') \cdot dr' \cdot d\phi' \tag{3-174}$$

where $\nu = 0, 1, 2, 3, \ldots$
β_m = positive roots of the transcendental equation 3-169; when $\nu = 0$, replace π by 2π

Solid Cylinder with $0 \le r \le b, 0 \le z < \infty$. Consider a semi-infinite solid cylinder of radius $r = b$ which extends in the z-direction from $z = 0$ to infinite as shown in Fig. 3-10. Initially the cylinder is at temperature $F(r, z)$ and for times $t > 0$ the boundary surfaces at $r = b$ and $r = z$ are kept at zero temperature.

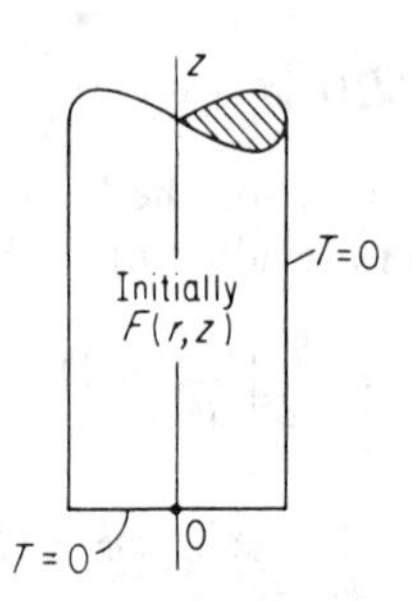

FIG. 3-10. Boundary conditions for a semi-infinite solid cylinder.

The boundary-value problem of heat conduction is given as

$$\frac{\partial^2 T}{\partial r^2} + \frac{1}{r}\frac{\partial T}{\partial r} + \frac{\partial^2 T}{\partial z^2} = \frac{1}{\alpha}\frac{\partial T}{\partial t} \quad \text{in} \quad 0 \le r \le b, 0 \le z < \infty, \text{ for } t > 0$$

$$T = 0 \quad \text{at} \quad r = b, \text{ for } t > 0$$

$$T = 0 \quad \text{at} \quad z = 0, \text{ for } t > 0$$

$$T = F(r, z) \quad \text{in} \quad 0 \le r \le b, 0 \le z \le \infty, \text{ for } t = 0 \tag{3-175}$$

where $T = T(r, z, t)$.

Assuming the separation of variables in the form

$$T(r, z, t) = \Gamma(t) \cdot R(r) \cdot Z(z) \tag{3-176}$$

the separation equations and their solutions are

$$\frac{d\Gamma}{dt} + \alpha(\beta^2 + \eta^2)\Gamma = 0 \quad \Gamma{:}e^{-\alpha(\beta^2+\eta^2)t} \tag{3-177a}$$

$$\frac{d^2R}{dr^2} + \frac{1}{r}\frac{dR}{dr} + \beta^2 R = 0 \quad R{:}J_0(\beta r) \quad 0 \le r \le b \tag{3-177b}$$

$$\frac{d^2Z}{dz^2} + \eta^2 Z = 0 \quad Z{:}\sin \eta z \quad 0 \le z < \infty \tag{3-177c}$$

The $Y_0(\beta r)$ function is excluded from the solution because the region includes $r = 0$. The $\cos \eta z$ function is excluded because of the boundary condition at $z = 0$. The separation parameters β and η are determined as follows.

The boundary condition at $r = b$ requires that

$$J_0(\beta b) = 0 \tag{3-178}$$

hence, β_m's constitute a discrete set of points which are the positive roots of the transcendental equation 3-178.

The parameter η assumes all values from $\eta = 0$ to infinity continuously since the region in the z direction extends to infinity.

Then the general solution of problem 3-175 is in the form

$$T(r, z, t) = \sum_{m=1}^{\infty} \int_{\eta=0}^{\infty} c_m(\eta) \cdot e^{-\alpha(\beta_m^2+\eta^2)t} \cdot J_0(\beta_m r) \cdot \sin \eta z \cdot d\eta \tag{3-179}$$

This solution should satisfy the initial condition, hence, for $t = 0$ it becomes

$$F(r, z) = \sum_{m=1}^{\infty} \int_{\eta=0}^{\infty} c_m(\eta) \cdot J_0(\beta_m r) \cdot \sin \eta z \cdot d\eta \tag{3-180}$$

We determine the unknown coefficients $c_m(\eta)$. The summation on the right-hand side of Eq. 3-180 is removed by operating both sides of this equation by

$$\int_{r'=0}^{b} r' \cdot J_0(\beta_m r') \cdot dr'$$

and by making use of the orthogonality of Bessel functions. Then Eq. 3-180 becomes

$$\frac{1}{N_r} \int_{r'=0}^{b} r' \cdot J_0(\beta_m r') \cdot F(r', z') \cdot dr' = \int_{\eta=0}^{\infty} c_m(\eta) \cdot \sin \eta z \cdot d\eta \tag{3-181}$$

where N_r is the norm for the r-separation, and for boundary condition of the first kind at $r = b$ it is given as

$$N_r = \int_{r'=0}^{b} r' \cdot J_0^2(\beta_m r') \cdot dr' = \frac{b^2}{2} [J_0'(\beta_m b)]^2 = \frac{b^2}{2} J_1^2(\beta_m b) \qquad (3\text{-}182)$$

Equation 3-181 is equivalent to representation of a function (i.e., the left-hand side of Eq. 3-181) in the semi-infinite interval $0 \leq z < \infty$ in terms of sine functions subject to a boundary condition of the first kind at $z = 0$. Such representation was considered in Chapter 2 and the unknown coefficient $c_m(\eta)$ is given as

$$c_m(\eta) = \frac{2}{\pi} \int_{z'=0}^{\infty} \sin \eta z' \cdot \left[\frac{1}{N_r} \int_{r'=0}^{b} r' \cdot J_0(\beta_m r') \cdot F(r', z') \cdot dr' \right] dz' \qquad (3\text{-}183)$$

Substituting Eqs. 3-182 and 3-183 into Eq. 3-179, the solution of the boundary-value problem of heat conduction becomes

$$T(r, z, t) = \frac{4}{\pi b^2} \sum_{m=1}^{\infty} e^{-\alpha \beta_m^2 t} \cdot \frac{J_0(\beta_m r)}{J_1^2(\beta_m b)}$$

$$\cdot \int_{\eta=0}^{\infty} e^{-\alpha \eta^2 t} \cdot \sin \eta z \cdot \sin \eta z' \cdot d\eta$$

$$\cdot \int_{z'=0}^{\infty} \int_{r'=0}^{b} \cdot r' J_0(\beta_m r') \cdot F(r', z') \cdot dr' \cdot dz' \qquad (3\text{-}184)$$

where the summation is taken over all eigenvalues β_m which are the positive roots of the transcendental equation 3-178.

3-7. NONHOMOGENEOUS BOUNDARY-VALUE PROBLEMS OF HEAT CONDUCTION INVOLVING MORE THAN ONE SPACE VARIABLE (SOLUTION WITH INTEGRAL TRANSFORM)

The heat conduction problem in the cylindrical coordinate system, in general, involves the following differential operator

$$D \equiv \frac{\partial^2 T}{\partial r^2} + \frac{1}{r} \frac{\partial T}{\partial r} + \frac{1}{r^2} \frac{\partial^2 T}{\partial \varphi^2} + \frac{\partial^2 T}{\partial z^2} \qquad (3\text{-}185)$$

In this section we examine the removal of the space variables r, φ, z from the differential operator 3-185 by the integral transform technique and illustrate the application with examples.

Removal of the z Variable. The range of the z variable may be finite, semi-infinite or infinite, and the boundary conditions associated with it may be of the first, second, or third kind. The removal of the z variable from the differential operator 3-185 is treated exactly the same as removal of the z variable from the differential operator in the Cartesian coordinate system. That is, the integral transform, the inversion formula, and the kernels are obtained from Table 2-1 when the range of z is finite, from Table 2-2 when the range of z is semi-infinite, and from Eq. 2-38 when the range of z is infinite. After removing the z variable as described in Chapter 2 the differential operator 3-185 becomes

$$\bar{D} \equiv \frac{\partial^2 \bar{T}}{\partial r^2} + \frac{1}{r}\frac{\partial \bar{T}}{\partial r} + \frac{1}{r^2}\frac{\partial^2 \bar{T}}{\partial \varphi^2} - \eta^2 \bar{T} + \left(\begin{array}{c}\text{Terms involving boundary}\\ \text{condition functions for}\\ \text{the } z \text{ variable}\end{array}\right) \tag{3-185a}$$

where η is the transform variable for the z variable.

Removal of the φ Variable. The range of the φ variable may be $0 \le \varphi \le \varphi_0$, where $\varphi_0 < 2\pi$ (i.e., portion of the circle), or $0 \le \varphi \le 2\pi$ (i.e., complete circle). These two cases require different treatment.

(i) $0 \le \varphi \le \varphi_0$ $(\varphi_0 < 2\pi)$: In this case the φ variable involves boundary conditions at $\varphi = 0$ and $\varphi = \varphi_0$, which may be of the first, second, or third kind. The removal of the φ variable from the differential operator 3-185a is treated exactly the same as the removal of the space variable from the differential operator in the Cartesian coordinate system for a finite region, and the integral transform, the inversion formula, and the kernels are obtained from Table 2-1. After removing the φ variable the differential operator 3-185a becomes

$$\bar{\bar{D}} \equiv \frac{\partial^2 \bar{\bar{T}}}{\partial r^2} + \frac{1}{r}\frac{\partial \bar{\bar{T}}}{\partial r} - \frac{\nu^2}{r^2}\bar{\bar{T}} - \eta^2 \bar{\bar{T}} + \left[\begin{array}{c}\text{Terms involving boundary condition}\\ \text{functions for the } \varphi \text{ variable and the}\\ \text{transform of the boundary condition}\\ \text{functions for the } z \text{ variable}\end{array}\right] \tag{3-185b}$$

where N is the transform variable for the φ variable.

(ii) $0 \le \varphi \le 2\pi$: In this case there is no prescribed boundary condition for the φ variable except the requirement that the temperature should be cyclic around the cylinder with a period of 2π. For this particular case the integral transform (i.e. Fourier transform) and the inversion formula with respect to the φ variable is defined as[5]

$$\text{(Integral transform)} \quad \bar{T}(\nu, \varphi) = \int_{\varphi'=0}^{2\pi} \cos \nu(\varphi - \varphi') \cdot T(\varphi') \cdot d\varphi' \tag{3-186a}$$

[5]The integral transform and the inversion formula as defined by equation 3-186 is obtained from the expansion of a function $F(\varphi)$ in the interval $-\pi \le \varphi \le \pi$ with Fourier series in the form

$$F(\varphi) = \frac{1}{2} A_0 + \sum_{\eta=1}^{\infty} (A_\eta \cos \eta\varphi + B_\eta \sin \eta\varphi) \tag{1}$$

$$\text{(Inversion formula)} \quad T(\varphi) = \frac{1}{\pi} \sum_{\eta=0}^{\infty} \bar{T}(\nu, \varphi) \tag{3-186b}$$

where $\nu = 0,1,2,3, \ldots$; and π should be replaced by 2π for $\nu = 0$. Then, the application of the integral transform 3-186a to the differential operator 3-185b removes from the differential operator 3-185b the partial derivative with respect to the φ variable and reduces it to the following form

$$\bar{\bar{D}} \equiv \frac{\partial^2 \bar{\bar{T}}}{\partial r^2} + \frac{1}{r}\frac{\partial \bar{\bar{T}}}{\partial r} - \frac{\nu^2}{r^2}\bar{\bar{T}} + \begin{bmatrix} \text{Terms involving the transform} \\ \text{of the boundary condition for} \\ \text{the } z \text{ variable} \end{bmatrix} \tag{3-185c}$$

Removal of the r Variable. After removing the space variables z and φ from the differential operator Eq. 3-185 as described above, the resulting doubly transformed differential operators 3-15b or 3-185c includes the

---

where the coefficients A_η and B_η are given as [8]

$$A_\eta = \frac{1}{\pi} \int_{-\pi}^{\pi} F(\varphi') \cdot \cos \eta \varphi' \cdot d\varphi' \quad B_\eta = \frac{1}{\pi} \int_{-\pi}^{\pi} F(\varphi') \cdot \sin \eta\varphi' \cdot d\varphi' \tag{2}$$

$\eta = 0,1,2,3, \ldots$

Substituting the coefficients Eq. 2 into Eq. 1

$$F(\varphi) = \frac{1}{2\pi} \int_{-\pi}^{\pi} F(\varphi') \cdot d\varphi' + \frac{1}{\pi} \sum_{\eta=1}^{\infty} \int_{-\pi}^{\pi} F(\varphi')[\cos \eta\varphi' \cdot \cos \eta\varphi' + \sin \eta\varphi \cdot \sin \eta\varphi']d\varphi'$$

or

$$F(\varphi) = \frac{1}{2\pi} \int_{-\pi}^{\pi} F(\varphi') \cdot d\varphi' + \frac{1}{\pi} \sum_{\eta=1}^{\infty} \int_{-\pi}^{\pi} F(\varphi') \cdot \cos \eta(\varphi - \varphi') \cdot d\varphi' \tag{3}$$

which is written in a more compact form as

$$F(\varphi) = \frac{1}{\pi} \sum_{\eta=0}^{\infty} \int_{-\pi}^{\pi} F(\varphi') \cdot \cos \eta(\varphi - \varphi') \cdot d\varphi' \tag{4}$$

where $\eta = 0,1,2,3,\ldots,$

and replace π ny 2π for $\eta = 0$.

The expansion given by Eq. 4 can be rewritten as the integral transform[l] and inversion formula for the function $F(\varphi)$ as

$$\begin{pmatrix}\text{Integral} \\ \text{transform}\end{pmatrix} \quad \bar{F}(\eta, \varphi) = \int_{\varphi'=-\pi}^{\pi} F(\varphi') \cdot \cos \eta(\varphi - \varphi') \cdot d\varphi \tag{5a}$$

$$\begin{pmatrix}\text{Inversion} \\ \text{formula}\end{pmatrix} \quad F(\varphi) = \frac{1}{\pi} \sum_{n=0}^{\infty} \bar{F}(\eta, \varphi) \tag{5b}$$

where, $\eta = 0,1,2,3,\ldots$; and replace π by 2π for $\eta = 0$.

The expansion given by Eq. 5 is valid if function $F(\varphi)$ is continuous in the interval $-\pi \le \varphi \le \pi$ and $F'(\varphi)$ is sectionally continuous in the same interval.

If the function $F(\varphi)$ is a periodic function of period 2π, that is if $F(\varphi + 2\pi) = F(\varphi)$, the series representation given by Eq. 5 is valid in any interval of length 2π, and the limits of the integral can be taken as $(0, 2\pi)$ as given in Eq. 3-186.

space variable r in the form

$$D \equiv \frac{\partial^2 T}{\partial r^2} + \frac{1}{r}\frac{\partial T}{\partial r} - \frac{\nu^2}{r^2} T \tag{3-187}$$

where the bars are omitted for convenience.

We examine the removal of the space variable r from the differential operator 3-187 by the application of integral transform (i.e. Hankel transform). The form of the integral transform, inversion formula and the kernels depends on the range of the r variable, that is whether it is finite or infinite. We examine below the ranges of r including $a \le r \le b$ (i.e. hollow cylinder), $0 \le r \le b$ (i.e. solid cylinder) and $0 \le r < \infty$ (i.e. an infinite region).

Finite Region $a \le r \le b$. We define the integral transform and inversion formula with respect to the r variable of T in the form (See Eq. 3-37.)

$$\begin{pmatrix}\text{Integral}\\ \text{transform}\end{pmatrix} \quad \bar{T}(\beta_m) = \int_a^b r' K_\nu(\beta_m, r') \cdot T(r') \cdot dr' \tag{3-188a}$$

$$\begin{pmatrix}\text{Inversion}\\ \text{formula}\end{pmatrix} \quad T(r) = \sum_{m=1}^{\infty} K_\nu(\beta_m, r) \cdot \bar{T}(\beta_m) \tag{3-188b}$$

The kernel $K_\nu(\beta, r)$ is the normalized eigenfunction of the following auxiliary differential equation:

$$\frac{d^2 K_\nu(\beta, r)}{dr^2} + \frac{1}{r}\frac{dK_\nu(\beta, r)}{dr} + \left(\beta^2 - \frac{\nu^2}{r^2}\right) K_\nu(\beta, r) = 0 \quad a \le r \le b \tag{3-189}$$

We take the integral transform of the differential operator 3-187 by applying the integral transform 3-188.

$$\bar{D}_\nu = \int_a^b r K_\nu(\beta_m, r)\left(\frac{\partial^2 T}{\partial r^2} + \frac{1}{r}\frac{\partial T}{\partial r} - \frac{\nu^2}{r^2} T\right) dr \tag{3-190a}$$

Integrating this by parts twice,

$$\bar{D}_\nu = \left[r\left(K_\nu \frac{\partial T}{\partial r} - T\frac{dK_\nu}{dr}\right)\right]_a^b + \int_a^b r\left(\frac{d^2 K_\nu}{dr^2} + \frac{1}{r}\frac{dK_\nu}{dr} - \frac{\nu^2}{r^2} K_\nu\right) \cdot T\, dr \tag{3-190b}$$

In view of the Eqs. 3-188a and 3-189, Eq. 3-190b becomes

$$\bar{D}_\nu = -\beta_m^2 \bar{T} + \left[r\left(K_\nu \frac{\partial T}{\partial r} - T\frac{dK_\nu}{dr}\right)\right]_a^b \tag{3-191}$$

where the terms with a bar denote the integral transform according to transform 3-188.

The terms inside the bracket in Eq. 3-191 are evaluated by making use of the boundary conditions at $r = a$ and $r = b$. For generality we assume boundary conditions at $r = a$ and $r = b$ are of the third kind and given in the form

$$-k_1 \frac{\partial T}{\partial r} + h_1 T = f_1 \quad \text{at} \quad r = a$$
$$k_2 \frac{\partial T}{\partial r} + h_2 T = f_2 \quad \text{at} \quad r = b \tag{3-192}$$

Then the boundary conditions for the auxiliary differential Eq. 3-189 are taken in the form

$$-k_1 \frac{dK_\nu}{dr} + h_1 K_\nu = 0 \quad \text{at} \quad r = a$$
$$k_2 \frac{dK_\nu}{dr} + h_2 K_\nu = 0 \quad \text{at} \quad r = b \tag{3-193}$$

The term inside the bracket in Eq. 3-191 is evaluated by means of Eqs. 3-192 and 3-193, then Eq. 3-191 becomes

$$\bar{D}_\nu = -\beta_m^2 \bar{T} + \left\{ a \cdot \frac{K_\nu(\beta_m, r)}{k_1}\bigg|_{r=a} \cdot f_1 + b \frac{K_\nu(\beta_m, r)}{k_2}\bigg|_{r=b} \cdot f_2 \right\} \tag{3-194}$$

where β_m's are the eigenvalues of the auxiliary Eq. 3-189 subject to the boundary conditions 3-193.

The kernels $K_\nu(\beta_m, r)$ and the eigenvalues β_m for the nine different combinations of boundary conditions at $r = a$ and $r = b$ are tabulated in Table 3-2. For the boundary condition of the first kind k_1 or k_2 or both are set equal to zero; for such cases the following changes should be made in Eq. 3-194.

When $k_1 = 0$, replace $\dfrac{K_\nu(\beta_m, r)}{k_1}\bigg|_{r=a}$ by $\dfrac{1}{h_1} \dfrac{dK_\nu(\beta_m, r)}{dr}\bigg|_{r=a}$

When $k_2 = 0$, replace $\dfrac{K_\nu(\beta_m, r)}{k_2}\bigg|_{r=b}$ by $-\dfrac{1}{h_2} \dfrac{dK_\nu(\beta_m, r)}{dr}\bigg|_{r=b}$

Finite Region $0 \le r \le b$. We define the integral transform and inversion formula with respect to the r variables of T in the form (See Eq. 3-24.)

$$\begin{pmatrix}\text{Integral}\\ \text{transform}\end{pmatrix} \quad \bar{T}(\beta_m) = \int_0^b r' \cdot K_\nu(\beta_m, r') \cdot T(r') \cdot dr' \tag{3-195a}$$

$$\begin{pmatrix}\text{Inversion}\\ \text{formula}\end{pmatrix} \quad T(r) = \sum_{m=1}^{\infty} K_\nu(\beta_m, r) \cdot \bar{T}(\beta_m) \tag{3-195b}$$

where the kernel $K_\nu(\beta, r)$ is the normalized eigenfunction of the differential Eq. 3-189 when $0 \le r \le b$. By applying the transform 3-195 we take integral transform of the differential operator 3-187 and obtain

$$\bar{D}_\nu = -\beta_m^2 \bar{T} + \left[r\left(K_\nu \frac{\partial T}{\partial r} - T\frac{dK_\nu}{dr} \right) \right]_0^b \tag{3-196}$$

It is to be noted that the terms in the bracket vanish at the lower limit; and can be evaluated at the upper limit by making use of the boundary condition at $r = b$. For generality, assuming a boundary condition at $r = b$ is of the third kind i.e.,

$$k_2 \frac{\partial T}{\partial r} + h_2 T = f_2 \quad \text{at} \quad r = b$$

the boundary condition for the auxiliary Prob. 3-189 is taken as

$$k_2 \frac{dK_\nu}{dr} + h_2 K_\nu = 0 \quad \text{at} \quad r = b$$

By making use of these boundary conditions, the terms inside the bracket in Eq. 3-196 is evaluated at the upper limit $r = b$, and we obtain

$$\bar{D}_\nu = -\beta_m^2 \bar{T} + b \cdot \left. \frac{K_\nu(\beta_m, r)}{k_2} \right|_{r=b} \cdot f_2 \tag{3-197}$$

The kernels $K_\nu(\beta_m, r)$ and the eigenvalues β_m for three different boundary conditions at $r = b$ are given in Table 3-1.

When the boundary condition at $r = b$ is of the first kind, we set $k_2 = 0$; in such cases the following change should be made in Eq. 3-197.

$$\text{When } k_2 = 0, \quad \text{replace} \quad \left. \frac{K_\nu(\beta_m, r)}{k_2} \right|_{r=b} \quad \text{by} \quad -\frac{1}{h_2} \left. \frac{dK_\nu(\beta_m, r)}{dr} \right|_{r=b}$$

Infinite Region $0 \le r < \infty$. We define integral transform and inversion formula with respect to the r variable of T in the form (See Eq. 3-44).

$$\begin{pmatrix}\text{Integral} \\ \text{transform}\end{pmatrix} \quad \bar{T}(\beta) = \int_0^\infty r' \cdot J_\nu(\beta r') \cdot T(r') \cdot dr'$$

$$\begin{pmatrix}\text{Inversion} \\ \text{formula}\end{pmatrix} \quad T(r) = \int_0^\infty \beta \cdot J_\nu(\beta r) \cdot \bar{T}(\beta) \cdot d\beta$$

In this case the kernal $J_\nu(\beta r)$ is the solution of the differential Eq. 3-189 for $0 \le r < \infty$. Taking the integral transform of the differential operator 3-187 by applying the above transform we obtain

$$\bar{D}_\nu = -\beta^2 \bar{T} + \left[r\left(K_\nu \frac{\partial T}{\partial r} - T\frac{dK_\nu}{dr} \right) \right]_0^\infty$$

The terms in the bracket vanish at the lower limit $r = 0$; if we assume further that the function T is such that the terms in the bracket vanish at the upper limit, $r \to \infty$, we obtain

$$\bar{D}_\nu = -\beta^2 \bar{T} \tag{3-198}$$

where β's assume all values from $\beta = 0$ to infinity continuously.

We now illustrate the application of the integral-transform technique to the solution of problems involving more than one space variable with the following examples.

Long Solid Cylinder $0 \le r \le b$, $0 \le \phi \le \phi_0$. Consider a long solid cylinder enclosed by the surface $r = b$ and the planes $\phi = 0$ and $\phi = \phi_0$. Solid is initially at temperature $F(r, \phi)$. For times $t > 0$ heat is generated within the solid at a rate of $g(r, \phi, t)$ Btu/hr ft^3, while the boundary surfaces at $\phi = 0$, $\phi = \phi_0$ and $r = b$ are kept at temperatures $f_1(r, t)$, $f_2(r, t)$ and $f_4(\phi, t)$ respectively. Figure 3-11 shows the geometry and the boundary conditions for the problem.

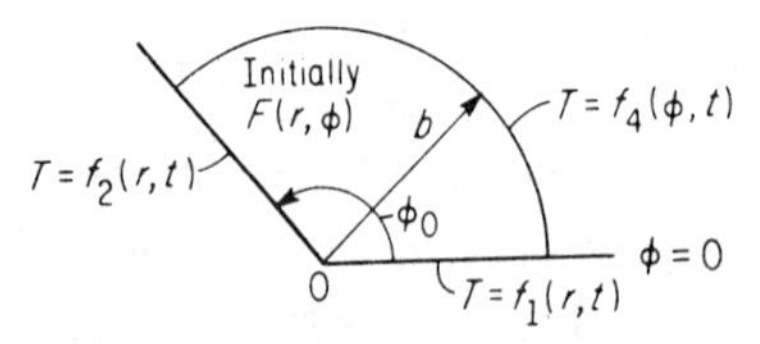

FIG. 3-11. Boundary conditions for a long cylinder enclosed by the surfaces $r = a$, $\phi = 0$, and $\phi = \phi_0$.

The boundary-value problem of heat conduction is given as

$$\frac{\partial^2 T}{\partial r^2} + \frac{1}{r}\frac{\partial T}{\partial r} + \frac{1}{r^2}\frac{\partial^2 T}{\partial \phi^2} + \frac{g(r, \phi, t)}{k} = \frac{1}{\alpha}\frac{\partial T}{\partial t}$$

$$\text{in} \quad 0 \le r \le b, 0 \le \phi \le \phi_0, \text{for } t > 0$$

$$\begin{aligned} T &= f_1(r, t) \quad \text{at} \quad \phi = 0, t > 0 \\ T &= f_2(r, t) \quad \text{at} \quad \phi = \phi_0, t > 0 \\ T &= f_4(\phi, t) \quad \text{at} \quad r = b, t > 0 \\ T &= F(r, \phi) \quad \text{in} \quad 0 \le r \le b, 0 \le \phi \le \phi_0, \text{for } t = 0 \end{aligned} \tag{3-199}$$

where $T \equiv T(r, \phi, t)$.

The partial derivative with respect to the ϕ-variable is removed by means of Fourier transform, and the partial derivative with respect to the r variable is removed by means of Hankel transform.

We define Fourier transform and the inversion formula of the temperature function $T(r, \phi, t)$ with respect to the ϕ variable in the range $0 \le \phi \le \phi_0$ as

$$\begin{pmatrix}\text{Integral}\\ \text{transform}\end{pmatrix} \quad \bar{T}(r, \nu, t) = \int_{\phi=0}^{\phi_0} K(\nu, \phi') \cdot T(r, \phi', t) \cdot d\phi' \tag{3-200a}$$

$$\begin{pmatrix}\text{Inversion}\\ \text{formula}\end{pmatrix} \quad T(r, \phi, t) = \sum_{\nu} K(\nu, \phi) \cdot \bar{T}(r, \nu, t) \tag{3-200b}$$

where the kernels $K(\nu, \phi)$ and the eigenvalues ν's are obtained from Table 2-1. For boundary conditions of the first kind at $\phi = 0$ and $\phi = \phi_0$ they are given as

$$K(\nu, \phi) = \sqrt{\frac{2}{\phi_0}} \sin \nu\phi \tag{3-200c}$$

where the eigenvalues ν are the positive roots of

$$\sin \nu\phi_0 = 0$$

$$\text{Or} \quad \nu = \frac{n\pi}{\phi_0} \quad \text{with} \quad n = 1, 2, 3, \ldots \tag{3-200d}$$

We take the integral transform of system 3-199 by applying the transform 3-200a.

$$\left(\frac{\partial^2 \bar{T}}{\partial r^2} + \frac{1}{r}\frac{\partial \bar{T}}{\partial r} - \frac{\nu^2}{r^2}\bar{T}\right) + \left.\frac{dK(\nu, \phi)}{d\phi}\right|_{\phi=0} \cdot f_1(r, t) - \left.\frac{dK(\nu, \phi)}{d\phi}\right|_{\phi=\phi_0} \cdot f_2(r, t) + \frac{\bar{g}(r, \nu, t)}{k}$$

$$= \frac{1}{\alpha}\frac{\partial \bar{T}}{\partial t} \quad \text{in} \quad 0 \le r \le b, \text{ for } t > 0 \tag{3-201}$$

$$\bar{T} = \bar{f}_4(\nu, t) \quad \text{at} \quad r = b, \text{ for } t > 0$$

$$\bar{T} = \bar{F}(r, \nu) \quad \text{in} \quad 0 \le r \le b, \text{ for } t = 0$$

where $\bar{T} \equiv \bar{T}(r, \nu, t)$, and quantities with bars refer to integral transforms according to Fourier transform 3-200a.

System 3-201 includes the differential operator

$$\frac{\partial^2 \bar{T}}{\partial r^2} + \frac{1}{r}\frac{\partial \bar{T}}{\partial r} - \frac{\nu^2}{r^2}\bar{T}, \quad 0 \le r \le b$$

In order to remove this differential operator from the system 3-201 we define the integral transform (Hankel transform) and the inversion formula of the temperature function $\bar{T}(r, \nu, t)$ with respect to the r variable in the range $0 \le r \le b$ as

$$\begin{pmatrix}\text{Integral}\\ \text{transform}\end{pmatrix} \quad \bar{\bar{T}}(\beta_m, \nu, t) = \int_{r'=0}^{b} r' K_\nu(\beta_m, r') \cdot \bar{T}(r', \nu, t) \cdot dr' \tag{3-202a}$$

$$\begin{pmatrix}\text{Inversion}\\ \text{formula}\end{pmatrix} \quad \bar{T}(r, \nu, t) = \sum_{m=1}^{\infty} K_\nu(\beta_m, r) \cdot \bar{\bar{T}}(\beta_m, \nu, t) \tag{3-202b}$$

where the kernel $K_\nu(\beta_m, r)$ and the eigenvalues β_m are obtained from Table 3-1. For boundary conditions of the first kind at $r = b$ they are given as

$$K_\nu(\beta_m, r) = \frac{\sqrt{2}}{b}\frac{J_\nu(\beta_m r)}{J'_\nu(\beta_m b)} \tag{3-202c}$$

and the eigenvalues β_m are the positive roots of the transcendental equation

$$J_\nu(\beta b) = 0 \tag{3-202d}$$

We take the integral transform of system 3-201 by applying the transform Eq. 3-202a. (See Eq. 3-197 for the integral transform of the differential operator with respect to the r variable for $0 \le r \le b$.)

$$-\beta_m^2 \bar{\bar{T}}(\beta_m, \nu, t) - b \left.\frac{dK_\nu(\beta_m, r)}{dr}\right|_{r=b} \cdot \bar{f}_4(\nu, t) + \left.\frac{dK(\nu, \phi)}{d\phi}\right|_{\phi=0} \cdot \bar{f}_1(\beta_m, t) - \left.\frac{dK(\nu, \phi)}{d\phi}\right|_{\phi=\phi_0} \cdot \bar{f}_2(\beta_m, t) + \frac{\bar{\bar{g}}(\beta_m, \nu, t)}{k} = \frac{1}{\alpha}\frac{d\bar{\bar{T}}(\beta_m, \nu, t)}{dt} \tag{3-203a}$$

$$\bar{\bar{T}}(\beta_m, \nu, t) = \bar{\bar{F}}(\beta_m, \nu) \quad \text{for} \quad t = 0 \tag{3-203b}$$

Equation 3-203 is rearranged as

$$\frac{d\bar{\bar{T}}(\beta_m, \nu, t)}{dt} + \alpha\beta_m^2 \bar{\bar{T}}(\beta_m, \nu, t) = A(\beta_m, \nu, t) \tag{3-204a}$$

$$\bar{\bar{T}}(\beta_m, \nu, t) = \bar{\bar{F}}(\beta_m, \nu) \quad \text{for} \quad t = 0 \tag{3-204b}$$

where

$$A(\beta_m, \nu, t) \equiv \frac{\alpha}{k}\bar{\bar{g}}(\beta_m, \nu, t) + \alpha\left\{\left.\frac{dK(\nu, \phi)}{d\phi}\right|_{\phi=0} \cdot \bar{f}_1(\beta_m, t) - \left.\frac{dK(\nu, \phi)}{d\phi}\right|_{\phi=\phi_0} \cdot \bar{f}_2(\beta_m, t) - b\left.\frac{dK_\nu(\beta_m, r)}{dr}\right|_{r=b} \cdot \bar{f}_4(\nu, t)\right\} \tag{3-204c}$$

Thus we reduced system 3-199 to an ordinary differential equation for the double transform of temperature subject to a prescribed initial condition as given by Eq. 3-204. Solution of Eq. 3-204 is easily determined and the resulting double transform of temperature is inverted successively by means of the inversion formulas 3-202b and 3-200b. Then the complete solution of the problem becomes

$$T(r, \phi, t) = \sum_\nu \sum_{m=1}^{\infty} K(\nu, \phi) \cdot K_\nu(\beta_m, r) \cdot e^{-\alpha\beta_m^2 t} \cdot \left[\bar{\bar{F}}(\beta_m, \nu) + \int_{t'=0}^{t} e^{\alpha\beta_m^2 t'} \cdot A(\beta_m, \nu, t') \cdot dt'\right] \tag{3-205}$$

$$A(\beta_m, \nu, t') = \frac{\alpha}{k}\bar{\bar{g}}(\beta_m, \nu, t') + \alpha\left\{\left.\frac{dK(\nu,\phi)}{d\phi}\right|_{\phi=0}\cdot\bar{f}_1(\beta_m, t') - \left.\frac{dK(\nu,\phi)}{d\phi}\right|_{\phi=\phi_0}\cdot\bar{f}_2(\beta_m, t') - b\cdot\left.\frac{dK_\nu(\beta_m, r)}{dr}\right|_{r=b}\cdot\bar{f}_4(\nu, t')\right\}$$

where
$$K(\nu,\phi) = \sqrt{\frac{2}{\phi_0}}\sin\nu\phi$$

$$K_\nu(\beta_m, r) = \frac{\sqrt{2}}{b}\frac{J_\nu(\beta_m r)}{J'_\nu(\beta_m b)}$$

$$\nu = \frac{n\pi}{\phi_0} \quad \text{with} \quad n = 1, 2, 3, \ldots$$

β_m's are positive roots of $J_\nu(\beta b) = 0$, and the transform of various functions are

$$\bar{\bar{g}}(\beta_m, \nu, t') = \int_{r'=0}^{b}\int_{\phi'=0}^{\phi_0} r'\cdot K_\nu(\beta_m, r')\cdot K(\nu, \phi')\cdot g(r', \phi', t')\cdot dr'\cdot d\phi'$$

$$\bar{\bar{F}}(\beta_m, \nu) = \int_{r'=0}^{b}\int_{\phi'=0}^{\phi_0} r'\cdot K_\nu(\beta_m, r')\cdot K(\nu, \phi')\cdot F(r', \phi')\cdot dr' d\phi'$$

$$\bar{f}_{1,2}(\beta_m, t') = \int_{r'=0}^{b} r'\cdot K_\nu(\beta_m, r')\cdot f_{1,2}(r', t')\cdot dr'$$

$$\bar{f}_4(\nu, t') = \int_{\phi'=0}^{\phi_0} K(\nu, \phi')\cdot f_4(\phi', t')\cdot d\phi'$$

We examine one special case of solution 3-205.

(a) A long solid cylinder $0 \le r \le b$, $0 \le \phi \le \phi_0$ is initially at temperature $F(r,\phi)$. For times $t > 0$ boundary surfaces at $\phi = 0$, $\phi = \phi_0$, $r = b$ are all kept at zero temperature, and there is no heat generation within the solid.

For this particular case we have

$$f_1(r,t) = f_2(r,t) = f_4(\phi,t) = g(r,\phi,t) = 0 \qquad \text{(3-206)}$$

Substituting Eq. 3-206 into the solution 3-205 we obtain

$$T(r,\phi,t) = \frac{4}{b^2\phi_0}\sum_{\nu}\sum_{m=1}^{\infty} e^{-\alpha\beta_m^2 t}\cdot\sin\nu\phi\cdot\frac{J_\nu(\beta_m r)}{J'^2_\nu(\beta_m b)} \cdot\int_{r'=0}^{b}\int_{\phi'=0}^{\phi_0} r'\cdot\sin\nu\phi'\cdot J_\nu(\beta_m r')\cdot F(r',\phi')\cdot dr'\cdot d\phi' \qquad \text{(3-207)}$$

where $\nu = n\pi/\phi_0$ with $n = 1, 2, 3, \ldots$, and β_m's are the positive roots of $J_\nu(\beta b) = 0$.

Semi-infinite Hollow Cylinder with a $a \le r \le b$, $0 \le z < \infty$. Consider a semi-infinite hollow cylinder of radius $a \le r \le b$, extending in the z-direction from $z = 0$ to infinity. Solid is initially at temperature $F(r, z)$. For times $t > 0$ heat is generated within the solid at a rate of $g(r, z, t)$ Btu/hr ft^3, while the boundary surfaces at $r = a$, $r = b$, $z = 0$ are kept at temperatures $f_1(z, t), f_2(z, t), f_3(r, t)$ respectively. Figure 3-12 shows the geometry and the boundary conditions for the problem. The boundary-

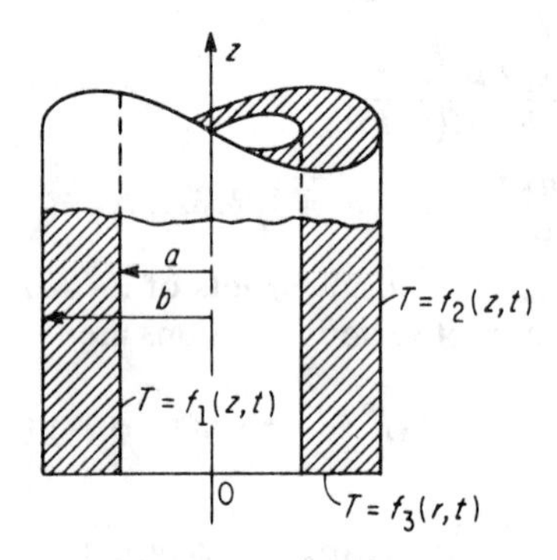

FIG. 3-12. Boundary conditions for a semi-infinite hollow cylinder.

value problem of heat conduction is given as

$$\frac{\partial^2 T}{\partial r^2} + \frac{1}{r}\frac{\partial T}{\partial r} + \frac{\partial^2 T}{\partial z^2} + \frac{g(r, z, t)}{k} = \frac{1}{\alpha}\frac{\partial T}{\partial t}$$

$$\text{in} \quad a \le r \le b,\ 0 \le z < \infty, \text{ for } t > 0$$

$$\begin{aligned} T &= f_1(z, t) \quad \text{at} \quad r = a,\ t > 0 \\ T &= f_2(z, t) \quad \text{at} \quad r = b,\ t > 0 \\ T &= f_3(r, t) \quad \text{at} \quad z = 0,\ t > 0 \\ T &= F(r, z) \quad \text{in} \quad a \le r \le b,\ 0 \le z < \infty,\ t = 0 \end{aligned} \tag{3-208}$$

where $T \equiv T(r, z, t)$.

The partial derivative with respect to the z-variable is removed by means of the Fourier transform and that with respect to r by means of the Hankel transform.

Since the range of the z variable is $0 \le z < \infty$, we define the Fourier transform and the inversion formula of the temperature function $T(r, z, t)$ with respect to the z-variable in the range $0 \le z < \infty$ as

$$\begin{pmatrix}\text{Integral} \\ \text{transform}\end{pmatrix} \quad \bar{T}(r, \eta, t) = \int_{z'=0}^{\infty} K(\eta, z') \cdot T(r, z', t) \cdot dz' \tag{3-209a}$$

$$\begin{pmatrix}\text{Inversion}\\ \text{formula}\end{pmatrix} \quad T(r,z,t) = \int_{\eta=0}^{\infty} K(\eta,z)\cdot\bar{T}(r,\eta,t)\cdot d\eta \tag{3-209b}$$

where the kernel $K(\eta, z)$ is obtained from Table 2-2. For the boundary condition of the first kind at $z = 0$ it is given as

$$K(\eta,z) = \sqrt{\frac{2}{\pi}}\sin\eta z \tag{3-209c}$$

We take the integral transform of system 3-208 by applying the transform Eq. 3-209a:

$$\begin{aligned}
&\frac{\partial^2\bar{T}}{\partial r^2} + \frac{1}{r}\frac{\partial\bar{T}}{\partial r} - \eta^2\bar{T} + \left.\frac{dK(\eta,z)}{dz}\right|_{z=0}\cdot f_3(r,t) + \frac{\bar{g}(r,\eta,t)}{k} = \frac{1}{\alpha}\frac{\partial\bar{T}}{\partial t}\\
&\qquad\qquad\qquad\qquad\qquad\qquad \text{in} \quad a \le r \le b,\ t > 0\\
&\qquad\qquad\quad \bar{T} = \bar{f}_1(\eta,r) \quad \text{at} \quad r = a,\ t > 0\\
&\qquad\qquad\quad \bar{T} = \bar{f}_2(\eta,t) \quad \text{at} \quad r = b,\ t > 0\\
&\qquad\qquad\quad \bar{T} = \bar{F}(r,\eta) \quad \text{in} \quad a \le r \le b,\ t = 0
\end{aligned} \tag{3-210}$$

where $\bar{T} \equiv \bar{T}(r,\eta,t)$.

System 3-210 includes a differential operator with respect to the r variable. We define the integral transform (i.e. Hankel Transform and inversion formula of the temperature function $\bar{T}(r,\eta,t)$ with respect to the r variable in the range $a \le r \le b$ as

$$\begin{pmatrix}\text{Integral}\\ \text{transform}\end{pmatrix} \quad \bar{\bar{T}}(\beta_m,\eta,t) = \int_{r'=a}^{b} r'\cdot K_0(\beta_m,r')\cdot\bar{T}(r',\eta,t)\cdot dr' \tag{3-211a}$$

$$\begin{pmatrix}\text{Inversion}\\ \text{formula}\end{pmatrix} \quad \bar{T}(r,\eta,t) = \sum_{m=1}^{\infty} K_0(\beta_m,r)\cdot\bar{\bar{T}}(\beta_m,\eta,t) \tag{3-211b}$$

The kernel $K_0(\beta_m, r)$ and eigenvalues β_m are obtained from Table 3-2 by setting $\nu = 0$. For boundary condition of the first kind at both boundaries they are given as (See Eq. 3-81.)

$$K_0(\beta_m,r) = \frac{\pi}{\sqrt{2}}\frac{\beta_m J_0(\beta_m b)\cdot Y_0(\beta_m b)}{\left[1 - \dfrac{J_0^2(\beta_m b)}{J_0^2(\beta_m a)}\right]^{1/2}}\left[\frac{J_0(\beta_m r)}{J_0(\beta_m b)} - \frac{Y_0(\beta_m r)}{Y_0(\beta_m b)}\right] \tag{3-211c}$$

β_m's are the positive roots of the following transcendental equation:

$$\frac{J_0(\beta a)}{J_0(\beta b)} - \frac{Y_0(\beta a)}{Y_0(\beta b)} = 0 \tag{3-211d}$$

We take the integral transform of system 3-210 by applying the transform Eq. 3-211a, and obtain

$$\frac{d\bar{\bar{T}}(\beta_m,\eta,t)}{dt} + \alpha(\beta_m^2 + \eta^2)\bar{\bar{T}}(\beta_m,\eta,t) = A(\beta_m,\eta,t) \tag{3-212a}$$

$$\bar{\bar{T}}(\beta_m,\eta,t) = \bar{\bar{F}}(\beta_m,\eta) \quad \text{for} \quad t = 0 \tag{3-212b}$$

where

$$A(\beta_m, \eta, t) \equiv \frac{\alpha}{k} \bar{\bar{g}}(\beta_m, \eta, t) + \alpha \left\{ a \left. \frac{dK_0(\beta_m, r)}{dr} \right|_{r=a} \cdot \bar{f}_1(\eta, t) - b \left. \frac{dK_0(\beta_m, r)}{dr} \right|_{r=b} \cdot \bar{f}_2(\eta, t) + \left. \frac{dK(\eta, z)}{dz} \right|_{z=0} \cdot \bar{f}_3(\beta_m, t) \right\} \qquad \text{(3-212c)}$$

Thus we reduced the partial differential equation of heat conduction to an ordinary differential equation for the double transform of temperature, $\bar{\bar{T}}(\beta_m, \eta, t)$, subject to a prescribed initial condition as given by Eq. 3-212, the solution of which is straightforward. When the resulting double transform of temperature is inverted successively by means of the inversion formulas 3-211b and 3-209b, the solution of the boundary-value problem of heat conduction is given as

$$T(r, z, t) = \int_{\eta=0}^{\infty} \sum_{m=1}^{\infty} K(\eta, z) \cdot K_0(\beta_m, r) \cdot e^{-\alpha(\beta_m^2 + \eta^2)t} \cdot \left[\bar{\bar{F}}(\beta_m, \eta) + \int_{t'=0}^{t} e^{\alpha(\beta_m^2 + \eta^2)t'} \cdot A(\beta_m, \eta, t') \cdot dt' \right] d\eta \qquad \text{(3-213)}$$

where

$$A(\beta_m, \eta, t') = \frac{\alpha}{k} \bar{\bar{g}}(\beta_m, \eta, t') + \alpha \left\{ a \left. \frac{dK_0(\beta_m, r)}{dr} \right|_{r=a} \cdot \bar{f}_1(\eta, t') - b \left. \frac{dK_0(\beta_m, r)}{dr} \right|_{r=b} \cdot \bar{f}_2(\eta, t') + \left. \frac{dK(\eta, z)}{dz} \right|_{z=0} \cdot \bar{f}_3(\beta_m, t') \right\}$$

$$K(\eta, z) = \sqrt{\frac{2}{\pi}} \sin \eta z$$

$$K_0(\beta_m, r) = \frac{\pi}{\sqrt{2}} \frac{\beta_m J_0(\beta_m b) \cdot Y_0(\beta_m b)}{\left[1 - \dfrac{J_0^2(\beta_m b)}{J_0^2(\beta_m a)} \right]^{1/2}} \left[\frac{J_0(\beta_m r)}{J_0(\beta_m b)} - \frac{Y_0(\beta_m r)}{Y_0(\beta_m b)} \right]$$

β_m's are positive roots of

$$\frac{J_0(\beta a)}{J_0(\beta b)} - \frac{Y_0(\beta a)}{Y_0(\beta b)} = 0$$

and transforms of various functions are defined as

$$\bar{\bar{g}}(\beta_m, \eta, t') = \int_{r'=a}^{b} \int_{z'=0}^{\infty} r' \cdot K_0(\beta_m, r') \cdot K(\eta, z') \cdot g(r', z', t') dr' dz'$$

$$\bar{\bar{F}}(\beta_m, \eta) = \int_{r'=a}^{b} \int_{z'=0}^{\infty} r' \cdot K_0(\beta_m, r') \cdot K(\eta, z') \cdot F(r', z') \cdot dr' \cdot dz'$$

$$\bar{f}_{1,2}(\eta, t') = \int_{z'=0}^{\infty} K(\eta, z') \cdot f_{1,2}(z', t') \cdot dz'$$

$$\bar{f}_3(\beta_m, t') = \int_{r'=a}^{b} r' \cdot K_0(\beta_m, r') \cdot f_3(r', t') \cdot dr'$$

3-8. STEADY-STATE PROBLEMS

Consider the steady-state boundary-value problem of heat conduction with heat generation within the solid given in the form

$$\nabla^2 T(r, \phi, z) + \frac{g(r, \phi, z)}{k} = 0 \quad \text{in region } R \tag{3-214a}$$

$$k_i \frac{\partial T(r, \phi, z)}{\partial n_i} + h_i T(r, \phi, z) = f_i \quad \text{on boundary surface } S_i \tag{3-214b}$$

We assume further that all h_i's do not vanish simultaneously.

Poisson's Eq. 3-214a may be reduced to Laplace's equation by defining a new dependent variable. That is, we define the new dependent variable $\theta(r, \phi, z)$ as

$$T(r, \phi, z) = \theta(r, \phi, z) + P(r, \phi, z) \tag{3-215}$$

and choose the function $P(r, \phi, z)$ such that system 3-214 reduces to the Laplace equation,

$$\nabla^2 \theta(r, \phi, z) = 0 \quad \text{in region } R \tag{3-216}$$

subject to transformed boundary conditions.

Table 3-3 gives $P(r, \phi, z)$ functions for transforming Poisson's equation to Laplace's equation in the cylindrical coordinate system for the case

TABLE 3-3

$P(r, \phi, z)$ FUNCTIONS IN THE CYLINDRICAL COORDINATE SYSTEM

[Poisson's equation $\nabla^2 T(r, \phi, z) = -\frac{g_0}{k}$ = constant, can be transformed into Laplace's equation $\nabla^2 \theta(r, \phi, z) = 0$ by defining the new variable $\theta(r, \phi, z)$ as $T(r, \phi, z) = \theta(r, \phi, z) + P(r, \phi, z)$.]

T is a function of	$P(r, \phi, z)$
r	$-\frac{g_0}{k} \frac{r^2}{4}$
r, ϕ	$-\frac{g_0}{k} \frac{r^2}{4}$
r, z	$-\frac{g_0}{k} \frac{r^2}{4} \left(\text{or } -\frac{g_0}{k} \frac{z^2}{2}\right)$
r, ϕ, z	$-\frac{g_0}{k} \frac{z^2}{2}$

of constant-heat generation, i.e.,

$$\frac{g_0}{k} = \text{const.}$$

The resulting Laplace equation, if it involves more than one nonhomogeneous boundary condition, is split up into a *set* of Laplace equations, each with one nonhomogeneous boundary condition as explained in the previous chapter. The simpler equations are solved either by the method of separation of variables or by the integral-transform technique.

In this section we examine solutions with integral-transform technique.

Long Solid Cylinder $0 \le r \le b, 0 \le \phi \le 2\pi$. A long solid cylinder $0 \le r \le b$, $0 \le \phi \le 2\pi$ is subjected to convective heat transfer at the boundary surface $r = b$ with a medium whose temperature varies around the circumference. There is no heat generation within the solid. The steady-state boundary-value problem of heat conduction is given as

$$\frac{\partial^2 T}{\partial r^2} + \frac{1}{r}\frac{\partial T}{\partial r} + \frac{1}{r^2}\frac{\partial^2 T}{\partial \phi^2} = 0 \quad \text{in} \quad 0 \le r \le b, 0 \le \phi \le 2\pi \tag{3-217a}$$

$$k_2 \frac{\partial T}{\partial r} + h_2 T = f_2(\phi) \quad \text{at} \quad r = b \tag{3-217b}$$

where $T \equiv T(r, \phi)$.

We examine the solution of this problem with two different approaches.

(1) We remove from the system 3-217 the second partial derivative with respect to the ϕ-variable by applying a Fourier transform, solve the resulting ordinary differential equation for the r variable and invert the transform of temperature by means of the inversion formula.

Since the range of ϕ-variable is $0 \le \phi \le 2\pi$ and that the temperature is cyclic with respect to the ϕ-variable with a period of 2π, we define the Fourier transform and the corresponding inversion formula of the temperature function $T(r, \phi)$ with respect to the ϕ-variable as (See Eq. 3-186.)

$$\begin{pmatrix}\text{Integral}\\ \text{transform}\end{pmatrix} \bar{T}(r, \nu, \phi) = \int_{\phi'=0}^{2\pi} \cos \nu(\phi - \phi') \cdot T(r, \phi') \cdot d\phi' \tag{3-218a}$$

$$\begin{pmatrix}\text{Inversion}\\ \text{formula}\end{pmatrix} \quad T(r, \phi) = \frac{1}{\pi}\sum_{\nu=0}^{\infty} \bar{T}(r, \nu) \tag{3-218b}$$

where $\nu = 0, 1, 2, 3, \ldots$, and π should be replaced by 2π when $\nu = 0$.

We take the integral transform of system 3-217 by applying the transform Eq. 3-218a.

$$\frac{d^2\bar{T}}{dr^2} + \frac{1}{r}\frac{d\bar{T}}{dr} - \frac{\nu^2}{r^2}\bar{T} = 0 \quad 0 \le r \le b \tag{3-219a}$$

$$k_2 \frac{d\bar{T}}{dr} + h_2 \bar{T} = \bar{f}_2(\nu) \quad \text{at} \quad r = b \tag{3-219b}$$

System 3-219 is now an ordinary differential equation, and the solution of which is in the form

$$\begin{aligned} \bar{T} &= c_1 r^\nu + c_2 r^{-\nu} \quad \text{for} \quad \nu = 1, 2, 3, \ldots \\ \bar{T} &= c_3 + c_4 \ln r \quad \text{for} \quad \nu = 0 \end{aligned} \tag{3-220}$$

The unknown coefficients are determined from the boundary conditions for the problem.

Since the temperature should remain finite at $r = 0$, we have

$$c_2 = c_4 = 0$$

Then solution 3-220 is given in the form

$$\bar{T} = c \cdot r^\nu \quad \text{for} \quad \nu = 0, 1, 2, 3, \ldots \tag{3-221}$$

The unknown coefficient c is determined from the boundary condition 3-219b:

$$\bar{T} = b \cdot \left(\frac{r}{b}\right)^\nu \cdot \frac{\bar{f}_2(\nu)}{k_2 \nu + h_2 b} \tag{3-222}$$

Inverting this result by means of inversion formula 3-218b, we obtain the solution as

$$T(r, \phi) = \frac{1}{\pi} \sum_{\nu=0}^{\infty} b \left(\frac{r}{b}\right)^\nu \cdot \frac{\bar{f}_2(\nu)}{k_2 \nu + h_2 b} \tag{3-223}$$

where $\nu = 0, 1, 2, \ldots$, and π should be replaced by 2π when $\nu = 0$.

The explicit form of solution 3-223 is

$$T(r, \phi) = \frac{1}{2\pi h_2} \int_{\phi'=0}^{2\pi} f_2(\phi') \cdot d\phi' + \frac{b}{\pi} \sum_{\nu=1}^{\infty} \left(\frac{r}{b}\right)^\nu \cdot \frac{1}{k_2 \nu + h_2 b} \int_{\phi'=0}^{2\pi} \cos \nu(\phi - \phi') \cdot f_2(\phi') \cdot d\phi' \tag{3-224}$$

(2) In this approach we remove from system 3-217 the partial derivatives with respect to both ϕ and r variables by applying a Fourier transform on ϕ and Hankel transform on r, and reduce the system to an algebraic expression for the double transform of temperature. The solution is obtained when the double transform is successively inverted by means of the inversion formulas.

The Fourier transform of system 3-217 with respect to the ϕ-variable by means of the transform Eq. 3-218 has been performed above and the result is given by Eq. 3-219.

We now define the Hankel transform and inversion formula of temperature function $\bar{T}(r, \nu)$ with respect to the space variable r in the range

$0 \le r \le b$ as

$$\binom{\text{Integral}}{\text{transform}} \quad \bar{\bar{T}} = \int_{r'=0}^{b} r' \cdot K_\nu(\beta_m, r') \cdot \bar{T} \cdot dr' \tag{3-225a}$$

$$\binom{\text{Inversion}}{\text{formula}} \quad \bar{T} = \sum_{m=1}^{\infty} K_\nu(\beta_m, r) \cdot \bar{\bar{T}} \tag{3-225b}$$

where the kernel $K_\nu(\beta_m, r)$ and eigenvalues β_m are obtained from Table 3-1. For the boundary condition of the third kind at $r = b$ they are given as

$$K_\nu(\beta_m, r) = \frac{\sqrt{2}}{b} \frac{1}{\left[\dfrac{H_2^2}{\beta_m^2} + \left(1 - \dfrac{\nu^2}{\beta_m^2 b^2}\right)\right]^{1/2}} \cdot \frac{J_\nu(\beta_m r)}{J_\nu(\beta_m b)} \tag{3-226}$$

and β_m's are the roots of the transcendental equation

$$\beta J_\nu'(\beta b) + H_2 J(\beta b) = 0 \tag{3-227}$$

We take the integral transform of system 3-219 by applying the Hankel transform 3-225a:

$$\bar{\bar{T}} = \frac{b}{\beta_m^2} \cdot \frac{K_\nu(\beta_m, r)}{k_2}\bigg|_{r=b} \cdot \bar{f}_2(\nu) \tag{3-228}$$

This double transform of temperature is inverted successively by means of the inversion formulas 3-225b and 3-218b, and the solution of the steady-state problem 3-217 becomes

$$T(r, \phi) = \frac{1}{\pi} \sum_{\nu=0}^{\infty} \sum_{m=1}^{\infty} K_\nu(\beta_m, r) \cdot \frac{b}{\beta_m^2} \cdot \frac{K_\nu(\beta_m, r)}{k_2}\bigg|_{r=b} \cdot \bar{f}_2(\nu) \tag{3-229}$$

where $\nu = 0, 1, 2, 3, \ldots$. π should be replaced by 2π when $\nu = 0$, and β_m's are positive roots of Eq. 3-220b.

$$\bar{f}_2(\nu) = \int_{\phi'=0}^{2\pi} \cos \nu(\phi - \phi') \cdot F(\phi') \cdot d\phi'$$

The explicit form of Eq. 3-229 is given as

$$\begin{aligned} T(r, \phi) = {} & \frac{1}{\pi k_2} \sum_{m=1}^{\infty} \frac{1}{b(H_2^2 + \beta_m^2)} \frac{J_0(\beta_m r)}{J_0(\beta_m b)} \int_{\phi'=0}^{2\pi} f_2(\phi') \cdot d\phi' \\ & + \frac{2b}{\pi k_2} \sum_{\nu=1}^{\infty} \sum_{m=1}^{\infty} \frac{1}{b^2(H_2^2 + \beta_m^2) - \nu^2} \frac{J_\nu(\beta_m r)}{J_\nu(\beta_m b)} \\ & \cdot \int_{\phi'=0}^{2\pi} \cos \nu(\phi - \phi') \cdot f_2(\phi') \cdot d\phi' \end{aligned} \tag{3-230}$$

Comparing the two solutions 3-224 and 3-230 we note that the former solution has one less summation than the latter solution and one obtains the following relation for the summation:

$$\left(\frac{r}{b}\right)^{\nu} = 2\,\frac{k_2\nu + h_2 b}{k_2}\sum_{m=1}^{\infty}\frac{1}{b^2(H_2^2+\beta_m^2)-\nu^2}\,\frac{J_\nu(\beta_m r)}{J_\nu(\beta_m b)}$$

3-9. TRANSIENT TEMPERATURE CHARTS

Heisler [9] prepared temperature–time charts for a long cylinder of radius b, which is initially at a uniform temperature T_i and for times $t > 0$ subjected to convection from the boundary surface $r = b$ into an environment at constant temperature T_e. Defining a dimensionless temperature θ as

$$\theta = \frac{T - T_e}{T_i - T_e}$$

The boundary-value problem of heat conduction is given as

$$\frac{1}{r}\frac{\partial}{\partial r}\left(r\frac{\partial \theta}{\partial r}\right) = \frac{1}{\alpha}\frac{\partial \theta}{\partial r} \quad \text{in} \quad 0 \le r \le b,\ t > 0 \tag{3-231a}$$

$$\frac{\partial \theta}{\partial r} = 0 \quad \text{at} \quad r = 0,\ t > 0 \tag{3-231b}$$

$$k\frac{\partial \theta}{\partial r} + h\theta = 0 \quad \text{at} \quad r = b,\ t > 0 \tag{3-231c}$$

$$\theta = 1 \quad \text{in} \quad 0 \le r \le b,\ t = 0 \tag{3-231d}$$

This problem is similar to that given by Eq. 3-45, and the solution of Eq. 3-231 is obtainable from the solution 3-56 by setting $T_0 = 1$.

Figure 3-13 shows the Heisler chart for the dimensionless temperature at the axis of the cylinder, $(T_0 - T_e)/(T_i - T_e)$, as a function of the dimensionless time $\alpha t/b^2$ for several different values of the parameter k/hb. Figure 3-14 is a position-correction chart which is used together with Fig. 3-13 to determine the temperature as a function of time at other positions in the cylinder. Temperature at any position r/b for any given value of k/hb and $\alpha t/b^2$ is determined by multiplying the value of $(T_0 - T_e)/(T_i - T_e)$ from Fig. 3-13 by the value of $(T - T_e)/(T_i - T_e)$ at the corresponding value of r/b and k/hb from Fig. 3-14.

3-10. SUMMARY OF HANKEL TRANSFORM OF PART OF THE LAPLACIAN IN THE CYLINDRICAL COORDINATE SYSTEM

The differential operator in the cylindrical coordinate system involving the r variable in the form

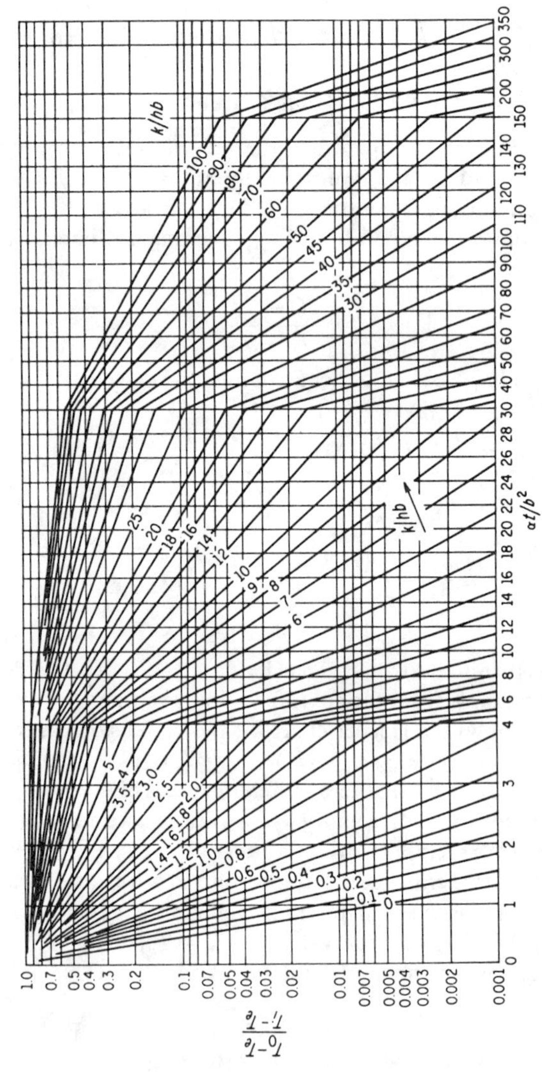

FIG. 3-13. Transient temperature T_0 at the axis of a long cylinder of radius b, subjected to convection at the boundary surface $r = b$. (From Heisler, Ref. 9.)

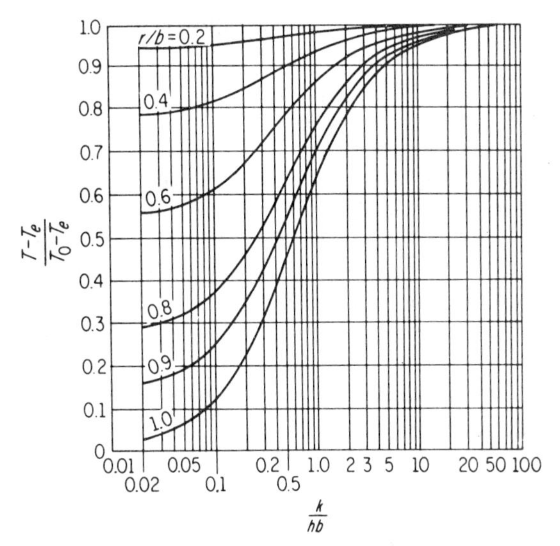

FIG. 3-14. Position-correction chart for use with Fig. 3-13. (From Heisler, Ref. 9.)

$$D \equiv \frac{\partial^2 T}{\partial r^2} + \frac{1}{r}\frac{\partial T}{\partial r} \tag{3-232}$$

can be removed by the application of an integral transform with respect to the r variable (i.e. Hankel transform of order $\nu = 0$). The transform, the inversion formula and the kernels to be used depend on the range of the r variable and the type of the boundary conditions. For a solid cylinder ($0 \le r \le b$) and a hollow cylinder ($a \le r \le b$) they are obtained respectively from Tables 3-1 and 3-2 by setting $\nu = 0$; and for an infinite region ($0 \le r < \infty$) they are obtained from Eq. 3-44 by setting $\nu = 0$.

The differential operator involving r and φ variables in the form

$$D \equiv \frac{\partial^2 T}{\partial r^2} + \frac{1}{r}\frac{\partial T}{\partial r} + \frac{1}{r^2}\frac{\partial^2 T}{\partial \varphi^2} \tag{3-233a}$$

can be removed by successive application of a Fourier transform with respect to the φ variable and a Hankel transform of order ν with respect to the r variable. When the range of the φ variable is $0 \le \varphi \le 2\pi$ the Fourier transform and the inversion formula are given by Eq. 3-186; when the range of the φ variable is $0 \le \varphi \le \varphi_0$, where $\varphi_0 < 2\pi$, the removal of the φ variable is treated exactly the same as that described in the Cartesian coordinate system for the finite regions and depending on the type of the boundary conditions associated with the φ variable, the integral transform, the inversion formula and the kernels are obtained from Table 2-1. Once the φ variable is removed the differential operator 3-233a includes the r variable in the form

$$D \equiv \frac{\partial^2 T}{\partial r^2} + \frac{1}{r}\frac{\partial T}{\partial r} - \frac{\nu^2}{r^2} \tag{3-233b}$$

where ν is the transform variable for the Fourier transform. The differential operator 3-233b can be removed by the application of a Hankel transform of order ν with respect to the r variable. The integral transform (i.e. Hankel transform of order ν), the inversion formula and the kernels are given by Table 3-1 for a solid cylinder ($0 \le r \le b$), by Table 3-2 for a hollow cylinder ($a \le r \le b$) and by Eq. 3-44 for an infinite region ($0 \le r < \infty$).

The differential operator involving the r and z variables in the form

$$D \equiv \frac{\partial^2 T}{\partial r^2} + \frac{1}{r}\frac{\partial T}{\partial r} + \frac{\partial^2 T}{\partial z^2} \tag{3-234a}$$

can be removed by successive application of a Fourier transform with respect to the z variable and a Hankel transform of order $\nu = 0$ with respect to the r variable. The removal of the z variable is treated exactly the same as that described in the Cartesian coordinate system. That is, the integral transform, the inversion formula, and the kernels for the z variable are obtained from Table 2-1 when the range of z is finite (i.e. $0 \le z \le L$), from Table 2-2 when the range of z is semiinfinite (i.e. $0 \le z < \infty$) and from Eq. 2-38 when the range of z is infinite (i.e. $-\infty < z < \infty$). Once the z variable is removed, the differential operator 3-234a includes the r variable in the form

$$D \equiv \frac{\partial^2 T}{\partial r^2} + \frac{1}{r}\frac{\partial T}{\partial r} \tag{3-234b}$$

which is of the same form as that given by Eq. 3-232, and can be removed by applying a Hankel transform of order $\nu = 0$ as described above.

Table 3-4 summarizes the removal of the differential operators 3-232, 3-233, and 3-234 given above for a solid cylinder ($0 \le r \le b$ and $0 \le r < \infty$), for a hollow cylinder ($a \le r \le b$) for the ranges of z finite, semi-infinite, and infinite when the range of the φ variable is $0 \le \varphi \le 2\pi$. For generality in the analysis the results are given for a boundary condition of the third kind, i.e.,

$$k_i \frac{\partial T}{\partial n_i} + h_i T = f_i \quad \text{on boundary } s_i \tag{3-235}$$

both for the r variable and the z variable. For a boundary condition of the first kind we set $k_i = 0$, and for such cases the term $\left.\frac{K}{k_i}\right|_i$ should be re-

TABLE 3-4

INTEGRAL TRANSFORM (i.e. HANKEL TRANSFORM) OF PART OF LAPLACIAN IN THE CYLINDRICAL COORDINATE SYSTEM

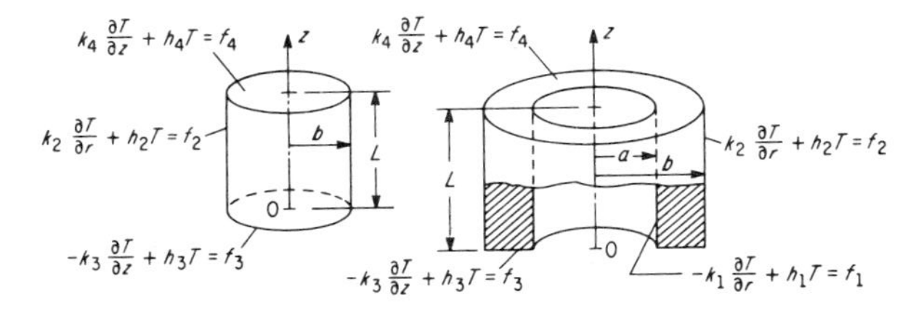

Boundary conditions for solid cylinder

Boundary conditions for hollow cylinder

Region	Definition of Integral Transform*	Definition of Inversion Formula*	Form of $\nabla^2 T$	Integral Transform of $\nabla^2 T$†
Solid Cylinder $0 \le r \le b$	$\bar{T} = \int_{r'=0}^{b} r' \cdot K_0(\beta_m, r') \cdot dr'$	$T = \sum_{m=1}^{\infty} K_0(\beta_m, r) \cdot \bar{T}$	$\frac{\partial^2 T}{\partial r^2} + \frac{1}{r}\frac{\partial T}{\partial r}$	$-\beta_m^2 \bar{T} + \left.\frac{K(\beta_m, r)}{k_2}\right\vert_{r=b} \cdot f_2$
$0 \le r < \infty$	$\bar{T} = \int_{r'=0}^{\infty} r' \cdot J_0(\beta r') \cdot dr'$	$T = \int_{\beta=0}^{\infty} \beta J_0(\beta r) \cdot \bar{T} \cdot d\beta$	$\frac{\partial^2 T}{\partial r^2} + \frac{1}{r}\frac{\partial T}{\partial r}$	$-\beta^2 \bar{T}$
$0 \le r \le b$ $0 \le \phi \le 2\pi$	$\bar{T} = \int_{r'=0}^{b} \int_{\varphi'=0}^{2\pi} Z \cdot T \cdot d\phi' \cdot dr'$ $Z \equiv r' \cdot K_\nu(\beta_m, r') \cdot \cos \nu(\varphi - \varphi')$	$T = \frac{1}{\pi} \sum_{\nu=0}^{\infty} \sum_{m=1}^{\infty} K_\nu(\beta_m, r) \cdot \bar{T}$ replace π by 2π for $\nu = 0$.	$\frac{\partial^2 T}{\partial r^2} + \frac{1}{r}\frac{\partial T}{\partial r} + \frac{1}{r^2}\frac{\partial^2 T}{\partial \varphi^2}$	$-\beta_m^2 \bar{T} + b \left.\frac{K_\nu(\beta_m, r)}{k_2}\right\vert_{r=b} \cdot \bar{f}_2(\nu)$
$0 \le r \le b$ $0 \le z \le L$	$\bar{T} = \int_{r'=0}^{b} \int_{z'=0}^{L} Z \cdot T \cdot dz' \cdot dr'$ $Z \equiv r' \cdot K_0(\beta_m, r') \cdot K(\eta_n, z')$	$T = \sum_{n=1}^{\infty} \sum_{m=1}^{\infty} Z \cdot \bar{T}$ $Z = K_0(\beta_m, r) \cdot K(\eta_n, z)$	$\frac{\partial^2 T}{\partial r^2} + \frac{1}{r}\frac{\partial T}{\partial r} + \frac{\partial^2 T}{\partial z^2}$	$-\beta_m^2 \bar{T} + b \left.\frac{K_0(\beta_m, r)}{k_2}\right\vert_{r=b} \cdot \bar{f}_2(\eta_n) - \eta_n^2 \bar{T}$ $+ \left.\frac{K(\eta_n, z)}{k_3}\right\vert_{z=0} \cdot \bar{f}_3(\beta_m) + \left.\frac{K(\eta_n, z)}{k_4}\right\vert_{z=L} \cdot \bar{f}_4(\beta_m)$
$0 \le r \le b$ $0 \le z < \infty$	$\bar{T} = \int_{r'=0}^{b} \int_{z'=0}^{\infty} Z \cdot T \cdot dz' \cdot dr'$ $Z \equiv r' \cdot K_0(B_m, r') \cdot K(\eta, z')$	$T = \int_{\eta=0}^{\infty} \sum_{m=1}^{\infty} Z \cdot T \cdot d\eta$ $Z \equiv K_0(\beta_m, r) \cdot K(\eta, z)$	$\frac{\partial^2 T}{\partial r^2} + \frac{1}{r}\frac{\partial T}{\partial r} + \frac{\partial^2 T}{\partial z^2}$	$-\beta_m^2 \bar{T} + b \left.\frac{K_0(\beta_m, r)}{k_2}\right\vert_{r=b} \cdot \bar{f}_2(\eta) - \eta^2 \bar{T}$ $+ \left.\frac{K(\eta, z)}{k_3}\right\vert_{z=0} \cdot \bar{f}_3(\beta_m)$

TABLE 3-4 (Continued)

Region	Definition of Integral Transform*	Definition of Inversion Formula*	Form of $\nabla^2 T$	Integral Transform of $\nabla^2 T$†
$0 \le r \le b$ $-\infty < z < \infty$	$\bar{T} = \int_{r'=0}^{b} \int_{z'=-\infty}^{\infty} Z \cdot T \cdot dz' \cdot dr'$ $Z \equiv r \cdot K_0(\beta_m, r') \cdot e^{i\eta z'}$	$T = \frac{1}{2\pi} \int_{\eta=-\infty}^{\infty} \sum_{m=1}^{\infty} Z \cdot \bar{T}\, d\eta$ $Z \equiv K_0(\beta_m, r) \cdot e^{-i\eta z}$	$\frac{\partial^2 T}{\partial r^2} + \frac{1}{r}\frac{\partial T}{\partial r} + \frac{\partial^2 T}{\partial z^2}$	$-\beta_m^2 \bar{T} + b \frac{K_0(\beta_m, r)}{k_2}\Big\vert_{r=b} \cdot \bar{f}_2(\eta) - \eta^2 \bar{T}$
$0 \le r < \infty$ $0 \le \varphi \le 2\pi$	$\bar{T} = \int_{r'=0}^{\infty} \int_{\varphi'=0}^{2\pi} Z \cdot T \cdot d\varphi' \cdot dr'$ $Z \equiv r' \cdot J_\nu(\beta r') \cdot \cos \nu(\varphi - \varphi')$	$T = \frac{1}{\pi} \sum_{\nu=0}^{\infty} \int_{\beta=0}^{\infty} \beta J_\nu(\beta r) \bar{T} d\beta$ replace π by 2π for $\nu = 0$.	$\frac{\partial^2 T}{\partial r^2} + \frac{1}{r}\frac{\partial T}{\partial r} + \frac{1}{r^2}\frac{\partial^2 T}{\partial \varphi^2}$	$-\beta^2 \bar{T}$
$0 \le r < \infty$ $0 \le z \le L$	$\bar{T} = \int_{r'=0}^{\infty} \int_{z'=0}^{L} Z \cdot T \cdot dz' \cdot dr'$ $Z \equiv r' \cdot J_0(\beta r') \cdot K(\eta_n, z')$	$T = \sum_{n=1}^{\infty} \int_{\beta=0}^{\infty} Z \cdot \bar{T} \cdot d\beta$ $Z \equiv \beta \cdot J_0(\beta r) \cdot K(\eta_n, z)$	$\frac{\partial^2 T}{\partial r^2} + \frac{1}{r}\frac{\partial T}{\partial r} + \frac{\partial^2 T}{\partial z^2}$	$-\beta^2 \bar{T} + \frac{K(\eta_n, z)}{k_3}\Big\vert_{z=0} \cdot \bar{f}_3(\beta)$ $+ \frac{K(\eta_n, z)}{k_4}\Big\vert_{z=L} \cdot \bar{f}_4(\beta) - \eta_n^2 \bar{T}$
$0 \le r < \infty$ $0 \le z < \infty$	$\bar{T} = \int_{r'=0}^{\infty} \int_{z'=0}^{\infty} Z \cdot T \cdot dz' \cdot dr'$ $Z \equiv r' \cdot J_0(\beta r') \cdot K(\eta, z')$	$T = \int_{\eta=0}^{\infty} \int_{\beta=0}^{\infty} Z \cdot \bar{T}\, d\beta\, d\eta$ $Z \equiv \beta \cdot J_0(\beta r) \cdot K(\eta, z)$	$\frac{\partial^2 T}{\partial r^2} + \frac{1}{r}\frac{\partial T}{\partial r} + \frac{\partial^2 T}{\partial z^2}$	$-\beta^2 \bar{T} + \frac{K(\eta, z)}{k_3}\Big\vert_{z=0} \cdot \bar{f}_3(\beta) - \eta^2 \bar{T}$
$0 \le r < \infty$ $-\infty < z < \infty$	$\bar{T} = \int_{r'=0}^{\infty} \int_{z'=-\infty}^{\infty} r' \cdot J_0(\beta r') \cdot e^{i\eta z'} \cdot T \cdot dz' \cdot dr'$	$T = \frac{1}{2\pi} \int_{\eta=-\infty}^{\infty} \int_{\beta=0}^{\infty} \beta \cdot J_0(\beta r) \cdot e^{-i\eta z} \cdot \bar{T} \cdot d\beta \cdot d\eta$	$\frac{\partial^2 T}{\partial r^2} + \frac{1}{r}\frac{\partial T}{\partial r} + \frac{\partial^2 T}{\partial z^2}$	$-\beta^2 \bar{T} - \eta^2 \bar{T}$
Hollow Cylinder				
$a \le r \le b$	$\bar{T} = \int_{r'=a}^{b} r' \cdot K_0(\beta_m, r') \cdot T \cdot dr'$	$T = \sum_{m=1}^{\infty} K_0(\beta_m, r) \cdot \bar{T}$	$\frac{\partial^2 T}{\partial r^2} + \frac{1}{r}\frac{\partial T}{\partial r}$	$-\beta_m^2 \bar{T} + a \frac{K_0(\beta_m, r)}{k_1}\Big\vert_{r=a} \cdot f_1$ $+ b \frac{K_0(\beta_m, r)}{k_2}\Big\vert_{r=b} \cdot f_2$
$a \le r \le b$ $0 \le \phi \le 2\pi$	$\bar{T} = \int_{r'=a}^{b} \int_{\varphi'=0}^{2\pi} Z \cdot T \cdot d\varphi' \cdot dr'$ $Z \equiv r' K_\nu(\beta_m, r') \cdot \cos \nu(\varphi - \varphi')$	$T = \frac{1}{\pi} \sum_{\nu=0}^{\infty} \sum_{m=1}^{\infty} K_\nu(\beta_m, r) \cdot \bar{T}$ replace π by 2π for $\nu = 0$.	$\frac{\partial^2 T}{\partial r^2} + \frac{1}{r}\frac{\partial T}{\partial r} + \frac{1}{r^2}\frac{\partial^2 T}{\partial \phi^2}$	$-\beta_m^2 \bar{T} + a \frac{K_\nu(\beta_m, r)}{k_1}\Big\vert_{r=a} \cdot \bar{f}_1(\nu)$ $+ b \frac{K_\nu(\beta_m, r)}{k_2}\Big\vert_{r=b} \cdot \bar{f}_2(\nu)$

$a \le r \le b$ $0 \le z \le L$	$\bar{\bar{T}} = \int_{r'=a}^{b} \int_{z'=0}^{L} Z \cdot T \cdot dz' \cdot dr'$ $Z \equiv r' K_0(\beta_m, r') \cdot K(\eta_n, z')$	$T = \sum_{n=1}^{\infty} \sum_{m=1}^{\infty} Z \bar{\bar{T}}$ $Z \equiv K_0(\beta_m, r) \cdot K(\eta_n, z)$	$\frac{\partial^2 T}{\partial r^2} + \frac{1}{r}\frac{\partial T}{\partial r} + \frac{\partial^2 T}{\partial z^2}$	$-\beta_m^2 \bar{\bar{T}} + a \left.\frac{K_0(\beta_m, r)}{k_1}\right\|_{r=a} \cdot \bar{\bar{f}}_1(\eta_n)$ $+ b \left.\frac{K_0(\beta_m, r)}{k_2}\right\|_{r=b} \cdot \bar{\bar{f}}_2(\eta_n) + \left.\frac{K(\eta_n, z)}{k_3}\right\|_{z=0} \cdot \bar{\bar{f}}_3(\beta_m)$ $+ \left.\frac{K(\eta_n, z)}{k_4}\right\|_{z=L} \cdot \bar{\bar{f}}_4(\beta_m) - \eta_n^2 \bar{\bar{T}}$
$a \le r \le b$ $0 \le z < \infty$	$\bar{\bar{T}} = \int_{r'=a}^{b} \int_{z'=0}^{\infty} Z \cdot T \cdot dz' \cdot dr'$ $Z \equiv r' K_0(\beta_m, r') \cdot K(\eta, z')$	$T = \int_{n=0}^{\infty} \sum_{m=1}^{\infty} Z \cdot \bar{\bar{T}} \cdot d\eta$ $Z \equiv K_0(\beta_m, r) \cdot K(\eta, z)$	$\frac{\partial^2 T}{\partial r^2} + \frac{1}{r}\frac{\partial T}{\partial r} + \frac{\partial^2 T}{\partial z^2}$	$-\beta_m^2 \bar{\bar{T}} + a \left.\frac{K_0(\beta_m, r)}{k_1}\right\|_{r=a} \cdot \bar{\bar{f}}_1(\eta) - \eta^2 \bar{\bar{T}}$ $+ b \left.\frac{K_0(\beta_m, r)}{k_2}\right\|_{r=b} \cdot \bar{\bar{f}}_2(\eta) + \left.\frac{K(\eta, z)}{k_3}\right\|_{z=0} \cdot \bar{\bar{f}}_3(\beta_m)$
$a \le r \le b$ $-\infty < z < \infty$	$\bar{\bar{T}} = \int_{r'=a}^{b} \int_{z'=-\infty}^{\infty} Z \cdot T \cdot dz' \cdot dr'$ $Z \equiv r' K_0(\beta_m, r') \cdot e^{i\eta z'}$	$T = \frac{1}{2\pi} \int_{\eta=-\infty}^{\infty} \sum_{m=1}^{\infty} Z \cdot \bar{\bar{T}} \cdot d\eta$ $Z \equiv K_0(\beta_m, r) \cdot e^{-i\eta z}$	$\frac{\partial^2 T}{\partial r^2} + \frac{1}{r}\frac{\partial T}{\partial r} + \frac{\partial^2 T}{\partial z^2}$	$-\beta_m^2 \bar{\bar{T}} + a \left.\frac{K_0(\beta_m, r)}{k_1}\right\|_{r=a} \cdot \bar{\bar{f}}_1(\eta)$ $+ b \left.\frac{K_0(\beta_m, r)}{k_2}\right\|_{r=b} \cdot \bar{\bar{f}}_2(\eta) - \eta^2 \bar{\bar{T}}$

*Kernels for the r variable are given in Tables 3-1 and 3-2. Kernels for the z-variable are given in Tables 2-1 and 2-2.

†For boundary condition of the first kind replace $\frac{K}{k_i}$ by $\frac{1}{h_i}\frac{dK}{dn_i}$, where $\frac{d}{dn_i}$ differentiation along the outward drawn normal to the boundary surface.

$\bar{f}(v), \bar{f}(\beta)$ and $\bar{f}(\eta)$ are integral transforms with respect to the φ, r and z variables respectively.

placed by $-\frac{1}{h_i}\frac{dK}{dn_i}\Big|_i$ in the last column of Table 3-4, where $\frac{d}{dn_i}$ denotes differentiation along the outward drawn normal to the boundary surface.

Tabulations similar to those given in Table 3-4 can easily be extended for the removal of the three-dimensional differential operator, i.e.,

$$D \equiv \frac{\partial^2 T}{\partial r^2} + \frac{1}{r}\frac{\partial T}{\partial r} + \frac{1}{r^2}\frac{\partial^2 T}{\partial \varphi^2} + \frac{\partial^2 T}{\partial z^2} \tag{3-236}$$

The space variables can be removed from the differential operator 3-236 by successive application of a Fourier transform with respect to the z variable, a Fourier transform with respect to the φ variable and a Hankel transform with respect to the r variable.

REFERENCES

1. G. N. Watson, *A Treatise on the Theory of Bessel Functions*, Oxford University Press, Cambridge, 1944.

2. N. W. McLachlan, *Bessel Functions for Engineers*, Clarendon Press, Oxford, 1961.

3. T. K. Sherwood and C. E. Reed, *Applied Mathematics in Chemical Engineering*, McGraw-Hill Book Company, 1939, p. 210.

4. Ian N. Sneddon, *Fourier Transforms*, McGraw-Hill Book Company, 1951, p. 52.

5. N. W. McLachlan, *Bessel Functions for Engineers*, Clarendon Press, Oxford, 1961, p. 32.

6. G. N. Watson, *A Treatise on the Theory of Bessel Functions*, Oxford University Press, Cambridge, 1944, p. 395.

7. *Ibid.*, p. 393.

8. R. V. Churchill, "Fourier Series and Boundary Value Problems," McGraw-Hill Book Company, 1963, pp. 78 and 106.

9. M. P. Heisler, "Transient Charts for Induction and Constant-Temperature Heating," *Trans. ASME*, Vol. 69 (April 1947), pp. 227–236.

PROBLEMS

1. Find a relation for one-dimensional, transient-temperature distribution in a long solid cylinder of radius $r = b$, which is initially at zero temperature and for times $t > 0$ the boundary surface $r = b$ is kept at a constant temperature T_0.

2. There is uniformly distributed volumetric-heat source of constant strength g_0 Btu/hr ft^3 inside a long solid cylinder of radius $r = b$. Initially the cylinder is at zero temperature, for times $t > 0$ the heat source releases its heat continuously at a constant rate while the boundary surface at $r = b$ is kept at constant temperature T_0. Find a relation for temperature distribution in the cylinder for times $t > 0$.

3. A long hollow cylinder of inner radius a and outer radius b is initially at zero temperature. For times $t > 0$ the boundary surface $r = a$ is kept insulated and that at $r = b$ is kept at constant temperature T_0. Determine the temperature distribution in the solid for times $t > 0$.

4. A long hollow cylinder $a \leq r \leq b$ is initially at temperature $F(r)$. For times $t > 0$ the boundaries at $r = a$ and $r = b$ are kept insulated. Determine the temperature distribution in the cylinder.

5. A long hollow cylinder $a \leq r \leq b$ is initially at temperature $F(r)$. For times $t > 0$ the boundary surface at $r = a$ kept at zero temperature and that at $r = b$ is subjected to convection into a medium at zero temperature. Find a relation for temperature distribution in the cylinder.

6. A long semicylinder $0 \leq r \leq b$, $0 \leq \phi \leq \pi$ is initially at temperature $F(r, \phi)$. For times $t > 0$ the boundary surfaces at $r = b$, $\phi = 0$ and $\phi = \pi$ are all kept at zero temperature. Determine the temperature distribution $T(r, \phi, t)$ in the cylinder for times $t > 0$.

7. A long solid cylinder of radius $r = b$ is initially at temperature $F(r, \phi)$ and for times $t > 0$ the boundary surface at $r = b$ is kept at zero temperature. Determine the temperature distribution $T(r, \phi, t)$ in the cylinder.

8. A finite solid cylinder $0 \leq r \leq b$, $-\pi \leq \phi \leq \pi$, $0 \leq z \leq L$ is initially at zero temperature. For times $t > 0$ heat is generated within the solid at a rate of $g(r, \phi, z)$ Btu/hr ft^3 while the boundaries are kept at zero temperature. Find the temperature distribution $T(r, \phi, z, t)$ in the cylinder.

9. A semi-infinite solid cylinder $0 \leq r \leq b$, $0 \leq z < \infty$ is initially at zero temperature. For times $t > 0$ the boundary surface at $z = 0$ is kept at zero temperature and that at $r = b$ is kept at a constant temperature T_0. Determine a relation for the temperature distribution $T(r, z, t)$ in the cylinder. Solve the same problem for the case when the boundary surface $r = b$ is subjected to convection to a medium at temperature T_∞.

10. A semi-infinite hollow cylinder $a \leq r \leq b$, $0 \leq z < \infty$ is initially at zero temperature. For times $t > 0$ heat is generated within the solid at a rate of $g(r, z)$ Btu/hr ft^3 while the boundary surfaces at $z = 0$ and $r = a$ are kept at zero temperature and that at $r = b$ dissipates heat by convection into a medium at zero temperature. Find a relation for temperature distribution $T(r, z, t)$ in the cylinder.

Evaluate the temperature distribution for the special case of $g_i(r, z, t)$ being an instantaneous volume heat source that releases its heat spontaneously at time $t = 0$.

11. A solid cylindrical disk of infinite radius, i.e., $0 \leq z \leq L$, $0 \leq r < \infty$ is initially at temperature $F(r, z)$. For times $t > 0$ the boundary surfaces $z = 0$ and $z = L$ are kept at zero temperature. Determine the temperature distribution $T(r, z, t)$ in the solid for times $t > 0$.

12. Determine the steady-state temperature distribution in a semi-infinite solid cylinder $0 \leq r \leq b$, $0 \leq z < \infty$ which is subjected to a temperature distribution $F(r)$ at the boundary surface $z = 0$ and to convection into a medium at zero temperature at the boundary surface $r = b$.

13. A long solid cylinder $0 \leq r \leq b$, $0 \leq \phi \leq \phi_0$ is initially at zero temperature. For times $t > 0$ heat is generated within the solid at a rate of $g(r, \phi)$ Btu/hr ft^3, while the boundary surfaces at $\phi = 0$ and $\phi = \phi_0$ are kept at zero temperature and the cylindrical surface at $r = b$ is subjected to convection with a medium at zero temperature. Find a relation for the temperature distribution $T(r, \phi, t)$ in the cylinder for times $t > 0$.

4

Heat Conduction in the Spherical Coordinate System

In this chapter we examine the solution of boundary-value problems of heat conduction in the spherical coordinate system. The method of separation of variables will be used to solve the homogeneous problems, integral-transform technique will be used to solve the nonhomogeneous problems. A brief discussion of Legendre's differential equation and of Legendre functions will be given.

4-1. SEPARATION OF HOMOGENEOUS DIFFERENTIAL EQUATION OF HEAT CONDUCTION

The differential equation of heat conduction in the spherical coordinate system with no heat generation within the solid is given as

$$\frac{\partial^2 T}{\partial r^2} + \frac{2}{r}\frac{\partial T}{\partial r} + \frac{1}{r^2 \sin\psi}\frac{\partial}{\partial \psi}\left(\sin\psi \frac{\partial T}{\partial \psi}\right) + \frac{1}{r^2 \sin^2\psi}\frac{\partial^2 T}{\partial \phi^2} = \frac{1}{\alpha}\frac{\partial T}{\partial t} \tag{4-1}$$

where $T \equiv T(r, \psi, \phi, t)$. This equation is put into a more convenient form by defining a new independent variable

$$\mu = \cos\psi \tag{4-2}$$

Equation 4-1 becomes

$$\frac{\partial^2 T}{\partial r^2} + \frac{2}{r}\frac{\partial T}{\partial r} + \frac{1}{r^2}\frac{\partial}{\partial \mu}\left[(1-\mu^2)\frac{\partial T}{\partial \mu}\right] + \frac{1}{r^2(1-\mu^2)}\frac{\partial^2 T}{\partial \phi^2} = \frac{1}{\alpha}\frac{\partial T}{\partial t} \tag{4-3}$$

where $T = T(r, \mu, \phi, t)$.

Equation 4-3 is separated by assuming the separation of the temperature function in the form

$$T(r, \mu, \phi, t) = \Gamma(t) \cdot R(r) \cdot M(\mu) \cdot \Phi(\phi) \tag{4-4}$$

Equation 4-3 becomes

$$\frac{1}{R}\frac{d^2 R}{dr^2} + \frac{2}{rR}\frac{dR}{dr} + \frac{1}{r^2}\frac{1}{M}\frac{d}{d\mu}\left[(1-\mu^2)\frac{dM}{d\mu}\right] + \frac{1}{r^2(1-\mu^2)}\frac{1}{\Phi}\frac{d^2\Phi}{d\phi^2} = \frac{1}{\alpha}\frac{1}{\Gamma}\frac{d\Gamma}{dt} \tag{4-5}$$

The only way Eq. 4-5 is satisfied is if each group of separated variables is set equal to a constant, and one way of performing this separation may be in the form

$$\frac{d\Gamma}{dt} + \alpha\lambda^2\Gamma = 0 \tag{4-6a}$$

$$\frac{d^2\Phi}{d\phi^2} + m^2\Phi = 0 \tag{4-6b}$$

$$\frac{d^2R}{dr^2} + \frac{2}{r}\frac{dR}{dr} + \left[\lambda^2 - \frac{n(n+1)}{r^2}\right]R = 0 \tag{4-6c}$$

$$\frac{d}{d\mu}\left[(1-\mu^2)\frac{dM}{d\mu}\right] + \left[n(n+1) - \frac{m^2}{1-\mu^2}\right]M = 0 \tag{4-6d}$$

where λ, m and n are the separation parameters.

Particular solutions of the above four separated equations are in the form

$$\Gamma : e^{-\alpha\lambda^2 t} \tag{4-7a}$$

$$\Phi : \sin m\phi, \cos m\phi \tag{4-7b}$$

$$R : (\lambda r)^{-1/2} \cdot J_{n+1/2}(\lambda r), (\lambda r)^{-1/2} \cdot Y_{n+1/2}(\lambda r) \tag{4-7c}$$

$$M : P_n^m(\mu), Q_n^m(\mu) \tag{4-7d}$$

Solution 4-7c is obtained from the differential equation 4-6c by defining a new variable

$$R = (\lambda r)^{-(1/2)} \cdot Z$$

Then Eq. 4-6c becomes

$$\frac{d^2Z}{dr^2} + \frac{1}{r}\frac{dZ}{dr} + \left[\lambda^2 - \frac{1}{r^2}\left(n + \frac{1}{2}\right)^2\right]Z = 0 \tag{4-8}$$

which is Bessel's differential equation of order $(n + \frac{1}{2})$ and its solutions are

$$Z : J_{n+1/2}(\lambda r), \quad Y_{n+1/2}(\lambda r)$$

The functions

$$j_n(x) = \sqrt{\frac{\pi}{2x}}\, J_{n+1/2}(x) \tag{4-9a}$$

$$y_n(x) \equiv \sqrt{\frac{\pi}{2x}}\, Y_{n+1/2}(x) \tag{4-9b}$$

are called the *spherical Bessel functions* of the first and second kinds [1]. Figures 4-1 and 4-2 show a plot of the spherical Bessel functions of the first and second kinds for $n = 0, 1, 2, 3$. It is to be noted that $y_n(x)$ func-

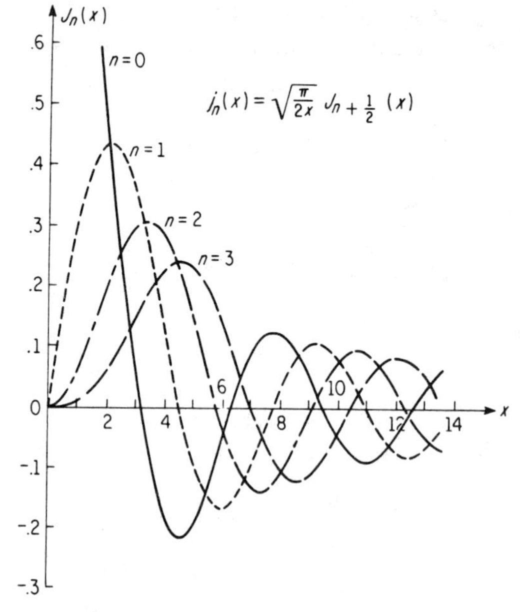

FIG. 4-1. Spherical Bessel functions of the first kind, $j_n(x)$ for $n = 0,1,2,3$. [*Handbook of Mathematical Functions*, National Bureau of Standards, *Applied Mathematics Series*, No. 55, Washington, D.C. (June 1964), p. 438.]

tions become infinite for $x = 0$, but $j_n(x)$ functions are zero for $n = 1, 2, 3$ and unity for $n = 0$ at $x = 0$.[1]

Equation 4-6d is called *Legendre's associated differential equation* and its solution includes two independent functions $P_n^m(\mu)$ and $Q_n^m(\mu)$ which are called *Legendre's associated functions of degree n and of order m* of the first and second kind respectively. When the order m is equal to zero, the functions are given as $P_n(\mu)$ and $Q_n(\mu)$ and called *Legendre functions of degree n* of the first and second kind respectively. A brief account of the properties of Legendre functions will be given in the next section.

We examine the separated solutions of the differential Eq. 4-3 for some special cases.

[1] $j_0(x) = \dfrac{\sin x}{x} \qquad j_1(x) = \dfrac{\sin x}{x^2} - \dfrac{\cos x}{x}$

$y_0(x) = -\dfrac{\cos x}{x} \qquad y_1(x) = \dfrac{\cos x}{x^2} - \dfrac{\sin x}{x}$

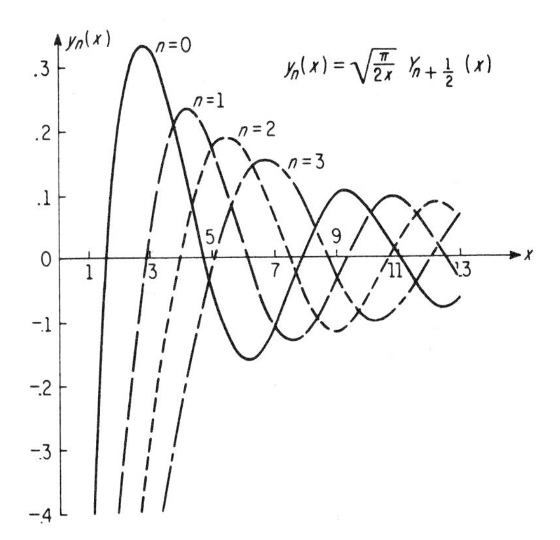

FIG. 4-2. Spherical Bessel functions of the second kind, $y_n(x)$ for $n = 0,1,2,3$. [*Handbook of Mathematical Functions*, National Bureau of Standards, *Applied Mathematics Series*, No. 55, Washington, D.C. (June 1964), p. 438.]

(1) Temperature is independent of the time variable. For this case we have

$$T \equiv T(r, \mu, \phi)$$

and the resulting separation equations and their particular solutions may be given in the form

$$\frac{d^2\Phi}{d\phi^2} + m^2\Phi = 0 \quad \Phi: \sin m\phi, \cos m\phi \tag{4-10a}$$

$$\frac{d^2R}{dr^2} + \frac{2}{r}\frac{dR}{dr} - \frac{n(n+1)}{r^2}R = 0 \quad R: r^n, r^{-(n+1)} \tag{4-10b}$$

$$\frac{d}{d\mu}\left[(1-\mu^2)\frac{dM}{d\mu}\right] + \left[n(n+1) - \frac{m^2}{1-\mu^2}\right]M = 0 \quad M: P_n^m(\mu), Q_n^m(\mu) \tag{4-10c}$$

It is to be noted that for this special case the differential Eq. 4-10b for the R separation, is an Euler-Cauchy type differential-equation solution of which include r^n and $r^{-(n+1)}$ functions. Differential equations for the Φ and M separations remain the same as in the previous case.

(2) Temperature is independent of ϕ-space variable—that is,

$$T \equiv T(r, \mu, t)$$

The resulting separation equations and their solutions may be given in the form

$$\frac{d\Gamma}{dt} + \alpha\lambda^2\Gamma = 0 \quad \Gamma : e^{-\alpha\lambda^2 t} \tag{4-11a}$$

$$\frac{d^2R}{dr^2} + \frac{2}{r}\frac{dR}{dr} + \left(\lambda^2 - \frac{n(n+1)}{r^2}\right)R = 0$$

$$R : (\lambda r)^{-(1/2)} J_{n+1/2}(\lambda r), (\lambda r)^{-(1/2)} Y_{n+1/2}(\lambda r) \tag{4-11b}$$

$$\frac{d}{d\mu}\left[(1-\mu^2)\frac{dM}{d\mu}\right] + n(n+1)M = 0 \quad M : P_n(\mu), Q_n(\mu) \tag{4-11c}$$

For this case the differential equation 4-11c for the M separation is Legendre's differential equation and its solutions are the Legendre functions.

(3) Temperature is a function of r and ψ space variables only,

$$T \equiv T(r, \mu)$$

The resulting separation equations and their solutions are in the form

$$\frac{d^2R}{dr^2} + \frac{2}{r}\frac{dR}{dr} - \frac{n(n+1)}{r^2} R = 0 \quad R : r^n, r^{-(n+1)} \tag{4-12a}$$

$$\frac{d}{d\mu}\left[(1-\mu^2)\frac{dM}{d\mu}\right] + n(n+1)M = 0 \quad M : P_n(\mu), Q_n(\mu) \tag{4-12b}$$

4-2. LEGENDRE FUNCTIONS AND LEGENDRE'S ASSOCIATED FUNCTIONS

In this section we examine some of the properties of Legendre functions and Legendre's associated functions that are peculiar to the solution of heat-conduction problems. For detailed treatment on Legendre functions the reader should refer to standard texts on this subject [2, 3, 4, 5].

Legendre Functions. Consider Legendre's differential equation

$$\frac{d}{d\mu}\left[(1-\mu^2)\frac{dM}{d\mu}\right] + n(n+1)M = 0, \quad -1 \le \mu \le 1 \tag{4-13}$$

When discussing the solutions of this equation we shall restrict our attention to the range of $-1 \le \mu \le 1$, because for heat-conduction problems in the spherical coordinate system the variable μ is related to the variable ψ by the relation

$$\mu = \cos\psi$$

Since the maximum range of ψ is from 0 to π, the corresponding range of μ, $1 \le \mu \le -1$, cover all the range in the ψ domain.

Legendre's differential equation in the range of $-1 \le \mu \le 1$ has two linearly independent solutions, $P_n(\mu)$ and $Q_n(\mu)$, which are called *Legendre functions of the first kind and second kind* respectively. For positive integer values of n (i.e., $n = 0, 1, 2, 3, \ldots$) Legendre functions of the first kind, $P_n(\mu)$, are polynomials which are continuous in the interval $-1 \le \mu \le 1$ and given as [Ref. 2, p. 86]

$$\begin{aligned}
P_0(\mu) &= 1 & P_4(\mu) &= \tfrac{1}{8}(35\mu^4 - 30\mu^2 + 3) \\
P_1(\mu) &= \mu & P_5(\mu) &= \tfrac{1}{8}(63\mu^5 - 70\mu^3 + 15\mu) \\
P_2(\mu) &= \tfrac{1}{2}(3\mu^2 - 1) & P_6(\mu) &= \tfrac{1}{16}(231\mu^6 - 315\mu^4 + 105\mu^2 - 5) \\
P_3(\mu) &= \tfrac{1}{2}(5\mu^3 - 3\mu) & P_7(\mu) &= \tfrac{1}{16}(429\mu^7 - 693\mu^5 + 315\mu^3 - 35\mu)
\end{aligned} \tag{4-14a}$$

The above polynomials are obtainable from Rodrigues' formula,

$$P_n(\mu) = \frac{1}{2^n \cdot n!} \frac{d^n}{d\mu^n} (\mu^2 - 1)^n \tag{4-14b}$$

Figure 4-3 shows a plot of first four of Legendre functions of the first kind; Appendix V gives the numerical values of the first seven of Legendre functions of the first kind.

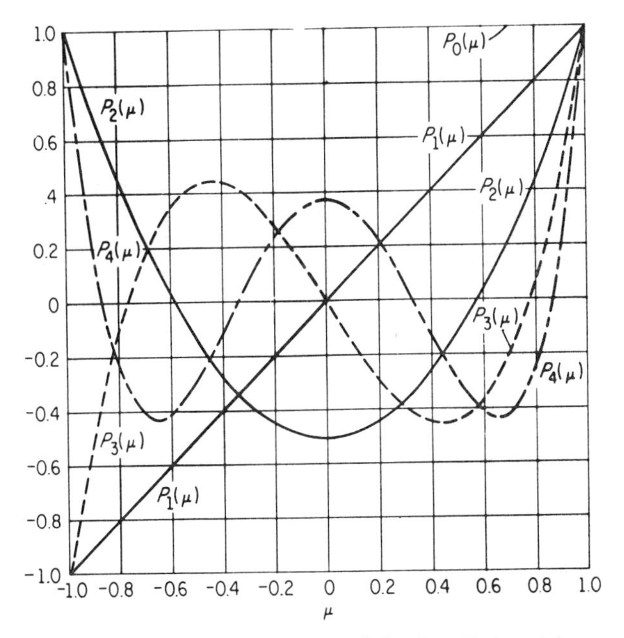

FIG. 4-3. Legendre functions of the first kind, $p_n(\mu)$ for $n = 0,1,2,3,4$.

For positive integer values of n the expressions for the Legendre functions of the second kind, $Q_n(\mu)$ are given in the form [Ref. 4, p. 135]

$$
\begin{aligned}
Q_0(\mu) &= \tfrac{1}{2}\ln\left(\frac{1+\mu}{1-\mu}\right) \\
Q_1(\mu) &= P_1(\mu)\cdot Q_0(\mu) - 1 \\
Q_2(\mu) &= P_2(\mu)\cdot Q_0(\mu) - \tfrac{3}{2}\mu \\
Q_3(\mu) &= P_3(\mu)\cdot Q_0(\mu) - \tfrac{5}{2}\mu^2 + \tfrac{2}{3} \\
Q_4(\mu) &= P_4(\mu)\cdot Q_0(\mu) - \tfrac{35}{8}\mu^3 + \tfrac{55}{24}\mu \\
Q_5(\mu) &= P_5(\mu)\cdot Q_0(\mu) - \tfrac{63}{8}\mu^4 + \tfrac{49}{8}\mu^2 - \tfrac{8}{15}
\end{aligned}
\tag{4-15}
$$

Figure 4-4 shows a plot of first three of the Legendre functions of the second kind. It is to be noted that the Legendre functions of the second kind are infinite series, which are convergent for $|\mu| < 1$ but becomes infinite as $\mu \to \pm 1$.

Legendre's Associated Functions. Legendre's associated differential equation is given in the form

$$
\frac{d}{d\mu}\left[(1-\mu^2)\frac{dM}{d\mu}\right] + \left[n(n+1) - \frac{m^2}{1-\mu^2}\right]M = 0 \quad -1 \le \mu \le 1 \tag{4-16}
$$

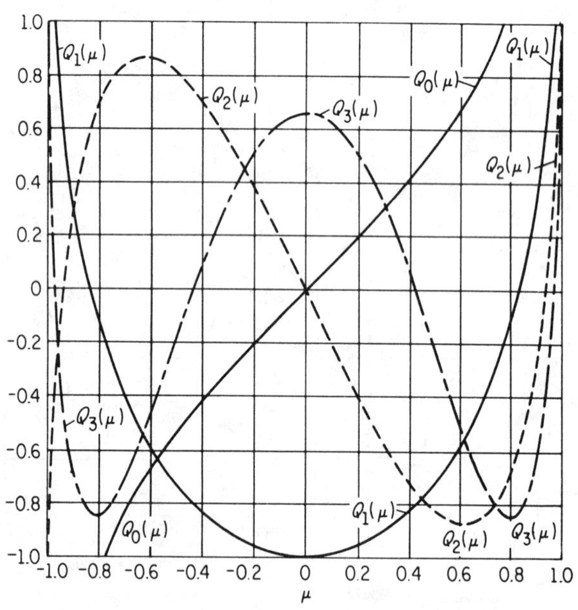

FIG. 4-4. Legendre functions of the second kind, $Q_n(\mu)$ for $n = 0,1,2,3$.

It can be shown that solutions of this differential equation are related to $P_n(\mu)$ and $Q_n(\mu)$ functions. We define a new variable

$$M = (1 - \mu^2)^{m/2} \cdot W \tag{4-17}$$

Then Eq. 4-16 becomes

$$(1 - \mu^2)\frac{d^2 W}{d\mu^2} - 2(m + 1)\mu\frac{dW}{d\mu} + (n - m)(n + m + 1)W = 0 \tag{4-18}$$

Consider Legendre's differential equation,

$$\frac{d}{d\mu}\left[(1 - \mu^2)\frac{dM_0}{d\mu}\right] + n(n + 1)M_0 = 0 \tag{4-19}$$

We differentiate Eq. 4-19 with respect to μ, m times,

$$(1 - \mu^2)\frac{d^{m+2} M_0}{d\mu^{m+2}} - 2(m + 1)\mu\frac{d^{m+1} M_0}{d\mu^{m+1}} + (n - m)(n + m - 1)\frac{d^m M_0}{d\mu^m} = 0 \tag{4-20}$$

Comparing Eqs. 4-20 and 4-18, we find

$$W = \frac{d^m M_0}{d\mu^m} \tag{4-21}$$

From Eqs. 4-17 and 4-21,

$$M = (1 - \mu^2)^{m/2} \cdot \frac{d^m M_0}{d\mu^m} \tag{4-22}$$

Equation 4-22 relates the solution of Legendre's associated differential equation to the solution of Legendre's differential equation, hence we have[2]

$$P_n^m(\mu) = (1 - \mu^2)^{m/2}\frac{d^m P_n(\mu)}{d\mu^m} \tag{4-23a}$$

$$Q_n^m(\mu) = (1 - \mu^2)^{m/2} \cdot \frac{d^m Q_n(\mu)}{d\mu^m} \tag{4-23b}$$

where m is a positive integer and $0 \leq m \leq n$.

[2]These relations are for $|\mu| \leq 1$. For the part of the complex plane outside the unit circle $|\mu| > 1$, the functions are defined as

$$P_n^m = (\mu^2 - 1)^{m/2}\frac{d^m P_n(\mu)}{d\mu^m}$$

$$Q_n^m = (\mu^2 - 1)^{m/2}\frac{d^m Q_n(\mu)}{d\mu^m}$$

In the process of relating the Legendre's associated functions to the Legendre functions, if we choose the new variable W as

$$M = (1 - \mu^2)^{-(m/2)} \cdot W \tag{4-24}$$

the resulting solutions are denoted by

$$P_n^{-m}(\mu) \quad \text{and} \quad Q_n^{-m}(\mu)$$

These solutions, however, are not independent of the $P_n^m(\mu)$ and $Q_n^m(\mu)$ functions.

4-3. THE INTEGRAL TRANSFORM (LEGENDRE TRANSFORM) AND THE INVERSION FORMULA

The partial derivatives with respect to the μ variable can be removed from the differential equation of heat conduction in the spherical coordinate system by applying an integral transform which is usually called the *Legendre transform*. Such transform and the corresponding inversion formula are constructed from the expansion of an arbitrary function in the interval $-1 \leq \mu \leq 1$ in an infinite series of Legendre polynomials.

In this section we examine the construction of the Legendre transform and the corresponding inversion formula for different ranges of μ and for different boundary conditions.

Region $-1 \leq \mu \leq 1$. Legendre functions of the first kind, $P_n(\mu)$, constitute an orthogonal set with respect to a weighting-function unity in the interval $-1 \leq \mu \leq 1$; that is, [Ref. 2, p. 92]

$$\int_{-1}^{1} P_n(\mu) \cdot P_{n'}(\mu)\, d\mu = \begin{cases} 0 & n \neq n' \\ \dfrac{2}{2n+1} & \text{for} \quad n = n' \end{cases} \tag{4-25}$$

We now consider expansion of an arbitrary function $F(\mu)$ in the interval $-1 \leq \mu \leq 1$ in an infinite series of $P_n(\mu)$ functions in the form

$$F(\mu) = \sum_{n=0}^{\infty} c_n \cdot P_n(\mu) \tag{4-26a}$$

The unknown coefficients c_n are immediately determined from the orthogonality property of $P_n(\mu)$ functions, and given as

$$c_n = \frac{\displaystyle\int_{-1}^{1} P_n(\mu') \cdot F(\mu') \cdot d\mu'}{N} \tag{4-26b}$$

where the norm N is

$$N = \int_{-1}^{1} P_n^2(\mu) \cdot d\mu = \frac{2}{2n+1} \tag{4-26c}$$

The expansion 4-26a becomes

$$F(\mu) = \sum_{n=0}^{\infty} \frac{P_n(\mu)}{\sqrt{N}} \int_{-1}^{1} \frac{P_n(\mu')}{\sqrt{N}} F(\mu') \cdot d\mu' \tag{4-27}$$

Expansion 4-27 is rewritten as the *inversion formula* for the *integral transform* (*Legendre transform*) of function $F(\mu)$ in the interval $-1 \le \mu \le 1$ as

$$\begin{pmatrix}\text{Inversion}\\ \text{formula}\end{pmatrix} \quad F(\mu) = \sum_{n=0}^{\infty} K_n(\mu)\, \bar{F}(n) \tag{4-28a}$$

$$\begin{pmatrix}\text{Integral}\\ \text{transform}\end{pmatrix} \quad \bar{F}(n) = \int_{\mu'=-1}^{1} K_n(\mu') \cdot F(\mu') \cdot d\mu' \tag{4-28b}$$

where the kernel is defined as

$$K_n(\mu) = \sqrt{\frac{2n+1}{2}}\, P_n(\mu) \tag{4-28c}$$

The above expansions are valid if function $F(\mu)$ is a sectionally continuous function on the interval $-1 \le \mu \le 1$ and has one-sided derivatives at each point μ at which $F(\mu)$ is discontinuous [6].

We now examine the expansion of an arbitrary function $F(\mu)$ in the interval $-1 \le \mu \le 1$ in terms of Legendre's associated functions of the first kind, $P_n^m(\mu)$. When n, n', m are positive integers and $n > m$, $n' > m$, we have [7]

$$\int_{-1}^{1} P_n^m(\mu) \cdot P_{n'}^m(\mu) \cdot d\mu = \begin{cases} 0 & \text{for } n \ne n' \\ \dfrac{2}{2n+1} \dfrac{(n+m)!}{(n-m)!} & \text{for } n = n' \end{cases} \tag{4-29}$$

Assuming a function $F(\mu)$ may be expanded in the interval $-1 \le \mu \le 1$ in an infinite series of $P_n^m(\mu)$ functions in the form

$$F(\mu) = \sum_{n}^{\infty} C_n \cdot P_n^m(\mu) \qquad n \ge m \tag{4-30a}$$

where the unknown coefficients C_n are determined by making use of the orthogonality of $P_n^m(\mu)$ functions as given by Eq. 4-29:

$$C_n = \frac{\displaystyle\int_{-1}^{1} P_n^m(\mu') \cdot F(\mu') \cdot d\mu'}{N} \tag{4-30b}$$

where the norm N is

$$N = \int_{-1}^{1} [P_n^m(\mu)]^2 \cdot d\mu = \frac{2}{2n+1} \frac{(n+m)!}{(n-m)!} \tag{4-30c}$$

The expansion given by Eq. 4-30 is rewritten as the *inversion formula* for the *intergral transform* (*Legendre transform*) of function $F(\mu)$ in the interval $-1 \leq \mu \leq 1$ as

$$\begin{pmatrix}\text{Inversion}\\ \text{formula}\end{pmatrix} F(\mu) = \sum_{n=0}^{\infty} K_n^m(\mu)\cdot \bar{F}(n,m) \tag{4-31a}$$

$$\begin{pmatrix}\text{Integral}\\ \text{transform}\end{pmatrix} \bar{F}(n,m) = \int_{\mu'=-1}^{1} K_n^m(\mu')\cdot F(\mu')\cdot d\mu' \tag{4-31b}$$

where the kernel is

$$K_n^m(\mu) = \sqrt{\frac{2n+1}{2}\frac{(n-m)!}{(n+m)!}}\, P_n^m(\mu). \tag{4-31c}$$

Region $0 \leq \mu \leq 1$. When the range of independent variable μ extends from 0 to 1, we have the following property of $P_n(\mu)$ functions [Ref. 7, p. 306, Example 2].

$$\int_{\mu=0}^{1} P_n(\mu)\cdot P_{n'}(\mu)d\mu$$

$$= 0 \quad \text{for } n \neq n' \text{ and } n - n' = \textit{even} \quad \text{(i.e., } n,n' \text{ both } \textit{even} \text{ or } n,n' \text{ both } \textit{odd}) \tag{4-32a}$$

$$= \frac{1}{2n+1} \quad \text{for} \quad n = n' \tag{4-32b}$$

$$= \frac{(-1)^{\frac{n+n'+1}{2}}}{2^{n+n'-1}\cdot(n-n')(n+n'+1)} \frac{n!\,n'!}{\left(\frac{n-1}{2}!\right)^2\left(\frac{n'}{2}!\right)^2} \quad \text{for } n = \text{odd},\ n' = \text{even} \tag{4-32c}$$

Therefore, Legendre polynomials constitute an orthogonal set in the interval $0 \leq \mu \leq 1$ if they are of *odd* degree or of *even* degree as shown by Eqs. 4-32a and 4-32b. By making use of this property an arbitrary function $F(\mu)$ is expanded in terms of *odd* Legendre polynomials or in terms of *even* Legendre polynomials. Since such expansions can be written as the inversion formula for the integral transform of function $F(\mu)$ in the interval $0 \leq \mu \leq 1$, we define the following *odd* and *even Legendre transforms* and the corresponding inversion formulas over the interval $0 \leq \mu \leq 1$ [8]. The odd Legendre transform and the inversion formula are defined as

$$\begin{pmatrix}\text{Inversion}\\ \text{formula}\end{pmatrix} F(\mu) = \sum_{n=0}^{\infty} K_{2n+1}(\mu)\cdot \bar{F}(2n+1) \tag{4-33a}$$

$$\begin{pmatrix}\textit{odd}\text{ Legendre}\\ \text{transform}\end{pmatrix} \bar{F}(2n+1) = \int_{\mu'=0}^{1} K_{2n+1}(\mu')\cdot F(\mu')\cdot d\mu' \tag{4-33b}$$

where

$$K_{2n+1}(\mu) = \sqrt{4n+3}\quad P_{2n+1}(\mu) \tag{4-33c}$$
$$n = 0,1,2,3,\ldots$$

The even Legendre transform and the inversion formula are defined as

$$\begin{pmatrix}\text{Inversion}\\ \text{formula}\end{pmatrix} F(\mu) = \sum_{n=0}^{\infty} K_{2n}(\mu)\cdot \bar{F}(2n) \tag{4-34a}$$

$$\begin{pmatrix}\textit{even}\text{ Legendre}\\ \text{transform}\end{pmatrix} \bar{F}(2n) = \int_{\mu'=0}^{1} K_{2n}(\mu')\cdot F(\mu')\cdot d\mu' \tag{4-34b}$$

where

$$K_{2n}(\mu) = \sqrt{4n+1}\quad P_{2n}(\mu) \tag{4-34c}$$
$$n = 0,1,2,3,\ldots$$

Table 4-1 gives a summary of the Legendre transforms and the corresponding inversion formulas.

4-4. REMOVAL OF PARTIAL DERIVATIVES WITH LEGENDRE TRANSFORM

In this section we examine the removal from the differential equation of heat conduction with the application of the Legendre transform of the following differential operators:

$$\frac{\partial}{\partial\mu}\left[(1-\mu^2)\frac{\partial T}{\partial\mu}\right] \quad \text{and} \quad \frac{\partial}{\partial\mu}\left[(1-\mu^2)\frac{\partial T}{\partial\mu}\right] + \frac{1}{1-\mu^2}\frac{\partial^2 T}{\partial\phi^2}$$

(1) Removal of the Differential Operator for the μ Variable. Consider the differential operator

$$D \equiv \frac{\partial}{\partial\mu}\left[(1-\mu^2)\frac{\partial T}{\partial\mu}\right] \tag{4-35}$$

Let $K_n(\mu)$ be the normalized Legendre function of the first kind satisfying Legendre's differential equation

$$\frac{d}{d\mu}\left[(1-\mu^2)\frac{dK_n}{d\mu}\right] + n(n+1)K_n = 0 \tag{4-36}$$

Taking $K_n(\mu)$ as the kernel we define the integral transform and the inversion formula of temperature function T with respect to the space

TABLE 4-1

LEGENDRE TRANSFORMS AND THE CORRESPONDING INVERSION FORMULAS.

Range of μ	Integral (Legendre) Transform	Inversion Formula	Kernel
$-1 \le \mu \le 1$	$\bar{F}(n) = \int_{-1}^{1} K_n(\mu') \cdot F(\mu') \cdot d\mu$	$F(\mu) = \sum_{n=0}^{\infty} K_n(\mu) \cdot \bar{F}(n)$	$K_n(\mu) = \sqrt{\frac{2n+1}{2}} P_n(\mu)$
$0 \le \mu \le 1$	$\bar{F}(2n+1) = \int_{0}^{1} K_{2n+1}(\mu') \cdot F(\mu') \cdot d\mu'$	$F(\mu) = \sum_{n=0}^{\infty} K_{2n+1}(\mu) \cdot \bar{F}(2n+1)$	$K_{2n+1}(\mu) = \sqrt{4n+3} \quad P_{2n+1}(\mu)$
$0 \le \mu \le 1$	$\bar{F}(2n) = \int_{0}^{1} K_{2n}(\mu') \cdot F(\mu') \cdot d\mu'$	$F(\mu) = \sum_{n=0}^{\infty} K_{2n}(\mu) \cdot \bar{F}(2n)$	$K_{2n}(\mu) = \sqrt{4n+1} \cdot P_{2n}(\mu)$
$-1 \le \mu \le 1$	$\bar{F}(n,m) = \int_{-1}^{1} K_n^m(\mu') \cdot F(\mu') \cdot d\mu'$	$F(\mu) = \sum_{n}^{\infty} K_n^m(\mu) \cdot \bar{F}(n,m)$	$K_n^m(\mu) = \sqrt{\frac{2n+1}{2} \cdot \frac{(n-m)!}{(n+m)!}} P_n^m(\mu)$

variable μ as

$$\begin{pmatrix}\text{Integral}\\ \text{transform}\end{pmatrix} \bar{T} = \int K_n(\mu')\cdot T\cdot d\mu' \tag{4-37a}$$

$$\begin{pmatrix}\text{Inversion}\\ \text{formula}\end{pmatrix} T = \sum_{n=0}^{\infty} K_n(\mu)\cdot \bar{T} \tag{4-37b}$$

The limits of integration depend on the range of μ and will be introduced later in the analysis.

By applying the integral transform Eq. 4-37a we take the integral transform of the differential operator 4-35:

$$\bar{D} = \int K_n(\mu)\cdot\frac{\partial}{\partial\mu}\left[(1-\mu^2)\frac{\partial T}{\partial\mu}\right]d\mu \tag{4-38}$$

Integrating Eq. 4-38 by parts twice, we obtain

$$\bar{D} = \left[(1-\mu^2)\left(K_n\frac{\partial T}{\partial\mu} - T\frac{dK_n}{d\mu}\right)\right]_{\substack{\text{lower}\\ \text{limit}}}^{\substack{\text{upper}\\ \text{limit}}} + \int T\,\frac{d}{d\mu}\left[(1-\mu^2)\frac{dK_n}{d\mu}\right]d\mu \tag{4-39}$$

In view of the relations 4-36 and 4-37a, the integral on the right-hand side is evaluated and Eq. 4-39 becomes

$$\bar{D} = \left[(1-\mu^2)\left(K_n\frac{\partial T}{\partial\mu} - T\frac{dK_n}{d\mu}\right)\right]_{\substack{\text{lower}\\ \text{limit}}}^{\substack{\text{upper}\\ \text{limit}}} - n(n+1)\,\bar{T} \tag{4-40}$$

Evaluation of the terms inside the large bracket depends on the range of μ variable and the boundary conditions. We examine the following cases.

The Range of $-1 \le \mu \le 1$. This is the case of complete sphere and the integral transform and the inversion formula are defined as

$$\begin{pmatrix}\text{Integral}\\ \text{transform}\end{pmatrix} \bar{T} = \int_{\mu'=-1}^{1} K_n(\mu')\cdot T\cdot d\mu' \tag{4-41a}$$

$$\begin{pmatrix}\text{Inversion}\\ \text{formula}\end{pmatrix} T = \sum_{n=0}^{\infty} K_n(\mu)\cdot \bar{T} \tag{4-41b}$$

where

$$K_n(\mu) = \sqrt{\frac{2n+1}{2}}\,P_n(\mu), \quad n = 0,1,2,3,\ldots \tag{4-41c}$$

In Eq. 4-40 the large bracket vanishes at the upper and lower limits (i.e., at $\mu = 1$ and $\mu = -1$), hence, the integral transform of the differential operator 4-35 becomes

$$\bar{D} = -n(n+1)\bar{T} \tag{4-42}$$

The range $0 \le \mu \le 1$, with boundary condition at $\mu = 0$ of the first kind. This is the case of a hemisphere as shown in Fig. 4-5a, with boundary condition at $\mu = 0$ is of the form

$$T|_{\mu=0} = \text{prescribed} \tag{4-43}$$

In Eq. 4-40 the terms in the large bracket vanish at the upper limit $\mu = 1$, but at the lower limit $\mu = 0$ they include an unknown quantity $\left.\frac{\partial T}{\partial \mu}\right|_{\mu=0}$ This unknown quantity can be removed if the kernel K_n is chosen with odd degrees, because Legendre functions of the first kind with odd degrees vanish at $\mu = 0$. Hence, replacing n by $2n + 1$ in Eq. 4-40 and noting that $K_{2n+1}(\mu)$ functions vanishes at $\mu = 0$ for $n = 0,1,2,3,\ldots$, the integral transform of the differential operator 4-35 becomes

$$\bar{D} = -(2n+1)(2n+2)\cdot\bar{T} + \left.\frac{dK_{2n+1}(\mu)}{d\mu}\right|_{\mu=0} T\Big|_{\mu=0}, \quad n = 0,1,2,3,\ldots \tag{4-44}$$

The integral transform (i.e. odd Legendre transform) and the inversion formula for the present case are defined as

$$\begin{pmatrix}\text{Integral}\\ \text{transform}\end{pmatrix} \quad \bar{T} = \int_{\mu'=0}^{1} K_{2n+1}(\mu')\cdot T\cdot d\mu' \tag{4-45a}$$

$$\begin{pmatrix}\text{Inversion}\\ \text{formula}\end{pmatrix} \quad T = \sum_{n=0}^{\infty} K_{2n+1}(\mu)\cdot\bar{T} \tag{4-45b}$$

where

$$K_{2n+1}(\mu) = \sqrt{4n+3}\cdot P_{2n+1}(\mu) \tag{4-45c}$$
$$n = 0,1,2,3,\ldots$$

The range $0 \le \mu \le 1$, the boundary condition at $\mu = 0$ is of the second kind. In this case the boundary condition at $\mu = 0$ is prescribed in the form

$$\left.\frac{\partial T}{\partial \mu}\right|_{\mu=0} = \text{prescribed} \tag{4-46}$$

as shown in Fig. 4-5b.

In Eq. 4-40 the bracket vanishes at the upper limit $\mu = 1$, but includes an unknown quantity $T|_{\mu=0}$ at the lower limit $\mu = 0$. This unknown quantity can be removed if we choose the kernel K_n with even degrees, because $dK_{2n}/d\mu$ vanishes at $\mu = 0$ for $n = 0,1,2,3,\ldots$; then the integral

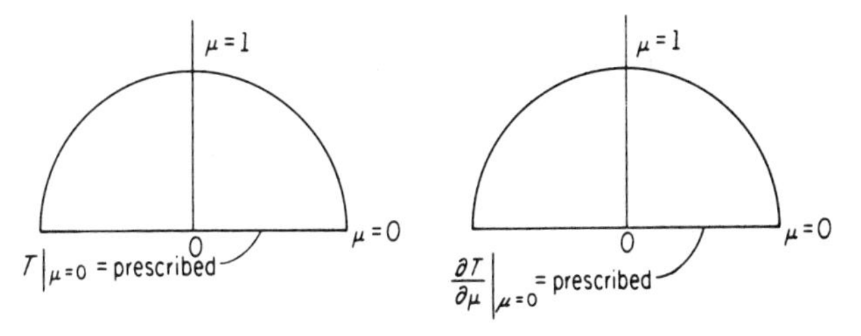

FIG. 4-5. Boundary conditions at the base ($\mu = 0$) of a hemisphere.

transform of the differential operator 4-35 becomes

$$\bar{D} = -2n(2n+1)\bar{T} - K_{2n}(\mu)\Big|_{\mu=0} \cdot \left[\frac{\partial T}{\partial \mu}\right]_{\mu=0}, \quad n = 0,1,2,3,\ldots \tag{4-47}$$

The integral transform (i.e. even Legendre transform) and the inversion formula are defined as

$$\begin{pmatrix}\text{Integral}\\ \text{transform}\end{pmatrix} \quad \bar{T} = \int_{\mu'=0}^{1} K_{2n}(\mu') \cdot T \cdot d\mu' \tag{4-48a}$$

$$\begin{pmatrix}\text{Inversion}\\ \text{formula}\end{pmatrix} \quad T = \sum_{n=0}^{\infty} K_{2n}(\mu) \cdot \bar{T} \tag{4-48b}$$

where (4-48c)

$$K_{2n}(\mu) = \sqrt{4n+1} \quad P_{2n}(\mu) \tag{4-48c}$$
$$n = 0,1,2,3,\ldots$$

(2) Removal of the Differential Operator for the μ and φ Variables. Consider the differential operator

$$D \equiv \frac{\partial}{\partial \mu}(1-\mu^2)\frac{\partial T}{\partial \mu} + \frac{1}{1-\mu^2}\frac{\partial^2 T}{\partial \varphi^2} \quad \text{in} \quad \begin{matrix}-1 \le \mu \le 1\\ 0 \le \varphi \le 2\pi\end{matrix} \tag{4-49}$$

We define an integral transform and the inversion formula with respect to the φ variable of the temperature function T as (see: Eq. 3-186)

$$\begin{pmatrix}\text{Integral}\\ \text{transform}\end{pmatrix} \quad \bar{T}(\mu, m, \varphi) = \int_{\varphi'=0}^{2\pi} \cos m(\varphi' - \varphi) \cdot T(\mu, \varphi') \cdot d\varphi' \tag{4-50a}$$

$$\begin{pmatrix}\text{Inversion}\\ \text{formula}\end{pmatrix} \quad T(\mu, \varphi) = \frac{1}{\pi}\sum_{m=0}^{\infty} \bar{T}(\mu, m, \varphi) \tag{4-50b}$$

where, $m = 0, 1, 2, 3, \ldots$, and π should be replaced by 2π for $m = 0$.

The application of the integral transform 4-50a removes from the differential operator 4-49 the φ variable and the result is

$$\bar{D} \equiv \frac{\partial}{\partial \mu}\left[(1-\mu^2)\frac{\partial \bar{T}}{\partial \mu}\right] - \frac{m^2}{1-\mu^2}\bar{T} \tag{4-51}$$

where, $\bar{T} \equiv \bar{T}(\mu, m, \varphi)$.

We define the integral transform and the inversion formula with respect to the μ variable of function $\bar{T}(\mu, m, \varphi)$ as

$$\begin{pmatrix}\text{Integral}\\ \text{transform}\end{pmatrix} \quad \bar{\bar{T}}(n, m, \varphi) = \int_{\mu'=-1}^{1} K_n^m(\mu') \cdot \bar{T}(\mu', m, \varphi) \cdot d\mu' \tag{4-52a}$$

$$\begin{pmatrix}\text{Inversion}\\ \text{formula}\end{pmatrix} \quad \bar{T}(\mu, m, \varphi) = \sum_{n}^{\infty} K_n^m(\mu) \cdot \bar{\bar{T}}(n, m, \varphi) \tag{4-52b}$$

where the kernel $K_n^m(\mu)$ is

$$K_n^m(\mu) = \sqrt{\frac{2n+1}{2}\frac{(n-m)!}{(n+m)!}}\, P_n^m(\mu) \quad \text{and} \quad n \geq m. \tag{4-52c}$$

The kernel $K_n^m(\mu)$ satisfies Legendre's associated differential equation

$$\frac{\partial}{\partial \mu}\left[(1-\mu^2)\frac{\partial K_n^m}{\partial \mu}\right] + \left[n(n+1) - \frac{m^2}{1-\mu^2}\right]K_n^m = 0 \tag{4-53}$$

We take integral transform of the differential operator 4-51 by applying the integral transform 4-52a.

$$\bar{\bar{D}} = \int_{-1}^{1} K_n^m \frac{\partial}{\partial \mu}\left[(1-\mu^2)\frac{\partial \bar{T}}{\partial \mu}\right]d\mu - \int_{-1}^{1}\frac{m^2}{1-\mu^2}K_n^m \bar{T} d\mu \tag{4-54a}$$

Integrating Eq. 4-54a by parts twice, we obtain

$$\bar{\bar{D}} = \left[(1-\mu^2)\left(K_n^m \frac{\partial \bar{T}}{\partial \mu} - \bar{T}\frac{dK_n^m}{d\mu}\right)\right]_{-1}^{1}$$

$$+ \int_{-1}^{1} \bar{T}\left\{\frac{\partial}{\partial \mu}\left[(1-\mu^2)\frac{dK_n^m}{d\mu}\right] - \frac{m^2}{1-\mu^2}K_n^m\right\}d\mu \tag{4-54b}$$

In Eq. 4-54b the terms in the first bracket vanish both at the upper and lower limits; when the integral is evaluated by making use of Eqs. 4-53 and 4-52a, then Eq. 4-54b simplifies to

$$\bar{\bar{D}} = -n(n+1) \cdot \bar{\bar{T}}(n, m, \varphi) \tag{4-54c}$$

Thus, by applying successively the integral transforms 4-50a and 4-52a, we removed from the differential operator 4-49 the partial derivatives with respect to the φ and μ variables, and reduced it to an algebraic relation as given by Eq. 4-54c.

4-5. ONE-DIMENSIONAL NONHOMOGENEOUS BOUNDARY-VALUE PROBLEMS OF HEAT CONDUCTION

A straightforward method for solving the one-dimensional boundary-value problems of heat conduction in the spherical coordinate system involving the space variable r, is to transform the problem into the cartesian coordinate system by means of a suitable transformation. Solution of the problem is then immcdiately obtained with the techniques discussed in Chapter 2. In this section we consider the application for a hollow sphere, a solid sphere, and infinite medium with spherical symmetry.

Hollow Sphere, $a \leq r \leq b$. Consider a hollow sphere, $a \leq r \leq b$, which is initially at temperature $F(r)$. For times $t > 0$ there is heat generation within the solid at a rate of $g(r)$ Btu/hr ft^3, while heat is dissipated by convection from the boundaries at $r = a$ and $r = b$ into a surrounding temperature of which varies with time. Figure 4-6 shows

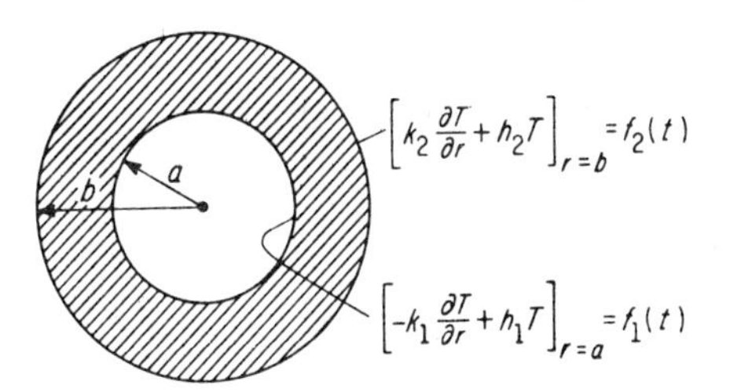

FIG. 4-6. Boundary conditions for a hollow sphere.

the geometry and the boundary conditions under consideration. The boundary-value problem of heat conduction is given as

$$\frac{1}{r} \frac{\partial^2}{\partial r^2} (rT) + \frac{g(r,t)}{k} = \frac{1}{\alpha} \frac{\partial T}{\partial t} \quad \text{in} \quad a \leq r \leq b,\ t > 0$$

$$-k_1 \frac{\partial T}{\partial r} + h_1 T = f_1(t) \quad \text{at} \quad r = a,\ t > 0$$

$$k_2 \frac{\partial T}{\partial r} + h_2 T = f_2(t) \quad \text{at} \quad r = b,\ t > 0 \tag{4-55}$$

$$T = F(r) \quad \text{in} \quad a \leq r \leq b,\ t = 0$$

where $T \equiv T(r, t)$.

The problem of sphere as given by Eq. 4-55 is readily transformed into a problem of slab by defining a new dependent variable U as

$$U = r \cdot T \tag{4-56}$$

Then system 4-54 becomes

$$\begin{aligned}
\frac{\partial^2 U}{\partial r^2} + \frac{r g(r,t)}{k} &= \frac{1}{\alpha}\frac{\partial U}{\partial t} \quad \text{in} \quad a \le r \le b,\ t > 0 \\
-k_1 \frac{\partial U}{\partial r} + \left(h_1 + \frac{k_1}{a}\right) U &= a \cdot f_1(t) \quad \text{at} \quad r = a,\ t > 0 \\
k_2 \frac{\partial U}{\partial r} + \left(h_2 - \frac{k_2}{b}\right) U &= b \cdot f_2(t) \quad \text{at} \quad r = b,\ t > 0 \\
U &= r \cdot F(r) \quad \text{in} \quad a \le r \le b,\ t > 0
\end{aligned} \tag{4-57}$$

Defining a new independent variable,

$$x = r - a \tag{4-58}$$

system 4-57 becomes:

$$\begin{aligned}
\frac{\partial^2 U}{\partial x^2} + \frac{(a+x) g(x,t)}{k} &= \frac{1}{\alpha}\frac{\partial U}{\partial t} \quad \text{in} \quad 0 \le x \le (b-a) \text{ for } t > 0 \\
-k_1 \frac{\partial U}{\partial x} + \left(h_1 + \frac{k_1}{a}\right) U &= a \cdot f_1(t) \quad \text{at} \quad x = 0,\ t > 0 \\
k_2 \frac{\partial U}{\partial x} + \left(h_2 - \frac{k_2}{b}\right) U &= b \cdot f_2(t) \quad \text{at} \quad x = b - a,\ t > 0 \\
U &= (x + a) \cdot F(x + a) \quad \text{in} \quad 0 \le x \le (b-a),\ t = 0
\end{aligned} \tag{4-59}$$

which is a problem of linear heat conduction in a slab $0 \le x \le (b - a)$ with heat generation within the solid and subject to a boundary condition of the third kind at both boundaries. Solution of this slab problem is readily obtained from the solution 2-73 given in Chapter 2, by changing the variables x and U to r and T respectively, and defining for convenience

$$H_1 \equiv \frac{h_1}{k_1} + \frac{1}{a}$$

$$H_2 \equiv \frac{h_2}{k_2} - \frac{1}{b}$$

The solution of the sphere problem 4-55 becomes

$$T(r,t) = \frac{1}{r}\sum_{m=1}^{\infty} e^{-\alpha\beta_m^2 t} \cdot K(\beta_m, r)\left[\int_{r'=a}^{b} r' \cdot F(r') \cdot K(\beta_m, r') \cdot dr' + \int_{t'=0}^{t} e^{\alpha\beta_m^2 t'} \cdot A(\beta_m, t') \cdot dt'\right] \tag{4-60a}$$

where

$$A(\beta_m, t') = \frac{\alpha}{k}\int_{r'=a}^{b} r' \cdot g(r', t') \cdot K(\beta_m, r') \cdot dr' + \alpha \left.\frac{K(\beta_m, r')}{k_1}\right|_{r'=a} \cdot a \cdot f_1(t') + \alpha \left.\frac{K(\beta_m, r')}{k_2}\right|_{r'=b} \cdot b \cdot f_2(t') \tag{4-60b}$$

$$K(\beta_m, r) = \sqrt{2}\,\frac{\beta_m \cos \beta_m(r-a) + H_1 \sin \beta_m(r-a)}{\left\{(\beta_m^2 + H_1^2)\left[(b-a) + \dfrac{H_2}{\beta_m^2 + H_2^2}\right] + H_1\right\}^{1/2}} \tag{4-61a}$$

where β_m's are the positive roots of the transcendental equation

$$\tan \beta(b-a) = \frac{\beta(H_1 + H_2)}{\beta^2 - H_1 H_2} \tag{4-61b}$$

(1) When solution 4-60 is applied to problems with boundary conditions of the first kind we set k_i equal to zero. For such cases the following changes should be made in Eq. 4-60b.

$$\text{When } k_1 = 0 \quad \text{replace} \quad \left.\frac{K(\beta_m, r)}{k_1}\right|_{r=a} \quad \text{by} \quad \frac{1}{h_1}\left.\frac{dK(\beta_m, r)}{dr}\right|_{r=a}$$

$$\text{When } k_2 = 0 \quad \text{replace} \quad \left.\frac{K(\beta_m, r)}{k_2}\right|_{r=b} \quad \text{by} \quad -\frac{1}{h_2}\left.\frac{dk(\beta_m, r)}{dr}\right|_{r=b}$$

(2) When there is no heat generation and the boundaries at $r = a$ and $r = b$ are insulated (i.e. h_1 and h_2 are zero) the solution should include the following term

$$\frac{3}{b^3 - a^3}\int_a^b r^2 \cdot F(r) \cdot dr$$

which is the average of the initial temperature distribution over the volume of the sphere and represents the equilibrium temperature after the transients have passed.

EXAMPLE 4-1. A hollow sphere is initially at temperature $F(r)$. For times $t > 0$ the boundary surfaces at $r = a$ and $r = b$ are kept at zero temperature.

The boundary conditions for this special case are both homogeneous boundary condition of the first kind. Setting $k_1 = k_2 = 0$, we have $H_1 = H_2 = \infty$, then the kernel $K(\beta_m, r)$ and the eigenvalues β_m are obtained from Eq. 4-61 as

$$K(\beta_m, r) = \sqrt{\frac{2}{b-a}} \sin \beta_m(r-a) \tag{4-62a}$$

and β_m's are the positive roots of $\sin \beta(b-a) = 0$ or given as

$$\beta_m = \frac{m\pi}{b-a} \qquad m = 1, 2, 3, \ldots \tag{4-62b}$$

Substituting Eq. 4-62 into solution 4-60 and taking $g(r, t) = f_1(t) = f_2(t) = 0$ we obtain

$$T(r, t) = \frac{2}{r(b - a)} \sum_{m=1}^{\infty} e^{-\alpha\beta_m^2 t} \cdot \sin \beta_m(r - a) \int_{r'=a}^{b} r' F(r') \cdot \sin \beta_m(r' - a) \cdot dr' \tag{4-63}$$

where

$$\beta_m = \frac{m\pi}{b - a}, \quad m = 1, 2, 3, \ldots.$$

EXAMPLE 4-2. A hollow sphere $a \leq r \leq b$ is initially at zero temperature. For times $t > 0$ heat is generated within the solid at a rate of $g(r, t)$ Btu/hr ft^3, while the boundaries are kept at zero temperature.

Since the boundary conditions are both of the first kind, the kernel $K(\beta_m, r)$ and the eigenvalues β_m are given by the Eq. 4-62. For this special case

$$f_1(t) = f_2(t) = F(r) = 0$$

Substituting these into Eq. 4-60 we obtain

$$T(r, t) = \frac{2}{r(b - a)} \frac{\alpha}{k} \sum_{m=1}^{\infty} e^{-\alpha\beta_m^2 t} \cdot \sin \beta_m(r - a) \int_{t'=0}^{t} e^{\alpha\beta_m^2 t'} dt' \cdot \int_{r'=a}^{b} r' g(r', t') \cdot \sin \beta_m(r' - a) \cdot dr' \tag{4-64}$$

where

$$\beta_m = \frac{m\pi}{b - a}, \quad m = 1, 2, 3, \ldots.$$

We examine some special cases of solution 4-64.

(a) The heat source is a spherical-surface heat source of radius r_1 (i.e., $a < r_1 < b$) which is situated concentrically inside the region. The heat source has a total strength of $g_{sph}(t)$ Btu/hr which is uniformly distributed over the spherical surface of the source and releases its heat continuously for times $t > 0$.

The spherical-surface heat source at $r = r_1$ of total strength $g_{sph}(t)$ Btu/hr is related to the volume heat source $g(r, t)$ Btu/hr ft^3 by

$$g(r', t') = \frac{g_{sph}(t')}{4\pi r'^2} \cdot \delta(r' - r_1) \tag{4-65}$$

Substituting Eq. 4-65 into Eq. 4-64 and performing the integration involving the delta function, we obtain

$$T(r, t) = \frac{1}{2\pi r r_1 (b - a)} \frac{\alpha}{k} \sum_{m=1}^{\infty} e^{-\alpha\beta_m^2 t} \cdot \sin \beta_m(r - a) \cdot \sin \beta_m(r_1 - a) \int_{t'=0}^{t} e^{\alpha\beta_m^2 t'} \cdot g_{sph}(t') \cdot dt' \tag{4-66}$$

where

$$\beta_m = \frac{m\pi}{b - a}, \quad m = 1,2,3,\ldots.$$

For a heat source with constant strength, i.e.,

$$g_{\text{sph}}(t') = g_{\text{sph}} = \text{const.}$$

the integration with respect to the time variable t' is performed and Eq. 4-66 becomes

$$T(r,t) = \frac{1}{2\pi r r_1 (b - a)} \frac{g_{\text{sph}}}{k} \sum_{m=1}^{\infty} \frac{1 - e^{-\alpha\beta_m^2 t}}{\beta_m^2} \cdot \sin\beta_m(r - a) \cdot \sin\beta_m(r_1 - a) \tag{4-67}$$

(b) The spherical surface-heat source at $r = r_1$ is an *instantaneous* heat source of total strength $g_{\text{sph},i}$ Btu, which is uniformly distributed over the surface of the sphere, and releases its heat spontaneously at time $t = \tau$.

The instantaneous spherical heat source of strength $g_{\text{sph},i}$ Btu, situated at r_1 and releasing its heat spontaneously at time τ, is related to the volume heat source $g(r,t)$ Btu/hr ft^3 by

$$g(r',t') = \frac{g_{\text{sph},i}}{4\pi r'^2} \cdot \delta(r' - r_1) \cdot \delta(t' - \tau) \tag{4-68}$$

Substituting Eq. 4-68 into Eq. 4-64 and performing the integrations we obtain

$$T(r,t) = \frac{1}{2\pi r r_1 (b - a)} \frac{\alpha g_{\text{sph},i}}{k} \sum_{m=1}^{\infty} e^{-\alpha\beta_m^2(t-\tau)} \cdot \sin\beta_m(r - a) \cdot \sin\beta_m(r_1 - a) \quad \text{for } t > \tau \tag{4-69}$$

where

$$\beta_m = \frac{m\pi}{b - a}$$

Defining for convenience the strength of the instantaneous surface-heat source as

$$S_{\text{sph},i} \equiv \frac{\alpha g_{\text{sph},i}}{k} \quad {}^\circ\text{F} \cdot \text{ft}^3 \tag{4-70}$$

then in Eq. 4-69 the term

$$\frac{1}{2\pi r r_1 (b - a)} \sum_{m=1}^{\infty} e^{-\alpha\beta_m^2(t-\tau)} \cdot \sin\beta_m(r - a) \cdot \sin\beta_m(r_1 - a) \quad \text{for } t > \tau \tag{4-71}$$

represents the temperature at r at time t due to an instantaneous spherical-surface heat source of radius r_1, of total strength $S_{\text{sph},i} = 1\,{}^\circ\text{F} \cdot \text{ft}^3$, releasing its heat spontaneously at time $t = \tau$ inside a hollow cylinder $a \leq r \leq b$ which is initially at zero temperature and for times $t > 0$ the boundaries of which are kept at zero temperature.

Solid Sphere $0 \le r \le b$. A solid sphere of radius $0 \le r \le b$ is initially at temperature $F(r)$. For times $t > 0$ heat is generated within the sphere at a rate of $g(r, t)$ Btu/hr ft³ and heat is dissipated by convection from the boundary at $r = b$ into a surrounding temperature of which varies with time. Figure 4-7 shows the geometry and the boundary conditions. The boundary-value problem of heat conduction is given as

$$\frac{1}{r}\frac{\partial^2}{\partial r^2}(rT) + \frac{g(r,t)}{k} = \frac{1}{\alpha}\frac{\partial T}{\partial t} \quad 0 \le r \le b, t > 0$$

$$k_2\frac{\partial T}{\partial r} + h_2 T = f_2(t) \quad \text{at } r = b, t > 0 \tag{4-72}$$

$$T \equiv F(r) \quad 0 \le r \le b, t = 0$$

where $T \equiv T(r, t)$.

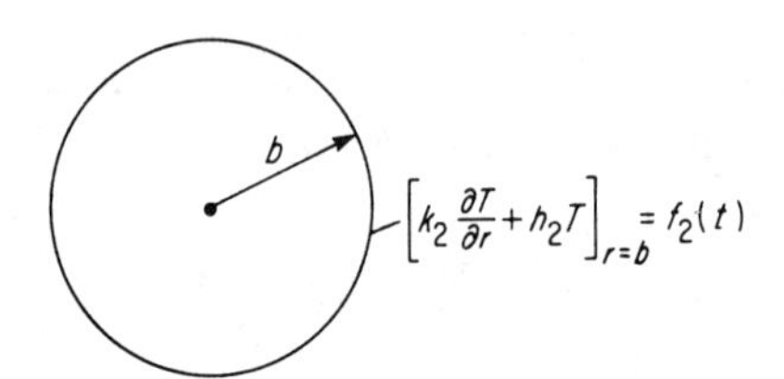

FIG. 4-7. Boundary condition for a solid sphere.

Defining a new dependent variable,

$$U = rT \tag{4-73}$$

system 4-72 becomes

$$\frac{\partial^2 U}{\partial r^2} + \frac{r \cdot g(r,t)}{k} = \frac{1}{\alpha}\frac{\partial U}{\partial t} \quad \text{in} \quad 0 \le r \le b, t > 0$$

$$k_2\frac{\partial U}{\partial r} + \left(h_2 - \frac{k_2}{b}\right)U = b \cdot f_2(t) \quad \text{at} \quad r = b, t > 0$$

$$U = 0 \quad \text{at} \quad r = 0, t > 0 \tag{4-74}$$

$$U = r \cdot F(r) \quad \text{at} \quad 0 \le r \le b, t = 0$$

which is a problem of linear heat flow in a slab $0 \le r \le b$, initially at temperature $r \cdot F(r)$ and for times $t > 0$ with heat generation within the solid at a rate of $r \cdot g(r, t)$ Btu/hr ft³ and subject to the boundary conditions shown in Eq. 4-74. Solution of this problem is immediately obtained from the general solution for a slab in Chapter 2. The kernels $K(\beta_m, r)$ and the eigenvalues β_m are obtained from Table 2-1 by replacing x by r and L by b.

Defining for convenience

$$H_2 \equiv \frac{h_2}{k_2} - \frac{1}{b}$$

the solution of system 4-72 is

$$T(r,t) = \frac{1}{r}\sum_{m=1}^{\infty} e^{-\alpha\beta_m^2 t}\cdot K(\beta_m, r)\cdot\left[\int_{r'=0}^{b} r'\cdot F(r')\cdot K(\beta_m, r')\cdot dr' + \int_{t'=0}^{t} e^{\alpha\beta_m^2 t'}\cdot A(\beta_m, t')\cdot dt'\right] \tag{4-75a}$$

where

$$A(\beta_m, t') = \frac{\alpha}{k}\int_{r'=0}^{b} r'\cdot g(r', t')\cdot K(\beta_m, r')\cdot dr' + \alpha\left.\frac{K(\beta_m, r)}{k_2}\right|_{r=b}\cdot b\cdot f_2(t') \tag{4-75b}$$

$$K(\beta_m, r) = \sqrt{2}\left[\frac{\beta_m^2 + H_2^2}{b(\beta_m^2 + H_2^2) + H_2}\right]^{1/2}\cdot\sin\beta_m r \tag{4-76a}$$

and β_m's are the positive roots of the transcendental equation[3]

$$\beta\cot\beta b + H_2 = 0 \tag{4-76b}$$

When applying the solution 4-75 to cases with the boundary condition at $r = b$ of the first kind (i.e., $k_2 = 0$) the following change should be made in Eq. 4-75b.

$$\left.\frac{K(\beta_m, r)}{k_2}\right|_{r=b} \quad \text{should be replaced by} \quad -\frac{1}{h_2}\left.\frac{dK(\beta_m, r)}{dr}\right|_{r=b}$$

When there is no heat generation and the boundary surface at $r = b$ is insulated, (i.e., $h_2 = 0$), then the solution includes term

$$\frac{3}{b^3}\int_{r=0}^{b} r^2\cdot F(r)\cdot dr$$

which is the average of the initial temperature distribution $F(r)$ over the volume of the sphere and represents the equilibrium temperature after the transients have passed.

We examine some special cases of solution 4-76.

[3]The roots of this transcendental equation are real if $bH_2 \geq -1$; for heat transfer problems this condition is satisfied because h_2 is always a positive quantity, or zero when the surface is insulated. Hence,

$$bH_2 = \frac{bh_2}{k_2} - 1 \geq -1.$$

(See Appendix I for the roots of Eq. 4-76b.)

EXAMPLE 4-3. Solid sphere of radius $r = b$ is initially at temperature $F(r)$. For times $t > 0$ heat is dissipated by convection from the boundary surface at $r = b$ into a surrounding at zero temperature.

For this case we have

$$g(r, t) = f_2(t) = 0$$

Substituting this in the solution 4-75 we obtain

$$T(r, t) = \frac{2}{r} \sum_{m=1}^{\infty} e^{-\alpha\beta_m^2 t} \cdot \frac{\beta_m^2 + H_2^2}{b(\beta_m^2 + H_2^2) + H_2} \cdot \sin \beta_m r \int_{r'=0}^{b} r' \cdot F(r') \cdot \sin \beta_m r' \cdot dr' \tag{4-77}$$

where β_m's are the positive roots of the transcendental equation

$$\beta \cot \beta b + H_2 = 0$$

where

$$H_2 \equiv \frac{h_2}{k_2} - \frac{1}{b}$$

EXAMPLE 4-4. A solid sphere of radius $r = b$ is initially at zero temperature. For times $t > 0$ there is heat generation within the sphere at a rate of $g(r, t)$ Btu/hr ft^3 while the boundary at $r = b$ is kept at zero temperature.

For this case the kernel is given as

$$K(\beta_m, r) = \sqrt{\frac{2}{b}} \sin \beta_m r$$

where

$$\beta_m = \frac{m\pi}{b} \qquad m = 1,2,3, \ldots$$

Substituting this kernel in solution 4-75 and taking $F(r) = f_2(t) = 0$, the solution of the present problem becomes

$$T(r, t) = \frac{2}{br} \frac{\alpha}{k} \sum_{m=1}^{\infty} e^{-\alpha\beta_m^2 t} \cdot \sin \beta_m r \int_{t'=0}^{t} e^{\alpha\beta_m^2 t'} \cdot dt' \cdot \int_{r'=0}^{b} r' \cdot g(r', t') \cdot \sin \beta_m r' \cdot dr' \tag{4-78}$$

where

$$\beta_m = \frac{m\pi}{b} \qquad m = 1,2,3, \ldots$$

We now examine some special cases of solution 4-78.

(a) Heat source is an instantaneous volume heat source of strength $g_i(r)$ Btu/ft^3 that releases its heat spontaneously at time $t = 0$.

Taking

$$g(r', t') = g_i(r') \cdot \delta(t' - 0)$$

and substituting this into Eq. 4-78 and performing integration with respect to the time variable t' we obtain

$$T(r,t) = \frac{2}{br}\frac{\alpha}{k}\sum_{m=1}^{\infty} e^{-\alpha\beta_m^2 t}\cdot \sin\beta_m r \cdot \int_{r'=0}^{b} r'\cdot g_i(r')\cdot \sin\beta_m r'\cdot dr' \tag{4-79}$$

(b) Heat source is an instantaneous point-heat source of strength g_{pi} Btu, situated at the center of the sphere and releases its heat spontaneously at time $t = 0$.

The instantaneous-point heat source is related to the volume-heat source by

$$g(r',t') = \frac{g_{pi}}{4\pi r'^2}\cdot\delta(t'-0)\cdot\delta(r'-0) \tag{4-80}$$

Substituting Eq. 4-80 into Eq. 4-78, and performing the integrations we obtain[4]

$$T(r,t) = \frac{1}{2rb^2}\frac{\alpha g_{pi}}{k}\sum_{m=1}^{\infty} e^{-\alpha\beta_m^2 t}\cdot m\cdot \sin\beta_m r \tag{4-81}$$

where

$$\beta_m = \frac{m\pi}{b}$$

(c) Heat source is an instantaneous spherical-surface heat source of radius r_1, of total strength $g_{\text{sph},i}$ Btu, situated concentrically inside the sphere, and releases its heat spontaneously at time $t = \tau$.

For this case the spherical instantaneous surface-heat source is related to the volume-heat source by

$$g(r',t') = \frac{g_{\text{sph},i}}{4\pi r'^2}\,\delta(r'-r_1)\cdot\delta(t'-\tau)$$

Substituting this into Eq. 4-78, we obtain

$$T(r,t) = \frac{1}{2\pi r r_1 b}\frac{\alpha g_{\text{sph},i}}{k}\sum_{m=1}^{\infty} e^{-\alpha\beta_m^2(t-\tau)}\cdot \sin\beta_m r\sin\beta_m r_1 \quad \text{for} \quad t>\tau \tag{4-82}$$

where

$$\beta_m = \frac{m\pi}{b}, \qquad m = 1, 2, 3, \ldots$$

Region $a \le r \le \infty$. We now examine the heat-conduction problem in a region extending from $r = a$ to $r = \infty$, that is an infinite region bounded internally with a sphere of radius $r = a$. Initially the region is at temperature $F(r)$, and for times $t > 0$ there is heat generation within the solid at a rate of $g(r,t)$ Btu/hr ft^3, while heat is dissipated by convection from the boundary at $r = a$ into a surrounding temperature of which varies with time. Figure 4-8 shows the region and the boundary condition under consideration.

[4]It is to be noted that when Eq. 4-80 is substituted into Eq. 4-78, the term $(\sin\beta_m r'/r')$ is finite as $r' \to 0$.

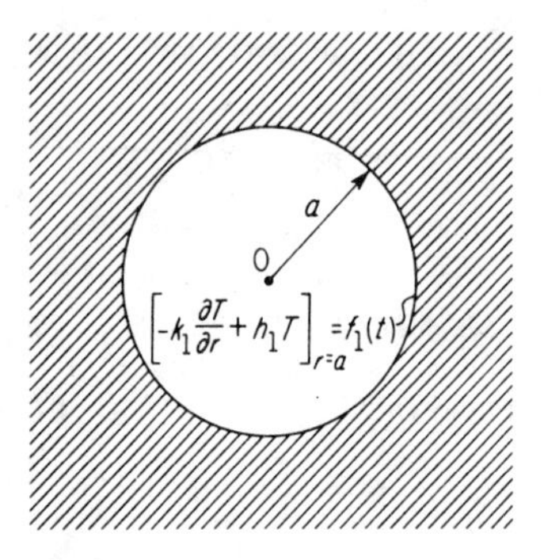

FIG. 4-8. Boundary condition for an infinite region bounded internally with a sphere of radius a.

The boundary-value problem of heat conduction is given as

$$\frac{1}{r}\frac{\partial^2}{\partial r^2}(rT) + \frac{g(r,t)}{k} = \frac{1}{\alpha}\frac{\partial T}{\partial t} \quad \text{in} \quad a \le r < \infty,\ t > 0$$

$$-k_1\frac{\partial T}{\partial r} + h_1T = f_1(t) \quad \text{at} \quad r = a,\ t > 0 \tag{4-83}$$

$$T = F(r) \quad \text{in} \quad a \le r \le \infty,\ t = 0$$

where $T \equiv T(r,t)$.

Defining new variables,

$$U = r\cdot T$$
$$x = r - a \tag{4-84}$$

The problem 4-83 is transformed into the following problem in the cartesian coordinate system:

$$\frac{\partial^2 U}{\partial x^2} + \frac{(a+x)g(x,t)}{k} = \frac{1}{\alpha}\frac{\partial U}{\partial t} \quad \text{in} \quad 0 \le x < \infty,\ t > 0$$

$$-k_1\frac{\partial U}{\partial x} + \left(h_1 + \frac{k_1}{a}\right)U = a\cdot f_1(t) \quad \text{at} \quad x = 0,\ t > 0 \tag{4-85}$$

$$U = (a+x)\cdot F(x+a) \quad \text{in} \quad 0 \le x < \infty,\ t = 0$$

where $U \equiv U(x,t)$.

System 4-85 is a heat-conduction problem for a semi-infinite region $0 \le x < \infty$ and the solution of which is obtained from the solution 2-128. After transforming the variables x to r and U to T according to 4-84, solution of the boundary-value problem of heat conduction becomes

$$T(r,t) = \frac{1}{r}\int_{\beta=0}^{\infty} e^{-\alpha\beta^2 t}\cdot K(\beta,r)\left[\int_{r'=a}^{\infty} r'F(r')\cdot K(\beta,r')\cdot dr' + \int_{t'=0}^{t} e^{\alpha\beta^2 t'}\cdot A(\beta,t')\cdot dt'\right]d\beta \qquad \text{(4-86a)}$$

where

$$A(\beta,t') = \frac{\alpha}{k}\int_{r'=a}^{\infty} r'\cdot g(r',t')\cdot K(\beta,r')\cdot dr' + \alpha\left.\frac{K(\beta,r')}{k_1}\right|_{r=a}\cdot a\cdot f_1(t') \qquad \text{(4-86b)}$$

and the kernel is

$$K = \sqrt{\frac{2}{\pi}}\,\frac{\beta\cos\beta(r-a) + H_1\sin\beta(r-a)}{\sqrt{\beta^2+H_1^2}}$$
$$H_1 = \frac{h_1}{k_1} + \frac{1}{a} \qquad \text{(4-87)}$$

When applying the solution 4-86 to the case with boundary condition of the first kind at $r = a$ (i.e., $k_1 = 0$) the following change should be made in Eq. 4-86b.

$$\text{Replace}\quad \left.\frac{K(\beta_m,r')}{k_1}\right|_{r=a} \quad\text{by}\quad \frac{1}{h_1}\left.\frac{dK(\beta_m,r)}{dr}\right|_{r=a}$$

We examine some special cases of solution 4-86.

EXAMPLE 4-5. An infinite region bounded internally with a sphere of radius $r = a$ is initially at zero temperature. For times $t > 0$ the boundary surface at $r = a$ is kept at temperature $f_1(t)$; but there is no heat generation within the solid. The boundary-value problem of heat conduction is given as

$$\frac{1}{r}\frac{\partial^2}{\partial r^2}(rT) = \frac{1}{\alpha}\frac{\partial T}{\partial t} \qquad a \le r < \infty,\ t > 0$$
$$T = f_1(t) \quad \text{at} \quad r = a,\ t > 0 \qquad \text{(4-88)}$$
$$T = 0 \qquad a \le r < \infty,\ t = 0$$

By comparing this problem with the problem 4-83 we note that

$$g(r,t) = F(r) = k_1 = 0$$
$$h_1 = 1 \qquad \text{(4-89a)}$$

The kernel $K(\beta, r)$ for the present problem is obtained from the generalized kernel 4-87 by setting $H_1 \to \infty$, and is given as

$$K(\beta,r) = \sqrt{\frac{2}{\pi}}\sin\beta(r-a) \qquad \text{(4-89b)}$$

The solution for the present problem is obtained by substituting Eqs. 4-89a and b into the solution 4-86; since $k_1 = 0$ we replace $\left.\frac{K(\beta, r)}{k_1}\right|_{r=a}$ by $\left.\frac{1}{h_1}\frac{dK(\beta, r)}{dr}\right|_{r=a}$ and obtain

$$T(r, t) = \frac{2\alpha a}{r\pi}\int_{t'=0}^{t} f_1(t')\cdot dt' \int_{\beta=0}^{\infty} e^{-\alpha\beta^2(t-t')}\cdot\beta\cdot\sin\beta(r-a)\cdot d\beta \tag{4-90}$$

The second integral on the right-hand side of Eq. 4-90 is evaluated by means of the relation [9]

$$\frac{2}{\pi}\int_{\beta=0}^{\infty} e^{-\alpha\beta^2(t-t')}\cdot\beta\cdot\sin\beta(r-a)\,d\beta = \frac{r-a}{\alpha(t-t')}\frac{1}{[4\pi\alpha(t-t')]^{1/2}}\,e^{-\frac{(r-a)^2}{4\alpha(t-t')}}$$

and we obtain

$$T(r, t) = \frac{a}{r}\int_{t'=0}^{t} f_1(t')\cdot\frac{r-a}{\sqrt{4\pi\alpha}(t-t')^{3/2}}\,e^{-\frac{(r-a)^2}{4\alpha(t-t')}}\,dt' \tag{4-91}$$

Equation 4-91 is simplified further by defining a new independent variable as

$$\xi = \frac{r-a}{\sqrt{4\alpha(t-t')}} \tag{4-92a}$$

and

$$d\xi = \frac{r-a}{2\sqrt{4\alpha}(t-t')^{3/2}}\,dt' \tag{4-92b}$$

Then solution 4-91 becomes

$$T(r, t) = \frac{2a}{r\sqrt{\pi}}\int_{\xi=\frac{r-a}{\sqrt{4\alpha t}}}^{\infty} e^{-\xi^2}\cdot f_1\left[t-\frac{(r-a)^2}{4\alpha\xi^2}\right]\cdot d\xi \tag{4-93}$$

For a constant surface temperature at $r = a$, i.e.,

$$f_1 = T_0 = \text{const.} \tag{4-94}$$

Equation 4-93 becomes

$$T(r, t) = \frac{2}{\sqrt{\pi}}\frac{a}{r}T_0\int_{\xi=\frac{r-a}{\sqrt{4\alpha t}}}^{\infty} e^{-\xi^2}\,d\xi \tag{4-95}$$

which is written as

$$\begin{aligned}\frac{T(r, t)}{T_0} &= \frac{a}{r}\operatorname{erfc}\left(\frac{r-a}{\sqrt{4\alpha t}}\right)\\ &= \frac{a}{r}\left[1-\operatorname{erf}\left(\frac{r-a}{\sqrt{4\alpha t}}\right)\right]\end{aligned} \tag{4-96}$$

4-6. HOMOGENEOUS BOUNDARY-VALUE PROBLEMS OF HEAT CONDUCTION INVOLVING MORE THAN ONE SPACE VARIABLE (SOLUTION WITH SEPARATION OF VARIABLES)

In this section we examine the solution of the homogeneous boundary-value problems of heat conduction involving more than one space variable with the method of the separation of variables.

Solid Sphere ($0 \leq r \leq b$, $-1 \leq \mu \leq 1$, $0 \leq \phi \leq 2\pi$). A solid sphere of radius $r = b$ is initially at temperature $F(r, \mu, \phi)$. For times $t > 0$ boundary surface at $r = b$ is kept at zero temperature.

The boundary-value problem of heat conduction is given as

$$\frac{\partial^2 T}{\partial r^2} + \frac{2}{r}\frac{\partial T}{\partial r} + \frac{1}{r^2}\frac{\partial}{\partial \mu}\left[(1-\mu^2)\frac{\partial T}{\partial \mu}\right] + \frac{1}{r^2(1-\mu^2)}\frac{\partial^2 T}{\partial \phi^2} = \frac{1}{\alpha}\frac{\partial T}{\partial t}$$

$$\begin{aligned} & \text{in the sphere,} \quad \text{for } t > 0 \\ T = 0 \quad & \text{at } r = b, \quad \text{for } t > 0 \\ T = F(r, \mu, \phi) \quad & \text{in the sphere,} \quad \text{for } t = 0 \end{aligned} \tag{4-97}$$

where $T \equiv T(r, \mu, \phi, t)$.

We assume the separation of variables in the form

$$T(r, \mu, \phi, t) = \Gamma(t) \cdot R(r) \cdot \Phi(\phi) \cdot M(\mu) \tag{4-98}$$

The resulting separated equations are given by Eq. 4-6; their particular solutions that will satisfy the problem are in the form

$$\Gamma: e^{-\alpha\lambda^2 t} \quad R: (\lambda r)^{-1/2} J_{n+1/2}(\lambda r) \quad \Phi: \begin{matrix}\sin m\phi \\ \cos m\phi\end{matrix} \quad M: P_n^m(\mu) \tag{4-99}$$

The $Q_n^m(\mu)$ and $(\lambda r)^{-1/2} Y_{n+1/2}(\lambda r)$ functions are excluded from the solutions because these functions have poles respectively at $\mu = \pm 1$ and $r = 0$.

Solutions 4-99 include the separation parameters λ, m, and n which are determined with the following considerations.

Since temperature is cyclic with a period of 2π, the value of Φ when increased by 2π should remain unaltered; hence, m's are taken as positive integers.

We choose n as an integer because it gives the degree of the Legendre polynomial. The parameter λ is determined from the boundary condition at $r = b$, hence λ's are the positive roots of the transcendental equation

$$J_{n+1/2}(\lambda b) = 0 \tag{4-100}$$

Hence the general solution of the problem is in the form[5]

$$T(r, \mu, \phi, t) = \sum_{n=0}^{\infty} \sum_{p=1}^{\infty} \sum_{m=0}^{n} e^{-\alpha\lambda_p^2 t} \cdot (\lambda_p r)^{-1/2} J_{n+1/2}(\lambda_p r)$$

$$\cdot P_n^m(\mu)[A_{mnp} \cos m\phi + B_{mnp} \sin m\phi] \qquad (4\text{-}101)$$

where λ_p's are the positive roots of Eq. 4-100.

Equation 4-101 should satisfy the initial condition for the problem, hence, for $t = 0$ it becomes

$$F(r, \mu, \phi) = \sum_{n=0}^{\infty} \sum_{p=1}^{\infty} \sum_{m=0}^{n} (\lambda_p r)^{-1/2} \cdot J_{n+1/2}(\lambda_p r)$$

$$\cdot p_n^m(\mu)[A_{mnp} \cos m\phi + B_{mnp} \sin m\phi] \qquad (4\text{-}102)$$

The unknown coefficients A_{mnp} and B_{mnp} are determined by making use of the orthogonality of the eigenfunctions. B_{mnp} is determined by operating both sides of Eq. 4-102 successively by

$$\int_0^{2\pi} \sin m\phi \cdot d\phi \qquad \int_{-1}^{1} P_n^m(\mu) d\mu \qquad \int_0^b r^{3/2} J_{n+1/2}(\lambda_p r) \cdot dr$$

and we obtain

$$B_{mnp} = \frac{1}{N_\phi \cdot N_\mu \cdot N_r} \int_{\phi'=0}^{2\pi} \int_{\mu'=-1}^{1} \int_{r'=0}^{b} \sin m\phi'$$

$$\cdot P_n^m(\mu') \cdot r'^{3/2} J_{n+1/2}(\lambda r') \cdot F(r', \phi', \mu') \cdot d\phi' \cdot dr' \cdot d\mu' \qquad (4\text{-}103)$$

where the norms N_ϕ, N_μ, and N_r are

$$N_\phi = \int_0^{2\pi} \sin^2 m\phi d\phi = \pi$$

$$N_\mu = \int_{\mu=-1}^{1} [P_n^m(\mu)]^2 d\mu = \frac{2}{2n+1} \frac{(n+m)!}{(n-m)!}$$

$$N_r = \int_0^b \lambda_p^{-1/2} \cdot r J_{n+1/2}^2(\lambda_p r) dr = \lambda_p^{-1/2}$$

$$\frac{b^2}{2} \left\{ J_{n+1/2}'^2(\lambda_p b) + \left[1 - \frac{(n+\frac{1}{2})^2}{\lambda_p^2 b^2}\right] J_{n+1/2}^2(\lambda_p b) \right\}$$

$$= \lambda_p^{-1/2} \frac{b^2}{2} J_{n+1/2}'^2(\lambda_p b)$$

A_{mnp} is obtained by operating both sides of Eq. 4-102 successively by

$$\int_0^{2\pi} \cos m\phi d\phi, \qquad \int_{-1}^{1} P_n^m(\mu)\, \mathrm{d}\mu \qquad \int_0^{b} r^{3/2} J_{n+1/2}(\lambda_p r) \cdot dr$$

[5]The summation over m is taken from $m = 0$ to n because of the requirement that $n \geq m$.

and we obtain

$$A_{mnp} = \frac{1}{N_\phi \cdot N_\mu \cdot N_r} \int_{\phi'=0}^{2\pi} \int_{\mu'=-1}^{1} \int_{r'=0}^{b} \cos m\phi'$$
$$\cdot P_n^m(\mu') \cdot r'^{3/2} J_{n+1/2}(\lambda r') \cdot F(r', \mu', \phi') \cdot d\phi' \cdot d\mu' \cdot dr' \qquad (4\text{-}104)$$

where the norms are

$$N_\phi = \int_0^{2\pi} \cos^2 m\phi d\phi = \begin{cases} \pi & \text{for } m = 1,2,3,\ldots \\ 2\pi & \text{for } m = 0 \end{cases}$$

and N_μ and N_r are exactly the same as those given above. Substituting A_{mnp} and B_{mnp} into the general solution 4-101 we obtain

$$T(r, \mu, \phi, t) = \frac{1}{\pi b^2} \sum_{n=0}^{\infty} \sum_{p=1}^{\infty} \sum_{m=0}^{n} e^{-\alpha \lambda_p^2 t}$$
$$\cdot (2n + 1) \frac{(n - m)!}{(n + m)!} \frac{r^{-1/2} \cdot J_{n+1/2}(\lambda_p r) \cdot P_n^m(\mu)}{J'^2_{n+1/2}(\lambda_p b)}$$

$$\cdot \int_{\phi'=0}^{2\pi} \int_{\mu'=1}^{1} \int_{r'=0}^{b} r'^{3/2} \cdot J_{n+1/2}(\lambda r') \cdot P_n^m(\mu')$$
$$\cdot \cos m(\phi - \phi') \cdot F(r', \mu', \phi') \cdot dr' \cdot d\phi' \cdot d\mu' \qquad (4\text{-}105)$$

where π should be replaced by 2π when $m = 0$,

λ_p's = positive roots of the transcendental Eq. 4-100

$n = 0,1,2,3,\ldots$

$m = 0,1,2,3,\ldots$

In Eq. 4-105 the $\cos m(\phi - \phi')$ term is obtained from the relation

$$\cos m(\phi - \phi') = \cos m\phi \cdot \cos m\phi' + \sin m\phi \cdot \sin m\phi'$$

4-7. NONHOMOGENEOUS BOUNDARY-VALUE PROBLEMS OF HEAT CONDUCTION INVOLVING MORE THAN ONE SPACE VARIABLE (SOLUTION WITH INTEGRAL TRANSFORM)

When the differential equation of heat conduction includes a differential operator in the form

$$D \equiv \frac{1}{r^2} \frac{\partial}{\partial r} \left(r^2 \frac{\partial T}{\partial r} \right) + \frac{1}{r^2} \frac{\partial}{\partial \mu} \left[(1 - \mu^2) \frac{\partial T}{\partial \mu} \right] \text{ in } \begin{matrix} 0 \le r \le b \\ -1 \le \mu \le 1 \end{matrix} \qquad (4\text{-}106)$$

it can be removed from the differential equation by applying an integral transform [10] called *Bessel-Legendre* transform. We now examine the construction of Bessel-Legendre transform and the corresponding inversion formula for the case of a solid sphere of radius $r = b$ subject to a boundary condition of the first kind at the surface $r = b$.

We consider the following *auxiliary* problem

$$\frac{1}{r^2}\frac{\partial}{\partial r}\left(r^2\frac{\partial Z}{\partial r}\right) + \frac{1}{r^2}\frac{\partial}{\partial \mu}\left[(1-\mu^2)\frac{\partial Z}{\partial \mu}\right] + \lambda^2 Z = 0 \quad \text{in} \quad \begin{array}{c} 0 \le r \le b, \\ -1 \le \mu \le 1 \end{array}$$

$$Z = 0 \quad \text{at} \quad r = b \tag{4-107}$$

where $Z \equiv Z(r, \mu)$.

Assuming the separation of the $Z(r, \mu)$ variable in the form

$$Z(r, \mu) = R(r) \cdot M(\mu)$$

Equation 4-107 becomes

$$\frac{1}{R}\frac{d}{dr}\left[r^2\frac{dR}{dr}\right] + \frac{1}{M}\frac{d}{d\mu}\left[(1-\mu^2)\frac{dM}{d\mu}\right] + \lambda^2 r^2 = 0 \tag{4-108}$$

This equation is satisfied if each group equals to a constant; then the resulting differential equations for the separated variables are

$$\frac{d}{d\mu}\left[(1-\mu^2)\frac{dM}{d\mu}\right] + n(n+1)M = 0 \qquad -1 \le \mu \le 1 \tag{4-109a}$$

$$\frac{d}{dr}\left(r^2\frac{dR}{dr}\right) + [\lambda^2 r^2 - n(n+1)]R = 0 \qquad 0 \le r \le b \tag{4-109b}$$

subject to

$$R\,|_{r=b} = 0 \text{ and } n = 0,1,2,3,\ldots \tag{4-109c}$$

Particular solutions of the differential Eq. 4-109 are

$$R : r^{-1/2} J_{n+1/2}(\lambda_{pn} r), \qquad M : P_n(\mu) \tag{4-110}$$

and λ_p's are the positive roots of

$$J_{n+1/2}(\lambda_n b) = 0 \tag{4-111}$$

since the solution R should satisfy the boundary condition 4-109c.

The eigenfunctions of the auxiliary Prob. 4-107 are chosen as

$$Z(r, \mu) = r^{-1/2} \cdot J_{n+1/2}(\lambda_{pn} r) \cdot P_n(\mu) \tag{4-112}$$

We assume that a function $F(r, \mu)$ may be represented in the interval $0 \le r \le b$, $-1 \le \mu \le 1$ in terms of the eigenfunctions 4-112 in the form

$$F(r, \mu) = \sum_{n=0}^{\infty}\sum_{p=1}^{\infty} c_{np}[r^{-1/2} J_{n+1/2}(\lambda_{pn} r) \cdot P_n(\mu)] \tag{4-113}$$

where $n = 0,1,2,3,\ldots$

λ_p's = positive roots of the transcendental Eq. 4-111.

The unknown coefficients c_{np} are determined by making use of the orthogonality of the eigenfunctions over the interval $0 \le r \le b$ and $-1 \le \mu \le 1$. We operate both sides of Eq. 4-113 successively by

$$\int_{r=0}^{b} r^{3/2} \cdot J_{n+1/2}(\lambda_{pn} r)\, dr \quad \text{and} \quad \int_{\mu=-1}^{1} P_n(\mu) \cdot d\mu$$

and we immediately obtain the unknown coefficients c_{np} as

$$c_{np} = \frac{1}{N_r N_\mu} \int_{r'=0}^{b} \int_{\mu'=-1}^{1} r'^{3/2} \cdot J_{n+1/2}(\lambda_{pn} r') \cdot P_n(\mu') \cdot F(r', \mu') \cdot dr' d\mu' \tag{4-114}$$

where the norms are[6]

$$N_r = \int_{r=0}^{b} r \cdot J_{n+1/2}^2(\lambda_{pn} r) \cdot dr = -\frac{b^2}{2} J_{n-1/2}(\lambda_{pn} b) \cdot J_{n+3/2}(\lambda_{pn} b) \tag{4-115a}$$

$$N_\mu = \int_{\mu=-1}^{1} P_n^2(\mu) \cdot d\mu = \frac{2}{2n+1} \tag{4-115b}$$

Substituting Eqs. 4-114 and 4-115 into the expansion 4-113, we obtain

$$F(r, \mu) = \sum_{n=0}^{\infty} \sum_{p=1}^{\infty} \frac{Z(r, \mu)}{\sqrt{N_r N_\mu}} \int_{r'=0}^{b} \int_{\mu'=-1}^{1} r'^2 \cdot \frac{Z(r', \mu')}{\sqrt{N_r N_\mu}} F(r', \mu') \cdot dr' d\mu' \tag{4-116}$$

where

$$Z(r, \mu) = r^{-1/2} J_{n+1/2}(\lambda_{pn} r) \cdot P_n(\mu)$$

Expansion Eq. 4-116 is written as the inversion formula for the integral transform (*Bessel-Legendre transform*) of function $F(r, \mu)$ in the interval $0 \le r \le b$, $-1 \le \mu \le 1$ as

$$\begin{pmatrix}\text{Inversion}\\ \text{formula}\end{pmatrix} F(r, \mu) = \sum_{n=0}^{\infty} \sum_{p=1}^{\infty} K_{np}(r, \mu) \cdot \bar{F}(n, p) \tag{4-117a}$$

$$\begin{pmatrix}\text{Integral}\\ \text{transform}\end{pmatrix} \bar{F}(n, p) = \int_{r'=0}^{b} \int_{\mu'=-1}^{1} r'^2 K_{np}(r', \mu') \cdot F(r', \mu') \cdot dr' d\mu' \tag{4-117b}$$

where the kernel $K_{np}(r, \mu)$ for the range of space variables $0 \le r \le b$, $-1 \le \mu \le 1$ and subject to the boundary condition of the first kind at

[6] McLachlain, Ref. 11.

$$\int_0^b r \cdot J_\nu^2(\lambda r) dr = \left[\frac{r^2}{2} \{ J_\nu^2(\lambda r) - J_{\nu-1}(\lambda r) \cdot J_{\nu+1}(\lambda r) \} \right]_0^b$$

for $\nu = n + 1/2$ and $J_{n+1/2}(\lambda r) = 0$, this relation becomes

$$\int_0^b r \cdot J_{n+1/2}^2(\lambda r) dr = -\frac{b^2}{2} J_{n-1/2}(\lambda b) \cdot J_{n+3/2}(\lambda b)$$

which is the same as Eq. 4-115a.

$r = b$ is given as

$$K_{np}(r, \mu) = \frac{r^{-1/2} J_{n+1/2}(\lambda_p r) \cdot P_n(\mu)}{\sqrt{N_r N_\mu}} \tag{4-118}$$

$$N_r N_\mu = -\frac{b^2}{2n+1} J_{n-1/2}(\lambda_p b) \cdot J_{n+3/2}(\lambda_p b)$$

The λ_p's are the positive roots of the transcendental equation

$$J_{n+1/2}(\lambda b) = 0$$

We now illustrate the application of Bessel-Legendre transform in the solution of heat-conduction problem in the spherical coordinate system.

Solid Sphere $0 \le r \le b$, $-1 \le \mu \le 1$. A solid sphere of radius $r = b$ is initially at temperature $F(r, \mu)$. For times $t > 0$ heat is generated within the solid at a rate of $g(r, \mu, t)$ Btu/hr ft^3 while the boundary surface is kept at temperature $f(\mu, t)$.

The boundary-value problem of heat conduction is given as

$$\frac{1}{r^2} \frac{\partial}{\partial r}\left(r^2 \frac{\partial T}{\partial r}\right) + \frac{1}{r^2} \frac{\partial}{\partial \mu}\left[(1-\mu^2)\frac{\partial T}{\partial \mu}\right] + \frac{g(r, \mu, t)}{k} = \frac{1}{\alpha}\frac{\partial T}{\partial t} \tag{4-119a}$$

$$\text{in} \quad 0 \le r \le b, \ -1 \le \mu \le 1, \text{ for } t > 0$$

$$T = f(\mu, t) \quad \text{at} \quad r = b, \qquad \text{for } t > 0 \tag{4-119b}$$

$$T = F(r, \mu) \quad \text{in} \quad 0 \le r \le b, \ -1 \le \mu \le 1, \text{ for } t = 0 \tag{4-119c}$$

where $T \equiv T(r, \mu, t)$.

To solve this problem we shall remove from the differential equation of heat conduction the partial derivatives with respect to the space variables r and μ by applying spherical Bessel-Legendre transform.

We define Bessel-Legendre transform and inversion formula of the temperature function $T(r, \mu, t)$ as

$$\begin{pmatrix}\text{Integral}\\ \text{transform}\end{pmatrix} \bar{T}(\lambda_{pn}, n, t) = \int_{r'=0}^{b} \int_{\mu'=-1}^{1} T(r', \mu', t) \cdot r'^2 \cdot K_{np}(r', \mu') \cdot dr' \cdot d\mu' \tag{4-120a}$$

$$\begin{pmatrix}\text{Inversion}\\ \text{formula}\end{pmatrix} T(r, \mu, t) = \sum_{n=0}^{\infty} \sum_{p=1}^{\infty} K_{np}(r, \mu) \cdot \bar{T}(\lambda_{pn}, n, t) \tag{4-120b}$$

The kernel $K_{np}(r, \mu)$ is defined as

$$K_{np}(r, \mu) = \frac{r^{-1/2} \cdot J_{n+1/2}(\lambda_{pn} r) \cdot P_n(\mu)}{\sqrt{N_r \cdot N_\mu}} \tag{4-121a}$$

$$N_r N_\mu = -\frac{b^2}{2n+1} \cdot J_{n-1/2}(\lambda_{pn} b) \cdot J_{n+3/2}(\lambda_{pn} b) \tag{4-121b}$$

The λ_{pn}'s are positive roots of

$$J_{n+1/2}(\lambda_n b) = 0. \tag{4-121c}$$

Where $K_{np}(r, \mu)$ is the normalized eigenfunctions of the following eigenvalue problem:

$$\frac{1}{r^2}\frac{\partial}{\partial r}\left[r^2\frac{\partial K}{\partial r}\right] + \frac{1}{r^2}\frac{\partial}{\partial \mu}\left[(1-\mu^2)\frac{\partial K}{\partial \mu}\right] + \lambda^2 K = 0 \quad \text{in} \quad 0 \le r \le b,$$

$$-1 \le \mu \le 1$$

$$K = 0 \quad \text{at} \quad r = b \qquad (4\text{-}122)$$

where $K \equiv K_{np}(r, \mu)$.

We take the integral transform of system 4-119 by applying the Bessel-Legendre transform Eq. 4-120a:

$$\int_{r=0}^{b}\int_{\mu=-1}^{1} r^2 K_{np}(r,\mu)\cdot\nabla^2 T(r,\mu,t)\cdot dr\cdot d\mu$$

$$+\frac{\bar{g}(\lambda_{pn}, n, t)}{k} = \frac{1}{\alpha}\frac{d\bar{T}(\lambda_{pn}, n, t)}{dt} \qquad (4\text{-}123a)$$

$$\bar{T}(\lambda_{pn}, n, t) = \bar{F}(\lambda_{pn}, n) \quad \text{for} \quad t = 0 \qquad (4\text{-}123b)$$

where

$$\nabla^2 \equiv \frac{1}{r^2}\frac{\partial}{\partial r}\left(r^2\frac{\partial}{\partial r}\right) + \frac{1}{r^2}\frac{\partial}{\partial \mu}\left[(1-\mu^2)\frac{\partial}{\partial \mu}\right]$$

and quantities with bars refer to integral transform according to transform Eq. 4-120a.

The integral transform of the Laplacian in Eq. 4-123a can be evaluated using the general relation 1-64 given in Chapter 1, (or it can be evaluated by performing the integration in Eq. 4-123a by parts twice and by making use of the auxiliary Eq. 4-122) and the result is given as

$$\int_{r=0}^{b}\int_{\mu=-1}^{1} r^2 K_{np}(r,\mu)\cdot\nabla^2 T(r,\mu,t)\cdot dr\cdot d\mu = -\lambda_{pn}^2\cdot\bar{T}(\lambda_{pn}, n, t)$$

$$-\int_{\mu'=-1}^{1} b^2\cdot f(\mu', t)\cdot\left[\frac{dK_{np}(r,\mu')}{dr}\right]_{r=b}\cdot d\mu' \qquad (4\text{-}124)$$

It is to be noted that in obtaining the result 4-124 from Eq. 1-64 we replaced

$$\left.\frac{K}{k_i}\right|_{r=b} \quad \text{by} \quad -\left.\frac{1}{h_i}\frac{dK}{dr}\right|_{r=b}$$

and chose $h_i = 1$.

Substituting Eq. 4-124 into Eq. 4-123 we obtain the following ordinary differential equation for the transform of temperature

$$\frac{d\bar{T}(\lambda_{pn}, n, t)}{dt} + \alpha\lambda_{pn}^2\bar{T}(\lambda_{pn}, n, t) = A(\lambda_{pn}, n, t) \qquad (4\text{-}125a)$$

$$\bar{T}(\lambda_{pn}, n, t)\,|_{t=0} = \bar{F}(\lambda_{pn}, n) \qquad (4\text{-}125b)$$

where

$$A(\lambda_{pn}, n, t) \equiv \frac{\alpha}{k}\,\bar{g}(\lambda_{pn}, n, t) - \int_{\mu'=-1}^{1} b^2 \cdot f(\mu', t) \cdot \left[\frac{dK_{np}(r, \mu')}{dr}\right]_{r=b} \cdot d\mu'$$

Solution of this ordinary differential equation subject to the prescribed initial condition is straightforward, and gives the Bessel-Legendre transform of temperature. When the transform of temperature is substituted into the inversion formula 4-120b we obtain the solution of the heat-conduction problem 4-119 as

$$T(r, \mu, t) = \sum_{n=0}^{\infty} \sum_{p=1}^{\infty} K_{np}(r, \mu) \cdot e^{-\alpha\lambda_{pn}^2 t} \cdot \left[\bar{F}(\lambda_{pn}, n) + \int_{t'=0}^{t} e^{\alpha\lambda_{pn}^2 t'} \cdot A(\lambda_{pn}, n, t') \cdot dt'\right] \tag{4-126a}$$

where

$$A(\lambda_{pn}, n, t') = \frac{\alpha}{k}\,\bar{g}(\lambda_{pn}, n, t') - b^2 \int_{\mu'=-1}^{1} f(\mu', t') \cdot \left[\frac{dK_{np}(r, \mu')}{dr}\right]_{r=b} \cdot d\mu' \tag{4-126b}$$

$$\bar{F}(\lambda_{pn}, n) = \int_{r'=0}^{b} \int_{\mu'=-1}^{1} F(r', \mu') \cdot r'^2 \cdot K_{np}(r', \mu') \cdot dr' \cdot d\mu' \tag{4-126c}$$

$$\bar{g}(\lambda_{pn}, n, t') = \int_{r'=0}^{b} \int_{\mu'=-1}^{1} g(r', \mu', t') \cdot r'^2 \cdot K_{np}(r', \mu') \cdot dr' \cdot d\mu' \tag{4-126d}$$

$$K_{np}(r, \mu) = \frac{r^{-1/2} J_{n+1/2}(\lambda_{pn} r) \cdot P_n(\mu)}{\sqrt{N_r N_\mu}} \tag{4-126e}$$

$$N_r N_\mu = -\frac{b^2}{2n+1} J_{n-1/2}(\lambda_{pn} b) \cdot J_{n+3/2}(\lambda_{pn} b) \tag{4-126f}$$

The λ_{pn}'s are the positive roots of the transcendental equation

$$J_{n+1/2}(\lambda_n b) = 0 \tag{4-126g}$$

We consider now one special case of solution 4-126.

(a) Sphere is initially at zero temperature, heat generation within the solid takes place at a constant and uniform rate, and the boundary-condition function is independent of time but proportional to μ.

For this special case the boundary condition, the initial condition and the heat-generation functions are given as

$$g(r, \mu, t) = g_0 = \text{const.} \tag{4-127a}$$

$$f(\mu, t) = c_0 \cdot \mu \tag{4-127b}$$

$$F(r, \mu) = 0 \tag{4-127c}$$

where c_0 = const. Substituting the functions defined in Eq. 4-127 into the solution 4-126 we obtain

$$T(r, \mu, t) = -\frac{1}{b^2} \sum_{n=0}^{\infty} \sum_{p=1}^{\infty} \frac{r^{-1/2} J_{n+1/2}(\lambda_{pn} r) \cdot P_n(\mu)}{J_{n-1/2}(\lambda_{pn} b) \cdot J_{n+3/2}(\lambda_{pn} b)} (2n + 1)$$

$$\cdot \left\{ \frac{\alpha}{k} g_0 \int_{r'=0}^{b} \int_{\mu'=-1}^{1} r'^{3/2} \cdot J_{n+1/2}(\lambda_{pn} r') \cdot P_n(\mu') \cdot dr' \cdot d\mu' - b^2 \cdot c_0 \right.$$

$$\left. \cdot \left[\frac{d}{dr} (r^{-1/2} J_{n+1/2}(\lambda_{pn} r) \right]_{r=b} \cdot \int_{\mu'=-1}^{1} \mu' \cdot P_n(\mu') \cdot d\mu' \right\}$$

$$\cdot e^{-\alpha \lambda_{pn}^2 t} \cdot \int_{t'=0}^{t} e^{\alpha \lambda_{pn}^2 t'} dt' \qquad (4\text{-}128)$$

where the λ_p's are the positive roots of the transcendental equation

$$J_{n+1/2}(\lambda_n b) = 0$$

In Eq. 4-128 various integrals and differentiations are evaluated as follows.

$$I_1 \equiv \int_{r'=0}^{b} r'^{3/2} J_{n+1/2}(\lambda_{pn} r') \cdot dr' \int_{\mu'=-1}^{1} P_n(\mu') \cdot d\mu'$$

$$= \begin{cases} 0 & \text{for n} = 1, 2, 3, \ldots \\ 2 \int_{r'=0}^{b} r'^{3/2} \cdot J_{1/2}(\lambda_{pn} r') \cdot dr' = 2 \dfrac{b^{3/2}}{\lambda_{p0}} J_{3/2}(\lambda_{p0} b) & \text{for } n = 0 \end{cases}$$

$$(4\text{-}129)$$

$$I_2 \equiv \int_{\mu'=-1}^{1} \mu' P_n(\mu') \cdot d\mu' = \begin{cases} 0 & \text{for } n = 0, 2, 3, 4, \ldots \\ 2/3 & \text{for } n = 1 \end{cases} \qquad (4\text{-}130)$$

$$I_3 \equiv e^{-\alpha \lambda^2 t} \int_{t'=0}^{t} e^{\alpha \lambda^2 t'} dt' = \frac{1 - e^{-\alpha \lambda^2 t}}{\alpha \lambda^2} \qquad (4\text{-}131)$$

$$I_4 \equiv \frac{d}{dr} [r^{-1/2} J_{n+1/2}(\lambda r)] \,|_{r=b} = \lambda b^{-1/2} \cdot J_{n-1/2}(\lambda b) \qquad (4\text{-}132)$$

Substituting Eq. 4-129 through Eq. 4-132 into Eq. 4-128, the solution becomes

$$T(r, \mu, t) = \sum_{p=1}^{\infty} \left[-\frac{2 g_0}{b^{1/2} k \lambda_{p0}^3} \cdot \frac{r^{-1/2} J_{1/2}(\lambda_{p0} r)}{J_{-1/2}(\lambda_{p0} b)} (1 - e^{-\alpha \lambda_{p0}^2 t}) \right.$$

$$\left. + \frac{2 c_0}{b^{1/2} \alpha \lambda_{p1}} \frac{r^{-(1/2)} J_{3/2}(\lambda_{p1} r) \cdot \mu}{J_{5/2}(\lambda_{p1} b)} (1 - e^{-\alpha \lambda_{p1}^2 t}) \right] \qquad (4\text{-}133)$$

where λ_{p0}'s = positive roots of $J_{1/2}(\lambda_0 b) = 0$
λ_{p1}'s = positive roots of $J_{3/2}(\lambda_1 b) = 0$, and summation is taken over all eigenvalues

4-8. STEADY-STATE PROBLEMS

The integral-transform technique can be applied to the solution of steady-state heat-conduction problems with or without the heat generation. The partial derivatives with respect to the space variables are removed and the problem is reduced to an ordinary differential equation for one of the space variables, which is then solved by one of the standard techniques.

Sometimes it is desirable to simplify the problem with heat generation (i.e., Poisson's equation) to the problem with no heat generation (i.e., Laplace's equation) by a suitable transformation of the dependent variable. For constant heat generation, Poisson's equation for the temperature function $T(r, \mu, \phi)$

$$\nabla^2 T(r, \mu, \phi) + \frac{g_0}{k} = 0 \tag{4-134}$$

can be transformed into the Laplace equation for the new variable $\theta(r, \mu, \phi)$

$$\nabla^2 \theta(r, \mu, \phi) = 0 \tag{4-135}$$

by defining the new variable $\theta(r, \phi, \mu)$ as

$$T(r, \mu, \phi) = \theta(r, \mu, \phi) - \frac{g_0}{k}\frac{r^2}{6} \tag{4-136}$$

and the transformation Eq. 4-136 is valid whenever temperature is a function of the space variable r.

In this section we examine the solution of steady-state problems with no heat generation within the solid with the integral-transform technique.

Solid Sphere ($0 \leq r \leq b$, $-1 \leq \mu \leq 1$, $0 \leq \phi \leq 2\pi$). We determine the steady-state temperature distribution in a solid sphere of radius $r = b$ with boundary surface at $r = b$ kept at temperature $f(\mu, \phi)$.

The boundary-value problem of heat conduction is

$$\frac{1}{r^2}\frac{\partial}{\partial r}\left(r^2\frac{\partial T}{\partial r}\right) + \frac{1}{r^2}\frac{\partial}{\partial \mu}\left[(1-\mu^2)\frac{\partial T}{\partial \mu}\right] + \frac{1}{r^2(1-\mu^2)}\frac{\partial^2 T}{\partial \phi^2} = 0$$

$$\text{in the sphere } 0 \leq r \leq b,\ -1 \leq \mu \leq 1,\ 0 \leq \phi \leq 2\pi \tag{4-137a}$$

$$T = f(\mu, \phi) \quad \text{at} \quad r = b \tag{4-137b}$$

where

$$T \equiv T(r, \mu, \phi)$$

We solve this problem by the application of integral transforms. The partial derivative with respect to the ϕ variable can be removed from the differential equation with Fourier transform. Since the range of independent variable ϕ is from 0 to 2π and the temperature function is cyclic with a period of 2π, we define *Fourier transform* and the corresponding inversion formula with respect to the ϕ-variable of the temperature function $T(r, \mu, \phi)$ as (See Eq. 3-186 or Eq. 4-50.)

$$\begin{pmatrix}\text{Integral}\\ \text{transform}\end{pmatrix} \quad \bar{T}(r, \mu, m) = \int_{\phi'=0}^{2\pi} \cos m(\phi - \phi') \cdot T(r, \mu, \phi') \cdot d\phi' \tag{4-138a}$$

$$\begin{pmatrix}\text{Inversion}\\ \text{formula}\end{pmatrix} \quad T(r, \mu, \phi) = \frac{1}{\pi} \sum_{m=0}^{\infty} \bar{T}(r, \mu, m) \tag{4-138b}$$

where $m = 0, 1, 2, 3, \ldots$, and π should be replaced by 2π when $m = 0$

We take the integral transform of system 4-137 by applying the transform Eq. 4-138a and remove the partial derivative with respect to the ϕ-variable,

$$\frac{\partial}{\partial r}\left(r^2 \frac{\partial \bar{T}}{\partial r}\right) + \frac{\partial}{\partial \mu}\left[(1 - \mu^2)\frac{\partial \bar{T}}{\partial \mu}\right] - \frac{m^2}{1 - \mu^2}\bar{T} = 0 \quad \text{in} \quad 0 \le r \le b, \; -1 \le \mu \le 1$$

$$\bar{T} = \bar{f}(\mu, m) \quad \text{at} \quad r = b \tag{4-139}$$

where $\bar{T} \equiv \bar{T}(r, \mu, m)$ and quantities with bars refer to integral transform according to transform Eq. 4-138a.

In system 4-139 the partial derivative with respect to the space variable μ is removed by applying a *Legendre transform*. We define Legendre transform and the corresponding inversion formula for the temperature function $\bar{T}(r, \mu, m)$ with respect to the μ variable in the range $-1 \le \mu \le 1$ as (See Eq. 4-52.)

$$\begin{pmatrix}\text{Integral}\\ \text{transform}\end{pmatrix} \quad \bar{\bar{T}}(r, n, m) = \int_{\mu=-1}^{1} K_n^m(\mu') \cdot \bar{T}(r, \mu', m) d\mu' \tag{4-140a}$$

$$\begin{pmatrix}\text{Inversion}\\ \text{formula}\end{pmatrix} \quad \bar{T}(r, \mu, m) = \sum_{n}^{\infty} K_n^m(\mu) \cdot \bar{\bar{T}}(r, n, m) \tag{4-140b}$$

where the kernel is

$$K_n^m(\mu) = \sqrt{\frac{2n+1}{2} \cdot \frac{(n-m)!}{(n+m)!}} \, P_n^m(\mu)$$

$$n = 0, 1, 2, 3, \ldots \qquad m \le n$$

We take the integral transform of system 4-139 by applying the transform Eq. 4-140a. (See Eq. 4-54 for the integral transform of the differential operator for the μ variable.

$$\frac{d}{dr}\left(r^2\frac{d\bar{\bar{T}}}{dr}\right) - n(n+1)\bar{\bar{T}} = 0 \quad \text{in} \quad 0 \le r \le b \tag{4-141a}$$

$$\bar{\bar{T}} = \bar{\bar{f}}(n,m) \quad \text{at} \quad r = b \tag{4-141b}$$

where $\bar{\bar{T}} \equiv \bar{\bar{T}}(r,n,m)$.

Thus we reduced the system of partial differential Eq. 4-137 to a second order ordinary differential equation as given by the Eq. 4-141, the particular solution of which is in the form

$$\bar{\bar{T}}(r,n,m): r^n,\ r^{-(n+1)} \tag{4-142}$$

The solution $r^{-(n+1)}$ should be excluded because the temperature should remain finite at $r = 0$. In view of the boundary condition Eq. 4-141b, the complete solution of system 4-141 is

$$\bar{\bar{T}}(r,n,m) = \left(\frac{r}{b}\right)^n \bar{\bar{f}}(n,m) \tag{4-143}$$

The double transform of temperature as given by Eq. 4-143 is inverted by applying successively the inversion formulas 4-140b and 4-138b. The solution of the boundary-value problem of heat conduction Eq. 4-137 becomes

$$T(r,\mu,\phi) = \frac{1}{\pi}\sum_{n=0}^{\infty}\sum_{m=0}^{n} K_n^m(\mu)\left(\frac{r}{b}\right)^n \int_{\mu'=-1}^{1} K_n^m(\mu')d\mu' \cdot \int_{\phi'=0}^{2\pi} f(\mu',\phi')\cdot\cos m(\phi-\phi')\cdot d\phi' \tag{4-144}$$

Substituting the value of the kernel $K_n^m(\mu)$, we obtain

$$T(r,\mu,\phi) = \frac{1}{\pi}\sum_{n=0}^{\infty}\sum_{m=0}^{n} \frac{2n+1}{2}\frac{(n-m)!}{(n+m)!}\left(\frac{r}{b}\right)^n P_n^m(\mu) \int_{\mu'=-1}^{1} P_n^m(\mu')d\mu' \cdot \int_{\phi'=0}^{2\pi} f(\mu',\phi')\cdot\cos m(\phi-\phi')\cdot d\phi' \tag{4-145}$$

where $m = 0,1,2,3,\ldots$

$n = 0,1,2,3,\ldots$, and π should be replaced by 2π when $m = 0$

Hemisphere ($0 \le r \le b$, $0 \le \mu \le 1$). We consider the steady-state temperature distribution in a hemisphere $0 \le r \le b$, $0 \le \mu \le 1$, the base of which, $\mu = 0$, is insulated and the hemispherical surface at $r = b$ is kept at a temperature $f(\mu)$. Figure 4-9 shows the boundary conditions for the problem. The boundary-value problem of heat conduction is given as

$$\frac{\partial}{\partial r}\left(r^2\frac{\partial T}{\partial r}\right) + \frac{\partial}{\partial \mu}\left[(1-\mu^2)\frac{\partial T}{\partial \mu}\right] = 0 \quad \text{in} \quad 0 \le r \le b, 0 \le \mu \le 1$$

$$\frac{\partial T}{\partial \mu} = 0 \quad \text{at} \quad \mu = 0 \tag{4-146}$$

$$T = f(\mu) \quad \text{at} \quad r = b$$

where $T \equiv T(r,\mu)$.

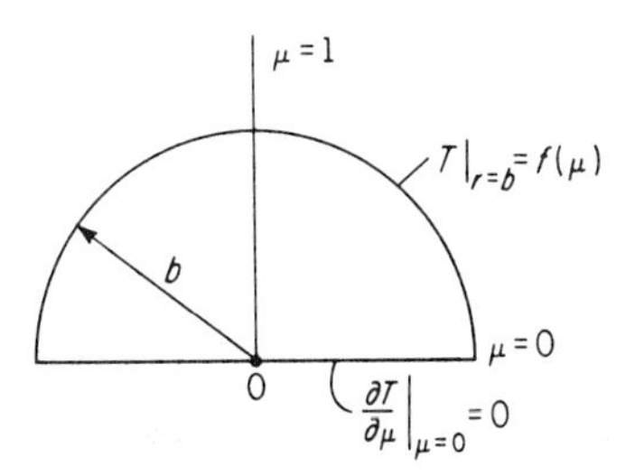

FIG. 4-9. Boundary conditions for a hemisphere.

The partial derivative with respect to μ variable can be removed by means of a Legendre transform. Since the range of space variable is $0 \leq \mu \leq 1$ and that the boundary condition at $\mu = 0$ is of the second kind we define an *even Legendre transform* and the corresponding inversion formula (see Eq. 4-34) of the temperature function $T(r, \mu)$ with respect to the μ variable as

$$\begin{pmatrix}\text{Integral}\\ \text{transform}\end{pmatrix} \quad \bar{T}(r, n) = \int_{\mu'=0}^{1} K_{2n}(\mu') \cdot T(r, \mu') d\mu' \tag{4-147a}$$

$$\begin{pmatrix}\text{Inversion}\\ \text{formula}\end{pmatrix} \quad T(r, \mu) = \sum_{n=0}^{\infty} K_{2n}(\mu) \cdot \bar{T}(r, n) \tag{4-147b}$$

where the kernel is

$$K_{2n}(\mu) = \sqrt{4n + 1} \quad P_{2n}(\mu)$$
$$n = 0,1,2,3, \ldots$$

We take the integral transform of system 4-146 by applying the transform Eq. 4-147a (see Eq. 4-47 for the transform of the Laplacian) and obtain

$$\frac{d}{dr}\left(r^2 \frac{d\bar{T}}{dr}\right) - 2n(2n + 1)\bar{T} = 0 \quad \text{in} \quad 0 \leq r \leq b \tag{4-148a}$$

$$\bar{T} = \bar{f}(n) \quad \text{at} \quad r = b \tag{4-148b}$$

where $\bar{T} \equiv \bar{T}(r, n)$.

System 4-148 is an ordinary differential equation, the solution of which remains finite within the region and subject to the boundary condition Eq. 4-148b is

$$\bar{T}(r, n) = \left(\frac{r}{b}\right)^{2n} \cdot \bar{f}(n) \tag{4-149}$$

The transform of temperature as given by Eq. 4-149 is inverted by means of inversion formula 4-147b,

$$T(r, \mu) = \sum_{n=0}^{\infty} K_{2n}(\mu) \left(\frac{r}{b}\right)^{2n} \int_{\mu'=0}^{1} f(\mu') \cdot K_{2n}(\mu') \cdot d\mu' \tag{4-150}$$

Substituting the kernel,

$$T(r,\mu) = \sum_{n=0}^{\infty} (4n+1)\left(\frac{r}{b}\right)^{2n} \cdot P_{2n}(\mu) \cdot \int_{\mu'=0}^{1} f(\mu') \cdot P_{2n}(\mu') \cdot d\mu' \tag{4-151}$$

where $n = 0,1,2,3,\ldots$.

4-9. TRANSIENT TEMPERATURE CHARTS

Heisler [12] prepared temperature-time charts for a solid sphere of radius $r = b_1$ initially at a uniform temperature T_i and for times $t > 0$ subjected to convection from the boundary surface at $r = b$ into an environment at constant temperature T_e.

Defining a dimensionless temperature θ as

$$\theta = \frac{T - T_e}{T_i - T_e}$$

the boundary-value problem of heat conduction is given as

$$\frac{1}{r^2}\frac{\partial}{\partial r}\left(r^2 \frac{\partial \theta}{\partial r}\right) = \frac{1}{\alpha}\frac{\partial \theta}{\partial r} \quad \text{in} \quad 0 \le r \le b, t > 0 \tag{4-152a}$$

$$k\frac{\partial \theta}{\partial r} + h\theta = 0 \quad \text{at} \quad r = b, t > 0 \tag{4-152b}$$

$$\theta = 1 \quad \text{in} \quad 0 \le r \le b, t = 0 \tag{4-152c}$$

Figure 4-10 shows the dimensionless center-temperature, $(T_0 - T_e)/(T_i - T_e)$ as a function of the dimensionless time $\alpha t/b^2$ for several different values of the parameter k/hb.

Figure 4-11 is a position-correction chart which is used in conjunction with Fig. 4-10 to determine temperature history at the interior regions of the sphere. Temperature at any position r/b for any given value of k/hb and $\alpha t/b^2$ is determined by multiplying the value of $(T_0 - T_e)/(T_i - T_e)$ from Fig. 4-10 by the value of $(T - T_e)/(T_0 - T_e)$ at the corresponding value of r/b and k/hb from Fig. 4-11.

4-10. SUMMARY OF LEGENDRE TRANSFORM OF PART OF THE LAPLACIAN IN THE SPHERICAL COORDINATE

The differential operator in the spherical coordinate system including the μ variable in the form

$$D \equiv \frac{\partial}{\partial \mu}\left[(1-\mu^2)\frac{\partial T}{\partial \mu}\right] \tag{4-153}$$

can be removed by the application of an integral transform with respect to the μ variable (i.e., Legendre transform). The transform, the inversion formula and the kernels to be used depend on the range of the μ variable

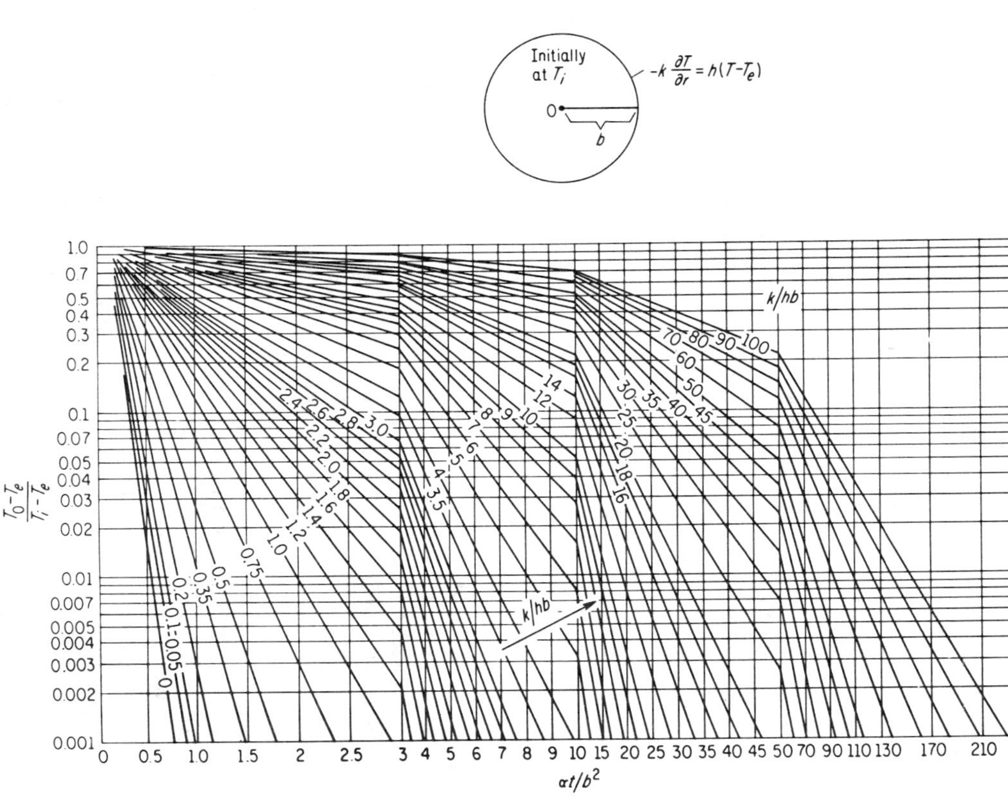

FIG. 4-10. Transient temperature T_0 at the center of a sphere of radius b, subjected to convection at the boundary surface $r = b$. (From Heisler, Ref. 12)

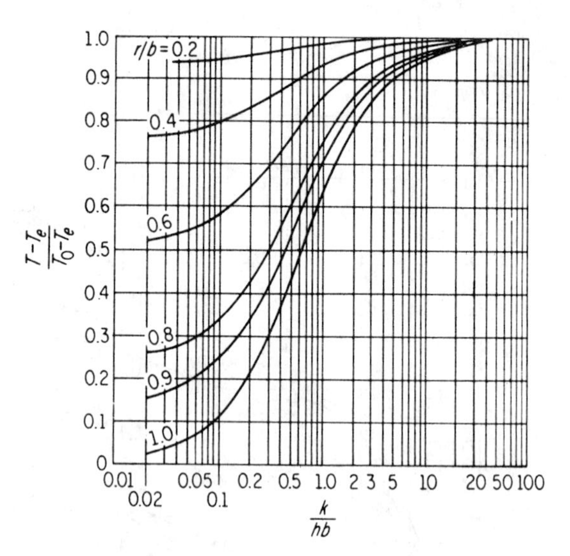

FIG. 4-11. Position-correction chart for use with Fig. 4-10. (From Heisler, Ref. 12.)

and the type of the boundary conditions. For $-1 \leq \mu \leq 1$, i.e., the complete sphere, they are given by Eq. 4-28. For $0 \leq \mu \leq 1$, i.e., the hemisphere, they are given by Eq. 4-33 when the boundary condition at $\mu = 0$ is of the first kind and by Eq. 4-34 when the boundary condition at $\mu = 0$ of the second kind.

The differential operator involving the μ and φ variables in the form

$$D \equiv \frac{\partial}{\partial \mu}\left[(1-\mu^2)\frac{\partial T}{\partial \mu}\right] + \frac{1}{1-\mu^2}\frac{\partial^2 T}{\partial \varphi^2} \tag{4-154a}$$

can be removed by successive application of a Fourier transform with respect to the φ variable and a Legendre transform with respect to the μ variable. When the range of φ variable is $0 \leq \varphi \leq 2\pi$, Fourier transform and the inversion formula are given by Eq. 3-186; when the range of φ is $0 \leq \varphi \leq \varphi_0(\varphi_0 < 2\pi)$ the removal of the φ variable is treated exactly the same as that described in the Cartesian coordinate system for finite regions and the integral transform, inversion formula and the kernels are obtained from Table 2-1. Once the φ variable is removed the differential operator 4-154a includes the μ variable in the form

$$D \equiv \frac{\partial}{\partial \mu}\left[(1-\mu^2)\frac{\partial T}{\partial \mu}\right] - \frac{m^2}{1-\mu^2}T \tag{4-154b}$$

TABLE 4-2

INTEGRAL TRANSFORM (i.e., LEGENDRE TRANSFORM) OF LAPLACIAN IN THE SPHERICAL COORDINATE SYSTEM

Region	Boundary Condition	Definition of Integral Transform*	Definition of Inversion Formula*	Form of $\nabla^2 T$	Integral Transform of $\nabla^2 T$
$-1 \le \mu \le 1$	Temperature-cyclic	$\bar{T} = \int_{-1}^{1} K_n(\mu') \cdot T \cdot d\mu'$	$T = \sum_{n=0}^{\infty} K_n(\mu) \cdot \bar{T}$	$\frac{\partial}{\partial \mu}\left[(1-\mu^2)\frac{\partial T}{\partial \mu}\right]$	$-n(n+1)\bar{T}$
$-1 \le \mu \le 1$ $0 \le \phi \le 2\pi$	Temperature-cyclic of period 2π	$\bar{T} = \int_{\phi=0}^{2\pi}\int_{-1}^{1} K_n^m(\mu') \cdot T \cdot \cos m(\phi - \phi') \cdot d\mu' \cdot d\gamma'$	$T = \frac{1}{\pi}\sum_{n=0}^{\infty}\sum_{m=0}^{n} K_n^m(\mu) \cdot \bar{T}$ replace π by 2π for $m = 0$	$\frac{\partial}{\partial \mu}\left[(1-\mu^2)\frac{\partial T}{\partial \mu}\right] + \frac{m^2}{1-\mu^2}\frac{\partial^2 T}{\partial \phi^2}$	$-n(n+1)\bar{T}$
$0 \le \mu \le 1$	$\left.\frac{\partial T}{\partial \mu}\right\|_{\mu=0} = f$	$\bar{T} = \int_{0}^{1} K_{2n}(\mu') \cdot T \cdot d\mu'$	$T = \sum_{n=0}^{\infty} K_{2n}(\mu) \cdot \bar{T}$	$\frac{\partial}{\partial \mu}\left[(1-\mu^2)\frac{\partial T}{\partial \mu}\right]$	$-2n(2n+1) \cdot \bar{T} - K_{2n}(\mu)\|_{\mu=0} \cdot f$
$0 \le \mu \le 1$	$T\|_{\mu=0} = f$	$\bar{T} = \int_{0}^{1} K_{2n+1}(\mu') \cdot T \cdot d\mu'$	$T = \sum_{n=0}^{\infty} K_{2n+1}(\mu) \cdot \bar{T}$	$\frac{\partial}{\partial \mu}\left[(1-\mu^2)\frac{\partial T}{\partial \mu}\right]$	$-(2n+1)(2n+2)\bar{T} + \left.\frac{dK_{2n+1}(\mu)}{d\mu}\right\|_{\mu=0} \cdot f$

TABLE 4-2 (Continued)

Region	Boundary Condition	Definition of Integral Transform*	Definition of Inversion Formula*	Form of $\nabla^2 T$	Integral Transform of $\nabla^2 T$
$-1 \le \mu \le 1$ $0 \le r \le b$	$T\mid_{r=b} = f(\mu)$	$\bar{T} = \int_{r'=0}^{b} \int_{\mu'=-1}^{1} r'^2 K_{np}(r', \mu') \cdot T dr' d\mu'$	$T = \sum_{n=0}^{\infty} \sum_{p=1}^{\infty} K_{np}(r, \mu) \cdot \bar{T}$	$\frac{1}{r^2} \frac{\partial}{\partial r}\left(r^2 \frac{\partial T}{\partial r}\right) +$ $\frac{1}{r^2} \frac{\partial}{\partial \mu}\left[(1-\mu^2)\frac{\partial T}{\partial \mu}\right]$	$-\lambda_{np}^2 \cdot T$ $- \int_{\mu'=-1}^{1} b^2 \cdot f(\mu')$ $\cdot \frac{dK_{np}(r, \mu')}{dr}\Big\vert_{r=b} \cdot d\mu'$

*Kernels are defined as:

$$K_n(\mu) = \sqrt{\frac{2n+1}{2}} P_n(\mu)$$

$$K_n^m(\mu) = \sqrt{\frac{2n+1}{2} \cdot \frac{(n-m)!}{(n+m)!}} P_n^m(\mu)$$

$m \le n$

$$K_{2n}(\mu) = \sqrt{4n+1} \cdot P_{2n}(\mu)$$

$$K_{2n+1}(\mu) = \sqrt{4n+3}\ P_{2n+1}(\mu)$$

$$K_{np}(r, \mu) = \frac{r^{-1/2} J_{n+1/2}(\lambda_{pn} r) \cdot P_n(\mu)}{\sqrt{N_r N_\mu}}$$

where $N_r N_\mu = -\frac{b^2}{2n+1} J_{n-1/2}(\lambda_{pn} b) \cdot J_{n+3/2}(\lambda_{pn} b)$

and the λ_{pn}'s are the positive roots of $J_{n+1/2}(\lambda_n b) = 0$

where m is the transform variable for the Fourier transform. When the range of μ is $-1 \leq \mu \leq 1$, the differential operator 4-154b can be removed by the application of Legendre transform given by Eq. 4-31.

The differential operator involving the r and μ variables in the form

$$D \equiv \frac{1}{r^2}\frac{\partial}{\partial r}\left(r^2\frac{\partial T}{\partial r}\right) + \frac{1}{r^2}\frac{\partial}{\partial \mu}\left[(1-\mu^2)\frac{\partial T}{\partial \mu}\right] \tag{4-155}$$

can be removed by the application of Bessel-Legendre transform given by Eq. 4-117 when the range of the space variables is $0 \leq r \leq b$, $-1 \leq \mu \leq 1$ and subject to a boundary condition of the first kind at $r = b$.

Table 4-2 summarizes the removal of the differential operators 4-153, 4-154 and 4-155 by the application of integral transform.

REFERENCES

1. Milton Abramowitz and Irene A. Stegun, "Handbook of Mathematical Functions," *National Bureau of Standards Applied Mathematic Series*, No. 55, Washington, D. C. (June 1964), p. 437.

2. T. M. MacRobert, *Spherical Harmonics*, Dover Publications, Inc., New York, 1948.

3. E. T. Whittaker and G. N. Watson, *A Course of Modern Analysis*, Cambridge University Press, London, 1958.

4. Parry Moon and D. E. Spencer, *Field Theory for Engineers*, D. Van Nostrand Co., Inc., New York, 1961, pp. 131–140.

5. William Elwood Byerly, *Fourier's Series and Spherical, Cylindrical, and Ellipsoidal Harmonics*, Dover Publications, Inc., New York, 1959.

6. Ruel V. Churchill, *Fourier Series and Boundary-Value Problems*, McGraw-Hill Book Company, New York, 1963, p. 213.

7. E. T. Whittaker and G. N. Watson, *A Course in Modern Analysis*, Cambridge University Press, London, 1931, p. 324.

8. C. J. Tranter, *Integral Transforms in Mathematical Physics*, John Wiley and Sons, Inc., New York, 1962, p. 97.

9. Ian N. Sneddon, *Fourier Transform*, McGraw-Hill Book Company, New York, 1951, p. 200.

10. S. Kaplan and G. Sonnemann, "The Helmholtz Transformation Theory and Application," *Research Report No. 2 of the Department of Mechanical Engineering*, University of Pittsburgh, 1961.

11. N. W. McLachlan, *Bessel Functions for Engineers*, Clarendon Press, Oxford, 1961, p. 110.

12. M. P. Heisler, "Transient Charts for Induction and Constant-Temperature Heating," *Trans. ASME*, Vol. 69 (April 1947), pp. 227–236.

PROBLEMS

1. Determine the one-dimensional transient temperature distribution in a solid sphere of radius $r = b$ which is initially at temperature T_i and for times $t > 0$ the boundary surface at $r = b$ is kept at constant temperature T_0.

2. A hollow sphere $a \leq r \leq b$ is initially at constant temperature T_i. For times $t > 0$ the inner-boundary surface at $r = a$ is kept at zero temperature and the outer-boundary surface $r = b$ is subjected to convection into a medium at zero temperature. Determine the temperature distribution in the solid for times $t > 0$.

3. Find the transient temperature distribution in a hollow sphere $a \leq r \leq b$ which is initially at temperature $F(r)$ and for times $t > 0$ the boundary surfaces at $r = a$ and $r = b$ are kept at zero temperature.

4. A solid sphere of radius $r = b$ is initially at temperature $F(r, \mu, \phi)$. For times $t > 0$ the boundary surface at $r = b$ is subjected to convection into a medium at zero temperature. Find a relation for the temperature distribution in the sphere.

5. Find the transient temperature distribution $T(r, \mu, t)$ in a solid sphere of radius $r = b$ which is initially at temperature $F(r,\mu)$ for times $t > 0$ the boundary surface at $r = b$ is kept at zero temperature.

6. A solid sphere of radius $r = b$ is initially at zero temperature. For times $t > 0$ heat is generated within the sphere at a rate of $g(r, \mu)$ Btu/hr ft^3 while the boundary surface at $r = b$ is kept at zero temperature. Find a relation for temperature distribution in the sphere for times $t > 0$.

7. Determine the steady-state temperature distribution $T(r, \mu)$ in a solid sphere of radius $r = b$ which is subjected to a temperature distribution $f(\mu)$ at the boundary surface $r = b$.

8. Determine the steady-state temperature distribution $T(r, \mu)$ in a hemisphere $0 \leq r \leq b$, $0 \leq \mu \leq 1$ which is subjected to zero temperature at the hemispherical surface $r = b$, and to a temperature distribution $f(r)$ at the base.

9. A hemisphere $0 \leq r \leq b$, $0 \leq \mu \leq 1$ is initially at a constant temperature T_i. For times $t > 0$ the base is kept insulated, and the hemispherical surface $r = b$ is subjected to convection with a medium at a constant temperature T_∞. Determine the temperature distribution in the solid for times $t > 0$.

10. Determine the steady-state temperature distribution in a solid sphere of radius $r = b$ in which heat is generated uniformly at a constant rate of g_0 Btu/hr ft^3 while the boundary surface at $r = b$ is subjected to convection into a medium at zero temperature.

5

Duhamel's Method and Use of Green's Functions in the Solution of Heat-Conduction Problems

Duhamel's method relates the solution of boundary-value problems of heat conduction with time-dependent boundary conditions and heat sources to the solution of a similar problem with time-independent boundary conditions and heat sources by means of a simple relation. Since it is often easier to obtain the solution of the latter problem, Duhamel's method is a useful tool for obtaining the solution of a problem with time-dependent boundary conditions and/or heat sources whenever the solution of a similar problem with time-independent boundary conditions and/or heat sources is available.

The solution of a boundary-value problem of heat conduction with distributed heat sources, nonhomogeneous boundary conditions and a prescribed initial condition can be represented in the integral form by means of a *Green's function* which is the solution of a similar problem for zero initial condition, homogeneous boundary conditions, and an instantaneous heat source of unit strength situated at a single location within the region. The method of Green's functions is useful whenever *Green's function* for the problem under consideration is available.

Later in this chapter it will be shown that the solutions with Duhamel's method and Green's function are related to the generalized solution 1-68 which was obtained by means of the integral-transform technique in the first chapter.

5-1. DUHAMEL'S METHOD

Consider a nonhomogeneous boundary-value problem of heat conduction for a region R with time-dependent heat sources and boundary conditions.

$$\nabla^2 T(\bar{r},t) + \frac{g(\bar{r},t)}{k} = \frac{1}{\alpha}\frac{\partial T(\bar{r},t)}{\partial t} \quad \text{in region } R,\ t > 0$$

$$k_i \frac{\partial T(\bar{r}_i,t)}{\partial n_i} + h_i T(\bar{r}_i,t) = f_i(\bar{r},t) \quad \text{on boundary } s_i,\ t > 0 \qquad (5\text{-}1)$$

$$T(\bar{r},t) = F(\bar{r}) \quad \text{in region } R,\ t = 0$$

We now consider the following auxiliary problem for the same region R,

$$\nabla^2\phi(\bar{r},\tau,t) + \frac{g(\bar{r},\tau)}{k} = \frac{1}{\alpha}\frac{\partial\phi(\bar{r},\tau,t)}{\partial t} \quad \text{in region } R,\ t > 0$$

$$k_i\frac{\partial\phi(\bar{r}_i,\tau,t)}{\partial n_i} + h_i\phi(\bar{r}_i,\tau,t) = f_i(\bar{r},\tau) \quad \text{on boundary } s_i,\ t > 0 \tag{5-2}$$

$$\phi(\bar{r},\tau,t) = F(\bar{r}) \quad \text{in region } R,\ t = 0$$

The physical significance of the auxiliary problem is that it represents the solution of the heat-conduction problem for the region R for *time-independent* boundary condition and heat generation; that is, in the auxiliary problem 5-2 the heat-generation term $g(\bar{r},t)$ is replaced by $g(\bar{r},\tau)$ and the boundary-condition function $f_i(\bar{r},t)$ is replaced by $f_i(\bar{r},\tau)$, which are the values of the time-dependent functions evaluated at a fixed time $t = \tau$. Therefore, function $\phi(\bar{r},\tau,t)$ represents the solution of system 5-1 for time-independent heat-generation and boundary-condition functions $g(\bar{r},\tau)$ and $f_i(\bar{r},\tau)$.

Assuming the solution $\phi(\bar{r},\tau,t)$ of the auxiliary system 5-2 is available, then the solution $T(\bar{r},t)$ of the system 5-1 is related to $\phi(\bar{r},\tau,t)$ function by the following simple relationship

$$T(\bar{r},t) = \frac{\partial}{\partial t}\int_{\tau=0}^{t}\phi(\bar{r},\tau,t-\tau)\cdot d\tau \tag{5-3a}$$

Performing the differentiation under the integral sign we obtain

$$T(\bar{r},t) = F(\bar{r}) + \int_{\tau=0}^{t}\frac{\partial}{\partial t}\phi(\bar{r},\tau,t-\tau)\cdot d\tau \tag{5-3b}$$

The relationship given by Eq. 5-3 is called *Duhamel's method* for the solution of boundary-value problems of heat conduction with time-dependent boundary conditions and heat sources in terms of the same problem, with time-independent boundary conditions and heat-generation functions. A proof of Duhamel's method is given by Bartels and Churchill [1] for a boundary condition of the first kind; a similar proof can easily be extended to the boundary condition of the third kind considered above. Proof of Duhamel's method is also given by Sneddon [2] and by Carslaw and Jaeger [3] and therefore it will not be reproduced here.

For a solid with zero initial temperature, i.e., $F(\bar{r}) = 0$, Eq. 5-3 simplifies to

$$T(\bar{r},t) = \int_{\tau=0}^{t}\frac{\partial}{\partial t}\phi(\bar{r},\tau,t-\tau)\cdot d\tau \tag{5-4}$$

We now examine some special cases of Duhamel's theorem:

The region R is initially at zero temperature. For times $t > 0$ all the boundary surfaces are subjected to convective-boundary condition with a medium at temperature $f(t)$ which varies only with time and there is no heat generation within the solid. The boundary-value problem of heat conduction is given as

$$\nabla^2 T(\bar{r},t) = \frac{1}{\alpha}\frac{\partial T(\bar{r},t)}{\partial t} \quad \text{in region } R,\ t > 0$$

$$k\frac{\partial T(\bar{r},t)}{\partial n_i} + hT(\bar{r},t) = h\cdot f(t) \quad \text{on boundaries, } t > 0 \tag{5-5}$$

$$T(\bar{r},t) = 0 \quad \text{in region } R,\ t = 0$$

For this special case we consider the following auxiliary problem:

$$\nabla^2 \phi(\bar{r},t) = \frac{1}{\alpha}\frac{\partial \phi(\bar{r},t)}{\partial t} \quad \text{in region } R,\ t > 0$$

$$k\frac{\partial \phi(\bar{r},t)}{\partial n_i} + h\phi(\bar{r},t) = h \quad \text{on boundaries, } t > 0 \tag{5-6}$$

$$\phi(\bar{r},t) = 0 \quad \text{in region } R,\ t = 0$$

which is similar to Prob. 5-5 except convection at the boundary is with a medium at temperature unity.

The solution of the boundary-value problem of heat conduction 5-5 in terms of the solution of the auxiliary problem 5-6 is given as

$$T(\bar{r},t) = \int_{\tau=0}^{t} f(\tau)\cdot\frac{\partial \phi(\bar{r},t-\tau)}{\partial t}\cdot d\tau \tag{5-7}$$

The result given by Eq. 5-7 is apparent from Eq. 5-4, because for this particular case we have

$$\phi(\bar{r},\tau,t) = f(\tau)\cdot\phi(\bar{r},t) \tag{5-8}$$

The result given by Eq. 5-7 can also be applied to the case of the solid which is initially at zero temperature, while for times $t > 0$ all the boundary surfaces are subjected to a uniform temperature $f(t)$ which varies only with time. For such a case the boundary-value problem of heat conduction 5-5 satisfies the boundary condition

$$T(\bar{r},t) = f(t) \quad \text{on boundaries, for } t > 0$$

and the auxiliary Prob. 5-6 satisfies the boundary condition

$$\phi(\bar{r},t) = 1 \quad \text{on boundaries, for } t > 0$$

5-2. A COMPARISON OF SOLUTIONS OBTAINED WITH DUHAMEL'S METHOD AND WITH INTEGRAL-TRANSFORM TECHNIQUE

Consider the boundary-value problem of heat conduction 1-53 and its general solution 1-68 which was obtained with the integral transform technique.

$$T(\bar{r}, t) = \sum_{m=1}^{\infty} e^{-\alpha\lambda_m^2 t} \cdot K(\lambda_m, \bar{r}) \left[\bar{F}(\lambda_m) + \int_{t'=0}^{t} e^{\alpha\lambda_m^2 t'} \cdot A(\lambda_m, t') \cdot dt' \right] \tag{5-9}$$

where

$$A(\lambda_m, t') = \frac{\alpha}{k} \bar{g}(\lambda_m, t') + \alpha \sum_{i=1}^{s} \int_{s_i} \frac{K(\lambda_m, \bar{r}')}{k_i} f_i(\bar{r}, t') \cdot ds_i$$

In deriving solution 5-9 it was assumed that both the heat-generation function $g(\bar{r}, t)$ and the boundary-condition function $f_i(\bar{r}, t)$ were functions of time.

We consider now a special case of the boundary-value problem of heat conduction 1-53: we assume that the heat-generation function and the boundary function are time-*independent* and choose them as $g(\bar{r}, \tau)$ and $f_i(\bar{r}, \tau)$ respectively which are the values of $g(\bar{r}, t)$ and $f_i(\bar{r}, t)$ functions evaluated at a fixed time $t = \tau$. The solution for this special case $\phi(\bar{r}, \tau, t)$ is obtained from the general solution 5-9 by replacing $A(\lambda_m, t')$ by $A(\lambda_m, \tau)$ and performing the integration with respect to the t' variable.

$$\phi(\bar{r}, \tau, t) = \sum_{m=1}^{\infty} K(\lambda_m, \bar{r}) \left[\bar{F}(\lambda_m) \cdot e^{-\alpha\lambda_m^2 t} + A(\lambda_m, \tau) \cdot \frac{1 - e^{-\alpha\lambda_m^2 t}}{\alpha\lambda_m^2} \right] \tag{5-10}$$

Equation 5-10 is the solution of heat conduction problem 1-53 for the case of time independent boundary conditions and heat generation within the solid. We now apply Duhamel's theorem to Eq. 5-10 in order to obtain the solution of the boundary value problem of heat-conduction 1-53 for the case of time-dependent boundary conditions and heat generation.

In Eq. 5-10 we replace the time variable t by $(t - \tau)$,

$$\phi(\bar{r}, \tau, t - \tau) = \sum_{m=1}^{\infty} K(\lambda_m, \bar{r}) \cdot \left[\bar{F}(\lambda_m) \cdot e^{-\alpha\lambda_m^2 (t-\tau)} + A(\lambda_m, \tau) \cdot \frac{1 - e^{-\alpha\lambda_m^2 (t-\tau)}}{\alpha\lambda_m^2} \right] \tag{5-11}$$

We integrate both sides of Eq. 5-11 with respect to the τ variable from $\tau = 0$ to $\tau = t$.

$$\int_{\tau=0}^{t} \phi(\bar{r}, \tau, t - \tau) d\tau = \sum_{m=1}^{\infty} K(\lambda_m, \bar{r}) \cdot \left[\bar{F}(\lambda_m) \cdot \frac{1 - e^{-\alpha\lambda_m^2 t}}{\alpha\lambda_m^2} + \int_{\tau=0}^{t} A(\lambda_m, \tau) \cdot \frac{1 - e^{-\alpha\lambda_m^2 (t-\tau)}}{\alpha\lambda_m^2} d\tau \right] \tag{5-12}$$

Differentiating both sides of Eq. 5-12 with respect to the time variable t,

$$\frac{\partial}{\partial t} \int_{\tau=0}^{t} \phi(\bar{r}, \tau, t - \tau) d\tau = \sum_{m=1}^{\infty} K(\lambda_m, \bar{r}) \cdot \left[\bar{F}(\lambda_m) \cdot e^{-\alpha\lambda_m^2 t} + \frac{\partial}{\partial t} \left(\int_{\tau=0}^{t} A(\lambda_m, \tau) \cdot \frac{1 - e^{-\alpha\lambda_m^2 (t-\tau)}}{\alpha\lambda_m^2} d\tau \right) \right] \tag{5-13}$$

Performing the differentiation under the integral sign, we obtain

$$\frac{\partial}{\partial t}\int_{\tau=0}^{t}\phi(\bar{r},\tau,t-\tau)\cdot d\tau = \sum_{m=1}^{\infty} e^{-\alpha\lambda_m^2 t}\cdot K(\lambda_m,\bar{r}) \cdot\left[\bar{F}(\lambda_m) + \int_{\tau=0}^{t} A(\lambda_m,\tau)\cdot e^{\alpha\lambda_m^2\tau}\cdot d\tau\right] \tag{5-14}$$

The right-hand side of Eq. 5-14 is the same as Eq. 5-9, that is

$$T(\bar{r},t) = \frac{\partial}{\partial t}\int_{\tau=0}^{t}\phi(\bar{r},\tau,t-\tau)\cdot d\tau \tag{5-15}$$

Thus, by Duhamel's theorem we related the solution 5-10 to the general solution 5-9.

5-3. APPLICATION OF DUHAMEL'S METHOD

Use of Duhamel's method in the solution of boundary-value problems of heat conduction with time-dependent boundary conditions and heat generation will be illustrated with examples.

Semi-infinite Solid $0 \le x < \infty$. A semi-infinite solid is initially at zero temperature. For times $t > 0$ the boundary surface at $x = 0$ is kept at a uniform temperature $f(t)$ which varies with time. There is no heat generation within the solid.

The boundary-value problem of heat conduction is given as

$$\frac{\partial^2 T(x,t)}{\partial x^2} = \frac{1}{\alpha}\frac{\partial T(x,t)}{\partial t} \quad \text{in} \quad 0 \le x < \infty, \text{ for } t > 0$$

$$T(x,t) = f(t) \quad \text{at} \quad x = 0, \text{ for } t > 0 \tag{5-16}$$

$$T(x,t) = 0 \quad \text{in} \quad 0 \le x < \infty, \text{ for } t = 0$$

To solve this problem with the special case of Duhamel's method we consider the following auxiliary problem:

$$\frac{\partial^2 \phi(x,t)}{\partial x^2} = \frac{1}{\alpha}\frac{\partial \phi(x,t)}{\partial t} \quad \text{in} \quad 0 \le x < \infty, \text{ for } t > 0$$

$$\phi(x,t) = 1 \quad \text{at} \quad x = 0, \text{ for } t > 0 \tag{5-17}$$

$$\phi(x,t) = 0 \quad \text{in} \quad 0 \le x < \infty, \text{ for } t = 0$$

Solution of this auxiliary problem is easily obtained and given as

$$\phi(x,t) = \text{erfc}\left(\frac{x}{\sqrt{4\alpha t}}\right) = \frac{2}{\sqrt{\pi}}\int_{\frac{x}{\sqrt{4\alpha t}}}^{\infty} e^{-\xi^2}d\xi \tag{5-18}$$

The solution $T(x,t)$ of the problem 5-16 is related to the solution $\phi(x,t)$ of the auxiliary problem 5-17 by means of the special case of Duhamel's theorem as

$$T(x,t) = \int_{\tau=0}^{t} f(\tau)\cdot\frac{\partial\phi(x,t-\tau)}{\partial t}\cdot d\tau \tag{5-19}$$

Since $\phi(x, t)$ is given by Eq. 5-18, we have

$$\frac{\partial \phi(x, t-\tau)}{\partial t} = \frac{2}{\sqrt{\pi}} \frac{\partial}{\partial t} \cdot \left[\int_{\frac{x}{\sqrt{4\alpha(t-\tau)}}}^{\infty} e^{-\xi^2} d\xi \right] = \frac{x}{\sqrt{4\pi\alpha(t-\tau)^3}} e^{-\frac{x^2}{4\alpha(t-\tau)}} \tag{5-20}$$

Substituting Eq. 5-20 into 5-19, the solution becomes

$$T(x, t) = \frac{x}{\sqrt{4\pi\alpha}} \int_{\tau=0}^{t} f(\tau) \cdot \frac{e^{-\frac{x^2}{4\alpha(t-\tau)}}}{(t-\tau)^{3/2}} d\tau \tag{5-21}$$

Semi-infinite Rectangular Strip $0 \le y \le b$, $0 \le x < \infty$. A semi-infinite rectangular strip $0 \le y \le b$, $0 \le x < \infty$ is initially at zero temperature. For times $t > 0$ the boundary surface at $y = b$ is kept at a temperature $f(t)$ which is a function of time and the remaining boundaries are kept at zero temperature. Figure 5-1 shows the boundary conditions

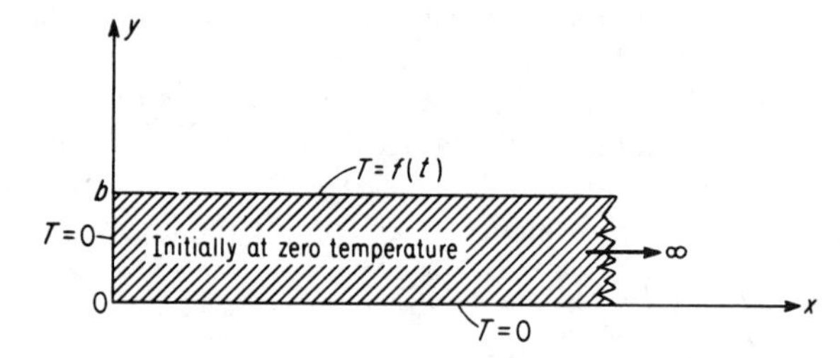

FIG. 5-1. Boundary conditions for a semi-infinite rectangular strip.

for the problem. The boundary-value problem of heat conduction is

$$\begin{aligned}
\frac{\partial^2 T}{\partial x^2} + \frac{\partial^2 T}{\partial y^2} &= \frac{1}{\alpha}\frac{\partial T}{\partial t} \quad \text{in} \quad 0 \le y \le b,\ 0 \le x < \infty,\ t > 0 \\
T &= 0 \quad \text{at} \quad x = 0,\ t > 0 \\
T &= 0 \quad \text{at} \quad y = 0,\ t > 0 \\
T &= f(t) \quad \text{at} \quad y = b,\ t > 0 \\
T &= 0 \quad \text{in} \quad 0 \le y \le b,\ 0 \le x < \infty,\ t = 0
\end{aligned} \tag{5-22}$$

where $T \equiv T(x, y, t)$.

This problem can be solved by Duhamel's method[1] if the solution of the above problem is known for a boundary condition $T = f(\tau) =$ con-

[1]For this case Duhamel's special case is also applicable; that is one chooses the boundary condition for the auxiliary problem as $\phi|_{y=b} = 1$ instead of $\phi|_{y=b} = f(\tau)$ and apply Duhamel's special case, Eq. 5-7.

stant at $y = b$. We consider the following auxiliary problem:

$$\frac{\partial^2 \phi}{\partial x^2} + \frac{\partial^2 \phi}{\partial y^2} = \frac{1}{\alpha}\frac{\partial \phi}{\partial t} \quad \text{in} \quad 0 \le y \le b,\ 0 \le x < \infty,\ t > 0$$

$$\begin{aligned} \phi &= 0 \quad \text{at} \quad x = 0,\ t > 0 \\ \phi &= 0 \quad \text{at} \quad y = 0,\ t > 0 \\ \phi = f(\tau) &= \text{const.} \quad \text{at} \quad y = b,\ t > 0 \\ \phi &= 0 \quad \text{in} \quad 0 \le y \le b,\ 0 \le x < \infty,\ t = 0 \end{aligned} \tag{5-23}$$

where $\phi \equiv \phi(x, y, \tau, t)$.

Solution of this auxiliary problem was considered in Chapter 2 and can be obtained from Eq. 2-229 by replacing T_0 with $f(\tau)$,

$$\phi(x, y, \tau, t) = \frac{4}{\pi b} f(\tau) \sum_{n=1}^{\infty} \sin \nu_n y \cdot \int_{\beta=0}^{\infty} \sin \beta x \, \frac{1 - e^{-\alpha(\beta^2+\nu_n^2)t}}{\beta^2 + \nu_n^2} \cdot (-1)^{n+1} \cdot \frac{\nu_n}{\beta}\, d\beta \tag{5-24}$$

where

$$\nu_n = \frac{n\pi}{b} \qquad n = 1, 2, 3, \ldots$$

According to Duhamel's formula 5-3b solution $T(x, y, t)$ of the Prob. 5-22 is related to the solution $\phi(x, y, t)$ of the auxiliary Prob. 5-23 by the relation

$$T(x, y, t) = \int_{\tau=0}^{t} \frac{\partial}{\partial t} \phi(x, y, \tau, t - \tau) \cdot d\tau \tag{5-25}$$

In view of the relation 5-24,

$$\frac{\partial \phi(x, y, \tau, t - \tau)}{\partial t} = \frac{4\alpha}{\pi b} \cdot f(\tau) \cdot \sum_{n=1}^{\infty} \sin \nu_n y \cdot \int_{\beta=0}^{\infty} \sin \beta x \cdot e^{-\alpha(\beta^2+\nu_n^2)(t-\tau)} \cdot \frac{\nu_n}{\beta} \cdot (-1)^{n+1} \cdot d\beta \tag{5-26}$$

Substituting 5-26 into 5-25, the solution of problem 5-22 becomes

$$T(x, y, t) = \frac{4\alpha}{\pi b} \sum_{n=1}^{\infty} (-1)^{n+1} \int_{\beta=0}^{\infty} e^{-\alpha(\beta^2+\nu_n^2)t} \cdot \nu_n \cdot \sin \nu_n y \int_{\tau=0}^{t} e^{\alpha(\beta^2+\nu_n^2)\tau} \cdot f(\tau) \cdot d\tau \, \frac{\sin \beta x}{\beta}\, d\beta \tag{5-27}$$

Long Solid Cylinder $0 \le r \le b$. A long solid cylinder of radius $r = b$ is initially at zero temperature. For times $t > 0$ heat is generated within the solid at a rate of $g(t)$ Btu/hr ft^3 which varies with time, while the boundary surface at $r = b$ is kept at zero temperature.

The boundary-value problem of heat conduction is

$$\frac{\partial^2 T}{\partial r^2} + \frac{1}{r}\frac{\partial T}{\partial r} + \frac{g(t)}{k} = \frac{1}{\alpha}\frac{\partial T}{\partial t} \quad \text{in} \quad 0 \le r \le b,\ t > 0$$

$$T = 0 \quad \text{at} \quad r = b,\ t > 0 \tag{5-28}$$

$$T = 0 \quad \text{in} \quad 0 \le r \le b,\ t = 0$$

where $T \equiv T(r, t)$.

We consider the following auxiliary problem for a time-independent heat-generation rate $g(\tau)$, which is the value of $g(t)$ at a fixed time $t = \tau$.

$$\frac{\partial^2 \phi}{\partial r^2} + \frac{1}{r}\frac{\partial \phi}{\partial r} + \frac{g(\tau)}{k} = \frac{1}{\alpha}\frac{\partial \phi}{\partial t} \quad \text{in} \quad 0 \le r \le b,\ t > 0$$

$$\phi = 0 \quad \text{at} \quad r = b,\ t > 0 \tag{5-29}$$

$$\phi = 0 \quad \text{in} \quad 0 \le r \le b,\ t = 0$$

where $\phi \equiv \phi(r, \tau, t)$.

Solution of this auxiliary problem is obtained from Eq. 3-100 as

$$\phi(r, \tau, t) = \frac{2}{b}\frac{g(\tau)}{k}\sum_{m=1}^{\infty}\frac{J_0(\beta_m r)}{J_1(\beta_m b)} \cdot \frac{1 - e^{-\alpha\beta_m^2 t}}{\beta_m^3} \tag{5-30}$$

where β_m's are the positive roots of the transcendental equation

$$J_0(\beta b) = 0$$

By Duhamel's method solution $T(r, t)$ of problem 5-28 is related to the solution of $\phi(r, \tau, t)$ of problem 5-29 by

$$T(r, t) = \int_{\tau=0}^{t} \frac{\partial}{\partial t}\phi(r, \tau, t - \tau)\cdot d\tau \tag{5-31}$$

Substituting 5-30 into 5-31, the solution becomes

$$T(r, t) = \frac{2}{b}\frac{\alpha}{k}\sum_{m=1}^{\infty} e^{-\alpha\beta_m^2 t} \cdot \frac{J_0(\beta_m r)}{\beta_m J_1(\beta_m b)} \int_{\tau=0}^{t} e^{\alpha\beta_m^2 \tau}\cdot g(\tau)\cdot d\tau \tag{5-32}$$

which is the same as Eq. 3-99, as it should be.

5-4. GREEN'S FUNCTIONS IN THE SOLUTION OF THREE-DIMENSIONAL BOUNDARY-VALUE PROBLEMS OF HEAT CONDUCTION FOR FINITE REGIONS

The use of Green's function in the solution of partial differential equations can be found in several Refs. [4, 5, 6, 7] and its application in the solution of heat-conduction problems is given by Carslaw and Jaeger. In this section we examine the use of Green's function in the solution of three-dimensional time-dependent boundary-value problems of heat conduction with distributed heat sources, nonhomogeneous boundary conditions and a prescribed initial condition. The cases of one and two-dimensional problems will be examined separately.

Consider the boundary-value problem of heat conduction for a three-dimensional bounded region R which is initially at temperature $F(\bar{r})$. For times $t > 0$ there is heat generation within the solid and the boundaries of which are subject to nonhomogeneous boundary condition of the third kind. For generality we assume that the heat generation and the boundary condition are both functions of position and time. We have

$$\nabla^2 T + \frac{g(\bar{r},t)}{k} = \frac{1}{\alpha}\frac{\partial T}{\partial t} \quad \text{in region } R,\ t > 0 \tag{5-33a}$$

$$k_i \frac{\partial T}{\partial n_i} + h_i T = f_i(\bar{r},t) \quad \text{on boundary } s_i,\ t > 0 \tag{5-33b}$$

$$T = F(\bar{r}) \quad \text{in region } R,\ t = 0 \tag{5-33c}$$

where $T \equiv T(\bar{r},t)$

$\partial/\partial n_i$ = denotes differentiation along the outward-drawn normal to the boundary surface s_i

We now consider the following auxiliary boundary-value problem

$$\nabla^2 G + \frac{1}{\alpha}\,\delta(\bar{r} - \bar{r}')\cdot\delta(t - \tau) = \frac{1}{\alpha}\frac{\partial G}{\partial t} \quad \text{in region } R,\ t > \tau \tag{5-34a}$$

$$k_i \frac{\partial G}{\partial n_i} + h_i G = 0 \quad \text{on boundary } s_i,\ t > \tau \tag{5-34b}$$

subject to the condition

$$G = 0 \quad \text{in region } R,\ t < \tau \tag{5-34c}$$

where $G \equiv G(\bar{r},\bar{r}' \mid t,\tau)$

$\delta(\bar{r} - \bar{r}')$ = three-dimensional delta function for the space variable, i.e., for the cartesian coordinate system—for example, it is $\delta(x - x')\cdot\delta(y - y')\cdot\delta(z - z')$

$\delta(t - \tau)$ = delta function for the time variable

The function $G(\bar{r},\bar{r}' \mid t,\tau)$ satisfying the auxiliary problem 5-34 is called *Green's function* for the boundary-value problem of heat conduction 5-33. It is to be noted that Green's function satisfies the *homogeneous boundary condition corresponding to the homogeneous part of the boundary condition satisfied by* $T(\bar{r},t)$. If the temperature function $T(\bar{r},t)$ in problem 5-33 satisfies *boundary condition of the first kind*, Green's function in the auxiliary problem 5-34 satisfies *homogeneous boundary condition of the first kind*, and so on.

Physical significance of the auxiliary problem 5-34 is as follows: $G(\bar{r},\bar{r}' \mid t,\tau)$ describes the temperature at $\bar{r}$ at time t due to an instantaneous point heat source of strength *unity*, i.e.,

$$S_{\text{pi}} = \frac{\alpha g_{\text{pi}}}{k} = 1, \quad {}^\circ\text{F}\cdot\text{ft}^3$$

where g_{pi} is in Btu, at the location $\bar{r}'$, releasing its heat spontaneously at time $t = \tau$ in the region R which is initially (i.e., for $t < \tau$) at zero temperature and the boundaries of which are subject to homogeneous boundary condition of the third kind.

Green's function $G(\bar{r}, \bar{r}' \mid t, \tau)$ as defined above satisfies the following reciprocity condition[4]

$$G(\bar{r}, \bar{r}' \mid t, \tau) = G(\bar{r}', \bar{r} \mid -\tau, -t) \tag{5-35}$$

where the function $G(\bar{r}, \bar{r}' \mid t, \tau)$ describes the temperature at $\bar{r}$ at a time t due to an impulsive point heat source introduced into the medium at $\bar{r}'$ at a time τ, and the function $G(\bar{r}', \bar{r} \mid -\tau, -t)$ describes the temperature at $\bar{r}'$ at a time $-\tau$ due to an impulsive point heat source introduced into the medium at $\bar{r}$ at a time $-t$.

We shall now obtain the solution of the boundary-value problem of heat conduction 5-33 for the temperature function $T(\bar{r}, t)$ in terms of the Green's function $G(\bar{r}, \bar{r}' \mid t, \tau)$ satisfying the system 5-34.

In the systems 5-33 and 5-34 we write the differential equations satisfied by T and G as a function of τ and $\bar{r}'$ variables. That is,

$$\nabla'^2 T + \frac{g(\bar{r}', \tau)}{k} = \frac{1}{\alpha} \frac{\partial T}{\partial \tau} \quad \text{in region } R,\ \tau < t \tag{5-36}$$

$$\nabla'^2 G + \frac{1}{\alpha} \delta(\bar{r}' - \bar{r}) \cdot \delta(\tau - t) = -\frac{1}{\alpha} \frac{\partial G}{\partial \tau} \quad \text{in region } R,\ \tau < t \tag{5-37}$$

The minus sign on the right-hand side of Eq. 5-37 results from the fact that the G function depends on t as a function of $t - \tau$. The Laplacian ∇'^2 in Eqs. 5-36 and 5-37 are with respect to the space variable $\bar{r}'$.

We multiply both sides of Eq. 5-36 by G and that of Eq. 5-37 by T and subtract the results.

$$(G\nabla'^2 T - T\nabla'^2 G) + \frac{g(\bar{r}', \tau)}{k} G - \frac{1}{\alpha} \delta(\bar{r}' - \bar{r}) \cdot \delta(\tau - t) \cdot T = \frac{1}{\alpha} \frac{\partial (GT)}{\partial \tau} \tag{5-38}$$

We integrate Eq. 5-38 with respect to the space variable $\bar{r}'$ over the region R and with respect to the time variable τ over the time interval $\tau = 0$ to t:

$$\int_{\tau=0}^{t} d\tau \int_R (G\nabla'^2 T - T\nabla'^2 G)\, d\bar{r}' + \frac{1}{k} \int_{\tau=0}^{t} d\tau \cdot \int_R g(\bar{r}', \tau) \cdot G \cdot d\bar{r}' - \frac{1}{\alpha} \int_{\tau=0}^{t} \delta(\tau - t)\, d\tau \cdot \int_R \delta(\bar{r}' - \bar{r}) \cdot T \cdot d\bar{r}' = \frac{1}{\alpha} \int_R [GT]_{\tau=0}^{t} \cdot d\bar{r}' \tag{5-39}$$

In eq. 5-39 various terms are evaluated as follows.

The volume integral is changed to surface integral by applying Green's theorem, that is

$$\int_R (G\nabla'^2 T - T\nabla'^2 G)\,d\bar{r}' = \int_{\text{surface}} \left(G\frac{\partial T}{\partial n_i} - T\frac{\partial G}{\partial n_i}\right) ds \qquad (5\text{-}40)$$

where $\partial/\partial n_i$ = differentiation along the outward-drawn normal to the boundary surface i.

Integrals involving delta functions are immediately evaluated, that is

$$\int_{\tau=0}^{t} \delta(\tau - t)\cdot d\tau \int_R \delta(\bar{r}' - \bar{r})\cdot T(\bar{r}', \tau)\cdot d\bar{r}' = T(\bar{r}, t) \qquad (5\text{-}41)$$

The term on the right-hand side is evaluated at the limits,

$$[GT]_{\tau=0}^{t} = -G\,|_{\tau=0}\cdot T\,|_{\tau=0} = -G\,|_{\tau=0}\cdot F(\bar{r}') \qquad (5\text{-}42)$$

since $[GT]$ vanishes for $\tau = t$, and $T = F(\bar{r}')$ for $\tau = 0$.

Substituting Eqs. 5-40, 5-41, and 5-42 into Eq. 5-39,

$$T(\bar{r}, t) = \int_R G\,|_{\tau=0}\cdot F(\bar{r}')\cdot d\bar{r}' + \frac{\alpha}{k}\int_{\tau=0}^{t} d\tau \int_R g(\bar{r}', \tau)\cdot G\cdot d\bar{r}'$$
$$+ \alpha\int_{\tau=0}^{t} d\tau \int_{\substack{\text{boundary}\\ \text{surface } i}} \left(G\frac{\partial T}{\partial n_i} - T\frac{\partial G}{\partial n_i}\right) ds \qquad (5\text{-}43)$$

In Eq. 5-43 the last term on the right-hand side is evaluated by making use of the boundary conditions for the heat-conduction and auxiliary problems. Multiplying Eq. 5-33b by G and 5-34b by T and subtracting the results, we obtain

$$\left[G\frac{\partial T}{\partial n_i} - T\frac{\partial G}{\partial n_i}\right] = \frac{G}{k_i}\bigg|_{\substack{\text{boundary}\\ \text{surface } i}} \cdot f_i \qquad (5\text{-}44)$$

Substituting 5-44 into 5-43, we have

$$T(\bar{r}, t) = \int_R G\,|_{\tau=0}\cdot F(\bar{r}')\cdot d\bar{r}' + \frac{\alpha}{k}\int_{\tau=0}^{t} d\tau \int_R g(\bar{r}', \tau)\cdot G\cdot d\bar{r}'$$
$$+ \alpha\int_{\tau=0}^{t} d\tau \sum_{i=1}^{s} \int_{\substack{\text{boundary}\\ \text{surface } i}} \frac{G}{k_i}\bigg|_i \cdot f_i\cdot ds_i \qquad (5\text{-}45)$$

where $G \equiv G(\bar{r}, \bar{r}' \mid t, \tau)$
$F(\bar{r})$ = initial-condition function
f_i = boundary-condition function for boundary surface i
g = heat-generation function
$i = 1, 2, \ldots, s$ number of continuous bounding surfaces of the region

Equation 5-45 gives the solution of the boundary-value problem of heat conduction 5-33 in terms of Green's function $G(\bar{r}, \bar{r}' \mid t, \tau)$ satisfying the system 5-34. Physical significance of the three different terms in the solution 5-45 is as follows.

The first term is for the effects of the initial temperature distribution, the second term for the effects of heat sources, and the third term for the effects of the boundary-condition functions.

In deriving Eq. 5-45, for generality, a boundary condition of the third kind is assumed for the boundaries. For a boundary condition of the first kind we set $k_i = 0$, and for such a case the following change should be made in Eq. 5-45.

$$\text{When} \quad k_i = 0, \quad \text{replace} \quad \left.\frac{G}{k_i}\right|_i \quad \text{by} \quad -\frac{1}{h_i}\left.\frac{\partial G}{\partial n_i}\right|_i \tag{5-46}$$

where $\partial/\partial n_i \equiv$ differentiation along the outward-drawn normal to the boundary surface i.

5-5. A COMPARISON OF SOLUTIONS OBTAINED WITH GREEN'S FUNCTION AND THE INTEGRAL-TRANSFORM TECHNIQUE

The boundary-value problem of heat conduction 5-33 which is solved in the previous section in terms of Green's function is similar to the boundary-value problem of heat conduction 1-53 which is solved in Chapter 1 with the integral-transform technique. It is instructive to compare the solutions obtained with these two different methods.

The solution with the integral-transform technique given by Eq. 1-68 is in the form

$$T(\bar{r}, t) = \sum_{m=1}^{\infty} e^{-\alpha\lambda_m^2 t} \cdot K(\lambda_m, \bar{r})$$

$$\cdot \left\{ \bar{F}(\lambda_m) + \int_{\tau=0}^{t} e^{\alpha\lambda_m^2 \tau} \left[\frac{\alpha}{k}\, \bar{g}(\lambda_m, \tau) + \alpha \sum_{i=1}^{s} \int_{s_i} \left.\frac{K(\lambda_m, \bar{r})}{k_i}\right|_{s_i} \cdot f_i ds_i \right] d\tau \right\} \tag{5-47}$$

Substituting the values of $\bar{F}$ and $\bar{g}$ in this relation, we obtain

$$T(\bar{r}, t) = \sum_{m=1}^{\infty} e^{-\alpha\lambda_m^2 t} \cdot K(\lambda_m, \bar{r}) \int_R K(\lambda_m, \bar{r}') \cdot F(\bar{r}') \cdot dr'$$

$$+ \sum_{m=1}^{\infty} e^{-\alpha\lambda_m^2 t} \cdot K(\lambda_m, \bar{r})$$

$$\cdot \int_{\tau=0}^{t} e^{\alpha\lambda_m^2 \tau} \cdot \frac{\alpha}{k} \int_R K(\lambda_m, \bar{r}') \cdot g(\bar{r}', \tau) \cdot d\tau \cdot d\bar{r}'$$

$$+ \alpha \sum_{m=1}^{\infty} e^{-\alpha\lambda_m^2 t} \cdot K(\lambda_m, \bar{r})$$

$$\cdot \int_{\tau=0}^{t} e^{\alpha\lambda_m^2 \tau} \left(\sum_{i=1}^{s} \int_{s_i} \left.\frac{K(\lambda_m, \bar{r})}{k_i}\right|_{s_i} \cdot f_i \cdot ds_i\; d\tau \right) \tag{5-48}$$

Equation 5-48 is now rearranged into a form which will be directly comparable with the solution 5-45 which was obtained using Green's function.

$$T(\bar{r},t) = \int_R \left[\sum_{m=1}^{\infty} e^{-\alpha\lambda_m^2 t} \cdot K(\lambda_m,\bar{r}) \cdot K(\lambda_m,\bar{r}') \cdot F(\bar{r}')\right] \cdot d\bar{r}' + \frac{\alpha}{k}\int_{\tau=0}^{t} d\tau$$
$$\cdot \int_R g(\bar{r}',\tau) \cdot \left[\sum_{m=1}^{\infty} e^{-\alpha\lambda_m^2(t-\tau)} \cdot K(\lambda_m,\bar{r}) \cdot K(\lambda_m,\bar{r}')\right] d\bar{r}' + \alpha \int_{\tau=0}^{t} d\tau \sum_{i=1}^{s}$$
$$\cdot \int_{\substack{\text{boundary}\\ \text{surface } i}} \left[\frac{1}{k_i}\sum_{m=1}^{\infty} e^{-\alpha\lambda_m^2(t-\tau)} \cdot K(\lambda_m,\bar{r}) \cdot K(\lambda_m,\bar{r}_i')\right] \cdot f_i \cdot ds_i \qquad (5\text{-}49)$$

By comparing Eq. 5-49 with 5-45 we note that Green's function is related to the three-dimensional kernels $K(\lambda_m,\bar{r})$ by the relation

$$G(\bar{r},\bar{r}' \mid t,\tau) = \sum_{m=1}^{\infty} e^{-\alpha\lambda_m^2(t-\tau)} \cdot K(\lambda_m,\bar{r}) \cdot K(\lambda_m,\bar{r}') \qquad (5\text{-}50a)$$

$$G(\bar{r},\bar{r}' \mid t,\tau)\big|_{\tau=0} = \sum_{m=1}^{\infty} e^{-\alpha\lambda_m^2 t} \cdot K(\lambda_m,\bar{r}) \cdot K(\lambda_m,\bar{r}') \qquad (5\text{-}50b)$$

5-6. GREEN'S FUNCTION IN THE SOLUTION OF ONE AND TWO-DIMENSIONAL BOUNDARY-VALUE PROBLEMS OF HEAT CONDUCTION FOR FINITE REGIONS

The solution of the boundary-value problems of heat conduction for one and two-dimensional finite regions can be expressed in terms of Green's functions following a similar procedure as discussed for the three-dimensional finite region. The results, however, are slightly different, and we now examine these two cases.

Two-dimensional Finite Region. Consider the following boundary-value problem of heat conduction for a two-dimensional finite region for the space variables x, y in the cartesian coordinate system:

$$\frac{\partial^2 T}{\partial x^2} + \frac{\partial^2 T}{\partial y^2} + \frac{g(x,y,t)}{k} = \frac{1}{\alpha}\frac{\partial T}{\partial t} \quad \text{in region } R,\ t > 0$$
$$k_i \frac{\partial T}{\partial n_i} + h_i T = f_i(\bar{r},t) \quad \text{on boundary } s_i,\ t > 0 \qquad (5\text{-}51)$$
$$T = F(x,y) \quad \text{in region } R,\ t = 0$$

where $T = T(x,y,t)$.

We consider the following auxiliary problem:

$$\frac{\partial^2 G}{\partial x^2} + \frac{\partial^2 G}{\partial y^2} + \frac{1}{\alpha}\delta(x - x')\delta(y - y')\delta(t - \tau)$$
$$= \frac{1}{\alpha}\frac{\partial G}{\partial t} \quad \text{in region } R,\ t > \tau$$
$$k_i \frac{\partial G}{\partial n_i} + h_i G = 0 \quad \text{on boundary } s_i,\ t > \tau \qquad (5\text{-}52)$$

subject to the condition

$$G = 0 \quad \text{in region } R,\ t < \tau$$

where $G \equiv G(x,y;x',y' \mid t,\tau)$ is the Green's function that describes the temperature at x, y at time t due to an instantaneous heat source of strength unity—i.e.,

$$S_{\text{Li}} = \frac{\alpha g_{\text{Li}}}{k} = 1, \quad {}^{\circ}\text{F} \cdot \text{ft}^2$$

where g_{Li} in Btu/ft at the location (x', y'), releasing its heat spontaneously at time $t = \tau$ in the two-dimensional finite region R which is initially (i.e., for $t < \tau$) at zero temperature and boundary surfaces of which are subject to homogeneous boundary conditions.

The solution of the boundary-value problem of heat conduction 5-51 is given in terms of the Green's function satisfying the auxiliary system 5-52 as

$$\begin{aligned} T(x,y,t) = & \iint_{\text{region}} G\,|_{\tau=0} \cdot F(x',y') \cdot dx' \cdot dy' \\ & + \frac{\alpha}{k} \int_{\tau=0}^{t} d\tau \iint_{\text{region}} g(x',y',\tau) \cdot G \cdot dx' dy' \\ & + \alpha \int_{\tau=0}^{t} d\tau \sum_{i=1}^{s} \int_{\substack{\text{boundary} \\ \text{path } i}} \frac{G}{k_i}\bigg|_i \cdot f_i \cdot dl \end{aligned} \tag{5-53}$$

where $G = G(x,y;x',y' \mid t,\tau)$. For boundary condition of the first kind we set $k_i = 0$; in such a case the following change should be made in Eq. 5-53.

$$\text{When} \quad k_i = 0, \quad \text{replace} \quad \frac{G}{k_i}\bigg|_i \quad \text{by} \quad -\frac{1}{h_i}\frac{\partial G}{\partial n_i}\bigg|_i$$

Green's function for use in the two-dimensional Prob. 5-53 is related to the kernels $K(\lambda_m, \bar{r})$ in a similar manner as given by Eq. 5-50, except for this case $K(\lambda_m, \bar{r})$ is a two-dimensional kernel and can be constructed from the one-dimensional kernels given in the previous chapters.

One-dimensional Finite Region. The boundary-value problem of heat conduction is given as

$$\begin{aligned} \frac{\partial^2 T}{\partial x^2} + \frac{g(x,t)}{k} &= \frac{1}{\alpha}\frac{\partial T}{\partial t} \quad \text{in region } R,\ t > 0 \\ k_i \frac{\partial T}{\partial n_i} + h_i T &= f_i(t) \quad \text{on boundary } i,\ t > 0 \\ T &= F(x) \quad \text{in region } R,\ t = 0 \end{aligned} \tag{5-54}$$

where $T \equiv T(x,t)$.

The corresponding auxiliary problem is

$$\frac{\partial^2 G}{\partial x^2} + \frac{1}{\alpha}\,\delta(x - x')\cdot\delta(t - \tau) = \frac{1}{\alpha}\frac{\partial G}{\partial t} \quad \text{in region } R,\ t > \tau$$

$$k_i \frac{\partial G}{\partial n_i} + h_i G = 0 \quad \text{on boundary } i,\ t > \tau \tag{5-55}$$

$$G = 0 \quad \text{in region } R,\ t < \tau$$

where $G \equiv G(x, x' \mid t, \tau)$ is the Green's function which describes temperature at x at time t due to an instantaneous heat source of strength unity, i.e.,

$$S_{si} = \frac{\alpha g_{si}}{k} = 1, \quad {}^\circ\text{F}\cdot\text{ft}$$

where g_{si} is in Btu/ft^2; situated at x', releasing its heat spontaneously at time $t = \tau$ in the one-dimensional finite region R, which is initially (i.e., $t < \tau$) at zero temperature and the boundaries of which are subject to homogeneous boundary conditions.

The solution of the boundary-value problem of heat conduction 5-51 is given in terms of the Green's function satisfying the auxiliary Prob. 5-55 as

$$T(x,t) = \int_R G\,|_{\tau=0}\cdot F(x')\cdot dx' + \frac{\alpha}{k}\int_{\tau=0}^{t} d\tau \cdot \int_R g(x',\tau)\cdot G\,dx' + \alpha\int_{\tau=0}^{t} d\tau \sum_{i=1} \frac{G}{k_i}\bigg|_i \cdot f_i \tag{5-56}$$

For boundary condition of the first kind (i.e., when $k_i = 0$) in Eq. 5-56 the term $\left.\frac{G}{k_i}\right|_i$ should be replaced by $-\left.\frac{1}{h_i}\frac{\partial G}{\partial n_i}\right|_i$.

5-7. APPLICATION OF GREEN'S FUNCTIONS

In the following examples the use of Green's functions in the solution of heat-conduction problems is illustrated.

A Rectangular Parallelepiped. A rectangular parallelepiped $0 \le x \le a$, $0 \le y \le b$, $0 \le z \le c$ is initially at temperature $F(x, y, z)$. For times $t > 0$ there is heat generation within the region at a rate of $g(x, y, z, t)$ Btu/hr ft^3, while the boundary surfaces are kept at zero temperature. Temperature distribution in the solid for times $t > 0$ is immediately determined by means of Eq. 5-45 provided that Green's function for the problem under consideration is known.

Green's function for a rectangular parallelepiped $0 \le x \le a$, $0 \le y \le b$, $0 \le y \le c$ subjected to homogeneous boundary condition of the first kind at all boundary surfaces is given by (see Eq. 2-254)

$$G(x, y, z, x', y', z' \mid t, \tau) = \frac{8}{abc} \sum_{m=1}^{\infty} \sum_{n=1}^{\infty} \sum_{p=1}^{\infty} e^{-\alpha(\beta_m^2 + \nu_n^2 + \eta_p^2)(t-\tau)} \cdot \sin \beta_m x' \cdot \sin \nu_n y' \cdot \sin \eta_p z' \cdot \sin \beta_m x \cdot \sin \nu_n y \cdot \sin \eta_p z \tag{5-57}$$

where $\beta_m = \frac{m\pi}{a} \quad m = 1, 2, 3, \ldots$

$$\nu_n = \frac{n\pi}{b} \quad n = 1, 2, 3, \ldots$$

$$\eta_p = \frac{p\pi}{c} \quad p = 1, 2, 3, \ldots$$

substituting Green's function 5-57 into Eq. 5-45, the solution of the present problem is given in the form

$$T(x, y, z, t) = \frac{8}{abc} \sum_{m=1}^{\infty} \sum_{n=1}^{\infty} \sum_{p=1}^{\infty} e^{-\alpha(\beta_m^2 + \nu_n^2 + \eta_p^2)t} \cdot \sin \beta_m x \cdot \sin \nu_n y \cdot \sin \eta_p z \cdot \int_{x'=0}^{a} \int_{y'=0}^{b} \int_{z'=0}^{c} F(x', y', z') \cdot \sin \beta_m x' \cdot \sin \nu_n y' \cdot \sin \eta_p z' \cdot dx' \cdot dy' \cdot dz' + \frac{\alpha}{k} \frac{8}{abc} \cdot \sum_{m=1}^{\infty} \sum_{n=1}^{\infty} \sum_{p=1}^{\infty} \sin \beta_m x \cdot \sin \nu_n y \cdot \sin \eta_p z \cdot \int_{\tau=0}^{t} d\tau \int_{x'=0}^{a} \int_{y'=0}^{b} \int_{z'=0}^{c} g(x', y', z', \tau) \cdot e^{-\alpha(\beta_m^2 + \nu_n^2 + \eta_p^2)(t-\tau)} \cdot \sin \beta_m x' \cdot \sin \nu_n y' \cdot \sin \eta_p z' \cdot dx' \cdot dy' \cdot dz' \tag{5-58}$$

The solution 5-58 is the same as solution 2-249 which was obtained with integral-transform technique.

A Long Hollow Cylinder $a \leq r \leq b$. A long hollow cylinder $a \leq r \leq b$ is initially at temperature $F(r)$. For times $t > 0$ heat is generated within the region at a rate of $g(r, t)$ Btu/hr ft^3, while the boundary surfaces at $r = a$ and $r = b$ are kept at zero temperatures.

Green's function for a hollow cylinder $a \leq r \leq b$ subjected to boundary condition of the first kind at the boundaries $r = a$ and $r = b$ is given (see Eq. 3-143 for $S_{cyl,i} = 1$) as

$$G(r, r' \mid t, \tau) = \frac{1}{2\pi} \sum_{m=1}^{\infty} e^{-\alpha\beta_m^2(t-\tau)} \cdot K_0(\beta_m r) \cdot K_0(\beta_m r') \tag{5-59}$$

where the kernel $K_0(\beta_m r)$ and the eigenvalues β_m are as defined by Eqs. 3-133 and 3-132 respectively.

Solution of the present problem is obtained by substituting the Green's function 5-59 into Eq. 5-56 and noting that the boundary-condition functions f_i's are zero. Since the region under consideration is in the form of concentric circles of radii a and b, in Eq. 5-56 the integrations over the cross section should be performed in the form

$$\int_a^b 2\pi r'\,dr'$$

Then the solution of the present problem is given in the form

$$T(r,t) = \sum_{m=1}^{\infty} e^{-\alpha\beta_m^2 t}\cdot K_0(\beta_m r)\cdot \int_{r'=a}^{b} r'\cdot K_0(\beta_m r')\cdot F(r')\cdot dr' + \frac{\alpha}{k}\sum_{m=1}^{\infty} e^{-\alpha\beta_m^2 t}\cdot K_0(\beta_m r)\cdot \int_{\tau=0}^{t} e^{\alpha\beta_m^2\tau}\cdot d\tau \int_{r'=a}^{b} r' \cdot K_0(\beta_m r')\cdot g(r',\tau)\cdot dr' \tag{5-60}$$

It is to be noted that the solution 5-60 is similar to the solution 3-128 of hollow-cylinder problem when the boundary-condition functions $\phi_1(t)$ and $\phi_2(t)$ in Eq. 3-128 are chosen as zero. Equation 3-128 was obtained with integral-transform technique.

REFERENCES

1. R. C. Bartels and R. V. Churchill, "Resolution of Boundary Problems by the Use of a Generalized Convolution," *Bull. Amer. Math. Soc.*, Vol. 48 (1942), pp. 276–282.

2. Ian N. Sneddon, *Fourier Transforms*, McGraw-Hill Book Company, New York, 1951, p. 164.

3. H. S. Carslaw and J. C. Jaeger, *Conduction of Heat in Solids*, Clarendon Press, Oxford, 1959, pp. 30–32.

4. P. M. Morse and H. Feshbach, *Methods of Theoretical Physics, Part I*, McGraw-Hill Book Company, New York, 1953, Chapter 7.

5. Ian N. Sneddon, *Partial Differential Equations*, McGraw-Hill Book Company, New York, 1957, pp. 167–196, 296–302.

6. John W. Dettman, *Mathematical Methods in Physics and Engineering*, McGraw-Hill Book Company, New York, 1962, Chapter 4.

7. R. Courant and D. Hilbert, *Methods of Mathematical Physics*, Vol. I, Interscience Publishers, Inc., New York, 1953, pp. 351–396.

8. H. S. Carslaw and J. C. Jaeger, *Conduction of Heat in Solids*, Clarendon Press, Oxford, 1959, pp. 353–386, 422–424.

PROBLEMS

1. Find the transient temperature distribution $T(x,t)$ in a semi-infinite solid $0 \le x < \infty$ which is initially at zero temperature and for times $t > 0$ the boundary surface at $x = 0$ is kept at constant temperature T_0. By making use of this solution and applying Duhamel's theorem determine the solution of the above problem when the boundary surface at $x = 0$ is kept at a temperature $f(t)$ which is a function of time.

2. Find the one-dimensional transient-temperature distribution $T(r, t)$ for a long solid cylinder of radius $r = b$, which is initially at zero temperature and for times $t > 0$ the boundary surface at $r = b$ is subjected to convection with a medium at constant temperature T_∞. Using this solution and by applying Duhamel's theorem determine the solution of the same problem when the boundary surface at $r = b$ is subjected to convection with a medium at a temperature $f(t)$ which varies with time.

3. Determine the one-dimensional transient temperature distribution $T(x, t)$ for a slab $0 \leq x \leq L$, which is initially at temperature $F(x)$ and for times $t > 0$ the boundary at $x = 0$ is subjected to heat flux $f(t)$ which varies with time and that at $x = L$ is kept at zero temperature, by making use of Duhamel's theorem and the solution of similar problems with constant heat flux at surface $x = 0$.

4. A point heat source is situated at a point x_1, y_1, z_1 in an infinite medium which is initially at zero temperature. For times $t > 0$ the point heat source releases its heat at a rate of $g(t)$ Btu/hr which varies with time. Using Duhamel's theorem and the solution of a similar problem with constant heat generation, determine temperature distribution in the solid.

5. Determine the one-dimensional transient-temperature distribution $T(r, t)$ for a solid sphere of radius $r = b$, which is initially at zero temperature and for times $t > 0$ the boundary surface at $r = b$ is kept at a temperature of unity. Using this solution, and by making use of Duhamel's theorem, determine the solution for the same problem when the boundary surface at $r = b$ is kept at a temperature $f(t)$ which varies with time.

6. Green's function for the heat-conduction problem in a semi-infinite region $0 \leq x < \infty$ which is subjected to a homogeneous boundary condition of the first kind at the boundary surface $x = 0$ is given as (see Eq. 2-150):

$$G(x, x' \mid t, \tau) = \frac{1}{\left[4\pi\alpha\,(t - \tau)\right]^{1/2}} \left[e^{-\frac{(x-x')^2}{4\alpha(t-\tau)}} - e^{-\frac{(x+x')^2}{4\alpha(t-\tau)}} \right]$$

Using this Green's function, determine the transient-temperature distribution $T(x, t)$ in a semi-infinite region $0 \leq x < \infty$, which is initially at temperature $F(x)$, for times $t > 0$ heat is generated within the solid at a rate of $g(x, t)$ Btu/hr ft^3 while the boundary surface at $x = 0$ is kept at zero temperature. Compare the resulting solution with Eq. 2-147.

7. Green's function for heat conduction in a rectangle $0 \leq x \leq a, 0 \leq y \leq b$ which is subjected to homogeneous boundary condition of the first kind at the boundaries is given as (see Eq. 2-216):

$$G(x, y, x', y' \mid t, \tau) = \frac{4}{ab} \sum_{m=1}^{\infty} \sum_{n=1}^{\infty} e^{-\alpha\lambda_{mn}^2 (t-\tau)}$$

$$\cdot \sin \beta_m x_1 \cdot \sin \nu_n y_1 \cdot \sin \beta_m x \cdot \sin \nu_n y$$

where $\lambda_{mn}^2 = \beta_m^2 + \nu_n^2$

$$\beta_m = \frac{m\pi}{a}, \quad m = 1, 2, 3, \ldots$$

$$\nu_n = \frac{n\pi}{b}, \quad n = 1, 2, 3, \ldots$$

Using the Green's function given above, determine the transient-temperature distribution in a rectangle $0 \le x \le a, 0 \le y \le b$ which is initially at temperature $F(x, y)$, for times $t > 0$ there is heat generation within the solid at a rate of $g(x, y, t)$ Btu/hr ft^3, while the boundary at $y = 0$ is kept at temperature $\phi(x, t)$ and the remaining boundaries are kept at zero temperature. Compare this solution with the solution given by Eq. 2-206.

8. Green's function for heat conduction in a long solid cylinder $0 \le r \le b$, subjected to homogeneous boundary condition of the third kind at the boundary surface $r = b$ is given as (see Eq. 3-118):

$$G(r, r' \mid t, \tau) = \frac{1}{\pi b^2} \sum_{m=1}^{\infty} e^{-\alpha\beta_m^2(t-\tau)} \cdot \frac{J_0(\beta_m r) \cdot J_0(\beta_m r')}{J_0^2(\beta_m b) + J_1^2(\beta_m b)}$$

where the β_m's are the positive roots of the transcendental equation

$$\frac{\beta J_1(\beta b)}{J_0(\beta b)} = \frac{h}{k}$$

Using this Green's function, determine the transient-temperature distribution $T(r, t)$ in a long solid cylinder of radius $r = b$, which is initially at temperature $F(r)$ and for times $t > 0$ there is heat generation within the solid at a rate of $g(r, t)$ Btu/hr ft^3 while the boundary surface at $r = b$ is subjected to convection with a medium at zero temperature. Compare the resulting solution with the solution given by Eq. 3-90.

9. Green's function for heat conduction in solid sphere of radius $r = b$, subjected to homogeneous boundary condition of the first kind at the boundary surface $r = b$ is given as (see Eq. 4-82 for $\alpha g/k = 1$):

$$G(r, r' \mid t, \tau) = \frac{1}{2\pi r r' b} \sum_{m=1}^{\infty} e^{-\alpha\beta_m^2(t-\tau)} \cdot \sin \beta_m r \cdot \sin \beta_m r'$$

where $\beta_m = \dfrac{m\pi}{b}, \quad m = 1, 2, 3, \ldots.$

Using this Green's function, determine the transient-temperature distribution $T(r, t)$ in a solid sphere of radius $r = b$, which is initially at zero temperature, and for times $t > 0$ there is heat generation within the solid at a rate of $g(r, t)$ Btu/hr ft^3 while the boundary surface at $r = b$ is kept at zero temperature. Compare the resulting solution with the solution given by Eq. 4-78.

6

Composite Regions

Temperature distribution in composite regions consisting of several layers has numerous applications in heat-transfer problems in rocket thrust chamber liners, the fuel elements for nuclear reactors, reentry bodies, and the like. The Laplace transform technique is used by Carslaw and Jaeger [1] to solve the transient boundary-value problems of heat conduction in solids consisting of many parallel layers. In principle the method solves the heat-transfer problem in any composite slab, but in practice if the number of layers is more than two, the inverse of the Laplace transform becomes quite difficult. The *adjoint*-solution technique which has been introduced by Goodman [2], provides a method of solution to a large class of heat conduction problems in composite slabs from the solution of but one adjoint problem. The primary disadvantage of the adjoint method is that only the solutions on the boundaries (i.e., interface) of the layers can be determined. In some practical problems, however, this restriction does not impose any serious limitation to the usefulness of the method. For example, if the temperature in a structure should be kept below a certain critical value and if the maximum temperature is known to occur on the boundary, then the method of adjoint solution is ideally suited for the problem. The adjoint method relates the solution of the heat-conduction problem for a multilayer region to the solution of one or more simple problems of a single region, each with zero initial temperature and a prescribed boundary condition.

Recently Tittle [3,4] introduced a technique for orthogonal expansion of functions over a one-dimensional, multilayer region. The method essentially is an extension of Sturm-Liouville problem to the case of one-dimensional multilayer region and it has the advantage over other analytical methods in that its application to the solution of the boundary-value problems of heat conduction is relatively simple, straightforward and similar to the ordinary expansion process in one-layer problems.

Bulavin and Kashcheev [5] used the method of separation of variables and of orthogonal expansion of functions over a one-dimensional multilayer region to solve the transient heat-conduction problem involving distributed volume heat sources in a multilayer region.

In this chapter we examine the methods of solution introduced by Goodman [2], Tittle [3], and Bulavin and Kashcheev [5].

6-1. THE ADJOINT-SOLUTION TECHNIQUE—BASIC CONCEPTS

Consider a composite slab consisting of m layers as shown in Fig. 6-1. We assume that the thermal properties for each layer are uniform but discontinuous at the interfaces. For no heat generation within the solid the differential equation of heat conduction for the ith layer is given as

$$\frac{\partial^2 T_i(x,t)}{\partial x^2} = \frac{1}{\alpha_i} \frac{\partial T_i(x,t)}{\partial t} \quad \text{in} \quad x_i \le x \le x_{i+1},\ t > 0 \tag{6-1a}$$

subject to

Prescribed boundary conditions at x_i and x_{i+1}, for $t > 0$. (6-1b)
Prescribed initial condition in $x_i \le x \le x_{i+1}$, for $t = 0$. (6-1c)

Since the region involves m layers, the problem will include $i = 1, 2, 3, \ldots, m$ differential equations of the form given by Eq. 6-1, coupled with each other through the boundary conditions at the common interfaces.

We now consider the following *adjoint* system

$$\frac{\partial^2 V_i(x,t)}{\partial x^2} = -\frac{1}{\alpha_i} \frac{\partial V_i(x,t)}{\partial t} \quad \text{in} \quad x_i \le x \le x_{i+1} \tag{6-2a}$$

subject to

Prescribed boundary conditions at x_i and x_{i+1} (6-2b)

and

Condition $V_i(x,t) = 0$ in $x_i \le x \le x_{i+1}$, for $t = t_0$ (6-2c)

The function $V_i(x,t)$ is said to be *adjoint* to $T_i(x,t)$, and the differential operator $\left(\alpha_i \frac{\partial^2}{\partial x^2} + \frac{\partial}{\partial t}\right)$ is said to be *adjoint* to the differential operator $\left(\alpha_i \frac{\partial^2}{\partial x^2} - \frac{\partial}{\partial t}\right)$.

It is to be noted that the function $V_i(x,t)$ is specified at the time $t = t_0$ whereas the temperature function $T_i(x,t)$ is specified at $t = 0$. Therefore, if we want to determine $V_i(x,t)$ over the time interval $0 - t_0$, it is necessary to integrate differential Eq. 6-2 backward. Therefore, it is much more convenient to define a new time variable τ in the form

$$\tau = t_0 - t \tag{6-3}$$

Then system 6-2 becomes

$$\frac{\partial^2 V_i(x,\tau)}{\partial x^2} = \frac{1}{\alpha_i} \frac{\partial V_i(x,\tau)}{\partial \tau} \quad \text{in} \quad x_i \le x \le x_{i+1} \quad \text{for} \quad \tau > 0 \tag{6-4a}$$

subject to

Prescribed boundary conditions at x_i and x_{i+1}, for $\tau > 0$ (6-4b)

and

Initial condition $V_i(x,\tau) = 0$ in $x_i \le x \le x_{i+1}$, for $\tau = 0$ (6-4c)

The adjoint-differential Eq. 6-4a is now similar in form to the differential equation of heat conduction 6-1a because the time variable τ runs forward but the boundary conditions 6-1b and 6-4b will not be the same. A judicious choice will be made for the boundary condition 6-4b for the reasons that will be apparent later in the analysis.

We now try to obtain the solution of the boundary-value problem of heat conduction 6-1 at the interfaces in terms of the solutions of the adjoint problem (i.e., V_i function).

We multiply the differential equation in the system 6-1a by $V_i(x, t)$ and 6-2a by $T_i(x, t)$ and subtract the results,

$$V_i \frac{\partial^2 T_i}{\partial x^2} - T_i \frac{\partial^2 V_i}{\partial x^2} = \frac{1}{\alpha_i} \frac{\partial}{\partial t} [V_i T_i] \tag{6-5}$$

We integrate Eq. 6-5 with respect to the space variable from $x = x_i$ to $x = x_{i+1}$ and with respect to the time variable from $t = 0$ to $t = t_0$.

$$\int_{t=0}^{t_0} dt \int_{x_i}^{x_{i+1}} \left(V_i \frac{\partial^2 T_i}{\partial x^2} - T_i \frac{\partial^2 V_i}{\partial x^2} \right) dx = \frac{1}{\alpha_i} \int_{x_i}^{x_{i+1}} \left\{ [V_i T_i]_{t=t_0} - [V_i T_i]_{t=0} \right\} dx \tag{6-6}$$

The integral on the left-hand side with respect to the space variable is evaluated by applying Green's theorem; on the right-hand side the first term under the integral vanishes, since

$$V_i |_{t=t_0} = 0$$

Then, Eq. 6-6 is written in the form

$$\alpha_i \int_{t=0}^{t_0} \left\{ \left[V_i \frac{\partial T_i}{\partial x} - T_i \frac{\partial V_i}{\partial x} \right]_{x_{i+1}} - \left[V_i \frac{\partial T_i}{\partial x} - T_i \frac{\partial V_i}{\partial x} \right]_{x_i} \right\} dt + \int_{x_i}^{x_{i+1}} [V_i T_i]_{t=0} dx = 0 \tag{6-7}$$

where $i = 1, 2, 3, \ldots, m$.

For the case of zero initial temperature we have $T_i |_{t=0} = 0$; then Eq. 6-7 reduces to

$$\int_{t=0}^{t_0} \left\{ \left[V_i \frac{\partial T_i}{\partial x} - T_i \frac{\partial V_i}{\partial x} \right]_{x_{i+1}} - \left[V_i \frac{\partial T_i}{\partial x} - T_i \frac{\partial V_i}{\partial x} \right]_{x_i} \right\} dt = 0 \qquad i = 1, 2, 3, \ldots, m \tag{6-8}$$

In the present analysis we restrict our attention to heat conduction problems in composite slabs with zero initial temperature, hence examine only the solution of system of integral equations described by Eq. 6-8.

For the $(i-1)$th layer Eq. 6-8 is written as

$$\int_{t=0}^{t_0}\left\{\left[k_{i-1}V_{i-1}\frac{\partial T_{i-1}}{\partial x}-k_{i-1}T_{i-1}\frac{\partial V_{i-1}}{\partial x}\right]_{x_i}-\left[k_{i-1}V_{i-1}\frac{\partial T_{i-1}}{\partial x}-k_{i-1}T_{i-1}\frac{\partial V_{i-1}}{\partial x}\right]_{x_{i-1}}\right\}\cdot dt=0 \tag{6-9}$$

and for the ith layer as

$$\int_{t=0}^{t_0}\left\{\left[k_iV_i\frac{\partial T_i}{\partial x}-k_iT_i\frac{\partial V_i}{\partial x}\right]_{x_{i+1}}-\left[k_iV_i\frac{\partial T_i}{\partial x}-k_iT_i\frac{\partial V_i}{\partial x}\right]_{x_i}\right\}dt=0 \tag{6-10}$$

Equations 6-9 and 6-10 are combined in the form

$$\int_{t=0}^{t_0}\left\{\left[k_{i-1}T_{i-1}\frac{\partial V_{i-1}}{\partial x}-k_{i-1}V_{i-1}\frac{\partial T_{i-1}}{\partial x}\right]_{x_{i-1}}+\left[k_{i-1}V_{i-1}\frac{\partial T_{i-1}}{\partial x}-k_iV_i\frac{\partial T_i}{\partial x}\right]_{x_i}+\left[k_iT_i\frac{\partial V_i}{\partial x}-k_{i-1}T_{i-1}\frac{\partial V_{i-1}}{\partial x}\right]_{x_i}+\left[k_iV_i\frac{\partial T_i}{\partial x}-k_iT_i\frac{\partial V_i}{\partial x}\right]_{x_{i+1}}\right\}dt=0 \tag{6-11}$$

$$i=2,3,\ldots,m$$

Equation 6-11 relates the interface values of the heat conduction problem Eq. 6-1 for zero initial temperature to the interface values of the adjoint problem. Therefore, if the solution of the adjoint problem 6-4 at the interfaces is known, then Eq. 6-11 provides a set of integral equations for the unknown interface temperatures. On the other hand, Eq. 6-11 includes certain terms which are not always known for any given problem, but such terms can be removed from the equation by judicious choice of the boundary conditions for the adjoint problem as illustrated with examples in the following sections.

6-2. ADJOINT SOLUTION FOR A MULTILAYER SLAB WITH PRESCRIBED TEMPERATURE AT THE OUTER SURFACES

Consider an m-layer slab which is initially at zero temperature and for times $t>0$ the outer surfaces at x_1 and x_{m+1} are kept at prescribed temperatures as shown in Fig. 6-1. To simplify the analysis we assume further that the layers are all in perfect thermal contact (i.e., continuity of heat flux and temperature at the interfaces).

The boundary-value problem of heat conduction is given as

$$\frac{\partial^2 T_i(x,t)}{\partial x^2}=\frac{1}{\alpha_i}\frac{\partial T_i(x,t)}{\partial t}\quad \text{in}\quad x_i\le x\le x_{i+1},\ \text{for } t>0 \tag{6-12}$$

$$i=1,2,3,\ldots,m$$

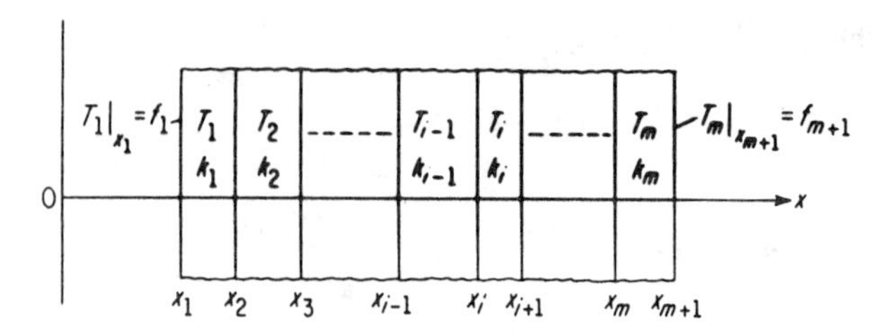

FIG. 6-1. A multilayer slab consisting of m parallel layers with prescribed temperature at the outer boundaries.

with the boundary conditions:

$$T_1(x_1, t) = f_1(t) \quad \text{at} \quad x = x_1 \tag{6-13a}$$

$$T_i(x_{i+1}, t) = T_{i+1}(x_{i+1}, t) \quad i = 1,2,3, \ldots, m - 1 \tag{6-13b}$$

$$k_i \frac{\partial T_i(x_{i+1}, t)}{\partial x} = k_{i+1} \frac{\partial T_{i+1}(x_{i+1}, t)}{\partial x}, \quad i = 1,2,3, \ldots, m - 1 \tag{6-13c}$$

$$T_m(x_{m+1}, t) = f_{m+1}(t) \quad \text{at} \quad x = x_{m+1} \tag{6-13d}$$

and the initial condition

$$T_i(x, t) = 0 \quad \text{in} \quad x_i \leq x \leq x_{i+1}, \text{for } t = 0 \quad i = 1,2,3, \ldots, m \tag{6-14}$$

The solution at the interfaces of the multilayer heat conduction problem described above by the system of Eqs. 6-12, 6-13 and 6-14 will be related to the solution of the system of equations described by Eq. 6-11. Various terms under the integral sign in Eq. 6-11 are simplified by judicious choice of the boundary conditions for the adjoint problem as described below. Choosing the boundary conditions for the adjoint problems as

$$V_{i-1}\big|_{x_i} = V_i\big|_{x_i} = 1 \tag{6-15}$$

and by making use of the perfect thermal contact condition at the interface, Eq. 6-13c, i.e.,

$$k_{i-1} \frac{\partial T_{i-1}}{\partial x}\bigg|_{x_i} = k_i \frac{\partial T_i}{\partial x}\bigg|_{x_i} \tag{6-16}$$

the terms in the second bracket in Eq. 6-11 vanish.

Choosing

$$V_{i-1}\bigg|_{x_{i-1}} = V_i\bigg|_{x_{i+1}} = 0 \tag{6-17}$$

the terms

$$\frac{\partial T_{i-1}}{\partial x}\bigg|_{x_{i-1}} \quad \text{and} \quad \frac{\partial T_i}{\partial x}\bigg|_{x_{i+1}}$$

are removed from Eq. 6-11.

Finally, by making use of the condition of continuity of temperature at the interface, Eq. 6-13b, i.e.,

$$T_{i-1}\Big|_{x_i} = T_i\Big|_{x_i} \tag{6-18}$$

Eq. 6-11 reduces to[1]

$$\int_{t=0}^{t_0}\left[\left(k_{i-1}T\frac{\partial V_{i-1}}{\partial x}\right)_{x_{i-1}} + \left(k_i T\frac{\partial V_i}{\partial x}\right)_{x_i} - \left(k_{i-1}T\frac{\partial V_{i-1}}{\partial x}\right)_{x_i} - \left(k_i T\frac{\partial V_i}{\partial x}\right)_{x_{i+1}}\right]dt = 0 \quad i = 2,3,\ldots,m \tag{6-19}$$

The subscripts on T have been omitted because they are no longer needed. Equation 6-19 is called *three-point temperature equation* because it relates the temperature at three successive boundaries, i.e., $T(x_{i-1},t)$, $T(x_i,t)$ and $T(x_{i+1},t)$.

Introducing the boundary conditions 6-13a and d, Eq. 6-19 yields the following $(m-1)$ simultaneous integral equations for the $(m-1)$ unknown interface temperatures.

For $i = 2$:

$$\int_{t=0}^{t_0}\left[k_1 f_1\left(\frac{\partial V_1}{\partial x}\right)_{x_1} + k_2\left(T\frac{\partial V_2}{\partial x}\right)_{x_2} - k_1\left(T\frac{\partial V_1}{\partial x}\right)_{x_2} - k_2\left(T\frac{\partial V_2}{\partial x}\right)_{x_3}\right]dt = 0 \tag{6-20}$$

For $i = 3,4,\ldots,(m-1)$:

$$\int_{t=0}^{t_0}\left[k_{i-1}\left(T\frac{\partial V_{i-1}}{\partial x}\right)_{x_{i-1}} + k_i\left(T\frac{\partial V_i}{\partial x}\right)_{x_i} - k_{i-1}\left(T\frac{\partial V_{i-1}}{\partial x}\right)_{x_i} - k_i\left(T\frac{\partial V_i}{\partial x}\right)_{x_{i+1}}\right]dt = 0 \tag{6-21}$$

[1] If the continuity of temperature, Eq. 6-18, were not used, the resulting equation would be

$$\int_{t=0}^{t_0}\left[\left(k_{i-1}T_{i-1}\frac{\partial V_{i-1}}{\partial x}\right)_{x_{i-1}} + \left(k_i T_i\frac{\partial V_i}{\partial x}\right)_{x_i} - \left(k_{i-1}T_{i-1}\frac{\partial V_{i-1}}{\partial x}\right)_{x_i} - \left(k_i T_i\frac{\partial V_i}{\partial x}\right)_{x_{i+1}}\right]dt = 0$$

This equation is called the *four-point* temperature equation because it relates the four temperatures at the boundaries, i.e., $T_{i-1}(x_{i-1},t)$, $T_i(x_{i-1},t)$, $T_{i-1}(x_i,t)$ and $T_i(x_{i+1},t)$. Since the continuity of temperature is not used as a boundary condition at the interface, we may prescribe any other boundary condition at the interface. One such boundary condition, for example, is the presence of contact resistance at the interface.

For $i = m$:

$$\int_{t=0}^{t_0}\left[k_{m-1}\left(T\frac{\partial V_{m-1}}{\partial x}\right)_{x_{m-1}} + k_m\left(T\frac{\partial V_m}{\partial x}\right)_{x_m} - k_{m-1}\left(T\frac{\partial V_{m-1}}{\partial x}\right)_{x_m} - k_m f_{m+1}\cdot\left(\frac{\partial V_m}{\partial x}\right)_{x_{m+1}}\right]dt = 0 \tag{6-22}$$

To solve the above system of $(m - 1)$ equations the *adjoint functions* V_i's are needed. The functions V_i and V_{i-1} are determined from the solution of system 6-4 subjected to boundary conditions 6-15 and 6-17; that is, V_{i-1} functions satisfy

$$\begin{aligned}\frac{\partial^2 V_{i-1}}{\partial x^2} &= \frac{1}{\alpha_{i-1}}\frac{\partial V_{i-1}}{\partial \tau} \quad \text{in} \quad x_{i-1} \le x \le x_i,\ \tau > 0\\ V_{i-1} &= 0 \quad \text{at} \quad x = x_{i-1},\ \tau > 0\\ V_{i-1} &= 1 \quad \text{at} \quad x = x_i,\ \tau > 0\\ V_{i-1} &= 0 \quad \text{in} \quad x_{i-1} \le x \le x_i,\ \tau = 0\end{aligned} \tag{6-23}$$

and the solution of which is

$$V_{i-1} = \frac{x - x_{i-1}}{x_i - x_{i-1}} + \frac{2}{\pi}\sum_{n=1}^{\infty}\frac{(-1)^n}{n}e^{-\alpha_{i-1}\frac{n^2\pi^2}{(x_i - x_{i-1})^2}\tau}\cdot\sin\frac{n\pi(x - x_{i-1})}{x_i - x_{i-1}}, \tag{6-24}$$

V_i functions satisfy

$$\begin{aligned}\frac{\partial^2 V_i}{\partial x^2} &= \frac{1}{\alpha_i}\frac{\partial V_i}{\partial \tau} \quad \text{in} \quad x_i \le x \le x_{i+1},\ \tau > 0\\ V_i &= 1 \quad \text{at} \quad x = x_i,\ \tau > 0\\ V_i &= 0 \quad \text{at} \quad x = x_{i+1},\ \tau > 0\\ V_i &= 0 \quad \text{in} \quad x_i \le x \le x_{i+1},\ \tau = 0\end{aligned} \tag{6-25}$$

and the solution of which is

$$V_i = 1 - \frac{x - x_i}{x_{i+1} - x_i} - \frac{2}{\pi}\sum_{n=1}^{\infty}\frac{1}{n}e^{-\alpha_i\frac{n^2\pi^2}{(x_{i+1} - x_i)^2}\tau}\cdot\sin\frac{n\pi(x - x_i)}{x_{i+1} - x_i} \tag{6-26}$$

Derivatives of V_{i-1} and V_i functions are evaluated as

$$\left.\frac{\partial V_{i-1}}{\partial x}\right|_{x_{i-1}} = \frac{1}{x_i - x_{i-1}} + \frac{2}{x_i - x_{i-1}}\sum_{n=1}^{\infty}(-1)^n e^{-\alpha_{i-1}\frac{\pi^2 n^2}{(x_i - x_{i-1})^2}\tau} \tag{6-27a}$$

$$\left.\frac{\partial V_i}{\partial x}\right|_{x_i} = -\frac{1}{x_i - x_{i-1}} - \frac{2}{x_{i+1} - x_{i-1}}\sum_{n=1}^{\infty}e^{-\alpha_i\frac{\pi^2 n^2}{(x_{i+1} - x_i)^2}\tau} \tag{6-27b}$$

$$\left.\frac{\partial V_{i-1}}{\partial x}\right|_{x_i} = \frac{1}{x_i - x_{i-1}} + \frac{2}{x_i - x_{i-1}} \sum_{n=1}^{\infty} e^{-\alpha_{i-1} \frac{\pi^2 n^2}{(x_i - x_{i-1})^2} \tau} \tag{6-27c}$$

$$\left.\frac{\partial V_i}{\partial x}\right|_{x_{i+1}} = -\frac{1}{x_{i+1} - x_i} - \frac{2}{x_{i+1} - x_i} \sum_{n=1}^{\infty} (-1)^n e^{\alpha_i \frac{\pi^2 n^2}{(x_{i+1} - x_i)^2} \tau} \tag{6-27d}$$

Substituting the derivatives of adjoint functions from 6-27 into Eqs. 6-20, 6-21 and 6-22, we obtain $(m - 1)$ simultaneous-integral equations for the $(m - 1)$ unknown interface temperatures. Since these integral equations are of convolution form with V_i functions playing the role of the kernel[2], i.e.,

$$\int_{t=0}^{t_0} T(x, t) \cdot V(x, t_0 - t) \cdot dt = 0 \tag{6-28}$$

they are most appropriately solved with an analog computer.

6-3. ADJOINT SOLUTION FOR A MULTILAYER SLAB FOR OTHER BOUNDARY CONDITIONS AT THE OUTER SURFACES

The choice of boundary conditions for the adjoint problem depends on the type of boundary conditions prescribed for the heat-transfer problem. To illustrate the procedure we examine a few such cases.

Prescribed Heat Flux at x_1 and Prescribed Temperature at x_{m+1}. For this case the boundary conditions at the outer surfaces of the m-layer slab is given as

$$k_1 \left.\frac{\partial T_1}{\partial x}\right|_{x_1} = \phi_1 \tag{6-29a}$$

$$T_m\Big|_{x_{m+1}} = f_{m+1} \tag{6-29b}$$

Figure 6-2 shows the boundary conditions and the geometry.

In this case the integral equations 6-21 and 6-22 are applicable for this problem, but the Eq. 6-20 should be modified because it involves the value of temperature at the boundary surface x_1 (i.e., $T\,|_{x_1} = f_1$) which is an unknown quantity for the present problem.

[2] In Ref. 6, in applying the adjoint-solution technique to transient heat-conduction problems in multilayer slabs the resulting integral equations were considered erroneously in the form

$$\int_{t=0}^{t_0} g(t) \cdot dt = 0$$

rather than in the convolution form as given by Eq. 6-28; and the integral equations were transformed to a set of algebraic equations. As pointed out by T. R. Goodman, the integral equations of the convolution form could not be reduced to algebraic equations.

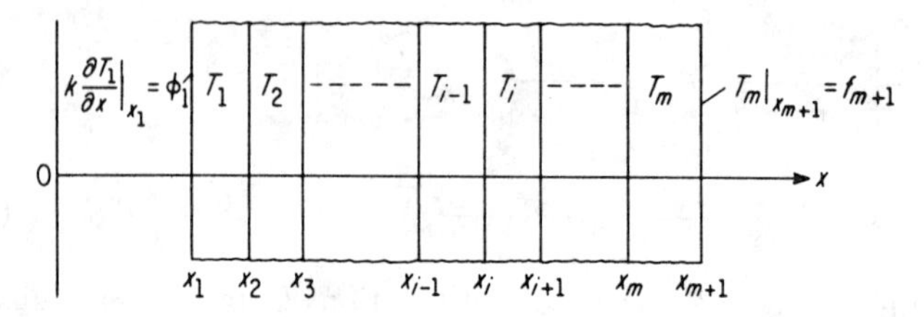

FIG. 6-2. A multilayer slab consisting of m parallel layers with prescribed heat flux at boundary x_1 and prescribed temperature at x_{m+1}.

An alternate relation that replaces Eq. 6-20 can be derived from Eq. 6-11, by setting $i = 2$ and choosing the boundary condition for the adjoint problem for V_1 as

$$\frac{\partial V_1}{\partial x}\bigg|_{x_1} = 0 \tag{6-30}$$

instead of $V_1 \,|_{x_1} = 0$ as chosen previously. The resulting alternate expression that replaces Eq. 6-20 is

$$\int_{t=0}^{t_0} \left[-\phi_1 \cdot V_1\bigg|_{x_1} + k_2 \left(T \frac{\partial V_2}{\partial x}\right)_{x_2} - k_1 \left(T \frac{\partial V_1}{\partial x}\right)_{x_2} - k_2 \left(T \frac{\partial V_2}{\partial x}\right)_{x_3} \right] dt = 0 \tag{6-31}$$

where the adjoint function V_1 is the solution of the following system:

$$\begin{aligned} \frac{\partial^2 V_1}{\partial x^2} &= \frac{1}{\alpha_1}\frac{\partial V_1}{\partial \tau} \quad \text{in} \quad x_1 \le x \le x_2,\ \tau > 0 \\ \frac{\partial V_1}{\partial x} &= 0 \quad \text{at} \quad x = x_1,\ \tau > 0 \\ V_1 &= 1 \quad \text{at} \quad x = x_2,\ \tau > 0 \\ V_1 &= 0 \quad \text{in} \quad x_1 \le x \le x_2,\ \tau = 0 \end{aligned} \tag{6-32}$$

Hence, Eq. 6-21, 6-22 and 6-31 are $(m - 1)$ simultaneous-integral equations for the $(m - 1)$ unknown interface temperatures.

Prescribed Heat Flux at x_{m+1} and Prescribed Temperature at x_1

$$T_1(x_1, t) = f_1 \tag{6-33a}$$

$$k_m \frac{\partial T_m(x_{m+1}, t)}{\partial x} = \phi_{m+1} \tag{6-33b}$$

Figure 6-3 shows the boundary conditions for this problem. The present problem differs from the problem in the previous section in that the boundary condition at the outer surface x_{m+1} is of the second kind. Therefore the integral Eqs. 6-20 and 6-21 are applicable for this problem, but Eq. 6-22 should be modified as it involves the temperature of the boundary surface at x_{m+1}, which is now an unknown quantity.

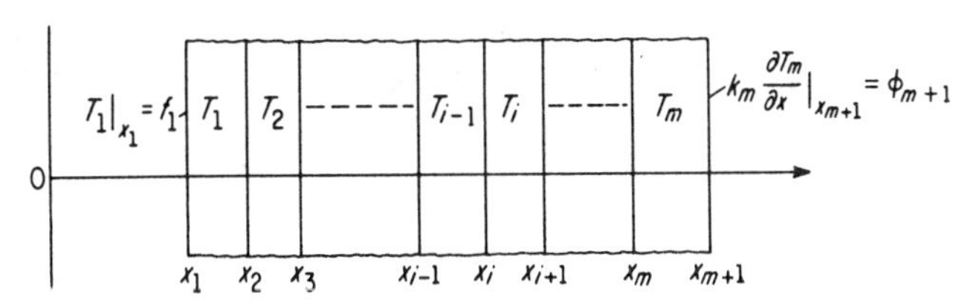

FIG. 6-3. A multilayer slab consisting of m parallel layers with prescribed temperature at boundary x_1 and prescribed heat flux at x_{m+1}.

An alternate Eq. to 6-22 can be derived from Eq. 6-11 by setting $i = m$ and choosing the boundary condition at x_{m+1} for the adjoint problem for V_m as

$$\left.\frac{\partial V_m}{\partial x}\right|_{x_{m+1}} = 0 \tag{6-34}$$

The resulting alternate expression that replaces Eq. 6-22 is

$$\int_{t=0}^{t_0} \left[k_{m-1} \left(T \frac{\partial V_{m-1}}{\partial x} \right)_{x_{m-1}} + k_m \left(T \frac{\partial V_m}{\partial x} \right)_{x_m} - k_{m-1} \left(T \frac{\partial V_{m-1}}{\partial x} \right)_{x_m} + \phi_{m+1} \cdot (V_m)_{x_{m+1}} \right] dt = 0 \tag{6-35}$$

where the adjoint function V_m is the solution of the following system:

$$\begin{aligned}
\frac{\partial^2 V_m}{\partial x^2} &= \frac{1}{\alpha} \frac{\partial V_m}{\partial \tau} \quad \text{in} \quad x_m \le x \le x_{m+1}, \tau > 0 \\
V_m &= 1 \quad \text{at} \quad x = x_m, \tau > 0 \\
\frac{\partial V_m}{\partial x} &= 0 \quad \text{at} \quad x = x_{m+1}, \tau > 0 \\
V_m &= 0 \quad \text{in} \quad x_m \le x \le x_{m+1}, \tau = 0
\end{aligned} \tag{6-36}$$

Hence, Eqs. 6-20, 6-21 and 6-35 provide $(m - 1)$ simultaneous-integral equations for the $(m - 1)$ unknown interface temperatures.

6-4. ADJOINT SOLUTION FOR A MULTILAYER SLAB AT STEADY STATE

The adjoint-solution technique discussed above for the time-dependent problems is readily applied to determining the steady-state temperature at the interfaces of a m-layer slab subject to a prescribed boundary condition at the external surfaces. To illustrate the procedure we examine the solution for a m-layer slab subject to prescribed temperatures at the outer-boundary surfaces, given as

$$T_1 = f_1 \quad \text{at} \quad x = x_1 \tag{6-37a}$$

$$T_m = f_{m+1} \quad \text{at} \quad x = x_{m+1} \tag{6-37b}$$

The geometry is similar to the one shown in Fig. 6-1. We assume further that there is perfect thermal contact between the layers at mutual interface.

Since there is no time-dependence, the three-point temperature equation 6-19 reduces to the following algebraic equation:

$$k_{i-1} T(x_{i-1}) \frac{dV_{i-1}(x_{i-1})}{dx} + k_i T(x_i) \frac{dV_i(x_i)}{dx} - k_{i-1} T(x_i) \frac{dV_{i-1}(x_i)}{dx}$$

$$-k_i T(x_{i+1}) \frac{dV_i(x_{i+1})}{dx} = 0, \quad i = 1,2,3,4, \ldots, m \tag{6-38}$$

The subscripts on T have been omitted because they are no longer needed.

Adjoint functions V_{i-1} and V_i satisfy the steady-state case of systems 6-23 and 6-25 respectively, and their solutions are given as

$$V_{i-1} = \frac{x - x_{i-1}}{x_i - x_{i-1}} \tag{6-39a}$$

$$V_i = \frac{x_{i+1} - x}{x_{i+1} - x_i} \tag{6-39b}$$

Substituting Eqs. 6-37 and 6-39 into Eq. 6-38 we obtain the following $(m - 1)$ simultaneous algebraic equations for the $(m - 1)$ unknown interface temperatures:

For $i = 2$:

$$\frac{k_1}{x_2 - x_1} f_1 - \left[\frac{k_2}{x_3 - x_2} - \frac{k_1}{x_2 - x_1}\right] T(x_2) + \frac{k_2}{x_3 - x_2} T(x_3) = 0 \tag{6-40}$$

For $i = 3,4, \ldots, m - 1$:

$$\frac{k_{i-1}}{x_i - x_{i-1}} T(x_{i-1}) - \left[\frac{k_i}{x_{i+1} - x_i} + \frac{k_{i-1}}{x_i - x_{i-1}}\right] T(x_i)$$

$$+ \frac{k_i}{x_{i+1} - x_i} T(x_{i+1}) = 0 \tag{6-41}$$

For $i = m$:

$$\frac{k_{m-1}}{x_m - x_{m-1}} T(x_{m-1}) - \left[\frac{k_m}{x_{m+1} - x_m} + \frac{k_{m-1}}{x_m - x_{m-1}}\right] T(x_m)$$

$$+ \frac{k_m}{x_{m+1} - x_m} \cdot f_{m+1} = 0 \tag{6-42}$$

The above $(m - 1)$ simultaneous algebraic equations are easily solved with a digital computer.

For other types of boundary conditions at the outer surfaces Eq. 6-40 and/or 6-42 should be modified in a similar manner as described previously.

6-5. ORTHOGONAL-EXPANSION TECHNIQUE OVER A MULTILAYER REGION

A relatively simple and straightforward method in the solution of boundary-value problems such as heat conduction in multilayer regions, based on the orthogonal-expansion technique of a function over a multilayer region, is introduced by Tittle [3]. The method is essentially an extension of ordinary orthogonality with respect to the usual weighting function (i.e., r in the cylindrical coordinate system) derived from the one-region Sturm-Liouville problem to the case of multilayer region by means of introduction of a new orthogonality factor called the *discontinuous-weighting function* [4]. Following Tittle [3], we examine the method of expansion of an arbitrary function $F(x)$ in an infinite series of eigenfunctions $X_n(x)$ over a one-dimensional composite region involving m layers (i.e., m layers of slabs, concentric cylinders or concentric spheres).

Consider a one-dimensional composite region involving m parallel layers. Let $T(x, t)$ be the temperature function that is to be be evaluated over the region $x_1 \leq x \leq x_{m+1}$, involving m parallel layers in which the thermal properties for each layer are uniform but discontinuous at the interfaces. We consider the case in which temperature function $T(x, t)$ is separated into space and time variables and represented in an infinite series in the form

$$T(x, t) = \sum_{n=1}^{\infty} A_n \cdot X_n(x) \cdot \Gamma(t) \quad \text{in } x_1 \leq x \leq x_{m+1} \tag{6-43}$$

where A_n are arbitrary constants, $X_n(x)$ are the eigenfunctions associated with the eigenvalue problem, and $\Gamma(t)$ is the solution of the separated equation for the time variable. In writing Eq. 6-43 it is assumed that all the boundary conditions including those of the interfaces and at the outer boundaries have been used, but the initial condition for the problem has not yet been applied.

Let $F(x)$ be the function that is to be satisfied by Eq. 6-43 at time $t = 0$ over the entire region $x_1 \leq x \leq x_{m+1}$ involving m layers. Then, for time $t = 0$ Eq. 6-43 becomes

$$F(x) = \sum_{n=1}^{\infty} A_n X_n(x) \quad x_1 \leq x \leq x_{m+1} \tag{6-44}$$

because the separation function (t) for the time variable is of exponential form. Equation 6-44 is an expansion of an arbitrary function $F(x)$ in an infinite series of the eigenfunctions $X_n(x)$ over a composite region involving m layers. In general, for a region involving more than one layer, the eigenfunctions $X_n(x)$ are not orthogonal with respect to the usual weighting function derived from a one-region Sturm-Liouville problem because the condition of the Sturm-Liouville problem is violated in that the coefficients of the differential equation (i.e., in the case of heat-conduction equation the thermal properties) are not continuous in a composite media [7].

However, an orthogonal set $G_n(x)$ can be constructed from the non-orthogonal set $X_n(x)$ by multiplying it in each layer i by the orthogonality factor W_i called the *discontinuous-weighting function* for layer i. Then the orthogonal set $G_n(x)$ is written as

$$G_n(x) = W_i \cdot X_n(x) \quad \text{in} \quad \text{layer } i \tag{6-45}$$

where the discontinuous-weighting functions W_i are derived from the properties of each layer and will be given later in the analysis.

We now express the initial-condition function $F(x)$ as a sum of m subfunctions $F_i(x)$ in the form

$$F(x) = \sum_{i=1}^{m} F_i(x) \tag{6-46}$$

where the subfunctions $F_i(x)$ are defined as

$$F_i(x) = F(x) \quad \text{in} \quad \text{layer } i \tag{6-47a}$$

$$F_i(x) = 0 \quad \text{in} \quad \text{all other layers} \tag{6-47b}$$

Each subfunction $F_i(x)$ is expanded into a series of orthogonal functions $G_n(x)$ in the form

$$F_i(x) = \sum_{n=1}^{\infty} B_{in} G_n(x) \tag{6-48}$$

Each expansion is carried out over the entire range of $x_1 \leq x \leq x_{m+1}$ spanning all m layers. The unknown coefficients B_{in} in the expansion 6-48 are determined by a generalized Fourier analysis over the entire range of m layers and is given in the form

$$B_{in} = \frac{\displaystyle\int_{m \text{ layers}} F_i(x) \cdot G_n(x) \cdot w(x) \cdot dx}{\displaystyle\int_{m \text{ layers}} [G_n(x)]^2 w(x) \cdot dx} \tag{6-49}$$

where $w(x)$ is the usual one-region Sturm-Liouville orthogonality function, e.g., r in cylindrical geometry.

In view of the relations 6-45, 6-46, 6-47, Eq. 6-49 becomes

$$B_{in} = \frac{W_i \displaystyle\int_{\text{layer } i} F(x) \cdot X_{in}(x) \cdot w(x) \cdot dx}{\displaystyle\sum_{i=1}^{m} W_i^2 \int_{\text{layer } i} X_{in}^2(x) \cdot w(x) \cdot dx} \tag{6-50}$$

In the series 6-48 the $G_n(x)$ function may be replaced by $W_i X_n(x)$—that is,

$$F_i(x) = \sum_{n=1}^{\infty} B_{in} W_i X_n(x) \tag{6-51}$$

by one-to-one correspondence of terms. In view of the relations 6-46 and 6-51, the unknown coefficient A_n in the expansion 6-44 becomes

$$A_n = \sum_{i=1}^{m} W_i B_{in} \tag{6-52}$$

Substituting B_{in} from Eq. 6-50 to Eq. 6-52, we obtain

$$A_n = \frac{\sum_{i=1}^{m} W_i^2 \int_{\text{layer } i} F(x) \cdot X_{in}(x) \cdot w(x) \cdot dx}{\sum_{i=1}^{m} W_i^2 \int_{\text{layer } i} X_{in}^2(x) \cdot w(x) \cdot dx} \tag{6-53}$$

Substituting Eq. 6-53 into Eq. 6-43, we obtain the solution for any layer j in the region $x_1 \le x \le x_{m+1}$ as

$$T_j(x, t) = \sum_{n=1}^{\infty} \frac{\sum_{i=1}^{m} W_i^2 \int_{\text{layer } i} F_i(x') \cdot X_{in}(x') \cdot w(x') \cdot dx'}{\sum_{i=1}^{m} W_i^2 \int_{\text{layer } i} X_{in}^2(x) \cdot w(x) \cdot dx} \cdot X_{jn}(x) \cdot \Gamma(t) \tag{6-54}$$

The solution of the boundary-value problem of heat conduction for a composite region involving m layers is now complete if the eigenfunctions $X_{in}(x)$ and the discontinuous weighting functions W_i for each region are known.

Tittle [4] has shown that the discontinuous-weighting function is *unchanged whether there is a perfect thermal contact or a linear contact resistance at the interface*. Also, the same applies to the boundary conditions at the outer surfaces—that is, whether it is of the boundary condition of the first, second, or third kind. The boundary conditions at interfaces and outer surfaces do affect the eigenfunctions and the eigenvalues.

It is to be noted that the solution 6-54 is very similar in form to the solutions obtained by the method of separation of variables for the one-layer regions, the only difference in this case being the introduction of the discontinuous-weighting function, W_i, and summation taken over the m layers.

We now examine the determination of the eigenfunctions $X_{in}(x)$, the eigenvalues β_{in} and the discontinuous-weighting functions W_i in the solution of heat-conduction problems for multilayer regions.

6-6. HOMOGENEOUS HEAT-CONDUCTION PROBLEM FOR A MULTILAYER REGION WITH PERFECT THERMAL CONTACT AT THE INTERFACES

Consider a one-dimensional composite region involving m parallel layers (i.e., slabs, concentric cylinders or spheres). We assume that there is perfect thermal contact between the layers at the interfaces (i.e., continuity of temperature and continuity of heat flux). The boundary conditions at the outer surfaces x_1 and x_{m+1} can be taken as any combinations of the boundary conditions of the first kind, the second and third kinds. We assume that the boundary conditions at the outer surfaces are both of the third kind. The boundary-value problem of heat conduction is in the form:

$$\nabla^2 T_i(x,t) = \frac{1}{\alpha_i}\frac{\partial T_i(x,t)}{\partial t} \quad \text{in} \quad x_i \le x \le x_{i+1},\ t > 0 \qquad i = 1,2,3,\ldots,m \tag{6-55}$$

The boundary conditions are

$$-k_1 \frac{\partial T_1(x_1,t)}{\partial x} + hT_1(x_1,t) = 0 \quad \text{at the outer surface } x_1,\ t > 0 \tag{6-56a}$$

$$T_i(x_{i+1},t) = T_{i+1}(x_{i+1},t) \quad \text{continuity of temperature at the interface, } i = 1,2,3,\ldots,(m-1) \text{ and } t > 0 \tag{6-56b}$$

$$k_i \frac{\partial T_i(x_{i+1},t)}{\partial x} = k_{i+1}\frac{\partial T_{i+1}(x_{i+1},t)}{\partial x} \quad \text{continuity of heat flux at the interface, } i = 1,2,3,\ldots,(m-1) \text{ and } t > 0 \tag{6-56c}$$

$$k_m \frac{\partial T_m(x_{m+1},t)}{\partial x} + hT_m(x_{m+1},t) = 0 \quad \text{at the outer surface } x_{m+1},\ t > 0 \tag{6-56d}$$

The initial condition is

$$T_i(x,0) = F_i(x) \quad \text{in} \quad x_i \le x \le x_{i+1} \qquad i = 1,2,3,\ldots,m \quad \text{for} \quad t = 0 \tag{6-57}$$

where

$$\nabla^2 \equiv \frac{1}{x^p}\frac{\partial}{\partial x}\left(x^p \frac{\partial}{\partial x}\right), \tag{6-58}$$

is the one-dimensional Laplace differential operator; p takes the values of 0, 1, 2 respectively for plates, cylinders, and spheres.

We seek the solution of the above boundary-value problem of heat conduction in the form

$$T_i(x,t) = \sum_{n=1}^{\infty} A_n \cdot X_{in}(x) \cdot e^{-\alpha_i \beta_{in}^2 t} \quad \text{in layer } i \qquad i = 1,2,3,\ldots,m \tag{6-59}$$

In Eq. 6-59 the eigenfunctions $X_{in}(x)$ satisfy the following system of eigenvalue problems.

$$\nabla^2 X_{in}(x) + \beta_{in}^2 X_{in}(x) = 0 \quad \text{in} \quad x_i \le x \le x_{i+1}, \quad i = 1,2,3,\ldots,m \tag{6-60}$$

where ∇^2 as defined by Eq. 6-58 and with the homogeneous boundary conditions

$$-k_1 \frac{dX_{1n}(x_1)}{dx} + hX_{1n}(x_1) = 0 \tag{6-61a}$$

$$X_{in}(x_{i+1}) = X_{i+1,n}(x_{i+1}) \qquad i = 1,2,3,\ldots,m-1 \tag{6-61b}$$

$$k_i \frac{dX_{in}(x_{i+1})}{dx} = k_{i+1} \frac{dX_{i+1,n}(x_{i+1})}{dx} \qquad i = 1,2,3,\ldots,m-1 \tag{6-61c}$$

$$k_m \frac{dX_{mn}(x_{m+1})}{dx} + hX_{mn}(x_{m+1}) = 0 \tag{6-61d}$$

The solution of the above eigenvalue problem is in the form

$$X_{in}(x) = C_{in}\phi_{in}(x) + D_{in}\psi_{in}(x) \quad x_i \le x \le x_{i+1}, \quad i = 1,2,3,\ldots,m \tag{6-62}$$

where the functions $\phi_{in}(x)$ and $\psi_{in}(x)$ are two linearly independent solutions of Eq. 6-60. Table 6-1 gives the functions $\phi_{in}(x)$ and $\psi_{in}(x)$ for plates, cylinders and spheres.

TABLE 6-1.
THE LINEARLY INDEPENDENT SOLUTIONS $\phi_{in}(x)$ AND $\psi_{in}(x)$ OF EQ. 6-60 FOR PLATES, CYLINDERS AND SPHERES

Geometry	$\phi_{in}(x)$	$\psi_{in}(x)$
Plate	$\cos(\beta_{in}x)$	$\sin(\beta_{in}x)$
Cylinder	$J_0(\beta_{in}x)$	$Y_0(\beta_{in}x)$
Sphere	$\frac{1}{x}\sin(\beta_{in}x)$	$\frac{1}{x}\cos(\beta_{in}x)$

Eigenfunctions 6-62 for $i = 1,2,3,\ldots,m$ include $2m$ unknown coefficients C_{in} and D_{in}. The $2m$ equations that are needed to determine the $2m$ unknown coefficients are obtained by substituting the eigenfunctions 6-62 into the boundary conditions 6-61 of the eigenvalue problem. The resulting $2m$ linear, homogeneous equations for determining the $2m$ arbitrary constants, C_{in} and D_{in}, are as follows.

At the outer boundary x_1:

$$h[C_{1n}\phi_{1n}(x_1) + D_{1n}\psi_{1n}(x_1)] - k_1[C_{1n}\phi'_{1n}(x_1) + D_{1n}\psi'_{1n}(x_1)] = 0$$

At the interfaces:

$i = 1$

$$C_{1n}\phi_{1n}(x_2) + D_{1n}\psi_{1n}(x_2) - C_{2n}\phi_{2n}(x_2) - D_{2n}\psi_{2n}(x_2) = 0$$

$$\frac{k_1}{k_2}\,[C_{1n}\phi'_{1n}(x_2) + D_{1n}\psi'_{1n}(x_2)] - C_{2n}\phi'_{2n}(x_2) - D_{2n}\psi'_{2n}(x_2) = 0$$

$i = 2$

$$C_{2n}\phi_{2n}(x_3) + D_{2n}\psi_{2n}(x_3) - C_{3n}\phi_{3n}(x_3) - D_{3n}\psi_{3n}(x_3) = 0$$

$$\frac{k_2}{k_3}[C_{2n}\phi'_{2n}(x_3) + D_{2n}\psi'_{2n}(x_3)] - C_{3n}\phi'_{3n}(x_3) - D_{3n}\psi'_{3n}(x_3) = 0 \qquad (6\text{-}63)$$

....

....

$i = m - 1$

$$C_{m-1,n}\cdot\phi_{m-1,n}(x_m) + D_{m-1,n}\cdot\psi_{m-1,n}(x_m) - C_{mn}\cdot\phi_{mn}(x_m) - D_{mn}\cdot\psi_{mn}(x_m) = 0$$

$$\frac{k_{m-1}}{k_m}\;[C_{m-1,n}\cdot\phi'_{m-1,n}(x_m) + D_{m-1,n}\cdot\psi'_{m-1,n}(x_m)]$$

$$- C_{mn}\phi'_{mn}(x_m) - D_{mn}\psi'_{mn}(x_m) = 0$$

At the outer boundary x_{m+1}:

$$h[C_{mn}\cdot\phi_{mn}(x_{m+1}) + D_{mn}\cdot\psi_{mn}(x_{m+1})]$$

$$+ k_m\;[C_{mn}\cdot\phi'_{mn}(x_{m+1}) + D_{mn}\cdot\psi'_{mn}(x_{m+1})] = 0$$

where

$$\psi' \equiv \frac{d\psi}{dx}, \quad \phi' \equiv \frac{d\phi}{dx}$$

Since the $2m$ equations in the above system 6-63 are all homogeneous equations, the only way the coefficients C_{in} and D_{in} are determined is by expressing them all in terms of one of them, say, C_{1n}, provided that C_{1n} is not zero. The resulting eigenfunctions $X_{in}(x)$ each will include the unknown coefficient C_{1n} as a product factor. When $X_{in}(x)$ are substituted in the solution 6-59 together with A_n as given by 6-53, the unknown coefficient C_{1n}, will cancel out.

Evaluation of the coefficients C_{in} and D_{in} from system 6-63 cannot yet be performed because the system includes the unknown eigenvalues β_{in}, which are determined by the following consideration.

Consider the interface at X_i between the layers $(i - 1)$ and i. Since there is no energy storage in the infinitesimal thickness of the interface, the time behavior of temperature at the interface X_i should be the same on either side of X_i. This condition is satisfied if we have

$$\alpha_{i-1}\beta^2_{(i-1)n} = \alpha_i\beta^2_{in} \quad \text{for} \quad i = 2, 3, \ldots, m \qquad (6\text{-}64)$$

By means of relation 6-64 it is possible to express all β_i's in terms of one of them, say, β_{1n}. An additional relationship that is needed to determine the unknown eigenvalue β_{1n} is obtained from the requirement that the system of $2m$ algebraic Eqs. 6-63 have nontrivial solution if the determinant of the coefficients C_{in} and D_{in} vanishes. By equating the determinant of the coefficients C_{in} and D_{in} to zero one obtains a transcendental equation, the positive roots of which gives the eigenvalues $\beta_{11} < \beta_{12} < \beta_{13} < \cdots < \beta_{1n} \le \cdots$. For each of these eigenvalues there are corresponding values of C_{in}, D_{in} and the eigenfunctions $X_{in}(x)$ which can be determined.

Solution 6-59 includes an additional unknown coefficient A_n, which is evaluated according to the relationship given by Eq. 6-53.

Summarizing, the solution of the boundary value problem of heat-conduction 6-55, 6-56 and 6-57 for an m-layer region is given in the form

$$T_j(x,t) = \sum_{n=1}^{\infty} \frac{\sum_{i=1}^{m} W_i^2 \int_{\text{layer } i} F_i(x') X_{in}(x')\, w(x')\,dx'}{\sum_{i=1}^{m} W_i^2 \int_{\text{layer } i} X_{in}^2(x')\, w(x')\,dx'} X_{jn}(x) e^{-\alpha_j \cdot \beta^2_{jn} t}$$

$$\text{for layer } j \quad j = 1, 2, 3, \ldots, m \tag{6-65}$$

Various terms in Eq. 6-65 are defined as follows:

$X_{in}(x)$ is the eigenfunction for the layer i and is determined for each region as described above. (See Eq. 6-62 and Table 6-1.)

$w(x)$ is the usual one-region Sturm-Liouville weighting function, i.e., $1, x, x^2$ for the rectangular, cylindrical and spherical-coordinate system respectively.

$F_i(x)$ is the initial-condition function for the layer i.

W_i is the *discontinuous-weighting function* for the layer i, and for time-dependent heat-conduction problems it is given as [3,4].

$$W_i = \left(\frac{k_i}{\alpha_i}\right)^{1/2} \tag{6-66}$$

It is to be noted that the discontinuous-weighting function W_i is not affected by the types of boundary conditions at interfaces and at outer surfaces. Such boundary conditions do effect the eigenvalues and eigenfunctions.

Eigenvalues β_{in} are related to each other by Eq. 6-64, and are determined from the solution of the transcendental equation obtained by equating to zero the determinant of the coefficients of C_{in} and D_{in} in system 6-63.

6-7. TWO-REGION CONCENTRIC CYLINDER WITH PERFECT THERMAL CONTACT AT THE INTERFACE

Transient heat-conduction problem in a two-region concentric solid cylinder, with perfect thermal contact at the interface and subjected to convection-boundary condition at the outer surface is solved by Tittle

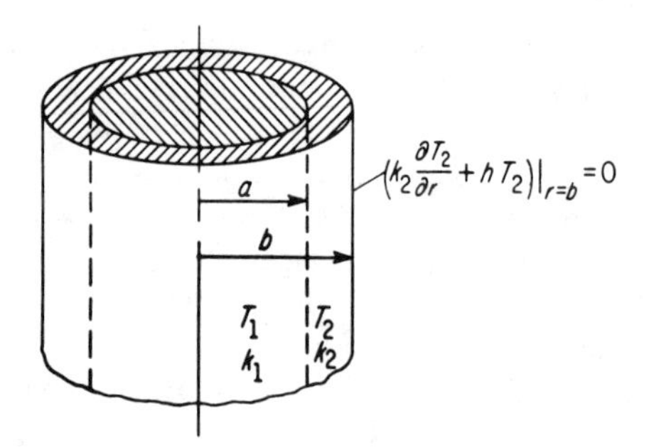

FIG. 6-4. Two-region solid cylinder with convective boundary condition at the outer surface and perfect thermal contact at the interface.

and Robinson [8] by applying the orthogonal-expansion technique over multilayer regions as described previously.

Figure 6-4 shows a two-region solid cylinder of outer radius $r = b$ with an inner core of radius $r = a$, with perfect thermal contact between the layers.

Let k_1, k_2 be the thermal conductivities of the inner and outer regions respectively and α_1, α_2 be the thermal diffusivities. The cylinders are initially at uniform temperature T_0 and for times $t \geq 0$ the boundary surface at $r = b$ dissipates heat by convection into a surrounding at zero temperature. There is no heat generation within the solid.

The boundary-value problem of heat conduction is given as

$$\begin{aligned}
&\frac{\partial^2 T_1}{\partial r^2} + \frac{1}{r}\frac{\partial T_1}{\partial r} = \frac{1}{\alpha_1}\frac{\partial T_1}{\partial t} \quad \text{in} \quad 0 \leq r \leq a,\ t > 0 \\
&\frac{\partial^2 T_2}{\partial r^2} + \frac{1}{r}\frac{\partial T_2}{\partial r} = \frac{1}{\alpha_2}\frac{\partial T_2}{\partial t} \quad \text{in} \quad a \leq r \leq b,\ t > 0
\end{aligned} \tag{6-67}$$

with boundary conditions

$$T_1 = T_2 \quad \text{at the interface } r = a,\ t > 0 \tag{6-68a}$$

$$k_1\frac{\partial T_1}{\partial r} = k_2\frac{\partial T_2}{\partial r} \quad \text{at the interface } r = a,\ t > 0 \tag{6-68b}$$

$$k_2\frac{\partial T_2}{\partial r} + h_2 T_2 = 0 \quad \text{at the outer boundary } r = b,\ t > 0 \tag{6-68c}$$

and the initial conditions

$$T_1 = T_0 \quad \text{in} \quad 0 \leq r \leq a,\ t = 0 \tag{6-69}$$

$$T_2 = T_0 \quad \text{in} \quad a \leq r \leq b,\ t = 0 \tag{6-70}$$

We seek the solution of this problem in the form

$$T_j(r,t) = \sum_{n=1}^{\infty} A_n \cdot X_{jn}(r) \cdot e^{-\alpha_j \beta_{jn}^2 t} \quad \text{in layer } j \qquad j = 1,2 \tag{6-71}$$

The eigenfunctions $X_{jn}(r)$ for the two-layer problem under consideration is given as (See Table 6-1.)

$$X_{1n}(r) = J_0(\beta_{1n} r) \quad \text{in layer 1, i.e., } 0 \le r \le a \tag{6-72}$$

$$X_{2n}(r) = B_n[J_0(\beta_{2n} r) + C_n Y_0(\beta_{2n} r)] \quad \text{in layer 2, i.e., } a \le r \le b \tag{6-73}$$

The $Y_0(\beta_{1n} r)$ function is excluded from the solution for region 1 because it includes the origin.

The eigenvalues β_{1n} and β_{2n} are related to each other according to Eq. 6-64 as

$$\alpha_1 \beta_{1n}^2 = \alpha_2 \beta_{2n}^2 \tag{6-74}$$

Now we have two unknown coefficients B_n, C_n and the eigenvalues β_{1n} (or β_{2n} since they are related) to be determined; three independent relations are needed to determine these three unknown quantities. The boundary conditions 6-68a, b, and c provide the required three independent relations.

C_n is evaluated by applying the condition 6-68c, that is

$$k_2 \beta_{2n}[J_1(\beta_{2n} b) + C_n Y_1(\beta_{2n} b)] = h_2[J_0(\beta_{2n} b) + C_n Y_0(\beta_{2n} b)]$$

or

$$C_n = \frac{h_2 J_0(\beta_{2n} b) - k_2 \beta_{2n} J_1(\beta_{2n} b)}{k_2 \beta_{2n} Y_1(\beta_{2n} b) - h_2 Y_0(\beta_{2n} b)} \tag{6-75}$$

B_n is evaluated by applying the condition 6-68a

$$J_0(\beta_{1n} a) = B_n[J_0(\beta_{2n} a) + C_n Y_0(\beta_{2n} a)]$$

or

$$B_n = \frac{J_0(\beta_{1n} a)}{J_0(\beta_{2n} a) + C_n Y_0(\beta_{2n} a)} \tag{6-76}$$

When the condition 6-68b is applied the following transcendental equation is obtained for evaluating the eigenvalues

$$k_1 \beta_{1n} J_1(\beta_{1n} a) = k_2 \beta_{2n} B_n[J_1(\beta_{2n} a) + C_n Y_1(\beta_{2n} a)] \tag{6-77}$$

The eigenvalues β_{1n} (or β_{2n}) are evaluated from the solution of this transcendental equation in which B_n and C_n are given by Eqs. 6-76 and 6-75 respectively, and β_{1n}, β_{2n} are related to each other by Eq. 6-74. A numerical method described in references [9,10] for solving transcendental equations may be used to evaluate the roots of the transcendental equation 6-77.

The solution 6-71 now involves only one unknown coefficient A_n, which can be evaluated according to the relationship given by Eq. 6-53.

For the cylinder problem under consideration various terms in Eq. 6-53 are given as

$$\left.\begin{aligned} W_i^2 &= \frac{k_i}{\alpha_i}, w = r, F_i = T_0 \quad i = 1,2 \\ X_{1n}(r) &= J_0(\beta_{1n} r) \\ X_{2n}(r) &= B_n [J_0(\beta_{2n} r) + C_n Y_0(\beta_{2n} r)] \end{aligned}\right\} \tag{6-78}$$

Substituting 6-78 into 6-53, the unknown coefficient A_n becomes

$$A_n = \frac{\dfrac{k_1}{\alpha_1}\displaystyle\int_0^a T_0 \cdot J_0(\beta_n r)\cdot r \cdot dr + \dfrac{k_2}{\alpha_2}\displaystyle\int_a^b T_0 \cdot B_n[J_0(\gamma_n r) + C_n Y_0(\gamma_n r)]\cdot r \cdot dr}{\dfrac{k_1}{\alpha_1}\displaystyle\int_0^a J_0^2(\beta_n r)\cdot r \cdot dr + \dfrac{k_2}{\alpha_2}\displaystyle\int_a^b B_n^2[J_0(\gamma_n r) + C_n Y_0(\gamma_n r)]^2 \cdot r \cdot dr} \tag{6-79}$$

The integrals in Eq. 6-79 are evaluated to yield

$$A_n = \frac{2T_0}{a\beta_n} \cdot \frac{J_1(\beta_n a) + \dfrac{k_2}{k_1}\left(\dfrac{\alpha_1}{\alpha_2}\right)^{1/2}\left[\dfrac{b}{a} Z_1(\gamma_n b) - Z_1(\gamma_n a)\right]}{J_0^2(\beta_n a) + J_1^2(\beta_n a) + \dfrac{k_2 \alpha_1}{k_1 \alpha_2}\left[\dfrac{b^2}{a^2}\{Z_0^2(\gamma_n b) + Z_1^2(\gamma_n b)\} - Z_0^2(\gamma_n a) - Z_1^2(\gamma_n a)\right]} \tag{6-80}$$

where

$$Z_p(x_n) \equiv B_n[J_p(x_n) + C_n Y_p(x_n)]$$

Summarizing, the solution of the problem described by Eqs. 6-67, 6-68, and 6-69 is given in the form

$$T_j(r, t) = \sum_{n=1}^{\infty} A_n \cdot X_{jn}(r) \cdot e^{-\alpha_j \beta_{jn}^2 t} \tag{6-71}$$

in layer j $\quad j = 1, 2$

where, $X_{jn}(r)$ is given by Eqs. 6-72 and 6-73, β_{1n} and β_{2n} are related to each other by Eq. 6-74, the coefficients C_n and B_n are given by Eqs. 6-75 and 6-76, β_{1n} (or β_{2n}) are the roots of Eq. 6-77, the coefficient A_n is given by Eq. 6-80.

The solution of the problem is now complete, and temperature distribution in the cylinder can be determined by evaluating the above results with a digital computer.

6-8. TWO-LAYER SLAB WITH CONTACT RESISTANCE AT THE INTERFACE

Transient-heat conduction in a two-layer slab with contact resistance at the interface has been solved by Moore [11] for several combinations of linear-boundary conditions at the outer surfaces. For simplicity, we consider here the case with boundary condition of the first kind at the outer boundaries.

Figure 6-5 shows a two-layer slab ($0 \le x \le L$) with contact resistance at the interface $x = a$ (i.e., $0 \le a \le L$). Initially the entire region is at a uniform temperature T_i and for times $t > 0$ the outer-boundary surfaces at $x = 0$ and $x = L$ are kept at constant temperatures T_0 and T_i respectively. The boundary-value problem of heat conduction is given as

$$\frac{\partial^2 T_1}{\partial x^2} = \frac{1}{\alpha_1}\frac{\partial T_1}{\partial t} \quad \text{in} \quad 0 \le x \le a, t > 0 \tag{6-81a}$$

$$\frac{\partial^2 T_2}{\partial x^2} = \frac{1}{\alpha_2}\frac{\partial T_2}{\partial t} \quad \text{in} \quad a \le x \le L, t > 0 \tag{6-81b}$$

The boundary conditions are

$$T_1 = T_0 \quad \text{at the outer boundary } x = 0, t > 0 \tag{6-82a}$$

$$k_1 \frac{\partial T_1}{\partial x} = k_2 \frac{\partial T_2}{\partial x} \quad \text{at the interface} \quad x = a, t > 0 \tag{6-82b}$$

$$-k_1 \frac{\partial T_1}{\partial x} = h(T_1 - T_2) \quad \text{at the interface} \quad x = a, t > 0 \tag{6-82c}$$

$$T_2 = T_i \quad \text{at the outer boundary } x = L, t > 0 \tag{6-82d}$$

where h is the contact conductance at the interface. The initial conditions are

$$T_1 = T_i \quad \text{in} \quad 0 \le x \le a, t = 0 \tag{6-83a}$$

$$T_2 = T_i \quad \text{in} \quad a \le x \le L, t = 0 \tag{6-83b}$$

For the above problem the boundary conditions at the outer-boundary surfaces are nonhomogeneous. If we express the solutions for each region

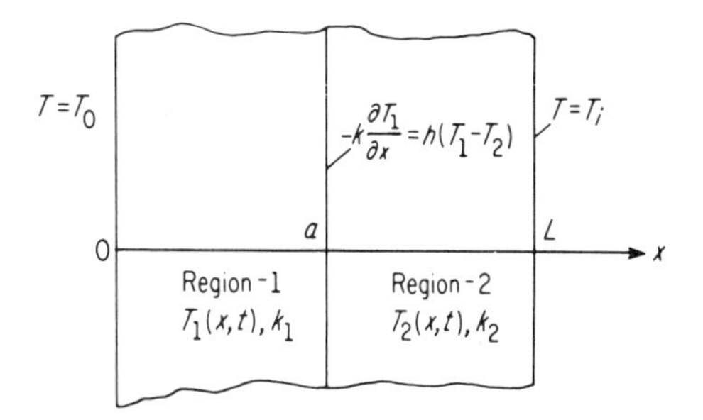

FIG. 6-5. Two-region slab with contact resistance at the interface and boundary condition of the first kind at the outer boundaries.

as the sum of the steady-state and transient solutions in the form

$$T_j(x, t) = T_j(x, \infty) + \sum_{n=1}^{\infty} A_n \cdot X_{jn}(x) \cdot e^{-\alpha_j \beta_{jn}^2 t} \quad \text{for region } j \tag{6-84}$$

where $j = 1$ for region $0 \le x \le a$

$j = 2$ for region $a \le x \le L$

then the steady-state solutions $T_j(x, \infty)$ are given in the form

$$T_1(x, \infty) = T_0 - \frac{T_0 - T_i}{a + \dfrac{k_1}{h} + \dfrac{k_1}{k_2}(L - a)} \cdot x \tag{6-85a}$$

$$T_2(x, \infty) = T_i + \frac{k_1}{k_2} \frac{T_0 - T_i}{a + \dfrac{k_1}{h} + \dfrac{k_1}{k_2}(L - a)} \cdot (L - x) \tag{6-85b}$$

The eigenfunctions $X_{jn}(x)$ are chosen in the form

$$X_{1n}(x) = \sin \beta_{1n} x + D_{1n} \cos \beta_{1n} x \tag{6-86a}$$

$$X_{2n}(x) = C_{2n} \sin \beta_{2n} x + D_{2n} \cos \beta_{2n} x \tag{6-86b}$$

The three unknown coefficients D_{1n}, C_{2n}, and D_{2n} will be determined from the three boundary conditions 6-82a, 6-82b, and 6-82d. The remaining boundary condition 6-82c will be used to establish a relationship for determining the eigenvalues. The procedure is as follows.

From the boundary condition at $x = 0$, i.e., 6-82a,

$$D_{1n} = 0 \tag{6-87}$$

From the boundary condition at $x = L$, i.e., Eq. 6-82d,

$$D_{2n} = -C_{2n} \cdot \tan \beta_{2n} L \tag{6-88}$$

Eigenfunctions 6-86 include two different eigenvalues β_{1n} and β_{2n}. Since there is no energy storage at the interface, the time behavior of temperature should be the same on either side of the interface. This condition is satisfied if we have (See Eq. 6-64.)

$$\alpha_1 \beta_{1n}^2 = \alpha_2 \beta_{2n}^2$$

or

$$\beta_{2n} = \sqrt{\frac{\alpha_1}{\alpha_2}} \cdot \beta_{1n} \tag{6-89}$$

In view of Eqs. 6-87 and 6-88, the eigenfunctions 6-86 become

$$X_{1n}(x) = \sin \beta_{1n} x \tag{6-90a}$$

$$X_{2n}(x) = C_{2n}[\sin \beta_{2n} x - \tan \beta_{2n} L \cdot \cos \beta_{2n} x] \tag{6-90b}$$

where the eigenvalues β_{1n} and β_{2n} are related by

$$\beta_{2n} = \sqrt{\frac{\alpha_1}{\alpha_2}} \beta_{1n}$$

The eigenfunction 6-90b now involves an unknown coefficient C_{2n}, which is determined by applying the boundary condition 6-82b at the interface. That is,

$$k_1 \sum_{n=1}^{\infty} A_n \beta_{1n} \cos \beta_{1n} a \cdot e^{-\alpha_1 \beta_{1n}^2 t}$$

$$= k_2 \sum_{n=1}^{\infty} A_n \beta_{2n} C_{2n} [\cos \beta_{2n} a + \tan \beta_{2n} L \cdot \sin \beta_{2n} a] \cdot e^{-\alpha_2 \beta_{2n}^2 t}$$

In view of the relation between the eigenvalues given by Eq. 6-89, this relation yields the value of C_{2n}, after some manipulation, as

$$C_{2n} = \frac{k_1}{k_2} \sqrt{\frac{\alpha_2}{\alpha_1}} \cdot \frac{\cos \beta_{1n} a}{\cos \beta_{2n} a + \tan \beta_{2n} L \cdot \sin \beta_{2n} a}$$

which may be written as

$$C_{2n} = \frac{k_1}{k_2} \sqrt{\frac{\alpha_2}{\alpha_1}} \cdot \frac{\cos \beta_{1n} a \cdot \cos \beta_{2n} a}{\cos [\beta_{2n}(L - a)]} \tag{6-91}$$

We now need a relation for determining the eigenvalues β_{1n} and the boundary condition 6-82c at the interface has not been used. By applying the boundary condition 6-82c we obtain

$$-k_1 \sum_{n=1}^{\infty} A_n \cdot \beta_{1n} \cdot \cos \beta_{1n} a \cdot e^{-\alpha_1 \beta_{1n}^2 t}$$

$$= h \left[\sum_{n=1}^{\infty} A_n \cdot \sin \beta_{1n} a \cdot e^{-\alpha_1 \beta_{1n}^2 t} \right]$$

$$- \sum_{n=1}^{\infty} A_n \cdot C_{2n} \cdot (\sin \beta_{2n} a - \tan \beta_{2n} L \cdot \cos \beta_{2n} a)\, e^{-\alpha_1 \beta_{1n}^2 t}$$

This relation is simplified by substituting the values of β_{2n} and C_{2n} from Eqs. 6-89 and 6-91 respectively, and after some algebraic and trigonometric manipulations we obtain the following transcendental equation for determining the eigenvalues β_{1n}.

$$-\beta_{1n} = \frac{h}{k_2} \sqrt{\frac{\alpha_2}{\alpha_1}} \cdot \tan \left[\beta_{1n}(L - a) \sqrt{\frac{\alpha_1}{\alpha_2}} \right] + \frac{h}{k_1} \tan \beta_{1n} a \tag{6-92}$$

Positive roots of this transcendental equation 6-92 yield the eigenvalues β_{1n}.

So far we determined the eigenfunctions and the eigenvalues for use in the solution of the problem given by Eq. 6-84, but the unknown coefficient A_n is yet to be determined. A_n is determined by means of the relation 6-53 and for the two-layer slab problem under consideration

Eq. 6-53 is written in the form

$$A_n = \frac{W_1^2 \int_0^a F_1(x)\cdot X_{1n}(x)\cdot w(x)\cdot dx + W_2^2 \int_a^L F_2(x)\cdot X_{2n}(x)\cdot w(x)\cdot dx}{W_1^2 \int_0^a X_{1n}^2(x)\cdot w(x)\cdot dx + W_2^2 \int_a^L X_{2n}^2(x)\cdot w(x)\cdot dx} \tag{6-93}$$

Various terms in Eq. 6-93 are given as follows.

The Sturm-Liouville weighting function $w(x)$ for the cartesian coordinate system is

$$w(x) = 1 \tag{6-94}$$

The discontinuous-weighting functions W_1 and W_2, from the relation 6-66 are

$$W_1 = \sqrt{\frac{k_1}{\alpha_1}}, \quad W_2 = \sqrt{\frac{k_2}{\alpha_2}} \tag{6-95}$$

The functions $F_1(x)$ and $F_2(x)$ are those functions which are to be expanded in an infinite series of the eigenfunctions for time $t = 0$ and are determined by substituting $T_j(x, t)\,|_{t=0} = T_i$ for $j = 1,2$ in the general solution 6-84. This yields,

$$\begin{aligned} T_i - T_1(x, \infty) &= \sum_{n=1}^{\infty} A_n \cdot X_{1n}(x) \quad \text{in} \quad 0 \le x \le a \\ T_i - T_2(x, \infty) &= \sum_{n=1}^{\infty} A_n \cdot X_{2n}(x) \quad \text{in} \quad a \le x \le L \end{aligned} \tag{6-96}$$

Hence the functions $F_1(x)$ and $F_2(x)$ are given as

$$F_1(x) = T_i - T_1(x, \infty) \tag{6-97a}$$

$$F_2(x) = T_i - T_2(x, \infty) \tag{6-97b}$$

where $T_1(x, \infty)$ and $T_2(x, \infty)$ are the steady-state solutions as defined by Eq. 6-85. It is to be noted that for the present problem with nonhomogeneous-boundary conditions at the outer surfaces the $F(x)$ functions are not merely the initial conditions for the region but include the steady-state solutions.

The eigenfunctions $X_{1n}(x)$ and $X_{2n}(x)$ for use in Eq. 6-93 are defined by Eq. 6-90.

Substituting the various quantities as defined above into Eq. 6-93, performing the integrations and simplifying the results, the unknown coefficient A_n is given as

$$A_n = \frac{2(T_i - T_0)}{\beta_{1n}a - \frac{1}{2}\sin(2\beta_{1n}a) + \frac{k_1}{k_2}\sqrt{\frac{\alpha_2}{\alpha_1}}\left\{\beta_{2n}(L-a) - \frac{1}{2}\sin\,[2\beta_{2n}(L-a)]\right\}} \tag{6-98}$$

where β_{2n} is related to β_{1n} by

$$\beta_{2n} = \beta_{1n} \sqrt{\frac{\alpha_1}{\alpha_2}}$$

Summarizing, the temperature distribution in the two-layer slab with contact resistance at the interface is given in the form

$$T_1(x,t) = T_1(x,\infty) + \sum_{n=1}^{\infty} A_n \cdot \sin \beta_{1n} x \cdot e^{-\alpha_1 \beta_{1n}^2 t} \quad \text{in } 0 \le x \le a \tag{6-99}$$

$$T_2(x,t) = T_2(x,\infty) + \sum_{n=1}^{\infty} A_n \cdot C_{2n}[\sin \beta_{2n} x - \tan \beta_{2n} L \cdot \cos \beta_{2n} x] \cdot e^{-\alpha_2 \beta_{2n}^2 t}$$
$$\text{in } a \le x \le L \tag{6-100}$$

where,

β_{1n} is related to β_{2n} by Eq. 6-89, C_{2n} is given by Eq. 6-91, β_{1n} (or β_{2n}) are the roots of Eq. 6-92, A_n is given by Eq. 6-98.

6-9. NONHOMOGENEOUS HEAT-CONDUCTION PROBLEM FOR A MULTILAYER REGION WITH PERFECT THERMAL CONTACT AT THE INTERFACES

The transient heat-conduction problem for multilayer symmetrical bodies (i.e., plates, cylinders, spheres) with heat generation within the solid has been solved by Bulavin and Kashcheev [5] by the method of separation of variables and by the construction of orthogonal expansion of functions over multilayer regions for the case of perfect thermal contact between the layers. The orthogonal-expansion technique and the orthogonality factor used for expansion is similar to that given by Tittle [3], but the treatment of the eigenvalue problem is somewhat different.

Figure 6-6 shows the geometry of one-dimensional composite region consisting of m parallel layers. It is assumed that the thermal properties of each layer are uniform but discontinuous at the interfaces between the layers which are in perfect thermal contact. The initial temperature

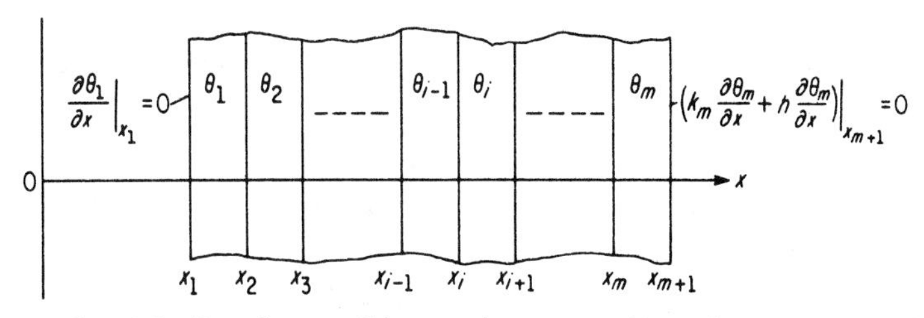

FIG. 6-6. Boundary conditions at the outer surfaces of an m-layer composite region with perfect thermal contact at the interfaces and heat generation within the layers.

distribution for each layer is prescribed, i.e., $T_i = F_i(x)$ for $t = 0$ and $i = 1,2,\ldots,m$. For times $t > 0$ there is heat generation within the solid at a rate of $g_i(x,t)$ Btu/hr ft^3, while the outer boundary surface at $x = x_{m+1}$ dissipates heat by convection into a surrounding temperature of which varies with time. The outer boundary at $x = x_1$ is kept insulated.

Defining:

$T_i(x,t)$ = temperature of the ith layer in $x_i \le x \le x_{i+1}$

$T_s(t)$ = temperature of the surrounding into which convection takes place from the outer surfaces

$c_i, k_i, \alpha_i, \rho_i$ = specific heat, thermal conductivity, thermal diffusivity, density respectively of the ith layer

The boundary-value problem of heat conduction is given as

$$\alpha_i \nabla^2 T_i(x,t) + \frac{\alpha_i}{k_i} g_i(x,t) = \frac{\partial T_i(x,t)}{\partial t} \quad \text{in} \quad x_i \le x \le x_{i+1}, t > 0 \quad i = 1,2,3,\ldots,m \tag{6-101}$$

with the boundary conditions

$$\frac{\partial T_1(x_1,t)}{\partial x} = 0 \qquad \text{at the outer surface } x_1$$

$$T_i(x_{i+1},t) = T_{i+1}(x_{i+1},t) \qquad \text{at the interface } x_{i+1} \quad i = 1,2,3,\ldots,m-1$$

$$k_i \frac{\partial T_i(x_{i+1},t)}{\partial x} = k_{i+1} \frac{\partial T_{i+1}(x_{i+1},t)}{\partial x} \qquad \text{at the interface } x_{i+1} \quad i = 1,2,3,\ldots,m-1 \tag{6-102}$$

$$-k_m \frac{\partial T_m(x_{m+1},t)}{\partial x} = h[T_m(x_{m+1},t) - T_s(t)] \text{ at the outer surface } x_{m+1}$$

and the initial condition

$$T_i(x,0) = F_i(x) \quad \text{in} \quad x_i \le x \le x_{i+1}, \quad i = 1,2,3,\ldots,m \tag{6-103}$$

where

$$\nabla^2 \equiv \frac{1}{x^p} \frac{\partial}{\partial x} \left(x^p \frac{\partial}{\partial x} \right); \tag{6-104}$$

is the one-dimensional Laplace differential operator; p takes the values of 0, 1, 2 respectively for plates, cylinders and spheres.

For convenience we define a new excess temperature $\theta_i(x,t)$ which is measured above the temperature of the surrounding $T_s(t)$—that is,

$$\theta_i(x,t) = T_i(x,t) - T_s(t), \quad i = 1,2,3,\ldots,m \tag{6-105}$$

In terms of the excess temperature $\theta_i(x,t)$ the above boundary-value problem of heat conduction becomes

$$\alpha_i \nabla^2 \theta_i(x,t) + \frac{\alpha_i}{k_i} g_i(x,t) - \frac{dT_s(t)}{dt} = \frac{\partial \theta_i(x,t)}{\partial t} \quad \text{in} \quad x_i \le x \le x_{i+1}, t > 0 \quad i = 1,2,3,\ldots,m \tag{6-106}$$

with the boundary conditions

$$\left.\begin{aligned}
&\frac{\partial \theta_1(x_1, t)}{\partial x} = 0 \text{ .... at the outer surface } x_1 \\
&\theta_i(x_{i+1}, t) = \theta_{i+1}(x_{i+1}, t) \text{ .... } i = 1,2,3, \ldots, m-1 \\
&k_i \frac{\partial \theta_i(x_{i+1}, t)}{\partial x} = k_{i+1} \frac{\partial \theta_{i+1}(x_{i+1}, t)}{\partial x} \text{ ... } i = 1,2,3, \ldots, m \\
&-k_m \frac{\partial \theta_m(x_{m+1}, t)}{\partial x} = h\theta_m(x_{m+1}, t) \text{ ...... at the outer surface } x_{m+1}
\end{aligned}\right\} \quad (6\text{-}107)$$

and the initial condition

$$\theta_i(x, 0) = F_i(x) - T_s(0) \equiv f_i(x) \quad \text{in} \quad x_i \le x \le x_{i+1}, \quad i = 1,2,3, \ldots, m \tag{6-108}$$

We seek the solution of the above problem in the form

$$\theta_i(x, t) = \sum_{n=1}^{\infty} X_{in}(x) \cdot \Gamma_n(t) \quad \text{in } x_i \le x \le x_{i+1} \tag{6-109}$$

where $\Gamma_n(t)$ is a function which is to be determined. $X_{in}(x)$ functions satisfy the following eigenvalue problem,[3]

$$\alpha_i \nabla^2 X_{in}(x) + \beta_n^2 X_{in}(x) = 0 \quad \text{in} \quad x_i \le x \le x_{i+1}, \quad i = 1,2, \ldots, m \tag{6-110}$$

with the boundary conditions

$$\frac{dX_{1n}(x_1)}{dx} = 0 \text{ .... at } x_1 \tag{6-111a}$$

$$X_{in}(x_{i+1}) = X_{i+1}(x_{i+1}) \text{ .... for } i = 1,2,3, \ldots, m-1 \tag{6-111b}$$

$$k_i \frac{dX_{in}(x_{i+1})}{dx} = k_{i+1} \frac{dX_{i+1,n}(x_{i+1})}{dx} \text{ .... for } i = 1,2,3, \ldots, m-1 \tag{6-111c}$$

$$-k_m \frac{dX_{mn}(x_{m+1})}{dx} = hX_{mn}(x_{m+1}) \text{ ...... at } x_{m+1} \tag{6-111d}$$

where β_m are the eigenvalues.

The solution of Eq. 6-110 is in the form

$$X_{in}(x) = C_{in}\phi_{in}(x) + D_{in}\psi_{in}(x) \tag{6-112}$$

where $\phi_{in}(x)$ and $\psi_{in}(x)$ are two linearly-independent solutions of differential equation 6-110. For the case of one-dimensional Laplace operator as defined by Eq. 6-104 the form of functions for $\phi_{in}(x)$ and $\psi_{in}(x)$ for plates, cylinders and spheres is shown in Table 6-2. Note the slight difference in the arguments of the functions given in Table 6-2 as compared with those given in Table 6-1.

[3]It is to be noted that the differential Eq. 6-110 chosen for the eigenvalue problem is slightly different from the eigenvalue Prob. 6-60 considered previously.

TABLE 6-2
LINEARLY-INDEPENDENT SOLUTIONS OF EQ. 6-90 FOR PLATES, CYLINDERS AND SPHERES.

Geometry	$\phi_{in}(x)$	$\psi_{in}(x)$
Plate	$\cos\left(\dfrac{\beta_n}{\sqrt{\alpha_i}}x\right)$	$\sin\left(\dfrac{\beta_n}{\sqrt{\alpha_i}}x\right)$
Cylinder	$J_0\left(\dfrac{\beta_n}{\sqrt{\alpha_i}}x\right)$	$Y_0\left(\dfrac{\beta_n}{\sqrt{\alpha_i}}x\right)$
Sphere	$\dfrac{1}{x}\sin\left(\dfrac{\beta_n}{\sqrt{\alpha_i}}x\right)$	$\dfrac{1}{x}\cos\left(\dfrac{\beta_n}{\sqrt{\alpha_i}}x\right)$

There are m solutions in the form of Eq. 6-112 for the m layers, hence we have $2m$ arbitrary constants (i.e., C_{in} and D_{in} for $i = 1,2, \ldots, m$) to be determined. The $2m$ equations that are needed to determine these $2m$ constants are obtained by substituting the eigenfunction 6-112 into the boundary condition 6-111. We obtain $2m$ linear homogeneous equations which are similar in form to the system 6-63, except for the first equation —that is, at the outer boundary x_1:

$$C_{1n}\phi'_{1n}(x_1) + D_{1n}\psi'_{1n}(x_1) = 0$$

At the interfaces:

$i = 1$

$$C_{1n}\,\phi_{1n}(x_2) + D_{1n}\psi_{1n}(x_2) - C_{2n}\phi_{2n}(x_2) - D_{2n}\psi_{2n}(x_2) = 0$$

$$\frac{h_1}{k_1}[C_{1n}\phi'_{1n}(x_2) + D_{1n}\psi'_{1n}(x_2)] - C_{2n}\phi'_{2n}(x_2) - D_{2n}\psi'_{2n}(x_2) = 0 \qquad (6\text{-}113)$$

$$\cdots\cdots\cdots\cdots\cdots\cdots$$

At the outer boundary x_{m+1}:

$$h[C_{mn}\phi_{mn}(x_{m+1}) + D_{mn}\psi_{mn}(x_{m+1})] + k_m[C_{mn}\phi'_{mn}(x_{m+1}) + D_{mn}\psi'_{mn}(x_{m+1})] = 0$$

where

$$\psi' \equiv \frac{d\psi}{dx}, \quad \phi' \equiv \frac{d\phi}{dx}$$

Functions ψ and ϕ are defined in Table 6-2. When the $2m$ homogeneous Eq. 6-113 are solved, one determines the $2m$ unknown coefficients in terms of one of them, say C_{1n}. Then, the resulting eigenfunctions $X_{in}(x)$ will all include the unknown coefficient C_{1n} as product and, as will be apparent later in the analysis, it will cancel out in the final solution. The system 6-113 contains the unknown eigenvalues β_n, which should be determined.

$$
D = \begin{vmatrix}
\underline{C_{1n}} & \underline{D_{1n}} & \underline{C_{2n}} & \underline{D_{2n}} & \underline{C_{3n}} & \underline{D_{3n}} & \underline{C_{m-1,n}} & \underline{D_{m-1,n}} & \underline{C_{m,n}} & \underline{D_{m,n}} \\
\phi'_{1n}(x_1) & \psi'_{1n}(x_1) & 0 & 0 & 0 & 0 \cdots\cdots 0 & 0 & 0 & 0 & 0 \\
\phi_{1n}(x_2) & \psi_{1n}(x_2) & -\phi_{2n}(x_2) & -\psi_{2n}(x_2) & 0 & 0 \cdots\cdots 0 & 0 & 0 & 0 & 0 \\
\frac{k_1}{k_2}\phi'_{1n}(x_2) & \frac{k_1}{k_2}\psi'_{1n}(x_2) & -\phi'_{2n}(x_2) & -\psi'_{2n}(x_2) & 0 & 0 \cdots\cdots 0 & 0 & 0 & 0 & 0 \\
0 & 0 & \phi_{2n}(x_3) & \psi_{2n}(x_3) & -\phi_{3n}(x_3) & -\psi_{3n}(x_3) \cdots 0 & 0 & 0 & 0 & 0 \\
\vdots & & \frac{k_2}{k_3}\phi'_{2n}(x_3) & \psi'_{2n}(x_3) & -\phi'_{3n}(x_3) & -\psi'_{3n}(x_3) \cdots 0 & 0 & 0 & 0 & 0 \\
\hdashline
0 & 0 & 0 & 0 & 0 & 0 \cdots\cdots 0 & \phi_{m-1,n}(x_m) & \psi_{m-1,n}(x_m) & -\phi_{mn}(x_m) & \psi_{mn}(x_m) \\
0 & 0 & 0 & 0 & 0 & 0 \cdots\cdots 0 & \frac{k_{m-1}}{k_m}\phi'_{m-1}(x_m) & \frac{k_{m-1}}{k_m}\psi'_{m-1,n}(x_m) & -\phi'_{mn}(x_m) & -\psi'_{mn}(x_m) \\
0 & 0 & 0 & 0 & 0 & 0 \cdots\cdots 0 & 0 & 0 & h\phi_{mn}(x_{m+1}) + k_m\phi'_{mn}(x_{m+1}) & h\psi_{mn}(x_{m+1}) + k_m\psi'_{mn}(x_{m+1})
\end{vmatrix} = 0 \qquad (6\text{-}114)
$$

Where $\psi' \equiv \frac{d\psi}{dx}$, $\phi' \equiv \frac{d\phi}{dx}$, and ψ, ϕ functions are as defined in Table 6-2.

An additional relation that is needed to determine the eigenvalues β_n. This relationship is obtained from the condition that if the homogeneous system 6-113 should have nontrivial solution the determinant of the coefficients should vanish. With this consideration we obtain the transcendental Eq. 6-114, the positive roots of which give the eigenvalues.

$$\beta_1 < \beta_2 < \beta_3 < \cdots < \beta_n < \cdots$$

For each of these eigenvalues there is the corresponding values of C_{in}, D_{in} and X_{in}. The coefficients C_{in}, D_{in} are determined from the solution of system 6-113.

We have yet to determine the unknown function $\Gamma(t)$ in the solution 6-109 and the procedure is as follows.

We consider the following orthogonality property [5] of the eigenfunctions $X_{in}(x)$ over the entire range of m layers (i.e., from x_1 to x_{m+1}):

$$\sum_{i=1}^{m} \frac{k_i}{\alpha_i} \int_{x_i}^{x} X_{in}(x) \cdot X_{in'}(x) \cdot x^p \cdot dx = \begin{cases} 0 & \text{for } n \neq n' \\ \text{const.} & \text{for } n = n' \end{cases} \tag{6-115}$$

It is to be noted that the term k_i/α_i is the same as the square of the discontinuous-weighting function, W_i^2, of Tittle given by Eq. 6-66 and the term x^p is the Sturm-Liouville weighting factor with $p = 0, 1, 2$ for the plates, cylinders and spheres respectively.

Assuming the heat-generation function $g_i(x, t)$ and the initial-condition function $F_i(x)$ satisfy Dirichlet's condition, we expand these functions over the entire range of m layers (i.e., from x_1 to x_{m+1}) in an infinite series of eigenfunctions $X_{in}(x)$ in the form

$$\frac{\alpha_i}{k_i} g_i(x, t) = \sum_{n=1}^{\infty} g_n^*(t) \cdot X_{in}(x) \tag{6-116}$$

$$f_i(x) = \sum_{n=1}^{\infty} f_n^* \cdot X_{in}(x) \tag{6-117}$$

Similarly, we expand unity in the form (this expansion will be needed later in the analysis):

$$1 = \sum_{n=1}^{\infty} I_n^* \cdot X_{in}(x) \tag{6-118}$$

The unknown functions $g_n^*(t)$, f_n^*, I_n^* in the above expansions are determined by multiplying both sides of Eqs. 6-116, 6-117, 6-118 by

$$\frac{k_i}{\alpha_i} X_{in'}(x) \cdot x^p$$

and integrating with respect to x from x_i to x_{i+1}, summing up the equalities over all values of i (i.e., $i = 1$ to m) and making use of the orthogonality condition given by Eq. 6-115. The results are

$$g_n^*(t) = \frac{\sum_{i=1}^{m} \int_{x_i}^{x_{i+1}} g_i(x,t) \cdot X_{in}(x) \cdot x^p \cdot dx}{N} \tag{6-119}$$

$$f_n^* = \frac{\sum_{i=1}^{m} \frac{k_i}{\alpha_i} \int_{x_i}^{x_{i+1}} f_i(x) \cdot X_{in}(x) \cdot x^p \cdot dx}{N} \tag{6-120}$$

$$I_n^* = \frac{\sum_{i=1}^{m} \frac{k_i}{\alpha_i} \int_{x_i}^{x_{i+1}} X_{in}(x) \cdot x^p \cdot dx}{N} \tag{6-121}$$

where the norm N is given as

$$N = \sum_{i=1}^{m} \frac{k_i}{\alpha_i} \int_{x_i}^{x_{i+1}} X_{in}^2(x) \cdot x^p \cdot dx \tag{6-122}$$

Substituting Eqs. 6-109, 6-116, 6-118 into Eq. 6-106 we obtain

$$\sum_{n=1}^{\infty} \Gamma_n(t) \cdot \alpha_i \nabla^2 X_{in}(x) + \sum_{n=1}^{\infty} g_n^*(t) \cdot X_{in}(x) - \sum_{n=1}^{\infty} I_n^* \cdot \frac{dT_s(t)}{dt} \cdot X_{in}(x)$$
$$= \sum_{n=1}^{\infty} \frac{d\Gamma_n(t)}{dt} \cdot X_{in}(x) \tag{6-123}$$

In view of Eq. 6-110, the term $\alpha_i \nabla^2 X_{in}(x)$ in Eq. 6-123 is replaced by $-\beta_n^2 X_{in}(x)$, and we have

$$\sum_{n=1}^{\infty} \left[\frac{d\Gamma_n(t)}{dt} + \beta_n^2 \Gamma_n(t) - g_n^*(t) + I_n^* \frac{dT_s(t)}{dt} \right] \cdot X_{in}(x) = 0 \tag{6-124}$$

from which we obtain the following ordinary differential equation for the function $\Gamma_n(t)$

$$\frac{d\Gamma_n(t)}{dt} + \beta_n^2 \Gamma_n(t) = g_n^*(t) - I_n^* \cdot \frac{dT_s(t)}{dt} \tag{6-125}$$

To solve Eq. 6-125 an initial condition is needed. For $t = 0$, Eq. 6-109 gives

$$f_i(x) = \sum_{n=1}^{\infty} X_{in}(x) \cdot \Gamma_n(0) \tag{6-126}$$

From Eqs. 6-117 and 6-126, we obtain

$$\Gamma_n(0) = f_n^* \tag{6-127}$$

Solution of differential Eq. 6-125 subject to the initial condition 6-127 gives

$$\Gamma_n(t) = e^{-\beta_n^2 t}\left[f_n^* + \int_{t'=0}^{t}\left\{g_n^*(t') - I_n^*\cdot\frac{dT_s(t')}{dt'}\right\}e^{\beta_n^2 t'}\cdot dt'\right] \tag{6-128}$$

Substituting Eq. 6-128 into Eq. 6-109, the solution of the boundary-value problem of heat conduction becomes

$$\theta_i(x,t) = \sum_{n=1}^{\infty} e^{-\beta_n^2 t}\cdot X_{in}(x)\left[f_n^* + \int_{t'=0}^{t}\left\{g_n^*(t') - I_n^*\frac{dT_s(t')}{dt'}\right\}e^{\beta_n^2 t'}dt'\right]$$

$$\text{in} \quad x_i \le x \le x_{i+1} \tag{6-129}$$

$$\text{and} \quad i = 1,2,3,\ldots,m$$

where

$$f^* = \frac{\displaystyle\sum_{i=1}^{m}\frac{k_i}{\alpha_i}\int_{x_i}^{x_{i+1}} f_i(x')\cdot X_{in}(x')\cdot x'^p\cdot dx'}{N} \tag{6-130a}$$

$$g^*(t') = \frac{\displaystyle\sum_{i=1}^{m}\int_{x_i}^{x_{i+1}} g_i(x',t')\cdot X_{in}(x')\cdot x'^p\cdot dx'}{N} \tag{6-130b}$$

$$I_n^* = \frac{\displaystyle\sum_{i=1}^{m}\frac{k_i}{\alpha_i}\int_{x_i}^{x_{i+1}} X_{in}(x')\cdot x'^p\cdot dx'}{N} \tag{6-130c}$$

and the norm N is

$$N = \sum_{i=1}^{m}\frac{k_i}{\alpha_i}\int_{x_i}^{x_{i+1}} X_{in}^2(x')\cdot x'^p\cdot dx' \tag{6-130d}$$

Hence the solution of the problem is formally complete.

6-10. TWO-REGION CONCENTRIC CYLINDER WITH HEAT GENERATION

We now illustrate the application of the above general solution for transient heat-conduction problem in a long two-layer cylinder with heat generation within the solid.

Figure 6-7 shows a cylinder of radius $r = b$ that contains an inner concentric cylindrical core of radius $r = a$. Let k_1, k_2 be the thermal conductivities respectively of the inner and outer cylinders and α_1, α_2 the thermal diffusivities. Initially the inner and outer regions are at temperatures $F_1(r)$ and $F_2(r)$ respectively. For times $t > 0$, heat is generated only in the inner cylindrical core at a rate of $g_1(t)$ Btu/hr ft^3 and it is dissipated

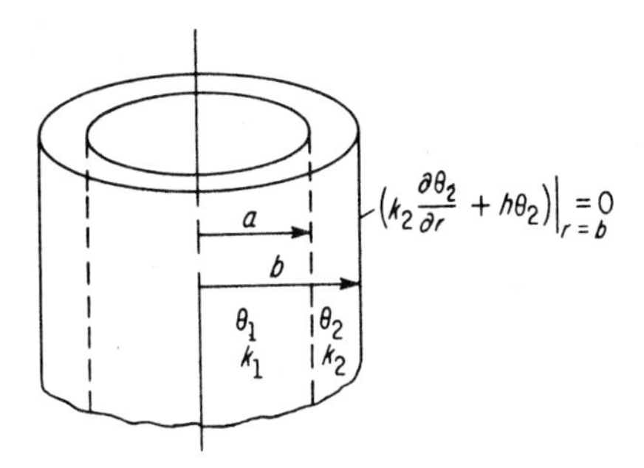

FIG. 6-7. Boundary condition at the outer surface for a two-region solid cylinder with perfect thermal contact at the interface and heat generation in the inner cylinder.

by convection from the boundary surface at $r = b$ into a surrounding at temperature $T_s(t)$ which varies with time.

Let

$T_1(r, t)$ = temperature distribution in the inner core, $0 \leq r \leq a$

$T_2(r, t)$ = temperature distribution in the outer cylinder, $a \leq r \leq b$

For convenience in the analysis we define an excess temperature $\theta_i(r, t)$ which is measured above the temperature of the surrounding $T_s(t)$ in the form

$$\theta_i(r, t) = T_i(r, t) - T_s(t), \quad i = 1,2 \tag{6-131}$$

In terms of the excess temperature $\theta_i(r, t)$ the boundary-value problem of heat conduction is given as

(6-132a)

$$\frac{\alpha_1}{r}\frac{\partial}{\partial r}\left(r\,\frac{\partial \theta_1(r,t)}{\partial r}\right) + \frac{\alpha_1}{k_1}\,g_1(t) - \frac{dT_s}{dt} = \frac{\partial \theta_1(r,t)}{\partial t} \quad \text{in} \quad 0 \leq r \leq a,\, t > 0$$

$$\frac{\alpha_2}{r}\frac{\partial}{\partial r}\left(r\,\frac{\partial \theta_2(r,t)}{\partial r}\right) - \frac{dT_s}{dt} = \frac{\partial \theta_2(r,t)}{\partial t} \quad \text{in} \quad a \leq r \leq b,\, t > 0$$

(6-132b)

with the boundary conditions

$$\theta_1(a, t) = \theta_2(a, t)$$

$$k_1\,\frac{\partial \theta_1(a,t)}{\partial r} = k_2\,\frac{\partial \theta_2(a,t)}{\partial r} \tag{6-133b}$$

$$-k_2\,\frac{\partial \theta_2(b,t)}{\partial r} = h\theta_2(b, t)$$

and the initial conditions

$$\theta_1(r, 0) = F_1(r) - T_s(0) \equiv f_1(r) \quad \text{in} \quad 0 \leq r \leq a \tag{6-134a}$$

$$\theta_2(r, 0) = F_2(r) - T_s(0) \equiv f_2(r) \quad \text{in} \quad a \leq r \leq b \tag{6-134b}$$

The solution of this problem is obtained from the general solution 6-129 as

$$\theta_i(r,t) = \sum_{n=1}^{\infty} e^{-\beta_n^2 t}\cdot X_{in}(x)\left[f_n^* + \int_{t'=0}^{t}\left\{g_n^*(t') - I_n^*\cdot\frac{dT_s(t')}{dt'}\right\}e^{\beta_n^2 t'}\cdot dt'\right] \tag{6-135}$$

in $0 \le r \le a$ for $i = 1$

in $a \le r \le b$ for $i = 2$

where

$$f_n^* = \frac{1}{N}\left[\frac{k_1}{\alpha_1}\int_0^a f_1(r')\cdot X_{1n}(r')\cdot r'\cdot dr' + \frac{k_2}{\alpha_2}\int_a^b f_2(r')\cdot X_{2n}(r')\cdot r'\cdot dr'\right]; \tag{6-136a}$$

$$g_n^*(t') = \frac{1}{N}g_1(t')\cdot\int_0^a X_{1n}(r')\cdot r'\cdot dr' \tag{6-136b}$$

$$I_n^* = \frac{1}{N}\left[\frac{k_1}{\alpha_1}\int_0^a X_{1n}(r')\cdot r'\cdot dr' + \frac{k_2}{\alpha_2}\int_a^b X_{2n}(r')\cdot r'\cdot dr'\right] \tag{6-136c}$$

and the norm N is

$$N = \frac{k_1}{\alpha_1}\int_0^a X_{1n}^2(r')\cdot r'\cdot dr' + \frac{k_2}{\alpha_2}\int_a^b X_{2n}^2(r')\cdot r'\cdot dr' \tag{6-136d}$$

We now determine the eigenfunctions and the eigenvalues for this problem. The eigenfunctions $X_{1n}(r)$ and $X_{2n}(r)$ satisfy the following eigenvalue problems.

$$\frac{\alpha_1}{r}\frac{d}{dr}\left[r\frac{dX_{1n}(r)}{dr}\right] + \beta_n^2 X_{1n}(r) = 0 \quad \text{in} \quad 0 \le r \le a \tag{6-137a}$$

$$\frac{\alpha_2}{r}\frac{d}{dr}\left[r\frac{dX_{2n}(r)}{dr}\right] + \beta_n^2 X_{2n}(r) = 0 \quad \text{in} \quad a \le r \le b \tag{6-137b}$$

with the boundary conditions

$$X_{1n}(a) = X_{2n}(a) \tag{6-138a}$$

$$k_1\frac{dX_{1n}(a)}{dr} = k_2\frac{dX_{2n}(a)}{dr} \tag{6-138b}$$

$$-k_2\frac{dX_{2n}(b)}{dr} = hX_{2n}(b) \tag{6-138c}$$

Solutions of differential Eqs. 6-137 subject to the boundary conditions 6-138 may be taken in the form

$$X_{1n}(r) = J_0\left(\frac{\beta_n}{\sqrt{\alpha_1}}r\right) \tag{6-139a}$$

$$X_{2n}(r) = C_{2n} J_0\left(\frac{\beta_n}{\sqrt{\alpha_2}} r\right) + D_{2n} Y_0\left(\frac{\beta_n}{\sqrt{\alpha_2}} r\right) \tag{6-139b}$$

It is to be noted that Y_0 function is excluded from the solution $X_{1n}(r)$ because the region includes the origin where the Y_0 function becomes infinite.

We have two unknown coefficients C_{2n}, D_{2n} and the eigenvalues β_n are to be determined, and the three boundary conditions given by Eq. 6-138 are sufficient to determine them.

The unknown coefficients C_{2n} and D_{2n} are determined from the boundary conditions 6-138a,b and given in the form

$$C_{2n} = \left[\frac{k_1\sqrt{\alpha_2}}{k_2\sqrt{\alpha_1}} J_1\left(\frac{\beta_n}{\sqrt{\alpha_1}} a\right) \cdot Y_0\left(\frac{\beta_n}{\sqrt{\alpha_2}} a\right) - J_0\left(\frac{\beta_n}{\sqrt{\alpha_1}} a\right) \cdot Y_1\left(\frac{\beta_n}{\sqrt{\alpha_2}} a\right)\right]\frac{1}{\Delta} \tag{6-140}$$

$$D_{2n} = \left[J_0\left(\frac{\beta_n}{\sqrt{\alpha_1}} a\right) \cdot J_1\left(\frac{\beta_n}{\sqrt{\alpha_2}} a\right) - \frac{k_1\sqrt{\alpha_2}}{k_2\sqrt{\alpha_1}} J_1\left(\frac{\beta_n}{\sqrt{\alpha_1}} a\right) \cdot J_0\left(\frac{\beta_n}{\sqrt{\alpha_2}} a\right)\right]\frac{1}{\Delta} \tag{6-141}$$

and Δ is the determinant of the coefficients.

The boundary condition 6-138c gives the following transcendental equation for determining the eigenvalues β_n

$$\frac{k_2}{hb} \cdot \frac{\beta b}{\sqrt{\alpha_2}} = \frac{C_{2n} J_0\left(\frac{\beta b}{\sqrt{\alpha_2}}\right) + D_{2n} Y_0\left(\frac{\beta b}{\sqrt{\alpha_2}}\right)}{C_{2n} J_1\left(\frac{\beta b}{\sqrt{\alpha_2}}\right) + D_{2n} Y_1\left(\frac{\beta b}{\sqrt{\alpha_2}}\right)} \tag{6-142}$$

Substituting the values of C_{2n} and D_{2n}, Eq. 6-142 is written in the form

$$\frac{\bar{\beta}}{Bi} = \frac{A}{B} \tag{6-143}$$

where

$$A = \left[J_1\left(\bar{\beta}\,\frac{a}{b}\right) \cdot Y_0(\bar{\beta}) - Y_1\left(\bar{\beta}\,\frac{a}{b}\right) \cdot J_0(\bar{\beta})\right] + \frac{k_1\sqrt{\alpha_2}}{k_2\sqrt{\alpha_1}}$$

$$\cdot \frac{J_1\left(\bar{\beta} \cdot \frac{a}{b}\,\sqrt{\frac{\alpha_2}{\alpha_1}}\right)}{J_0\left(\bar{\beta}\,\frac{a}{b}\,\sqrt{\frac{\alpha_2}{\alpha_1}}\right)} \cdot \left[Y_0\left(\bar{\beta}\,\frac{a}{b}\right) \cdot J_0(\bar{\beta}) - J_0\left(\bar{\beta}\,\frac{a}{b}\right) \cdot Y_0(\bar{\beta})\right]$$

$$B = \left[J_1\left(\bar{\beta}\,\frac{a}{b}\right) \cdot Y_1(\bar{\beta}) - Y_1\left(\bar{\beta}\,\frac{a}{b}\right) \cdot J_1(\bar{\beta})\right]$$

$$+ \frac{k_1\sqrt{\alpha_2}}{k_2\sqrt{\alpha_1}} \frac{J_1\left(\bar{\beta} \cdot \frac{a}{b}\,\sqrt{\frac{\alpha_2}{\alpha_1}}\right)}{J_0\left(\bar{\beta}\,\frac{a}{b}\,\sqrt{\frac{\alpha_2}{\alpha_1}}\right)} \cdot \left[Y_0\left(\bar{\beta}\,\frac{a}{b}\right) \cdot J_1(\bar{\beta}) - J_0\left(\bar{\beta}\,\frac{a}{b}\right) \cdot Y_1(\bar{\beta})\right]$$

where

$$\bar{\beta} = \frac{\beta b}{\sqrt{\alpha_2}}$$

$$Bi = \frac{hb}{k_2}$$

The transcendental Eq. 6-143 can be solved using a digital computer. Integrations in Eqs. 6-136b, 6-136c, and 6-136d may be performed and explicit relations obtained for the functions $g_n^*(t)$, I_n^*, and the norm N.

REFERENCES

1. H. S. Carslaw and J. C. Jaeger, *Conduction of Heat in Solids*, Clarendon Press, Oxford, 1959, pp. 319–326.

2. Theodore R. Goodman, "The Adjoint Heat-Conduction Problems for Solids," ASTIA-AD 254–769, (AFOSR-520), April 1961.

3. C. W. Tittle, "Boundary-Value Problems in Composite Media: Quasi-Orthogonal Functions," *Journal of Applied Physics*, Vol. 36, No. 4 (1965), pp. 1486–1488.

4. Personal communication with C. W. Tittle.

5. P. E. Bulavin and V. M. Kashcheev, "Solution of Nonhomogeneous Heat-Conduction Equation for Multilayered Bodies," *International Chemical Engineering*, Vol. 5, No. 1 (1965), pp. 112–115.

6. Charles W. Bouchillon, "Unsteady Heat Conduction in Composite Slabs," ASME Paper 64–WA/HT-13, 1964.

7. Ruel V. Churchill, *Fourier Series and Boundary-Value Problems*, McGraw-Hill Book Company, New York, 1963, p. 70.

8. C. W. Tittle and Virgil L. Robinson, "Analytical Solution of Conduction Problems in Composite Media," ASME paper 65–WA–HT–52, 1965.

9. L. S. Allen and C. W. Tittle, "Some Functions in the Theory of Neutron Logging," *Journal of the Graduate Research Center*, Southern Methodist University, Vol. 33, No. 1 (1964), pp. 33–54.

10. Harry Lee Beach, "The Application of the Orthogonal Expansion Technique to Conduction Heat Transfer Problems in Multilayer Cylinders," M.S. Thesis, Mech. and Aerospace Eng., N.C. State University, Raleigh, N.C., (1967), App. C and D.

11. Clifford J. Moore, "Heat Transfer Across Surfaces in Contact: Studies of Transients in One-Dimensional Composite Systems," Southern Methodist University, Mechanical Engineering Department, Dallas, Texas. Ph.D. Dissertation, 1967.

PROBLEMS

1. A three-layer slab has outer surfaces at $x = x_1$ and $x = x_4$ and interfaces at $x = x_2$ and $x = x_3$. The properties of the solid are uniform within each layer but discontinuous at the interfaces where there is a perfect thermal contact between the layers. Initially all the layers are at zero temperature. For times $t > 0$

the outer-boundary surface at $x = x_4$, is kept at a prescribed temperature,

$$T_3(x_4, t) = f_4$$

and the outer-boundary surface at $x = x_1$ is subjected to convection boundary condition of the form

$$\left[-k_1 \frac{\partial T(x,t)}{\partial x} + h_1 T(x,t)\right]_{x=x_1} = f_1$$

Using the adjoint-solution technique formulate the integral equations and the adjoint problems for determining the temperatures at the interfaces $x = x_2$ and $x = x_3$.

2. A three-layer concentric hollow cylinder has inner and outer-boundary surfaces at $r = r_1$ and $r = r_4$ respectively and interfaces at r_2 and r_3. The properties of the solid are uniform within each layer but discontinuous at the interfaces where there is perfect thermal contact between the layers. Initially all the layers are at zero temperature. For times $t > 0$ the boundary surfaces r_1 and r_4 are kept at constant temperatures f_1 and f_4 respectively, i.e.,

$$T_1(r_1, t) = f_1, \quad T_3(r_4, t) = f_4$$

Using the adjoint-solution technique formulate the integral equations and the adjoint problems for determining temperature at the interfaces $r = r_2$ and $r = r_3$.

3. A two-layer slab has outer surfaces at $x = 0$ and $x = L$, and the interface at $x = a$. The properties of the solid are uniform within each layer but discontinuous at the interface where there is a contact resistance described by a boundary condition in the form

$$-k_1 \frac{\partial T_1}{\partial x} = h_1(T_1 - T_2) \quad \text{at} \quad x = a$$

Initially the entire solid is at zero temperature. For times $t > 0$ the outer boundaries are subjected to convection heat transfer in the form

$$-k_1 \frac{\partial T_1}{\partial x} + hT_1 = f_0 \quad \text{at} \quad x = 0$$

$$k_2 \frac{\partial T_2}{\partial x} + hT_2 = f_2 \quad \text{at} \quad x = x_2$$

Using the technique of orthogonal expansion over multilayer regions determine a relation for the transient-temperature distributions $T_1(x, t)$ and $T_2(x, t)$ in the layers 1 and 2.

4. Consider a two-layer solid cylinder $0 \le r \le b$, with interface at $r = a$ (i.e., $0 < a < b$). The properties of the solid are uniform within each layer but discontinuous at the interface where there is perfect thermal contact between the cylinders. Initially the entire solid is at zero temperature. For times $t > 0$ the outer boundary surface $r = b$ is subjected to a constant temperature T_b. Using the orthogonal-expansion technique over multilayer regions determine a relation for the temperature distribution $T(r, t)$ in both layers for times $t > 0$.

5. A two-layer solid cylinder $0 \le r \le b$ has an interface at $r = a$ (i.e., $0 < a < b$). The properties of the solid are uniform within each layer but discontinuous at the interface where there is perfect thermal contact. Initially the entire region is at zero temperature. For times $t > 0$ there is heat generation within the inner cylinder (i.e., $0 \leq r \leq a$) at a constant rate of g Btu/hr ft^3 while the outer boundary surface $r = b$ is kept at a constant temperature T_b. Find a relationship for the temperature distribution in the cylinders for times $t > 0$.

6. A two-region solid sphere $0 \leq r \leq b$ has perfect thermal contact between the layers at the interface $r = a$ (i.e., $0 < a < b$). Properties of the solid are uniform within each layer but discontinuous at the interface. Initially both regions are at zero temperature. For times $t > 0$ the outer boundary surface at $r = b$ is kept at a constant temperature T_b. Using the orthogonal-expansion technique over multilayer regions determine a relationship for temperature distribution in the spheres for times $t > 0$.

7. Solve Prob. 6 by assuming a contact resistance between the spheres at the interface $r = a$.

8. The determinant 6-114 can be used for evaluating the eigenvalues for the one-dimensional, n-layer heat-conduction problem 6-106 subjected to the boundary conditions as described by 6-107. If the boundary condition at the outer surface x_1 were of the third kind instead of the second kind, determine the resulting changes that should be made in the determinant 6-114.

7

Approximate Methods in the Solution of Heat-conduction Problems

An approximate method in the solution of transient heat-conduction problems is the *integral method* which has recently received considerable attention; it is capable of handling nonlinear problems of heat conduction such as problems with temperature-dependent thermal conductivities, or with boundaries involving some power of temperature as in the case of fourth-power law of radiation, or with boundaries involving change of phase (i.e., solidification, melting, ablation). The use of integral methods in the solution of transient-heat-conduction problems was first introduced by Goodman [1]. The method is analogous to the momentum and energy-integral methods of von Kármán and Pohlhausen which are used in the solution of the boundary-layer momentum and heat-transfer problems in fluid mechanics [2]. An excellent treatment of integral methods in the solution of various types of heat-conduction problems and a bibliography is given by Goodman [3].

The method is simple and straightforward, easily applicable to both linear and nonlinear one-dimensional transient-boundary-value problems of heat conduction for certain boundary conditions. The results are approximate, but several solutions obtained with this method when compared with the exact solutions have confirmed that the accuracy is acceptable for engineering purposes.

Another approximate method which provides a way of solving heat-conduction problems in irregular geometries is the *method of Galerkin.*

In this chapter we examine the application of these methods to the solution of heat-conduction problems.

7-1. THE INTEGRAL METHOD—GENERAL CONSIDERATIONS

When the differential equation of heat conduction is solved with the techniques described in the previous chapters, the resulting solution satisfies the problem at each differential element over the entire range of space variables considered; but, such solutions are rather difficult to obtain when the problem is nonlinear or involves complicated boundary conditions. An approximate but powerful method in the solution of one-dimensional, time-dependent boundary-value problems of heat conduction is to solve them in such a manner that the solution will satisfy the problem not at every differential element in the region but only on the *average* over the region considered. The "integral method" is based on such a concept and the basic steps to be followed in the solution of one-dimensional, time-dependent differential equation of heat conduction subject to some prescribed boundary and uniform initial conditions are summarized below.

(1) The differential equation of heat conduction is integrated over a phenomenological distance $\delta(t)$, called the *thermal layer*, in order to remove from the differential equation derivatives with respect to the space variable. The thermal layer in the heat-conduction problem is defined as the distance from the origin beyond which the initial temperature distribution within the region remains unaffected by the applied boundary condition, and hence there is no heat flow in the region beyond $\delta(t)$. The concept of the thermal layer is similar to that of the boundary-layer thickness in fluid mechanics. The various terms resulting from the integration are evaluated by making use of the boundary and initial conditions for the problem and from the definition of the thermal layer. The resulting equation is called the *heat-balance integral*.

(2) A suitable profile is assumed for the temperature distribution over the thermal layer. A profile which is frequently used is the polynomial approximation[1] and the unknown coefficients associated with the polynomials are evaluated by means of the natural and derived conditions at each end of the thermal layer. The resulting temperature profile is a function of the space variable x and the thermal-layer thickness $\delta(t)$.

(3) When this temperature profile is substituted into the heat-balance integral and integrations with respect to the space variable x are performed, one obtains an ordinary differential equation for the thermal-layer thickness $\delta(t)$ with t as the independent variable.

(4) Solution of this ordinary differential equation subject to an initial condition [i.e., in most cases $\delta(t) = 0$ for $t = 0$] gives a relation for $\delta(t)$ as a function of the time variable.

(5) Substituting the value of $\delta(t)$ thus determined into the temperature profile chosen, one obtains a relation for the temperature distribution as a function of time and space variables, and the solution of the problem is complete.

To illustrate the procedure discussed above we examine first a simple problem of heat conduction for a semi-infinite region.

Semi-infinite Solid $x > 0$. Consider a semi-infinite solid $0 \leq x < \infty$ which is initially at a uniform temperature T_i. For times $t > 0$ the boundary surface at $x = 0$ is kept at a constant temperature T_0. The

[1]Trigonometric or exponential approximations may also be chosen if one wishes so.

boundary-value problem of heat conduction is given as

$$\frac{\partial^2 T(x,t)}{\partial x^2} = \frac{1}{\alpha}\frac{\partial T(x,t)}{\partial t} \quad \text{in} \quad 0 \le x < \infty,\ t > 0 \tag{7-1a}$$

$$T(x,t) = T_0 \quad \text{at} \quad x = 0,\ t > 0 \tag{7-1b}$$

$$T(x,t) = T_i \quad \text{in} \quad 0 \le x < \infty,\ t = 0 \tag{7-1c}$$

The solution of this problem can easily be obtained by directly solving the differential equation of heat conduction with the methods discussed in the previous chapters. We now examine the solution of this problem with the integral methods in order to illustrate the procedure and to compare the approximate and the exact solutions.

Figure 7-1 shows the geometry, the boundary condition, and the location of the thermal layer $\delta(t)$ at any instant t. For all practical purposes the region $x \ge \delta(t)$ is at the initial temperature T_i and there is no heat flux beyond $x \ge \delta(t)$.

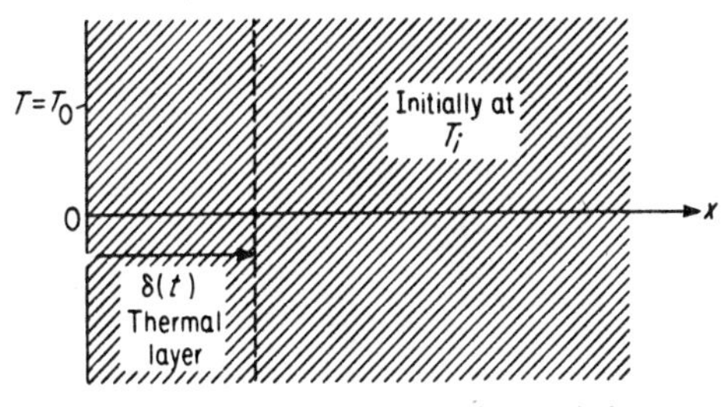

FIG. 7-1. A semi-infinite region and the "thermal layer."

The basic steps in the solution are as follows.

(1) We integrate differential equation of heat conduction 7-1a with respect to the space variable from $x = 0$ to $x = \delta(t)$.

$$\left.\frac{\partial T}{\partial x}\right|_{x=\delta(t)} - \left.\frac{\partial T}{\partial x}\right|_{x=0} = \frac{1}{\alpha}\int_{x=0}^{\delta(t)} \frac{\partial T}{\partial t}\, dx \tag{7-2}$$

When the integral on the right-hand side is rearranged, Eq. 7-2 becomes

$$\left.\frac{\partial T}{\partial x}\right|_{x=\delta} - \left.\frac{\partial T}{\partial x}\right|_{x=0} = \frac{1}{\alpha}\left[\frac{\partial}{\partial t}\left(\int_{x=0}^{\delta} T\, dx\right) - \left.T\right|_{x=\delta} \cdot \frac{d\delta}{dt}\right] \tag{7-3}$$

where $\delta \equiv \delta(t)$. By the definition of thermal layer we have

$$\left.\frac{\partial T}{\partial x}\right|_{x=\delta} = 0 \quad \text{and} \quad T\,|_{x=\delta} = T_i \tag{7-4}$$

Substituting Eq. 7-4 into Eq. 7-3,

$$-\alpha \left.\frac{\partial T}{\partial x}\right|_{x=0} = \frac{\partial}{\partial t}\left[\int_{x=0}^{\delta} T \cdot dx - T_i \cdot \delta\right] \tag{7-5}$$

For convenience we define

$$\theta \equiv \int_{x=0}^{\delta} T\,dx \tag{7-6}$$

Equation 7-5 becomes

$$-\alpha \left.\frac{\partial T}{\partial x}\right|_{x=0} = \frac{\partial}{\partial t}\left[\theta - T_i \cdot \delta\right] \tag{7-7}$$

Equation 7-7 is called the *heat-balance integral* for the problem considered.

(2) We assume a polynomial approximation for the temperature profile across the thermal layer. A second, third, and fourth-degree polynomial has been examined by Reynolds and Dolton [4]. Here we consider a third-degree polynomial approximation in the form

$$T = a + bx + cx^2 + dx^3 \qquad 0 \le x \le \delta(t) \tag{7-8}$$

Four conditions are needed to evaluate the four unknown coefficients. Three of these conditions are given as[2]

$$T|_{x=0} = T_0 \qquad T|_{x=\delta} = T_i \qquad \left.\frac{\partial T}{\partial x}\right|_{x=\delta} = 0 \tag{7-9}$$

The fourth condition may be derived by making use of the differential equation 7-1a. Since at $x = 0$ we have $T = T_0 =$ constant, the derivative of temperature with respect to the time variable vanishes at $x = 0$, and differential equation 7-1a yields

$$\left.\frac{\partial^2 T}{\partial x^2}\right|_{x=0} = 0 \tag{7-10}$$

By means of the four conditions given by Eq. 7-9 and 7-10 the four unknown coefficients in Eq. 7-8 are evaluated and the profile becomes

$$\frac{T - T_i}{T_0 - T_i} = 1 - \frac{3}{2}\frac{x}{\delta} + \frac{1}{2}\left(\frac{x}{\delta}\right)^3 \tag{7-11}$$

which is a function of x and δ.

(3) Substituting the profile 7-11 into the heat-balance integral 7-7 and performing the operations indicated, we obtain the following first-order ordinary differential equation for the thermal-layer thickness δ,

$$4\alpha = \delta \cdot \frac{d\delta}{dt} \tag{7-12}$$

subject to the initial condition

$$\delta = 0 \quad \text{for} \quad t = 0 \tag{7-13}$$

(4) Solution of Eq. 7-12 subject to the initial condition 7-13 gives the thermal-layer thickness as a function of time as

$$\delta = \sqrt{8\alpha t} \tag{7-14}$$

[2]If a second-degree polynomial approximation is assumed, these three conditions are sufficient to determine the three unknown coefficients in the polynomial.

(5) Substituting Eq. 7-14 into the profile 7-11 we obtain the temperature distribution as a function of the space and time variables in the form

$$\frac{T - T_i}{T_0 - T_i} = 1 - \frac{3}{2}\frac{x}{\sqrt{8\alpha t}} + \frac{1}{2}\left(\frac{x}{\sqrt{8\alpha t}}\right)^3 \qquad 0 \le x \le \delta \qquad (7\text{-}15)$$

In the above analysis the fourth condition 7-10 was derived by evaluating the differential equation of heat conduction at $x = 0$. The fourth condition could also be derived by evaluating the differential equation of heat conduction 7-1a at $x = \delta$; in such case the fourth derived condition would be in the form

$$\left.\frac{\partial^2 T}{\partial x^2}\right|_{x=\delta} = 0 \qquad (7\text{-}16)$$

When the four unknown coefficients in Eq. 7-8 are determined by using the three conditions 7-9 together with the fourth condition 7-16, the resulting temperature profile becomes

$$\frac{T - T_i}{T_0 - T_i} = \left(1 - \frac{x}{\delta}\right)^3 \qquad (7\text{-}17)$$

When this temperature profile is substituted into the heat-balance integral Eq. 7-7 and the resulting ordinary differential equation is solved, one obtains the thermal-layer thickness as

$$\delta = \sqrt{24\alpha t} \qquad (7\text{-}18)$$

and the corresponding temperature distribution becomes

$$\frac{T - T_i}{T_0 - T_i} = \left(1 - \frac{x}{\sqrt{24\alpha t}}\right)^3 \qquad 0 \le x \le \delta \qquad (7\text{-}19)$$

which is different in form from that given by Eq. 7-15, although in both cases a third-degree polynomial approximation is used. Some ambiguity always exists in the accuracy of the final result. Question concerning which of these two results is more accurate cannot be answered without comparing these approximate solutions with the exact result. However, it is encouraging to know that the error involved in the solutions with approximate integral method is usually small and may be acceptable for most engineering purposes. Furthermore, the results are rather insensitive as to the initial choice of the form of profile. A choice of a fourth-degree polynomial approximation may improve the result slightly and for practical purposes there is no gain in accuracy by choosing profiles higher than fourth degree.

We now examine the solution of the above problem by means of a fourth-degree polynomial and then compare these approximate solutions with the exact result.

A Fourth-Degree Polynomial Approximation. In this case we assume a temperature profile in the form

$$T = a + bx + cx^2 + dx^3 + ex^4 \qquad (7\text{-}20)$$

The five unknown coefficients are evaluated by applying the following five conditions:

$$T\big|_{x=0} = T_0 \qquad T\big|_{x=\delta(t)} = T_i \qquad \left.\frac{\partial T}{\partial x}\right|_{x=\delta(t)} = 0$$

$$\left.\frac{\partial^2 T}{\partial x^2}\right|_{x=0} = 0 \qquad \left.\frac{\partial^2 T}{\partial x^2}\right|_{x=\delta(t)} = 0 \tag{7-21}$$

Then the temperature profile becomes

$$\frac{T - T_i}{T_0 - T_i} = 1 - 2\left(\frac{x}{\delta}\right) + 2\left(\frac{x}{\delta}\right)^3 - \left(\frac{x}{\delta}\right)^4 \qquad 0 \le x \le \delta(t) \tag{7-22}$$

Substituting this temperature distribution into the heat-balance integral Eq. 7-7 and proceeding as before, we obtain

$$\delta(t) = \sqrt{\frac{40}{3}\,\alpha t}$$

and the corresponding temperature distribution within the solid becomes

$$\frac{T - T_i}{T_0 - T_i} = 1 - 2\left(\frac{x}{\sqrt{\frac{40}{3}\alpha t}}\right) + 2\left(\frac{x}{\sqrt{\frac{40}{3}\alpha t}}\right)^3 - \left(\frac{x}{\sqrt{\frac{40}{3}\alpha t}}\right)^4$$

$$0 \le x \le \delta(t) \tag{7-23}$$

Comparison with the Exact Solution. The exact solution of the boundary-value problem of heat conduction 7-1 is straightforward and is given as

$$\frac{T - T_i}{T_0 - T_i} = 1 - \operatorname{erf}\left(\frac{x}{\sqrt{4\alpha t}}\right) \tag{7-24}$$

Figure 7-2 shows a comparison of approximate temperature distributions based on a second-, third- and fourth-degree polynomial approximation with the exact solution 7-24. The agreement between the exact and approximate solutions is very good for the three different polynomial approximations considered above. The agreement is better in the regions near the boundary surface $x = 0$, but deviation increases in the regions away from the boundary surface. Higher-degree polynomial approximation agrees better with the cxact solution; however, the simpler second-degree polynomial approximation seems to yield reasonably good results which may be considered sufficiently accurate for most engineering applications.

The heat flux at the boundary surface $x = 0$ is a quantity of interest for the heat-conduction problem considered above. The heat flux at the boundary surface $x = 0$ for the various temperature profiles considered above may be expressed in the form

$$q = -k\left.\frac{\partial T}{\partial x}\right|_{x=0} = C \cdot \frac{k(T_0 - T_i)}{\sqrt{\alpha t}}$$

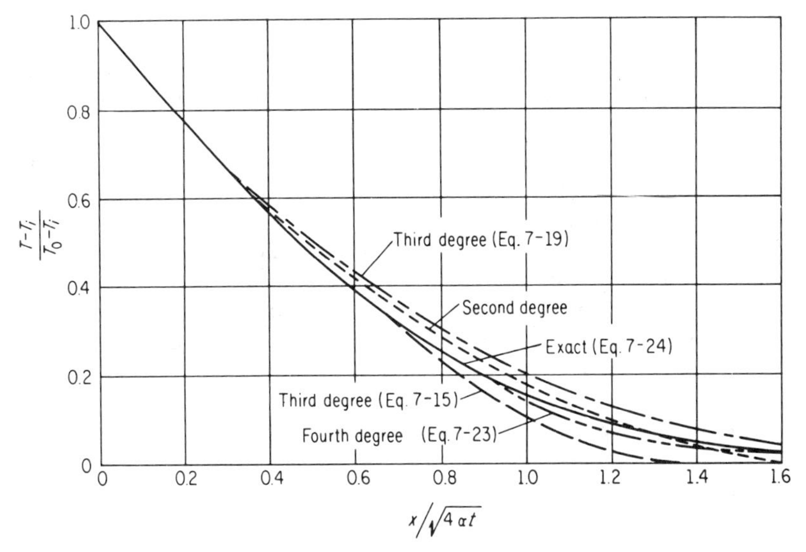

FIG. 7-2. Comparison of temperature distribution obtained from exact and integral solutions for a semi-infinite solid ($0 \leq x < \infty$).

or

$$\frac{q}{k(T_0 - T_i)/\sqrt{\alpha t}} = C \tag{7-25}$$

Table 7-1 gives the values of the constant C as calculated from the exact and approximate temperature profiles and compares the error involved.

TABLE 7-1

ERROR INVOLVED IN THE SURFACE HEAT FLUX

Temperature Profile	C (as Defined by Eq. 7-25)	Percent Error Involved
Exact (Eq. 7-24)	$\frac{1}{\sqrt{\pi}} = 0.565$	0
Cubic approximation (Eq. 7-15)	$\frac{3}{2\sqrt{8}} = 0.530$	6
Cubic approximation (Eq. 7-19)	$\frac{3}{\sqrt{24}} = 0.612$	8
Fourth-degree approximation (Eq 7-23)	$\frac{2}{\sqrt{\frac{40}{3}}} = 0.548$	3

7-2. PROBLEMS IN ONE-DIMENSIONAL FINITE REGION

In the previous section we examined the application of integral methods to the solution of heat-conduction problem for a semi-infinite region, in which the thermal-layer thickness $\delta(t)$ could increase indefinitely because there was no boundary in the region $x > 0$. For the problem of heat conduction in a finite region, say, in a region $0 \leq x \leq L$, the treatment of the problem is similar to that for a semi-infinite region so long as the thermal-layer thickness $\delta(t)$ is less than the thickness of the region, L. As soon as $\delta(t) = L$, the thermal layer has no physical significance and a different treatment of the problem is needed. To illustrate the procedure we examine the application of integral method in the solution of one-dimensional, time-dependent boundary-value problem of heat conduction for a slab of finite thickness.

Slab of Thickness, L. A slab of thickness $0 \leq x \leq L$ is initially at temperature T_i. For times $t > 0$ the boundary surface at $x = 0$ is kept at a constant temperature T_0 and the boundary surface at $x = L$ is kept insulated. The solution of this problem has been treated by Reynolds and Dolton. [4]

The boundary-value problem of heat conduction is given as

$$\frac{\partial^2 T(x,t)}{\partial x^2} = \frac{1}{\alpha}\frac{\partial T(x,t)}{\partial t} \quad \text{in} \quad 0 \leq x \leq L,\ t > 0 \tag{7-26a}$$

$$T(x,t) = T_0 \quad \text{at} \quad x = 0,\ t > 0 \tag{7-26b}$$

$$\frac{\partial T(x,t)}{\partial x} = 0 \quad \text{at} \quad x = L,\ t > 0 \tag{7-26c}$$

$$T(x,t) = T_i \quad \text{in} \quad 0 \leq x \leq L,\ t = 0 \tag{7-26d}$$

Figure 7-3 shows the thermal-layer thickness $\delta(t)$ for $\delta(t) < L$. We integrate the differential equation of heat conduction 7-26a over the thermal-layer thickness as before and obtain the following *heat-balance integral*:

$$\left.\frac{\partial T}{\partial x}\right|_{x=\delta} - \left.\frac{\partial T}{\partial x}\right|_{x=0} = \frac{1}{\alpha}\frac{\partial}{\partial t}[\theta - T_i \cdot \delta] \tag{7-27}$$

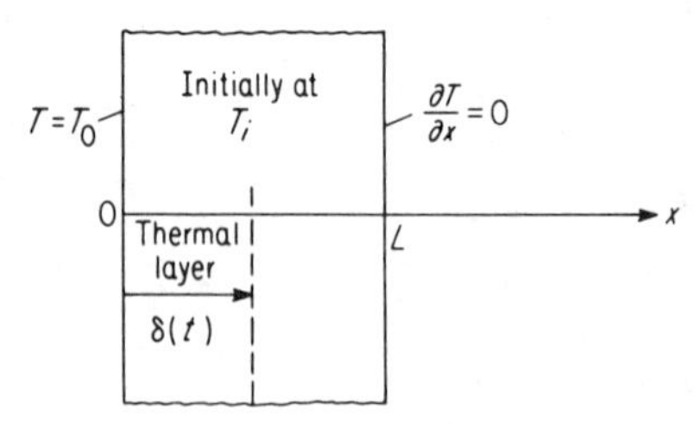

FIG. 7-3. A slab of thickness L and the "thermal layer."

where

$$\theta = \int_0^\delta T \cdot dx$$

Since the region is finite we treat the problem in *two stages*: (1) The *first stage*, during which the thermal-layer thickness δ is less than the slab thickness L (i.e., $\delta \leq L$); (2) The *second stage*, during which δ exceeds the slab thickness L.

The First Stage. The treatment of the problem during the first stage is exactly the same as that discussed in the previous section for a semi-infinite region since the thermal layer has not yet reached the boundary surface $x = L$. The heat-balance integral 7-27 becomes

$$-\alpha \left.\frac{\partial T}{\partial x}\right|_{x=0} = \frac{\partial}{\partial t}[\theta - T_i \cdot \delta] \tag{7-28}$$

which is the same as Eq. 7-7.

Choosing a cubic profile and the auxiliary conditions as discussed in the previous section the temperature distribution may be taken in the form:

$$\frac{T - T_i}{T_0 - T_i} = 1 - \frac{3}{2}\left(\frac{x}{\delta}\right) + \frac{1}{2}\left(\frac{x}{\delta}\right)^3 \tag{7-29a}$$

where

$$\delta = \sqrt{8\alpha t} \tag{7-29b}$$

Equation 7-29 is valid in $0 \leq x \leq \delta$, for

$$\delta \leq L \quad \text{or} \quad t \leq t_L \tag{7-30}$$

where t_L is defined as the time required for the thermal layer to reach the boundary surface $x = L$; the value of t_L is determined by substituting $\delta = L$ in Eq. 7-29b, yielding

$$t_L = \frac{L^2}{8\alpha} \tag{7-31}$$

The Second Stage. For times $t > t_L$, the thickness of the thermal layer exceeds the thickness of the plate, L, hence the thermal layer has no physical significance and the determination of temperature distribution within the slab requires different treatment.

We assume a cubic profile of the form

$$T = a + bx + cx^2 + dx^3 \qquad 0 \leq x \leq L \tag{7-32}$$

and consider the following three auxiliary conditions:

$$T|_{x=0} = T_0 \qquad \left.\frac{\partial T}{\partial x}\right|_{x=L} = 0 \qquad \left.\frac{\partial^2 T}{\partial x^2}\right|_{x=0} = 0 \tag{7-33}$$

The first two of these conditions are the boundary conditions for the slab and the last one is a derived condition which is obtained by evaluating the differential equation of heat conduction at $x = 0$.

Equation 7-32 involves four unknown coefficients but we have chosen only three auxiliary conditions because the heat-balance integral provides the fourth relation. We express three of the coefficients in terms of one of them, say b, and the resulting profile is in the form

$$T = T_0 + bL\left(\frac{x}{L}\right) - \frac{1}{3}\,bL\left(\frac{x}{L}\right)^3 \qquad 0 \le x \le L \tag{7-34a}$$

It is to be noted that the coefficient b is a function of time. By subtracting from both sides of this equation T_i and defining a new variable

$$\eta(t) = \frac{bL}{T_0 - T_i}$$

Eq. 7-34a is written in the form

$$\frac{T - T_i}{T_0 - T_i} = 1 + \eta(t)\cdot\left(\frac{x}{L}\right) - \frac{1}{3}\,\eta(t)\cdot\left(\frac{x}{L}\right)^3 \tag{7-34b}$$

Now the problem is reduced to that of determining the unknown coefficient $\eta(t)$ and the heat-balance integral is used to determine $\eta(t)$. In the second stage, since the region extends from $x = 0$ to $x = L$, the heat balance integral Eq. 7-27 becomes

$$-\alpha\left.\frac{\partial T}{\partial x}\right|_{x=0} = \frac{\partial}{\partial t}\,[\theta - T_i\cdot L] \tag{7-35}$$

where

$$\theta = \int_{x=0}^{L} T\,dx$$

Substituting the temperature profile 7-34b into the heat-balance integral 7-35 and performing the operations indicated we obtain the following first-order ordinary differential equation for the unknown coefficient $\eta(t)$.

$$-\alpha\,\frac{\eta(t)}{L^2} = \frac{5}{12}\,\frac{d\eta(t)}{dt} \tag{7-36a}$$

or

$$\frac{d\eta(t)}{dt} + \frac{12}{5}\,\frac{\alpha}{L^2}\,\eta(t) = 0 \tag{7-36b}$$

In order to solve Eq. 7-36 an initial condition is needed. This initial condition is obtained from the temperature profile 7-34b by applying the condition that temperature should be equal to T_i at $x = L$ at time $t = t_L$, yielding

$$\eta(t) = -\frac{3}{2} \quad \text{for} \quad t = t_L = \frac{L^2}{8\alpha} \tag{7-37}$$

The solution of the differential equation 7-36b subject to the initial condition 7-37 gives

$$\eta(t) = -\frac{3}{2}\,e^{-\left(\frac{12}{5}\frac{\alpha}{L^2}t - \frac{3}{10}\right)} \tag{7-38}$$

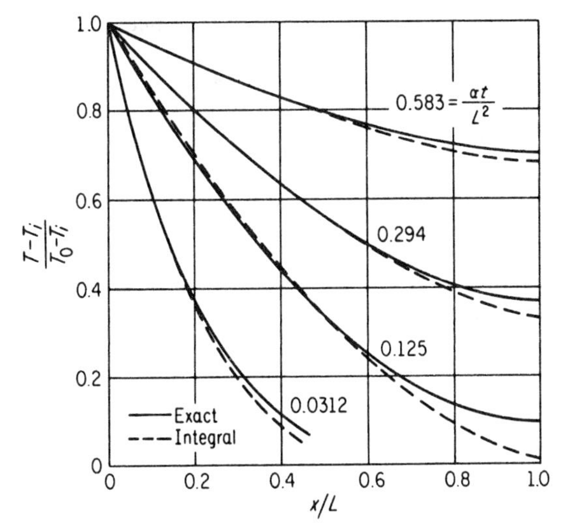

FIG. 7-4. Comparison of temperature distribution from exact and approximate integral solutions for a slab of thickness L. (W. C. Reynolds and T. A. Dolton, Department of Mechanical Engineering, *Report No. 36*, Stanford University, Stanford, Calif., September 1958.)

Substituting Eq. 7-38 into Eq. 7-34b, the temperature distribution in the slab for times $t > t_L$ is obtained.

Figure 7-4 shows a comparison of the approximate and the exact solutions. The agreement is quite good in the regions close to the boundary surface $x = 0$, but deviation from the exact solution is larger at the boundary surface $x = L$.

7-3. PROBLEMS WITH CYLINDRICAL AND SPHERICAL SYMMETRY

In solving the heat-conduction problem with the integral-transform technique in the cartesian system we used a polynomial approximation to represent the temperature profile. Lardner and Pohle [5] investigated the application of the integral method to the solution of the heat-conduction problem in regions with cylindrical and spherical symmetry and have shown that polynomial representation alone of the temperature profile would be inaccurate as an approximation. This is to be expected, since the volume into which the heat diffuses does not remain the same for equal increments of r in the cylindrical and spherical coordinate system, hence a modification in the temperature profile assumed is necessary.

Lardner and Pohle suggested the following modifications for the temperature profiles.

In the cylindrical coordinate system the temperature profile may be chosen in the form

$$T = (\text{polynomial in } r) \cdot \ln r \tag{7-39}$$

and in the spherical coordinate system the form

$$T = \frac{\text{polynomial in } r}{r} \tag{7-40}$$

We illustrate the use of the modified temperature profiles with the following examples.

Cylindrical Hole Exposed to a Constant Heat Flux [5]. Consider a region exterior to a cylindrical hole of radius $r = b$ and extending to infinity. Initially the region is at zero temperature and for times $t > 0$ the surface of the cylindrical hole is exposed to a constant heat flux. Figure 7-5 shows the geometry and the boundary condition and the

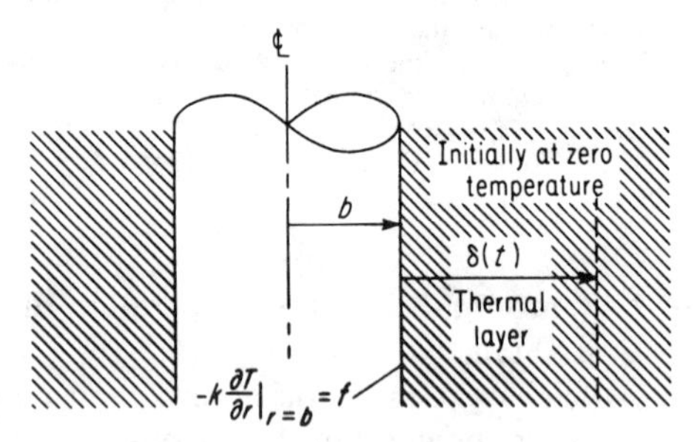

FIG. 7-5. Region exterior to a cylindrical hole of radius $r = b$ and the "thermal layer."

thermal-layer thickness δ. Assuming cylindrical symmetry, the boundary-value problem of heat conduction is given as

$$\frac{1}{r}\frac{\partial}{\partial r}\left(r\frac{\partial T}{\partial r}\right) = \frac{1}{\alpha}\frac{\partial T}{\partial t} \quad \text{in} \quad b \le r < \infty, t > 0 \tag{7-41a}$$

$$-k\frac{\partial T}{\partial r} = f \quad \text{at} \quad r = b, t > 0 \tag{7-41b}$$

$$T = 0 \quad \text{in} \quad b \le r < \infty, t = 0 \tag{7-41c}$$

where $T \equiv T(r, t)$. We multiply both sides of the differential equation of heat conduction 7-41a by r and integrate it over the thermal-layer thickness $\delta(t)$, i.e., from $r = b$ to $r = b + \delta(t)$, and obtain

$$\left[r\frac{\partial T}{\partial r}\right]_{b+\delta(t)} - \left[r\frac{\partial T}{\partial r}\right]_b = \frac{1}{\alpha} \cdot \left[\frac{\partial}{\partial t}\int_b^{b+\delta(t)} T\cdot r\cdot dr - (T\cdot r)\bigg|_{b+\delta(t)} \cdot \frac{d\delta(t)}{dt}\right] \tag{7-42}$$

which simplifies to the following heat-balance integral:

$$-\left[r\frac{\partial T}{\partial r}\right]_b = \frac{1}{\alpha}\frac{d\theta}{dt} \tag{7-43a}$$

where

$$\theta = \int_b^{b+\delta(t)} r\cdot T\cdot dr \tag{7-43b}$$

In order to solve Eq. 7-43 a profile should be assumed for the temperature distribution over the thermal layer. In choosing the profile we shall examine two different cases: (1) a polynomial approximation alone, and (2) a polynomial approximation modified by the term $\ln r$ as given by Eq. 7-39.

Polynomial Approximation. We assume a second-degree polynomial approximation in the form

$$T = a_1 + a_2 r + a_3 r^2 \quad \text{in} \quad b \le r \le b + \delta(t) \tag{7-44}$$

The three unknown coefficients a_1, a_2, a_3 are determined from the following three conditions:

$$-k\frac{\partial T}{\partial r}\bigg|_b = f \qquad -k\frac{\partial T}{\partial r}\bigg|_{b+\delta(t)} = 0 \qquad T\,|_{b+\delta(t)} = 0 \tag{7-45}$$

Then the temperature profile becomes

$$T = \frac{1}{2}\frac{fb/k}{\delta/b}\left[1 + \frac{\delta}{b} - \frac{r}{b}\right]^2 \quad \text{in} \quad b \le r \le (b + \delta) \tag{7-46}$$

Substituting the profile 7-46 into the heat-balance integral 7-43 and performing the operations, we obtain

$$b = \frac{1}{24\alpha}\frac{d}{dt}(\delta^3 + 4b\delta^2) \tag{7-47}$$

with

$$\delta = 0 \quad \text{for} \quad t = 0$$

The solution of this differential equation results in the following relation for the thermal-layer thickness:

$$\frac{24\alpha}{b^2}\cdot t = \left(\frac{\delta}{b}\right)^3 + 4\left(\frac{\delta}{b}\right)^2 \tag{7-48}$$

Equation 7-48 together with Eq. 7-46 gives the temperature distribution within the solid.

When the cylinder is very large as compared with the thickness of the thermal layer (i.e., $b > \delta$), Eq. 7-48 is approximated by

$$\frac{24\alpha}{b^2}\, t \cong 4\left(\frac{\delta}{b}\right)^2$$

or

$$\delta = \sqrt{6\alpha t} \quad \text{for} \quad b > \delta \tag{7-49}$$

which is the thermal thickness that would be obtained for a semi-infinite region ($0 \le x < \infty$) in the cartesian coordinate system with a constant heat flux applied at $x = 0$.

The temperature at the boundary surface $x = 0$ is obtained from the temperature profile 7-46 by substituting $r = b$,

$$T\,\big|_{r=b} = \frac{f}{2k}\,\delta \tag{7-50}$$

where δ is given by Eq. 7-48.

Figure 7-6 shows a comparison of the surface temperature T_s at $r = b$ evaluated from the approximate Eq. 7-50 and the exact solutions; the

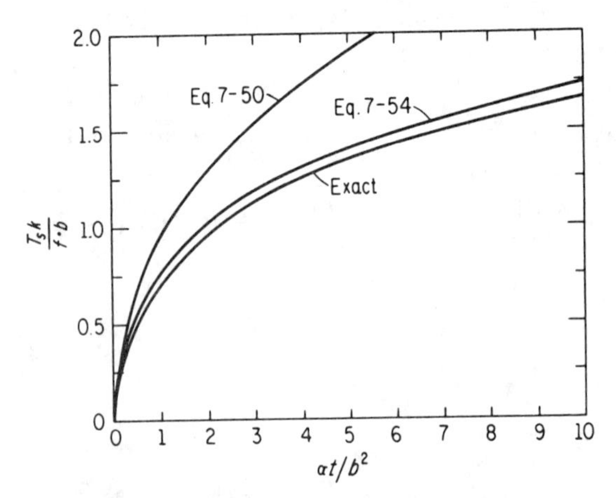

FIG. 7-6. Comparison of surface temperature from exact and approximate integral solutions for a region exterior to a cylindrical hole of radius b, with zero initial temperature and constant heat flux at the surface. [T. J. Lardner and F. V. Pohle, *ASME Trans., J. Applied Mechanics* (1961), p. 311.]

agreement is not good. This is because a polynomial approximation is used for representing the temperature profile in the cylindrical region.

Polynomial Approximation Multiplied by a Logarithmic Term. We consider a second-degree polynomial approximation multiplied by a logarithmic term in the form

$$T = (a_1 + a_2 r + a_3 r^2) \cdot \ln r \tag{7-51}$$

It can be verified that this profile when chosen in the following form satisfies conditions considered above:

$$T = -\frac{fb/k}{\delta/b} \cdot \left[1 + \frac{\delta}{b} - \frac{r}{b}\right]^2 \frac{\ln\left(\dfrac{r/b}{1 + \delta/b}\right)}{2 \ln\left(1 + \dfrac{\delta}{b}\right) + \delta/b} \tag{7-52}$$

The modified profile given by Eq. 7-52 involves a logarithmic term which is multiplied by multiplier of the second-degree polynomial profile 7-46.

Substituting 7-52 into the heat-balance integral 7-43 we obtain the following relation for the thermal-layer thickness $\delta(t)$,

$$\frac{\alpha t}{b^2} = -\frac{(72\eta^2 - 96\eta + 36) \ln \eta - 13\eta^4 + 36\eta^2 - 32\eta + 9}{144(\eta - 1) \cdot (2 \ln \eta + \eta - 1)} \tag{7-53}$$

where

$$\eta = 1 + \frac{\delta}{b}$$

Equation 7-52 together with Eq. 7-53 gives the temperature distribution in the solid.

The surface temperature at $r = b$ is obtained by setting in Eq. 7-52 r equal to b—that is,

$$T\big|_{r=b} = \frac{fb}{k} \cdot \frac{\delta}{b} \cdot \frac{\ln\left(1 + \dfrac{\delta}{b}\right)}{2 \ln\left(1 + \dfrac{\delta}{b}\right) + \delta/b} \tag{7-54}$$

The surface temperature 7-54 is also included in Fig. 7-6. It can be seen that the approximate solution based on the second-degree profile modified with a logarithmic term agrees closely with the exact solution.

Problems with Spherical Symmetry. Differential equation of heat conduction in the spherical coordinate system with spherical symmetry is in the form

$$\frac{1}{r} \frac{\partial^2}{\partial r^2} (rT) = \frac{1}{\alpha} \frac{\partial T}{\partial t} \tag{7-55}$$

Defining a new dependent variable,

$$U = rT \tag{7-56}$$

Eq. 7-55 becomes

$$\frac{\partial^2 U}{\partial r^2} = \frac{1}{\alpha} \frac{\partial U}{\partial t} \tag{7-57}$$

which is now a problem in the cartesian coordinate system and we choose a polynomial profile for U, i.e.,

$$U = (\text{polynomial in } r) \tag{7-58}$$

Changing the dependent variable from U to T according to 7-56 we have

$$T = \frac{\text{polynomial in } r}{r} \tag{7-59}$$

which is the form of the temperature profile proposed by Lardner and Pohle for problems involving spherical symmetry.

Since the differential equation of heat conduction 7-55 can be transformed into the cartesian coordinate system by means of the transformation 7-56, it is convenient to solve the problem in the Cartesian coordinate system by using a polynomial approximation and then transform the resulting solution to the spherical coordinate system.

7-4. PROBLEMS INVOLVING HEAT GENERATION

The integral method has been applied to solve transient heat-conduction problems involving heat generation within the solid [3, 4]. To illustrate the application we examine the problem of semi-infinite region with time-dependent heat generation.

Semi-infinite Solid, $x \geq 0$. A semi-infinite solid is initially at a constant temperature T_i. For times $t > 0$ heat is generated within the solid at a rate of $g(t)$ Btu/hr ft^3 while the boundary at $x = 0$ is kept at a constant temperature T_0. The boundary-value problem of heat conduction is given as

$$\frac{\partial^2 T}{\partial x^2} + \frac{g(t)}{k} = \frac{1}{\alpha}\frac{\partial T}{\partial t} \quad \text{in} \quad 0 \leq x < \infty, t > 0 \tag{7-60a}$$

$$T = T_0 \quad \text{at} \quad x = 0, t > 0 \tag{7-60b}$$

$$T = T_i \quad \text{in} \quad 0 \leq x < \infty, t = 0 \tag{7-60c}$$

where $T \equiv T(x, t)$.

We integrate Eq. 7-60 from $x = 0$ to $x = \delta(t)$,

$$\left.\frac{\partial T}{\partial x}\right|_{x=\delta} - \left.\frac{\partial T}{\partial x}\right|_{x=0} + \int_0^\delta \frac{g(t)}{k}\,dx = \frac{1}{\alpha}\left[\frac{\partial}{\partial t}\int_0^\delta T\,dx - T\,|_\delta \cdot \frac{d\delta}{dt}\right] \tag{7-61}$$

The various terms in Eq. 7-61 are

$$\left.\frac{\partial T}{\partial x}\right|_{x=\delta} = 0 \tag{7-62a}$$

$$\int_0^\delta \frac{g(t)}{k}\,dx = \frac{\delta}{k}\cdot g(t) \tag{7-62b}$$

$$T\,|_{\delta} = T_i + \frac{1}{\rho c_p}\int_0^t g(t)\cdot dt \tag{7-62c}$$

The result 7-62a is obtained by using the definition of thermal layer, and 7-62c is obtained from the differential equation of heat conduction (i.e. by evaluating Eq. 7-60 at $x = \delta$ and then integrating it from $t = 0$ to t). Substituting Eq. 7-62 into Eq. 7-61,

$$-\alpha \left.\frac{\partial T}{\partial x}\right|_{x=0} + \frac{1}{\rho c_p}\cdot\delta\cdot g(t) = \frac{\partial}{\partial t}\int_0^{\delta} T\cdot dx - T_i\frac{d\delta}{dt} - \frac{1}{\rho c_p}\cdot\frac{d\delta}{dt}\cdot\int_0^t g(t)\cdot dt \tag{7-63}$$

Defining

$$\theta = \int_0^{\delta} T\cdot dx \tag{7-64a}$$

$$G = \int_0^t g(t)\cdot dt \tag{7-64b}$$

Eq. 7-63 is written in the form

$$-\alpha\left.\frac{\partial T}{\partial x}\right|_{x=0} = \frac{\partial}{\partial t}\left[\theta - \frac{G\delta}{\rho c_p} - T_i\cdot\delta\right] \tag{7-65}$$

Equation 7-65 is the heat-balance integral for the problem under consideration.

We assume a cubic profile in the form

$$T = a_1 + a_2 x + a_3 x2 + a_4 x3 \tag{7-66}$$

and choose the four conditions needed to evaluate the unknown coefficient as

$$T\,|_{x=0} = T_0 \tag{7-67a}$$

$$\left.\frac{\partial T}{\partial x}\right|_{x=\delta} = 0 \tag{7-67b}$$

$$T\,|_{x=\delta} = T_i + \frac{G}{\rho c_p} \tag{7-67c}$$

$$\left.\frac{\partial^2 T}{\partial x^2}\right|_{x=\delta} = 0 \tag{7-67d}$$

Then the temperature profile is in the form

$$T = T_0 + \left[1 - \left(1 - \frac{x}{\delta}\right)^3\right]\cdot F \quad \text{in} \quad 0 \le x \le \delta \tag{7-68}$$

where

$$F \equiv (T_i - T_0) + \frac{G}{\rho c_p}.$$

Substituting Eq. 7-68 into the heat-balance integral 7-65 we obtain the following differential equation:

$$12\alpha F^2 = (F\delta)\frac{d(F\delta)}{dt} \tag{7-69a}$$

with

$$\delta = 0 \quad \text{for} \quad t = 0 \tag{7-69b}$$

Solution of Eq. 7-69 gives the following relation for the thermal-layer thickness:

$$\delta^2 = 24\alpha \frac{\int_0^t F^2 dt}{F^2} \tag{7-70}$$

where

$$F = (T_0 - T_i) + \frac{G}{\rho c_p}$$

$$G = \int_0^t g(t) \cdot dt$$

Equation 7-68 together with the value of δ evaluated from 7-70 gives the temperature distribution in the solid.

For no heat generation within the solid, Eq. 7-68 becomes

$$\frac{T - T_i}{T_0 - T_i} = \left(1 - \frac{x}{\delta}\right)^3 \tag{7-71}$$

and the relation 7-70 for the thermal-layer thickness reduces to

$$\delta = \sqrt{24\alpha t} \tag{7-72}$$

It is to be noted that Eqs. 7-71 and 7-72 are the same as Eqs. 7-17 and 7-18 respectively, derived previously for a semi-infinite region with no heat generation.

In the above example, we considered the case in which heat generation was a function of time. The problems involving heat generation, which is a function of position, can also be handled with the integral method.

7-5. PROBLEMS INVOLVING NONLINEAR BOUNDARY CONDITIONS

In the previous sections we examined the use of integral methods in the solution of transient, one-dimensional boundary-value problems of heat conduction in which both the differential equation and the boundary conditions were linear, and demonstrated that the resulting approximate solutions closely agree with the exact solutions.

The advantage of the integral method is that it can also handle nonlinear problems. In this section we examine the solution of problems in which the differential equation of heat conduction is linear but the boundary conditions are nonlinear. For example, in heat-conduction problems associated with transient radiation cooling of spacecraft components or in high-speed flights of rockets, surface temperatures are so high that the fourth-power radiation law should be considered. The boundary conditions for such problems are nonlinear. The integral method has been applied to the solution of transient heat-conduction problems with nonlinear boundary conditions [3,6,7]. We illustrate the solution of such problems with the following examples.

Semi-infinite Solid, $x \geq 0$, with Arbitrary Surface Heat Flux. Goodman [6] has provided a simple means of integrating the transient, one-dimensional differential equation of heat conduction for a semi-infinite region with nonlinear boundary conditions by applying the integral method. Consider a semi-infinite solid $x \geq 0$ which is initially at temperature T_i. For times $t > 0$ the boundary surface at $x = 0$ is subjected to a heat flux which is a prescribed function of surface temperature and time. Figure 7-7 shows the geometry and the boundary condition. The boundary-value problem of heat conduction is given as

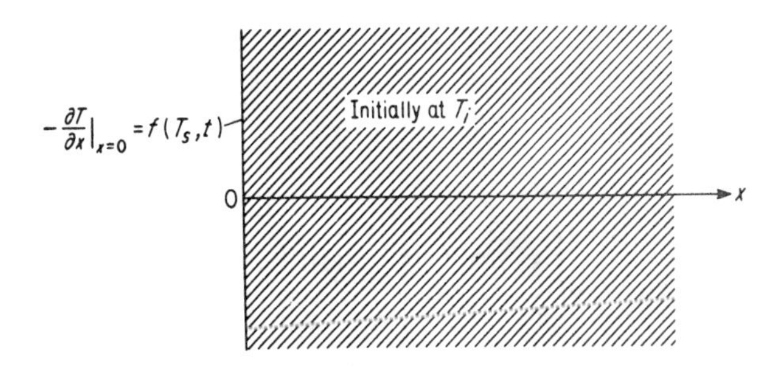

FIG. 7-7. Semi-infinite region with prescribed surface heat flux which is a function of surface temperature and time.

$$\frac{\partial^2 T}{\partial x^2} = \frac{1}{\alpha}\frac{\partial T}{\partial t} \quad \text{in} \quad 0 \leq x < \infty, t > 0 \tag{7-73a}$$

$$-\frac{\partial T}{\partial x} = f(T_s, t) \quad \text{at} \quad x = 0, t > 0 \tag{7-73b}$$

$$T = T_i \quad \text{in} \quad 0 \leq x < \infty, t = 0 \tag{7-73c}$$

where $T \equiv T(x, t)$. The boundary-condition function $f(T_s, t)$ is a function of T_s, the boundary surface temperature at $x = 0$.

Integrating the differential equation of heat conduction over the thermal layer, $\delta(t)$, we have

$$\left.\frac{\partial T}{\partial x}\right|_0^\delta = \frac{1}{\alpha}\left[\frac{\partial}{\partial t}\int_0^\delta T\cdot dx - T\,|_\delta\cdot\frac{d\delta}{dt}\right] \tag{7-74}$$

In view of the conditions

$$\left.\frac{\partial T}{\partial x}\right|_{x=\delta} = 0 \qquad T\,|_{x=\delta} = T_i \qquad -\left.\frac{\partial T}{\partial x}\right|_{x=0} = f(T_s, t) \tag{7-75}$$

Eq. 7-74 becomes

$$\alpha\cdot f(T_s, t) = \frac{d}{dt}[\theta - T_i\cdot\delta] \tag{7-76a}$$

where

$$\theta = \int_{x=0}^{\delta} T\cdot dx \tag{7-76b}$$

Equation 7-6 is the heat-balance integral for the problem considered.

We choose a cubic approximation in the form

$$T = a_1 + a_2x + a_3x^2 + a_4x^3 \tag{7-77}$$

Four auxiliary conditions are needed to determine these four unknown coefficients. Three of the conditions are given by Eq. 7-75 and the fourth condition is taken as

$$\left.\frac{\partial^2 T}{\partial x^2}\right|_{x=\delta} = 0 \tag{7-78}$$

Then the temperature profile becomes

$$T - T_i = \frac{\delta\cdot f(T_s, t)}{3}\left(1 - \frac{x}{\delta}\right)^3 \qquad 0 \le x \le \delta \tag{7-79}$$

A relationship is obtained between the boundary-surface temperature T_s and the thermal-layer thickness δ by setting x equal to zero, in Eq. 7-79;

$$T_s - T_i = \frac{\delta\cdot f(T_s, t)}{3} \tag{7-80}$$

By combining Eqs. 7-79 and 7-80, the temperature profile is expressed in the form

$$\frac{T - T_i}{T_s - T_i} = \left(1 - \frac{x}{\delta}\right)^3 \tag{7-81}$$

Substituting Eq. 7-81 into the heat-balance integral 7-76, performing the operations indicated, and eliminating δ from the result by means of the Eq. 7-80, we obtain the following first-order, ordinary differential equation for the surface temperature T_s,

$$\frac{4}{3}\,\alpha\cdot f(T_s, t) = \frac{d}{dt}\left[\frac{(T_s - T_i)^2}{f(T_s, t)}\right] \tag{7-82a}$$

with

$$T_s = T_i \quad \text{for} \quad t = 0 \tag{7-82b}$$

Equation 7-82 should be integrated numerically if the boundary-condition function $f(T_s, t)$ depends on both the surface temperature and the time. It is to be noted that the integration of Eq. 7-82 requires less effort than the integration of the original differential equation with finite-difference technique. Furthermore, for some special cases considered below Eq. 7-82 can be integrated.

Case 1. The boundary-condition function is independent of time but is a function of surface temperature only. Taking $f(T_s, t) = f(T_s)$, Eq. 7-82 becomes

$$\frac{4}{3}\,\alpha \cdot f(T_s) = \frac{d}{dT_s}\left[\frac{(T_s - T_i)^2}{f(T_s)}\right] \cdot \frac{dT_s}{dt}$$

or

$$\frac{4}{3}\,\alpha = \frac{2(T_s - T_i)\cdot f(T_s) - f'(T_s)\cdot(T_s - T_i)^2}{f^3(T_s)} \cdot \frac{dT_s}{dt} \tag{7-83a}$$

with

$$T_s = T_i \quad \text{for} \quad t = 0 \tag{7-83b}$$

Integrating Eq. 7-83,

$$\frac{4}{3}\,\alpha t = \int_{T_i}^{T_s} \frac{2(T_s - T_i)\cdot f(T_s) - f'(T_s)\cdot(T_s - T_i)^2}{f^3(T_s)} \cdot dT_s \tag{7-84}$$

We now illustrate the application of Eq. 7-84 to the solution of heat-conduction problem with a fourth-power radiation boundary condition with the following example.

EXAMPLE. Consider a semi-infinite region $x \geq 0$ which is initially at temperature T_i. For times $t > 0$ the boundary surface at $x = 0$ is dissipating heat according to a fourth-power radiation law into a surrounding at a constant temperature T_∞. The boundary-value problem of heat conduction is

$$\frac{\partial^2 T}{\partial x^2} = \frac{1}{\alpha}\frac{\partial T}{\partial t} \quad \text{in} \quad 0 \leq x < \infty, t > 0 \tag{7-85a}$$

$$k\frac{\partial T}{\partial x} = \sigma\epsilon(T_s^4 - T_\infty^4) \quad \text{at} \quad x = 0, t > 0 \tag{7-85b}$$

$$T = T_i \quad \text{in} \quad 0 \leq x < \infty, t = 0 \tag{7-85c}$$

where ϵ is the emissivity of the surface, σ is the Stefan-Boltzmann constant, T_s is the boundary surface temperature at $x = 0$, and all temperatures are the absolute temperatures.

Comparing the boundary condition for the present problem with the boundary condition for the problem 7-73b, we have

$$f(T_s) = -\frac{\sigma\epsilon}{k}(T_s^4 - T_\infty^4) \tag{7-86}$$

TABLE 7-2

NUMERICAL EVALUATION OF EQ. 7-87 FOR $\frac{T_\infty}{T_i} = 0.05$

$\frac{T_s}{T_i}$	1.0	0.9	0.8	0.7	0.6	0.5	0.4
$\left(\frac{\sigma\epsilon}{k}\right)^2 \cdot T_i^6 \cdot \alpha t$	0.00	15.60×10^{-3}	14.95×10^{-2}	91.50×10^{-2}	50.00×10^{-1}	31.55×10^{0}	23.55×10^{1}

Substituting Eq. 7-86 into solution 7-84 we obtain

$$\frac{4}{3}\left(\frac{\sigma\epsilon}{k}\right)^2 \alpha t = \int_{T_i}^{T_s} \left[2\frac{T_s - T_i}{(T_s^4 - T_\infty^4)^2} - \frac{4T_s^3(T_s - T_i)}{(T_s^4 - T_\infty^4)^3} \right] dT_s$$

or

$$\left(\frac{\sigma\epsilon}{k}\right)^2 T_i^6 \cdot \alpha \cdot t = \frac{3}{2} \int_{\xi'=1}^{\xi} \left[\frac{\xi' - 1}{(\xi'^4 - U^4)^2} - 2\frac{\xi'^3(\xi' - 1)^2}{(\xi'^4 - U^4)^3} \right] d\xi' \tag{7-87}$$

where

$$\xi = \frac{T_s}{T_i}, \quad U = \frac{T_\infty}{T_i}$$

Integration on the right-hand side of Eq. 7-87 does not have the solution in elementary functions, but it can be evaluated numerically. Schneider [7] evaluated this integral numerically for $T_\infty / T_i = 0.05$ and his results are shown in Table 7-2.

For the special case of $T_\infty = 0$, that is radiation taking place into a sink at zero absolute temperature (or for the cases when $U < \xi$), Eq. 7-87 can be integrated and we obtain

$$\left(\frac{\sigma\epsilon}{k}\right)^2 T_i^6 \alpha t = \frac{3}{2\xi^8} \left[\frac{1}{6}\xi^2 - \frac{3}{7}\xi + \frac{1}{4} \right] + \frac{1}{56} \tag{7-88}$$

where

$$\xi = \frac{T_s}{T_i}$$

The problem of heat conduction in a semi-infinite solid with the fourth-power law of thermal radiation from the boundary surface into a medium at zero absolute temperature has been solved numerically by Jaeger [8] and results are presented graphically. When the integral-method solution 7-88 is compared with Jaeger's numerical results it is impossible to detect any difference [6].

7-6. PROBLEMS INVOLVING TEMPERATURE-DEPENDENT THERMAL PROPERTIES

When the thermal properties of the solid vary significantly with temperature or when the range of temperature involved is so large that the thermal properties cannot be treated as constant throughout the region, it becomes necessary to include in the analysis the variation of thermal properties with temperature. The approximate integral methods are used in the solution of one-dimensional, transient-boundary-value problem of heat conduction in which thermal properties are temperature dependent [3,9,10,11,12]. In this section we examine the solution of a nonlinear differential equation of heat conduction with linear boundary conditions.

Semi-infinite Solid $0 \le x < \infty$. Goodman [9] solved the transient heat-conduction problem for a semi-infinite solid $0 \le x < \infty$ with variable thermal properties both for the cases of prescribed surface temperature and prescribed surface heat flux. The region is assumed to be at zero temperature initially. We examine here the case with prescribed surface heat flux. The boundary-value problem of heat conduction is

$$\frac{\partial}{\partial x}\left(k\,\frac{\partial T}{\partial x}\right) = \rho c_p \frac{\partial T}{\partial x} \quad \text{in} \quad 0 \le x < \infty, t > 0 \tag{7-89a}$$

$$-k\,\frac{\partial T}{\partial x} = f(t) \quad \text{at} \quad x = 0, t > 0 \tag{7-89b}$$

$$T = 0 \quad \text{in} \quad 0 \le x < \infty, t = 0 \tag{7-89c}$$

where $T \equiv T(x, t)$, and we assume that the properties k, ρ, c_p are all dependent upon temperature.

By applying the Goodman transformation [9]

$$U = \int_0^T \rho c_p \cdot dT \tag{7-90}$$

system 7-89 transforms into

$$\frac{\partial}{\partial x}\left[\alpha \cdot \frac{\partial U}{\partial x}\right] = \frac{\partial U}{\partial t} \quad \text{in} \quad 0 \le x < \infty, t > 0 \tag{7-91a}$$

$$-\frac{\partial U}{\partial x} = \frac{f(t)}{\alpha_s} \quad \text{at} \quad x = 0, t > 0 \tag{7-91b}$$

$$U = 0 \quad \text{in} \quad 0 \le x < \infty, t = 0 \tag{7-91c}$$

where subscript s refers to the condition at the boundary surface $x = 0$, and that $\alpha \equiv \alpha(U)$.

We integrate the differential equation of heat conduction 7-91a over the thermal layer $\delta(t)$:

$$\alpha \cdot \frac{\partial U}{\partial x}\bigg|_{x=0}^{x=\delta} = \frac{\partial}{\partial t}\left[\int_0^\delta U \cdot dx - U\,|_\delta \cdot \delta\right] \tag{7-92}$$

In view of the conditions

$$\frac{\partial U}{\partial x}\bigg|_{x=\delta} = 0 \qquad U\,|_\delta = 0 \qquad \alpha_s \cdot \frac{\partial U}{\partial x}\bigg|_{x=0} = -f(t) \tag{7-93}$$

Eq. 7-92 becomes

$$f(t) = \frac{d\theta}{dt} \tag{7-94}$$

where

$$\theta = \int_0^\delta U \cdot dx,$$

which is the heat-balance integral for the problem under consideration.

We choose a cubic profile for the temperature function $U(x)$ in the form

$$U(x) = a_1 + a_2 x + a_3 x^2 + a_4 x^3 \tag{7-95}$$

The four unknown coefficients are determined from the following four conditions:

$$U\big|_{x=\delta} = 0 \qquad \left.\frac{\partial U}{\partial x}\right|_{x=\delta} = 0 \qquad \left.\frac{\partial U}{\partial x}\right|_{x=0} = -\frac{f(t)}{\alpha_s} \qquad \left.\frac{\partial^2 U}{\partial x^2}\right|_{x=\delta} = 0 \tag{7-96}$$

The first three of these conditions are the natural conditions for the problem and the last one is the derived condition. The corresponding temperature profile is in the form

$$U = \frac{\delta \cdot f(t)}{3\alpha_s}\left(1 - \frac{x}{\delta}\right)^3 \qquad 0 \le x \le \delta \tag{7-97}$$

Substituting the profile 7-97 into the heat-balance integral 7-94 we obtain the following ordinary differential equation for the thermal-layer thickness δ:

$$\frac{d}{dt}\left[\frac{\delta^2 \cdot f(t)}{12\alpha_s}\right] = f(t) \tag{7-98a}$$

with

$$\delta = 0 \quad \text{for } t = 0 \tag{7-98b}$$

The solution of Eq. 7-98 is

$$\delta = \left[\frac{12\,\alpha_s}{f(t)} \int_0^t f(t') \cdot dt'\right]^{1/2} \tag{7-99}$$

In this equation α_s is the thermal diffusivity evaluated at the temperature of the boundary surface $x = 0$. Since the boundary-surface temperature is not yet known, Eq. 7-99 can not be used to evaluate δ. To avoid this difficulty we relate the boundary-surface temperature U_s to the thermal-layer thickness δ; such relationship is obtained by substituting $x = 0$ in Eq. 7-97, yielding

$$U_s = \frac{\delta \cdot f(t)}{3\alpha_s} \tag{7-100}$$

Eliminating δ between Eqs. 7-99 and 7-100, we obtain the following relation for U_s:

$$U_s\sqrt{\alpha_s} = \left[\frac{4}{3} \cdot f(t) \int_0^t f(t') \cdot dt'\right]^{1/2} \tag{7-101}$$

where α is given as a function of temperature at the onset of the problem. When the value of α_s, given as a function of U_s, is substituted in Eq. 7-101, the resulting expression is a transcendental equation for U_s, the solution of which gives the value of U_s as a function of time. Knowing U_s as a function of time, the value of α_s is also known as a function of time.

The solution of the problem now becomes complete, because the thermal-layer thickness is evaluated as a function of time from Eq. 7-99 and the temperature U from Eq. 7-97. With U known, the actual temperature distribution T is obtained through the transformation 7-90.

7-7. PROBLEMS INVOLVING MELTING AND SOLIDIFICATION

Problems of heat conduction involving melting or solidification are complicated because the interface between the solid and liquid phases moves as the latent heat is absorbed or liberated at the interface. The location of the moving interface is not known a priori, and the thermal properties of solid and liquid are different. Examples are found in the making of ice, the freezing of food, and the solidification or melting of metals in the casting process. Such problems are nonlinear; exact solutions are extremely difficult. One important case for which the exact solution exists is the solidification (or melting) of a semi-infinite region as a result of one-dimensional heat flow. This problem involves a semi-infinite region $x \geq 0$ which is initially at a constant temperature T_i greater than the melting temperature. For times $t > 0$ the boundary surface at $x = 0$ is kept at zero temperature and the liquid starts to solidify, with the solidification front moving in the positive x direction. This problem is discussed by Neumann and a generalization of Neumann's solution is given by Carslaw and Jaeger [13]. For other boundary conditions, for finite regions, or for spheres and cylinders no exact solutions exist.

The integral method as applied to the solution of heat-conduction problems involving change of phase provides a useful means for obtaining solutions to such complicated problems. In this section we illustrate the application of this method to the solution of one-dimensional melting problems following Goodman [1].

Application of the integral method to the treatment of the heat-conduction problem involving two-dimensional solidification front in a square is given by Poots [14].

The problem of heat conduction involving change of phase is different from the usual heat-conduction problems in that the interface between the solid and liquid phases is moving and the boundary condition at this interface requires special treatment. We give below a brief discussion of the boundary condition at the moving interface.

Boundary Condition at the Moving Interface. Figure 7-8a shows the interface between the solid and liquid phases for a one-dimensional *solidification* problem in which a liquid which is initially at a temperature higher than the fusion temperature and contained in the region $x \geq 0$ is solidified as a result of cooling applied at the boundary surface at $x = 0$. As latent heat of fusion is removed from the fluid by cooling, it changes to the solid phase and the solid–liquid interface moves in the positive x direction.

Let $S(t)$ be the location of the solid–liquid interface at any instant of time t as measured from the origin $x = 0$. Subscripts s and l denote the conditions respectively in the solid and liquid phase. One natural boundary condition at the solid–liquid interface is the continuity of temperature

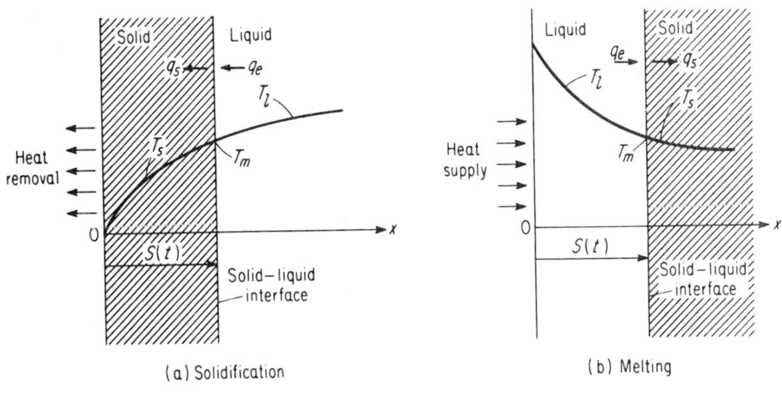

FIG. 7-8. Solid–liquid interface for one-dimensional solidification and melting problems.

—that is,

$$T_s = T_l \quad \text{at} \quad x = S(t) \tag{7-102}$$

Another boundary condition for the interface is derived by writing an energy-balance equation for the interface. For the solidification problem under consideration the heat of fusion released at the interface as a result of solidification should be removed from the interface as a result of cooling applied at $x = 0$. Assuming no storage of heat at the interface, the conservation of energy reduces to a simple energy-balance equation at the interface:

$$\begin{pmatrix}\text{Net energy removed from}\\ \text{interface per unit area}\\ \text{and per unit time in}\\ \text{direction of the cooled face}\end{pmatrix} = \begin{pmatrix}\text{Energy released at}\\ \text{interface per unit area per unit}\\ \text{time as a result of release of}\\ \text{latent heat during solidification}\end{pmatrix}$$

If q_s and q_l are heat fluxes respectively at the solid and liquid sides of the interface pointing in the direction of the cooled face, we have

$$q_s - q_l = \rho L \frac{dS(t)}{dt}$$

where L = latent heat of fusion
ρ = density (no subscript is written for density because density of the liquid and solid phases are assumed to be equal)

Then for the *solidification* problem shown in Fig. 7-8a, the boundary condition at the interface becomes

$$\left[k_s \frac{\partial T_s}{\partial x} - k_l \frac{\partial T_l}{\partial x}\right]_{x=S(t)} = \rho L \frac{dS(t)}{dt} \tag{7-103}$$

Figure 7-8b shows the solid–liquid interface for a *melting* problem in which a solid that is initially at a temperature lower than the fusion temperature and is contained in the region $x \geq 0$ is melted as a result of heating applied at the boundary $x = 0$. As the solid melts, the solid–liquid interface will move in the positive x direction. In this case the latent heat of fusion absorbed at the interface during melting is supplied to the interface as a result of heating applied at $x = 0$. By writing an energy-balance equation for the interface it can be shown that the boundary condition 7-103 is also applicable to the *melting* problem shown in Fig. 7-8b.

Several special cases of the boundary condition 7-103 are of interest. Referring to the problem of solidification shown in Fig. 7-8a, if it is assumed that the liquid phase is initially at the *fusion* temperature, then there is no temperature gradient within the liquid and the boundary condition 7-103 for this case reduces to

$$k_s \left. \frac{\partial T_s}{\partial x} \right|_{x=S(t)} = \rho L \frac{dS(t)}{dt}$$

In the case of the melting problem shown in Fig. 7-8b, if we assume that the solid phase is initially at the fusion temperature then there is no temperature gradient in the solid phase, and the boundary condition 7-103 reduces to

$$-k_l \left. \frac{\partial T_l}{\partial x} \right|_{x=S(t)} = \rho L \frac{dS(t)}{dt}$$

The boundary condition as given by Eq. 7-103 is a nonlinear one; its nonlinearity becomes apparent with the following consideration.

At the solid–liquid interface the continuity of temperature requires that temperatures at the solid and liquid phases should be equal to the melting temperature, T_m:

$$T_s(x, t) = T_m = T_l(x, t) \quad \text{at} \quad x = S(t)$$

Differentiating this relation, and considering that the melting temperature T_m stays constant, we obtain

$$\frac{\partial T_s}{\partial x} dx + \frac{\partial T_s}{\partial t} dt = 0 = \frac{\partial T_l}{\partial x} dx + \frac{\partial T_l}{\partial t} dt \quad \text{at} \quad x = S(t)$$

or

$$\frac{dS(t)}{dt} = -\frac{\partial T_s/\partial t}{\partial T_s/\partial x} = -\frac{\partial T_l/\partial t}{\partial T_l/\partial x} \tag{7-104}$$

Substituting $\frac{dS(t)}{dt}$ from Eq. 7-104 into 7-103, we obtain

$$\left[k_s \frac{\partial T_s}{\partial x} - k_l \frac{\partial T_l}{\partial x} \right]_{x=S(t)} = -\rho L \frac{\partial T_s/\partial t}{\partial T_s/\partial x} = -\rho L \frac{\partial T_l/\partial t}{\partial T_l/\partial x} \tag{7-105}$$

The nonlinearity of the boundary condition is now apparent.

Melting of Semi-infinite Solid with Constant Surface Temperature at $x = 0$. Goodman [1] applied the integral method to solve the heat-transfer problem associated with the melting of a semi-infinite solid $x \geq 0$. To simplify the problem it is assumed that the solid is initially at the melting temperature T_m, and for times $t > 0$, a constant temperature T_0 greater than the melting temperature (i.e., $T_0 > T_m$) is applied at the boundary surface at $x = 0$.

As a result of the applied temperature T_0 the solid melts and the solid–liquid interface moves in the positive x direction. This problem is similar to the solidification problem studied by Stefan [15] in which a liquid initially at the melting temperature and contained in the region $x \geq 0$ is solidified by the application of a constant temperature lower than the melting temperature at the boundary surface $x = 0$.

We define a thermal-layer thickness $\delta(t)$ beyond which the temperature gradient in the liquid phase is zero. For the present problem the location of the thermal layer is identical with the location of the solid–liquid interface—that is,

$$\delta(t) = S(t)$$

Since the solid is initially at the melting temperature there is no temperature gradient in the solid beyond the solid–liquid interface. The temperature gradient exists only in the liquid phase, and neglecting the effects of any fluid motion resulting from natural convection we assume that heat transfer in the liquid phase takes place solely by conduction. The boundary-value problem of heat conduction for the liquid phase is given in the form

$$\frac{\partial^2 T_l}{\partial x^2} = \frac{1}{\alpha}\frac{\partial T_l}{\partial t} \quad \text{in} \quad 0 \leq x \leq S(t), t > 0 \tag{7-106a}$$

$$T_l = T_0 \quad \text{at} \quad x = 0, t > 0 \tag{7-106b}$$

$$T_l = T_m \quad \text{at} \quad x = S(t), t > 0 \tag{7-106c}$$

$$-k_l\frac{\partial T_l}{\partial x} = \rho L\frac{dS(t)}{dt} \quad \text{at} \quad x = S(t), t > 0 \tag{7-106d}$$

The boundary condition 7-106d at the interface is obtained from Eq. 7-103 by taking $\partial T_s/\partial x$ equal to zero. No differential equation is needed for the solid phase because it is assumed to be at the melting temperature throughout.

In the following analysis the subscript l will be omitted from the system 7-106, since there is only one phase (i.e., liquid phase) in which temperature variation is considered.

We integrate the differential equation of heat conduction over the thermal-layer thickness; since for the present problem $\delta(t) = S(t)$, integration is performed from $x = 0$ to $x = S(t)$.

$$\left.\frac{\partial T}{\partial x}\right|_{x=S(t)} - \left.\frac{\partial T}{\partial x}\right|_{x=0} = \frac{1}{\alpha}\frac{d}{dt}\int_0^{S(t)} T dx - T\,|_{x=S(t)} \cdot S(t) \tag{7-107}$$

In view of the boundary conditions 7-106c and 7-106d, Eq. 7-107 becomes

$$-\alpha\left[\frac{\rho L}{k}\frac{dS(t)}{dt} + \left.\frac{\partial T}{\partial x}\right|_{x=0}\right] = \frac{d}{dt}\left[\theta - T_m \cdot S(t)\right] \tag{7-108a}$$

where

$$\theta = \int_0^{S(t)} T \cdot dx \tag{7-108b}$$

Equation 7-108 is the heat-balance integral for the melting problem considered.

We choose a second-degree polynomial approximation for the temperature profile in the liquid phase in the form

$$T = a + b(x - S) + c(x - S)^2 \tag{7-109}$$

where $S \equiv S(t)$. The three unknown coefficients are determined from the following three conditions:

$$T(x,t)\,|_{x=0} = T_0 \tag{7-110a}$$

$$T(x,t)\,|_{x=S(t)} = T_m \tag{7-110b}$$

$$-k\left.\frac{\partial T(x,t)}{\partial x}\right|_{x=S(t)} = \rho L\frac{dS(t)}{dt} \tag{7-110c}$$

However, the last of these conditions is not suitable for this purpose because if it is used to determine the unknown constants the resulting temperature profile will involve $dS(t)/dt$ term; when such a profile is substituted into the heat-balance integral, a second-order, ordinary differential equation will be obtained instead of the usual first-order equation. To avoid this difficulty Goodman derived another condition to replace the third condition 7-110c. The procedure to obtain this condition is as follows.

Differentiating the boundary condition 7-110b with respect to t,

$$\frac{\partial T}{\partial x}\cdot\frac{dS(t)}{dt} + \frac{\partial T}{\partial t} = 0 \tag{7-111}$$

The boundary condition 7-106d is written in the form

$$-\frac{\partial T}{\partial x} = A\frac{dS(t)}{dt} \tag{7-112a}$$

where

$$A = \frac{\rho L}{k} \tag{7-112b}$$

and properties refer to liquid-phase properties. Eliminating $dS(t)/dt$ between 7-111 and 7-112,

$$\left(\frac{\partial T}{\partial x}\right)^2 = A\frac{\partial T}{\partial t} \tag{7-113}$$

the $\partial T/\partial t$ term can be eliminated from this result by means of the differential equation of heat conduction 7-106a, and we obtain

$$\left(\frac{\partial T}{\partial x}\right)^2 = \alpha A \frac{\partial^2 T}{\partial x^2} \quad \text{at} \quad x = S(t) \tag{7-114}$$

We now use the conditions 7-110a, 7-110b, and 7-114 to determine the three unknown constants in Eq. 7-109; the result is

$$T - T_m = b(x - S) + c(x - s)^2 \tag{7-115a}$$

where

$$b = \frac{A\alpha}{S}[1 - (1 + \mu)^{1/2}] \tag{7-115b}$$

$$c = \frac{bS + (T_0 - T_m)}{S^2} \tag{7-115c}$$

and

$$A = \frac{\rho L}{k} \tag{7-115d}$$

$$\mu = \frac{2(T_0 - T_m)}{A\alpha} \tag{7-115e}$$

Substituting the temperature profile 7-115 into the heat-balance integral 7-108 and performing the operations indicated, we obtain the following ordinary differential equation for $S(t)$, the location of the liquid–solid interface,

$$S\frac{dS}{dt} = 6\alpha \frac{1 - (1 + \mu)^{1/2} + \mu}{5 + (1 + \mu)^{1/2} + \mu} \tag{7-116a}$$

with

$$S = 0 \quad \text{for} \quad t = 0 \tag{7-116b}$$

The solution of the differential Eq. 7-116 is

$$S = K\sqrt{t} \tag{7-117a}$$

where

$$K = \left[12\alpha \frac{1 - (1 + \mu)^{1/2} + \mu}{5 + (1 + \mu)^{1/2} + \mu}\right]^{1/2} \tag{7-117b}$$

Figure 7-9 shows a comparison of the constant K as evaluated from the approximate solution 7-117 and from the exact solution. The agreement is close. Goodman [1] also tried a cubic approximation for the temperature profile, and the result with cubic approximation agreed even much closer with the exact result than with the second-degree profile.

To simplify the analysis in the present problem it is assumed that the solid is at the melting temperature. In most practical problems the solid is not at the melting temperature but that there is temperature gradient in

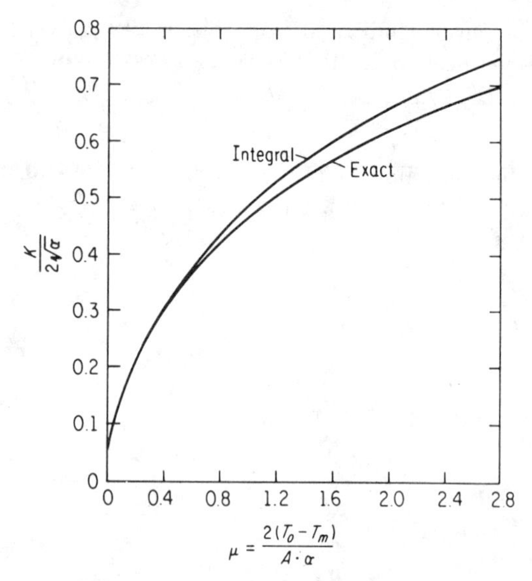

FIG. 7-9. Comparison of exact and approximate integral solutions for the melting constant K for a semi-infinite region. [T. R. Goodman, *Trans. ASME*, Vol. 80 (1958), pp. 335–342.]

the solid. The integral method is capable of taking into account such a two-region problem, but the resulting equations are more complicated.

7-8. PROBLEMS INVOLVING ABLATION

During high-speed reentry conditions into the earth's atmosphere, the aerodynamic heating is so high and the flight time is sufficiently short that a method of absorbing the heat generated by aerodynamic heating is to allow a portion of the solid surface (i.e., heat shield) to melt or vaporize. This is known as *ablation*. The peak heat flux may range, depending on conditions of reentry, from 100 to 10,000 Btu/ft^2 sec and the duration of reentry from 50 to 100 sec respectively [16]. For a given heat flux at the boundary surface, the temperature distribution within the solid before the start of ablation can be calculated by the well-known methods. As soon as the ablation starts, the problem becomes more complex mathematically, since it then involves simultaneous prediction of the rate of melting, the total amount of material melted, and the transient-temperature distribution within the solid. To simplify the analysis in such problems it is assumed that the ablated material is completely removed from the surface as a gas as soon as it is ablated. There will therefore be no liquid phase over the surface of ablation.[3]

Figure 7-10 shows the location of the boundary surface $x = S(t)$ of a solid on which ablation takes place as a result of aerodynamic heating of intensity $f(t)$ Btu/hr ft^2. As a result of ablation the boundary surface will

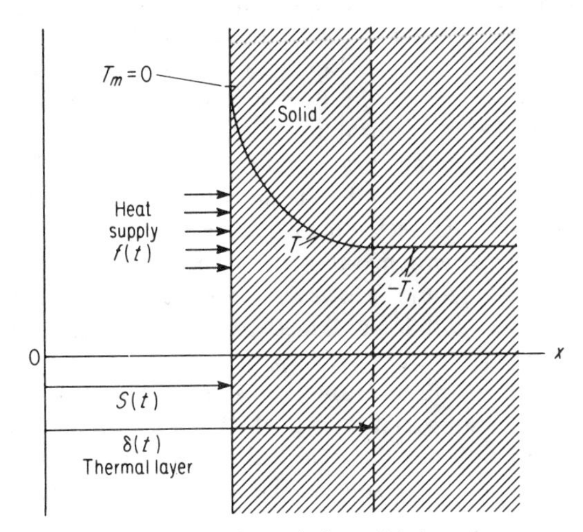

FIG. 7-10. Location of the solid boundary and "thermal layer" during ablation of one-dimensional semi-infinite solid.

move in the positive x direction. Assuming no liquid phase exists on the solid surface, the boundary condition at this surface is given as

$$\begin{pmatrix}\text{Net heat flux entering}\\\text{boundary surface}\end{pmatrix} = \begin{pmatrix}\text{Energy absorbed as a result}\\\text{of melting per unit time per unit}\\\text{area of boundary surface}\end{pmatrix}$$

$$f(t) - q_s = \rho L \frac{dS(t)}{dt}$$

$$f(t) + k_s \frac{\partial T}{\partial x}\bigg|_{x=S(t)} = \rho L \frac{dS(t)}{dt} \tag{7-118}$$

where L = latent heat of fusion.

[3]The assumption of the boundary surface being completely free from the ablated material is not always correct. For example, during the ablation of many materials containing organic matter such as plastic or resins, a porous charred layer is formed on the surface [17]. This layer attains a thickness of several millimeters in the quasi-steady state. For some shield materials which melt before they vaporize the presence of the liquid layer over the solid surface cannot be ignored, and the liquid layer can be subject to a strong body force due to acceleration or deceleration of the body. The solution of such problems involves coupling of the energy equation with the equations of motion; the solution is extremely difficult and treatment of such problems is beyond the scope of the present work.

Landau [18] gave an exact treatment of the ablation problem on the assumption that ablated material is removed immediately on formation. He examined a slab with one face insulated while heat flows in through the other face at a rate of $f(t)$ Btu/hr ft^2, and a problem for the semi-infinite region. Goodman [1] used the integral method for the ablation of a semi-infinite region with constant heat flux at the boundary surface, and Altman [16] examined a similar case with time-dependent heat flux. In the following example we consider the case examined by Altman.

Ablation for a Semi-Infinite Region with Variable Surface Heat Flux. Consider the heat-transfer problem for a semi-infinite solid $0 \leq x < \infty$ resulting from aerodynamic heating at a rate of $f(t)$ Btu/hr ft^2 applied to the boundary surface at $x = 0$. The heat-transfer problem for times before the boundary-surface temperature reaches the melting temperature can be obtained by several methods as well as by straightforward application of the integral method. However, after the start of ablation the solution of this problem with integral method requires special consideration.

Let $t = 0$ be the time at which the boundary surface at $x = 0$ reaches the melting temperature and ablation starts with complete removal of ablated material as soon as it is formed. For convenience in the analysis, the melting temperature of the solid will be taken as zero—i.e., $T_m = 0$—and the temperature of the solid at the start of the aerodynamic heating taken as $-T_i$.

Figure 7-10 shows the location of the boundary surface $x = S(t)$ at time t after the boundary surface has reached the melting temperature. The one-dimensional boundary-value problem of heat conduction for the solid during ablation is given as

$$\frac{\partial^2 T}{\partial x^2} = \frac{1}{\alpha}\frac{\partial T}{\partial t} \quad \text{in} \quad x \geq S(t),\ t > 0 \tag{7-119a}$$

$$f(t) + k\frac{\partial T}{\partial x} = \rho L \frac{dS(t)}{dt} \quad \text{at} \quad x = S(t),\ t > 0 \tag{7-119b}$$

The initial condition for this problem is the condition at the beginning of ablation, which is obtained from the solution of the heat-conduction problem for a semi-infinite region subject to a prescribed heat flux at the boundary surface $x = 0$. This will be discussed later in the analysis.

The purpose of the present analysis is to determine the location of the boundary surface as a function of time after the start of ablation (i.e., to determine the total amount ablated) and the temperature distribution in the solid during the ablation. Once the location of the boundary surface is determined the velocity of the boundary surface, which is the rate of ablation, is easily determined.

As a first step in the calculations we integrate the differential equation of heat conduction 7-119a from $x = S(t)$ to $x = \delta(t)$, where $\delta(t)$ is the thermal-layer thickness as shown in Fig. 7-10.

$$\alpha \left[\left. \frac{\partial T}{\partial x} \right|_{\delta(t)} - \left. \frac{\partial T}{\partial x} \right|_{S(t)} \right] = \frac{d}{dt} \left[\int_{S(t)}^{\delta(t)} T\,dx - T\,|_{\delta(t)} \cdot \delta(t) + T\,|_{S(t)} \cdot S(t) \right] \tag{7-120}$$

In view of the conditions,

$$T\,|_{\delta(t)} = -T_i \tag{7-121a}$$

$$\left. \frac{\partial T}{\partial x} \right|_{\delta(t)} = 0 \tag{7-121b}$$

$$T|_{S(t)} = 0 \tag{7-121c}$$

$$f(t) = -k \left. \frac{\partial T}{\partial x} \right|_{S(t)} + \rho L \frac{dS(t)}{dt} \tag{7-121d}$$

Equation 7-120 becomes

$$\frac{d}{dt} \left[\theta + T_i \cdot \delta(t) + \alpha \frac{\rho L}{k} S(t) \right] = \frac{\alpha f(t)}{k} \tag{7-122a}$$

where

$$\theta = \int_{S(t)}^{\delta(t)} T\,dx \tag{7-122b}$$

Equation 7-122 is the heat-balance integral for the ablation problem under consideration.

We choose a fourth-degree polynomial in the form

$$T = a + b \frac{x - S}{\delta - S} + c \left(\frac{x - S}{\delta - S} \right)^2 + d \left(\frac{x - S}{\delta - S} \right)^3 + e \left(\frac{x - S}{\delta - S} \right)^4 \tag{7-123}$$

where

$$S \equiv S(t) \qquad \delta \equiv \delta(t)$$

The five unknown coefficients are determined from the following five conditions:

$$T\,|_{x=S} = 0 \qquad T\,|_{x=\delta} = -T_i \qquad \left. \frac{\partial T}{\partial x} \right|_{x=\delta} = 0 \tag{7-124a}$$

$$\left. \frac{\partial^2 T}{\partial x^2} \right|_{x=\delta} = 0 \qquad \left. \frac{\partial^3 T}{\partial x^3} \right|_{x=\delta} = 0 \tag{7-124b}$$

The first three of these conditions, Eq. 7-124a, are the natural conditions for the problem. The last two conditions, Eq. 7-124b, are derived from the differential equation of heat conduction.

After solving for the unknown constants, the temperature profile 7-123 becomes

$$\frac{T}{T_i} = -4 \frac{x - S}{\delta - S} + 6 \left(\frac{x - S}{\delta - S} \right)^2 - 4 \left(\frac{x - S}{\delta - S} \right)^3 + \left(\frac{x - S}{\delta - S} \right)^4 \tag{7-125}$$

It is to be noted that the temperature profile 7-125 involves two unknown parameters, namely S and $(\delta - S)$.

Substituting the temperature profile 7-125 into the heat-balance integral 7-122 and performing the operations indicated, we obtain

$$\frac{d}{dt}(\delta - S) + 5\left(1 + \frac{\alpha \rho L}{kT_i}\right)\frac{dS}{dt} = \frac{5\alpha f(t)}{kT_i} \qquad (7\text{-}126)$$

This equation involves the unknown term dS/dt, which is evaluated by substituting the temperature profile 7-125 into the boundary condition 7-119b. The result is

$$\frac{dS}{dt} = \frac{1}{\rho L}\left[f(t) - \frac{4kT_i}{\delta - S}\right] \qquad (7\text{-}127)$$

Substituting 7-127 into 7-126 we obtain the following first-order ordinary differential equation for the unknown variable $(\delta - S)$,

$$\frac{d}{dt}(\delta - S) - \frac{20kT_i}{\rho L}\left(1 + \frac{\alpha \rho L}{kT_i}\right)\cdot\frac{1}{\delta - S} = -\frac{5f(t)}{\rho L} \qquad (7\text{-}128)$$

For convenience, this equation is written in the form

$$\frac{dy}{dt} - \frac{A}{y} = -B \cdot f(t) \qquad (7\text{-}129)$$

where

$$y = \delta - S$$

$$A = \frac{20kT_i}{\rho L}\left(1 + \frac{\alpha \rho L}{kT_i}\right)$$

$$B = \frac{5}{\rho L}$$

Equation 7-129 shows that the unknown variable y is a function of the parameters $A, B, f(t)$ and the initial value of y at time $t = 0$, i.e., at the start of ablation. The parameters A and B are constants for a given material, and the heat-flux function $f(t)$ is prescribed. For the initial value of y, i.e.,

$$y\,|_{t=0} = [\delta - S]_{t=0} \qquad (7\text{-}129\text{a})$$

we take $S\,|_{t=0} = 0$ and $\delta\,|_{t=0}$ as the value of the thermal-layer thickness at the initiation of ablation. $\delta\,|_{t=0}$ is evaluated with straightforward application of the integral method to the heating problem preceding ablation.

Once $y(t)$ is determined from the solution of the differential Eq. 7-129, the thickness of the material ablated over a time interval $t = 0$ to t is obtained by integrating Eq. 7-127—that is,

$$S = \int_0^t \frac{1}{\rho L}\left[f(t') - \frac{4kT_i}{y(t')}\right]dt' \qquad (7\text{-}130)$$

We now examine the solution of Eq. 7-129 for one special case.

(1) We assume that the applied heat flux is constant. Taking $f(t) = f =$ constant, Eq. 7-129 becomes

$$\frac{dt}{dy} = \frac{y}{A - Bfy} \tag{7-131a}$$

with

$$y = \frac{4kT_i}{f} \quad \text{for} \quad t = 0 \tag{7-131b}$$

The initial condition chosen for this problem is the value of the thermal-layer thickness at the initiation of ablation which is evaluated by using a fourth-degree temperature profile for the heating problem with constant heat flux f at the boundary.

For convenience we write Eq. 7-131 in dimensionless form as

$$\frac{d\tau}{d\xi} = \frac{1}{5}\,\frac{\xi}{4(1 + \nu) - \xi} \tag{7-132a}$$

with

$$\xi = 4 \quad \text{for} \quad \tau = 0 \tag{7-132b}$$

where

$$\xi = \frac{f}{kT_i}\,y = \frac{f}{kT_i}(\delta - S)$$

$$\tau = \frac{f^2}{kT_i\rho L}\,t$$

$$\nu = \frac{\alpha\rho L}{kT_i}$$

Integrating Eq. 7-132 from $\xi = 4$ to ξ, we obtain

$$\tau = \frac{1}{5}\int_4^\xi \frac{\xi}{4(1 + \nu) - \xi}\,d\xi \tag{7-133}$$

When the integration is performed

$$\tau = -\frac{1}{5}\left[(\xi - 4) + 4(1 + \nu)\cdot \ln \frac{4(1 + \nu) - \xi}{4\nu}\right] \tag{7-134}$$

The thickness of the material ablated is obtained by integrating Eq. 7-127. First we write Eq. 7-127 in dimensionless form, using the above notation as

$$d\bar{S} = \frac{\xi - 4}{\xi}\,d\tau \tag{7-135a}$$

with

$$\bar{S} = 0 \quad \text{for} \quad \tau = 0 \tag{7-135b}$$

where

$$\bar{S} = \frac{f}{kT_i} S$$

$$\xi = \frac{f}{kT_i} y = \frac{f}{kT_i} (\delta - S)$$

Eliminating $d\tau$ from Eq. 7-135 by means of Eq. 7-132, we obtain

$$\frac{d\bar{S}}{d\xi} = \frac{1}{5} \frac{\xi - 4}{4(1 + \nu) - \xi} \tag{7-136a}$$

with

$$\bar{S} = 0 \quad \text{for} \quad \xi = 4 \tag{7-136b}$$

Integration of Eq. 7-136 yields the relation for the material ablated as

$$\bar{S} = -\frac{1}{5} \left[(\xi - 4) + 4\nu \ln \frac{4(1 + \nu) - \xi}{4\nu} \right] \tag{7-137}$$

Equations 7-134 and 7-137 are coupled relations for evaluating the material ablated, i.e., the location of the boundary surface.

(2) When the applied heat flux $f(t)$ in Eq. 7-129 is nonconstant, then Eq. 7-129 should be integrated numerically.

7-9. METHOD OF GALERKIN IN THE SOLUTION OF STEADY, TWO-DIMENSIONAL HEAT-CONDUCTION PROBLEMS

The approximate method introduced by Galerkin for solving boundary-value problems has found wider application in recent years. In this section we present a brief description of Galerkin's method in the solution of steady heat-conduction problems in two-dimensional finite regions. For fundamental treatment of this method and the mathematical proofs and application to problems in other fields the reader should refer to books by Kantorovich and Krylov [19], and by Mikhlin [20].

The basic idea of Galerkin's method will now be illustrated with the following heat conduction problem subject to linear, homogenous boundary conditions.

Consider the steady-state heat-conduction problem in a two-dimensional, homogeneous, finite region with internal heat generation in the solid. Assuming the boundaries of the region are subjected to linear, homogeneous boundary conditions (i.e., boundary condition of the first, second, or third kind), the boundary-value problem of heat conduction is given in the form[4]

$$L\left[T(x, y)\right] \equiv \frac{\partial^2 T}{\partial x^2} + \frac{\partial^2 T}{\partial y^2} + \frac{g}{k} = 0 \text{ in region } R \tag{7-138a}$$

[4] It is to be noted that the method is applicable to equations of elliptic, hyperbolic, and parabolic types.

subjected to

$$T = 0 \quad \text{or} \quad \frac{\partial T}{\partial n} = 0, \quad \text{or} \quad \frac{\partial T}{\partial n} + HT = 0 \qquad \text{(7-138b)}$$

on the boundary S of region R, (but boundary conditions are not all of the second kind) where $\partial/\partial n$ = differentiation along the outward-drawn normal to the boundary.

The basis of Galerkin's method of solution of system 7-138 is that of seeking an approximate solution to system 7-138 in the form

$$T_n(x, y) = \sum_{i=1}^{n} c_i \cdot \phi_i(x, y) \qquad \text{(7-139)}$$

where $T_n(x, y)$ = approximate profile, i.e., n-term approximation
$\phi_i(x, y)$ = set of known functions chosen at onset of problem such that they satisfy boundary conditions 7-138b of the problem
c_i = unknown coefficients to be determined by Galerkin's method

which satisfies the system 7-138 as closely as possible.

Galerkin's method may be stated as [Ref. 19, p. 262]: In order that $T_n(x, y)$ be the solution of the differential equation 7-138a, it is necessary that $L[T_n(x, y)]$ be identically equal to zero; and this requirement, if $L[T_n(x, y)]$ is considered to be continuous, is equivalent to the requirement of orthogonality of the expression $L[T_n(x, y)]$ to all functions of the system $\phi_i(x, y)$. Since we have n functions of $\phi_i(x, y)$, i.e., $i = 1, 2, 3, \ldots, n$, we obtain n orthogonality conditions, from which n algebraic equations are obtained for determining the n unknown coefficients $c_1, c_2, c_3, \ldots, c_n$. After determining c_is from the solution of the algebraic system, these coefficients are substituted into Eq. 7-139 and an approximate solution is obtained for the temperature distribution in the region.

The foregoing statement is demonstrated as follows.

We substitute the approximate temperature profile 7-139 into the differential equation 7-138a,

$$L\left[\sum_{i=1}^{n} c_i \cdot \phi_i(x, y)\right] = 0 \qquad \text{(7-140)}$$

We multiply Eq. 7-140 by $\phi_j(x, y)$ and integrate over the region R,

$$\iint_R L\left[\sum_{i=1}^{n} c_i \cdot \phi_i(x, y)\right] \cdot \phi_j(x, y) \cdot dx \cdot dy = 0 \quad \text{for} \quad j = 1, 2, 3, \ldots, n \qquad \text{(7-141)}$$

The system 7-141 yields a system of n algebraic equations for determining the n unknown coefficients $c_1, c_2, \ldots, c_n$. When the unknown coefficients

$c_1, c_2, \ldots, c_n$ are determined from the solution of system 7-141 and substituted into Eq. 7-139, it gives the nth approximate solution for the problem.

Galerkin's method as described above provides a simple and straightforward method for obtaining approximate solution to steady-state heat-conduction problems. It is applicable to problems in regular or arbitrarily shaped finite regions, provided that the $\phi_i(x, y)$ functions satisfying the boundary conditions for the problem are found at the onset of the problem. In the case of problems involving nonhomogeneous boundary conditions it is possible to define a new independent variable, say $U(x, y)$, that will reduce the nonhomogeneous boundary conditions to homogeneous ones.

We now examine the construction of $\phi_i(x, y)$ functions.

Construction of $\phi_i(x, y)$ Functions. The construction of $\phi_i(x, y)$ functions for approximate representation of temperature profile in the form as given by 7-139 requires special consideration because the profile 7-139 should satisfy both the differential equation and the boundary conditions of system 7-138. The functions $\phi_i(x, y)$ are usually chosen as different combination of polynomials or trigonometric functions that satisfy the boundary conditions of problem 7-138. We therefore examine the construction of $\phi_i(x, y)$ functions for different types of boundary conditions.

Boundary Condition of the First Kind. Consider the case when the boundary condition for the differential equation 7-138a is of the form

$$T(x, y) = 0 \quad \text{on the boundary } S \text{ of the region } R \tag{7-142}$$

Suppose we choose a function $w(x, y)$ which is continuous in the closed region R, have continuous derivatives $\dfrac{\partial w}{\partial x}$ and $\dfrac{\partial w}{\partial y}$, and satisfies the boundary condition 7-142, i.e.,

$$w(x, y) = 0 \quad \text{on the boundary } S \text{ of the region } R \tag{7-143}$$

Then, the system of functions $\phi_i(x, y)$ may be chosen as the system consisting of the products of $w(x, y)$ and various powers of x and y in the form

$$\begin{aligned}
\phi_1(x, y) &= w(x, y) \\
\phi_2(x, y) &= x \cdot w(x, y) \\
\phi_3(x, y) &= y \cdot w(x, y) \\
\phi_4(x, y) &= x^2 \cdot w(x, y) \\
\phi_5(x, y) &= x \cdot y \cdot w(x, y) \\
&\cdots\cdots\cdots\cdots
\end{aligned} \tag{7-144}$$

It is apparent that the functions 7-144 also satisfy the boundary condition for the problem (i.e., vanish on the boundary).

The form of $w(x, y)$ functions is closely related to the contour bounding the region R. We examine below determination of $w(x, y)$ functions from the knowledge of the equations for the contour of the boundary for various geometries.

(a) *Regions having a single continuous boundary contour.* Consider a two-dimensional finite region R the boundary of which has a contour that satisfies an equation of the form

$$F(x, y) = 0 \tag{7-145}$$

where the function $F(x, y)$ is continuous together with its partial derivatives $\partial F/\partial x$ and $\partial F/\partial y$. In such cases the function $w(x, y)$ may be chosen as

$$w(x, y) = \pm F(x, y) \tag{7-146}$$

For example, consider a circular region of radius R and with center at the origin. The contour of this region satisfies the equation

$$F(x, y) = R^2 - x^2 - y^2 = 0 \tag{7-147}$$

Hence the function $w(x, y)$ is chosen as

$$w(x, y) = R^2 - x^2 - y^2 \tag{7-148}$$

(b) *Regions having boundary contour in the form of a convex polygon.* Consider a finite region whose boundary contour is a convex polygon having equations for its sides in the form

$$\begin{aligned} a_1 x + b_1 y + d_1 &= 0 \\ a_2 x + b_2 y + d_2 &= 0 \\ \cdots\cdots&\cdots\cdots \\ a_m x + b_m y + d_m &= 0 \end{aligned} \tag{7-149}$$

For such cases the function $w(x, y)$ is chosen in the form

$$w(x, y) = \pm(a_1 x + b_1 y + d_1)\ldots(a_m x + b_m y + d_m) \tag{7-150}$$

which vanishes at every point of the boundary and satisfies the boundary condition for the problem.

We illustrate this with the following examples.

Figure 7-11a shows a rectangular region $(-a, a;\ -b, b)$ whose sides have equations in the form

$$a - x = 0 \qquad a + x = 0 \qquad b - y = 0 \qquad b + y = 0$$

Hence the $w(x, y)$ function for use in this region, according to Eq. 7-150, is given as

$$w(x, y) = (a^2 - x^2)(b^2 - y^2) \tag{7-151}$$

Figure 7-11b shows a triangular region having its apex at the origin and its base parallel to the y-axis. The equations for the sides of this triangle are given in the form

$$y - \alpha x = 0 \qquad y + \beta x = 0 \qquad L - x = 0$$

and the corresponding $w(x,y)$ function is given as

$$w(x,y) = (y - \alpha x)\cdot(y + \beta x)\cdot(L - x) \tag{7-152}$$

For a right-angle triangle having two of its sides lying on the $0x$ and $0y$ axis as shown in Fig. 7-11c, $w(x,y)$ function is given in the form

$$w(x,y) = x\cdot y\cdot\left(1 - \frac{x}{a} - \frac{y}{b}\right) \tag{7-153}$$

The above considerations for combining the equations for the contours is applicable in the case of regions bounded by curves. For example, Fig. 7-11d shows a crescent-shaped closed region formed by two circles of radii R_1 and R_2. Equations for the contours of these two circles are given as

$$R_1^2 - x^2 - y^2 = 0 \qquad R_2^2 - (x - L)^2 - y^2 = 0$$

Hence the corresponding $w(x,y)$ function is given in the form

$$w(x,y) = (R_1^2 - x^2 - y^2)\cdot[R_2^2 - (x - L)^2 - y^2] \tag{7-154}$$

(c) *Regions forming nonconvex polygons.* For two-dimensional finite regions whose boundary contour forms nonconvex polygon, the construction of $w(x,y)$ function is more involved. The reader may refer to the book by Kantorovich and Krylov for treatment of such cases. [Ref. 19 pp. 277–279]

Other Boundary Conditions. The construction of $w(x,y)$ for finite regions subjected to a homogeneous boundary condition of the first kind is generally a relatively simple matter as described above. When the boundaries are subjected to homogeneous boundary condition of the second or third kind (but not all of the second kind), i.e.

$$\frac{\partial T}{\partial n} = 0$$

or

$$\frac{\partial T}{\partial n} + HT = 0$$

on the boundaries of the region R, determination of $\phi(x,y)$ function that will satisfy the boundary conditions is generally difficult for the irregular geometries.

However, one may consider that $\phi_i(x,y)$ functions may be taken as the products of powers of x and y such that the approximate temperature profile chosen as a polynomial in the form

$$T_n(x,y) = c_1 + c_2x + c_3y + c_4x^2 + c_5xy + \cdots + c_ny^m \tag{7-155}$$

This polynomial is then subjected to the boundary conditions prescribed for the region; namely, Eq. 7-155 is required to satisfy the boundary conditions for the region and with this requirement some of the coefficients c_i's are eliminated. The resulting relation gives an approximate temperature profile that satisfies the boundary conditions for the region. The remaining unknown coefficients are determined by Galerkin's method.

7-10. APPLICATION OF GALERKIN'S METHOD

We now illustrate the application of Galerkin's method to the solution of steady heat-conduction problems in finite regions with examples.

EXAMPLE. Consider a rectangular region $(-a, a; -b, b)$ shown in Fig. 7-11a subjected to zero temperature at the boundaries and with heat generation within the solid at a uniform rate of g Btu/hr ft^3. The steady-state temperature distribution in the region satisfies the following system.

$$\frac{\partial^2 T}{\partial x^2} + \frac{\partial^2 T}{\partial y^2} + \frac{g}{k} = 0$$

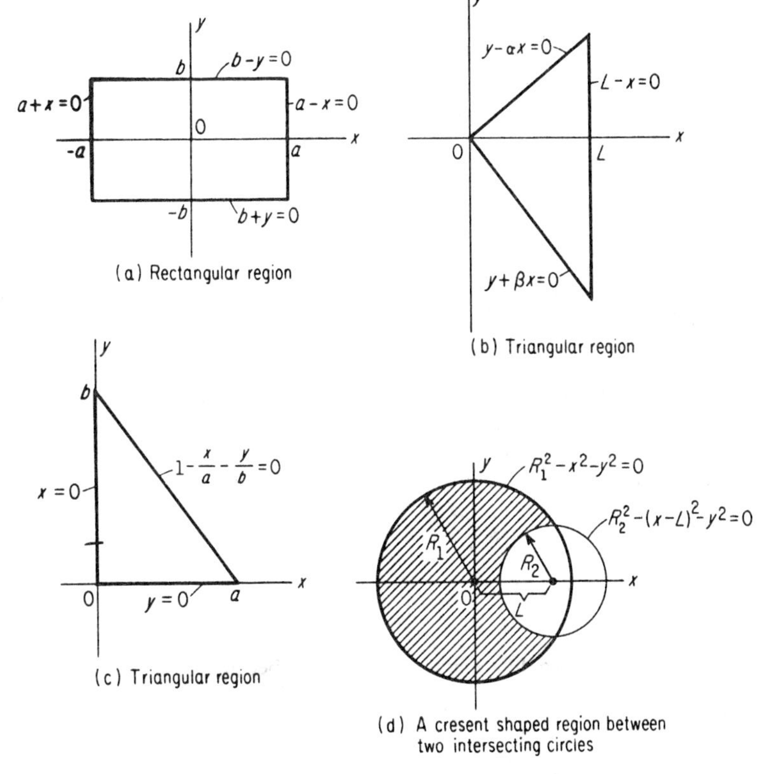

FIG. 7-11. Regions having boundary contour in the form of a convex polygon and a region bounded by two circles.

in the region

$$T = 0 \quad \text{at} \quad x = \pm a \tag{7-156}$$
$$T = 0 \quad \text{at} \quad y = \pm b$$

The exact solution of this system is easily obtainable by the standard techniques. In order to illustrate the procedure and demonstrate the accuracy of the result, we present an approximate solution of this problem by Galerkin's method.

For the rectangular region subjected to homogeneous boundary condition of the first kind at the boundaries the $w(x, y)$ function is chosen as (See Eq. 7-151)

$$w(x, y) = (a^2 - x^2)(b^2 - y^2) \tag{7-157}$$

In view of the symmetry involved in the present problem, the system of functions $\phi_i(x, y)$ is chosen as the product of $w(x, y)$ and *even* powers of x and y in the form

$$\begin{aligned}
\phi_1(x, y) &= (a^2 - x^2)(b^2 - y^2) \\
\phi_2(x, y) &= x^2 \cdot (a^2 - x^2)(b^2 - y^2) \\
\phi_3(x, y) &= y^2 \cdot (a^2 - x^2)(b^2 - y^2) \\
\phi_4(x, y) &= x^2 y^2 \cdot (a^2 - x^2)(b^2 - y^2) \\
&\vdots \\
\phi_n(x, y) &= x^{2i} \cdot y^{2j} \cdot (a^2 - x^2)(b^2 - y^2)
\end{aligned} \tag{7-158}$$

Hence, the general form of the approximate profile for temperature is taken in the form

$$T_n(x, y) = (a^2 - x^2)(b^2 - y^2) \cdot [c_1 + c_2 x^2 + c_3 y^2 + c_4 x^2 y^2 + \cdots + c_n x^{2i} y^{2j}] \tag{7-159}$$

To simplify the problem we consider the solution with only *one-term* approximations—namely, we choose the temperature profile, by taking $n = 1$, in the form

$$T_1(x, y) = c_1 \cdot (a^2 - x^2)(b^2 - y^2) \equiv c_1 \cdot \phi_1(x, y) \tag{7-160}$$

and we wish to determine the unknown coefficient c_1 by Galerkin's method.

The general relation for determining the unknown coefficients c_i's is given by Eq. 7-141; for the present profile with $n = 1$ and for the rectangular region under consideration it becomes

$$\int_{x=-a}^{a} \int_{y=-b}^{b} \left[c_1 \frac{\partial^2 \phi_1}{\partial x^2} + c_1 \frac{\partial^2 \phi_1}{\partial y^2} + \frac{g}{k} \right] \cdot \phi_1 \cdot dx \cdot dy = 0 \tag{7-161}$$

where

$$\phi_1 = (a^2 - x^2)(b^2 - y^2).$$

After we have performed differentiations, Eq. 7-161 becomes

$$4 \int_{x=0}^{a} \int_{y=0}^{b} \left[-2c_1 (a^2 - x^2)(b^2 - y^2)^2 - 2c_1 (b^2 - y^2)(a^2 - x^2)^2 + \frac{g}{k} (a^2 - x^2)(b^2 - y^2) \right] dx \cdot dy = 0$$

Performing the integrations and solving for c_1, we obtain

$$c_1 = \frac{5}{8}\frac{g/k}{a^2 + b^2} \tag{7-162}$$

Hence the one-term approximation for the temperature distribution is given in the form

$$T_1(x, y) = \frac{5}{8}\frac{g/k}{a^2 + b^2}(a^2 - x^2)(b^2 - y^2) \tag{7-163}$$

It is of interest to compare this approximate solution with the exact solution, which is given as

$$T(x, y) = \frac{g}{k}\left[\frac{a^2 - x^2}{2} - 2a^2 \sum_{n=0}^{\infty} \frac{(-1)^n}{\beta_n^3} \frac{\cosh\left(\beta_n \frac{y}{b}\right) \cdot \cos\left(\beta_n \frac{x}{a}\right)}{\cosh\left(\beta_n \frac{b}{a}\right)}\right] \tag{7-164}$$

where

$$\beta_n = \frac{(2n + 1)\pi}{2}$$

We compare the center temperature (i.e., at $x = 0$, $y = 0$) for the particular case of a square region (i.e., $a = b$). Equations 7-163 and 7-164 become

$$T_1(0, 0) = \frac{5}{16}\frac{ga^2}{k} = 0.3125\frac{ga^2}{k} \tag{7-163a}$$

$$T(0, 0) = \frac{ga^2}{k}\left[\frac{1}{2} - 2\sum_{n=0}^{\infty}\frac{(-1)^n}{\beta_n^3 \cdot \cosh\beta_n}\right] = 0.293\frac{ga^2}{k} \tag{7-164a}$$

Thus the error involved with one term temperature approximation is less than 7 percent. The accuracy of the approximate solution would have been improved if two or more terms were used.

For a two-term approximation the temperature profile may be chosen in the form

$$\begin{aligned} T_2(x, y) &= c_1\phi_1(x, y) + c_2 \cdot \phi_2(x, y) \\ &= (c_1 + c_2x^2)(a^2 - x^2)(b^2 - y^2) \end{aligned} \tag{7-165}$$

Proceeding as before, we substitute 7-165 into 7-141 and obtain two algebraic equations for determining the unknown coefficients c_1 and c_2.

REFERENCES

1. T. R. Goodman, "The Heat Balance Integral and Its Application to Problems Involving a Change of Phase," *ASME Trans.* Vol. 80, No. 2 (1958), pp. 335–342.

2. H. Schlichting, *Boundary Layer Theory*, Chapter 12, McGraw-Hill Book Company, New York, 1955.

3. T. R. Goodman, "Application of Integral Methods to Transient Nonlinear Heat Transfer," in T. F. Irvine and J. P. Hartnett, (eds.), *Advances in Heat Transfer*, Vol. 1, Academic Press, New York, 1964.

4. W. C. Reynolds and T. A. Dolton, *The Use of Integral Methods in Transient Heat Transfer Analysis*, Department of Mechanical Engineering, Report No. 36, Stanford University, Stanford, Calif., Sept. 1, 1958.

5. T. J. Lardner and F. B. Pohle, "Application of Heat Balance Integral to the Problems of Cylindrical Geometry," *ASME Trans., J. Applied Mechanics* (1961), pp. 310–312.

6. T. R. Goodman, "The Heating of Slabs with Arbitrary Heat Inputs," *J. Aerospace Sciences*, Vol. 26 (1959), pp. 187–188.

7. P. J. Schneider, "Radiation Cooling of Finite Heat Conducting Solids," *J. Aerospace Sciences*, Vol. 27 (1960), pp. 546–549.

8. J. C. Jaeger, "Conduction of Heat in Solids with a Power Law of Heat Transfer at Its Surface," *Cambridge Phil. Soc. Proc.*, Vol. 46, Part 4 (October 1950), p. 634.

9. T. R. Goodman, "Heat Balance Integral—Further Considerations and Refinements," *ASME J. Heat Transfer*, Vol. 83 (1961), pp. 83–86.

10. J. C. Y. Koh, "One-Dimensional Heat Conduction with Arbitrary Heating Rate and with Variable Properties," *J. Aerospace Sciences*, Vol. 28 (1961), pp. 989–990.

11. Kwang-Zu Yang, "Calculation of Unsteady Heat Conduction in a Single Layer and Composite Finite Slabs with and without Property Variations by an Improved Integral Procedure," *International Developments in Heat Transfer 1, Proceedings*, (1961), pp. 18–27.

12. K. T. Yang, "Transient Conduction in a Semi-infinite Solid with Variable Thermal Conductivity," *Trans. ASME*, Vol. 80 (1958), pp. 146–147.

13. H. S. Carslaw and J. C. Jaeger, *Conduction of Heat in Solids*, Chapter XI, Clarendon Press, Oxford, 1959.

14. G. Poots, "An Approximate Treatment of Heat Conduction Problem Involving a Two-Dimensional Solidification Front," *Int. J. Heat Mass Transfer*, Vol. 5, (1962), pp. 339–348.

15. L. R. Ingersol, O. J. Zobel and A. C. Ingersol, *Heat Conduction with Engineering and Geological Application*, McGraw-Hill Book Company, New York, 1948, pp. 194–196.

16. Manfred Altman, "Some Aspects of the Melting Solution for a Semi-infinite Slab," *Chemical Engineering Progress Symposium Series*, Vol. 57, Buffalo, 1961, pp. 16–23.

17. R. J. Barriault and J. Yos, "Analysis of the Ablation of Plastic Heat Shields That Form a Charred Surface Layer," *ARS J.* (September 1960), pp. 823–829.

18. H. G. Landau, "Heat Conduction in a Melting Solid," *Quart. Appl. Math.*, Vol. 8 (1950), pp. 81–94.

19. L. V. Kantorovich and V. I. Krylov, *Approximate Methods of Higher Analysis*, John Wiley & Sons, Inc., New York, 1964, pp. 258–309.

20. S. G. Mikhlin, *Variational Methods in Mathematical Physics*, The Macmillan Company, New York, 1964, Chapter IX.

PROBLEMS

1. Using a third-degree polynomial approximation, determine temperature distribution in a semi-infinite solid $0 \leq x < \infty$ which is initially at temperature T_i and for times $t > 0$ the boundary surface at $x = 0$ is subjected to convective boundary condition described by the equation

$$k \frac{\partial T}{\partial x} + hT = f_1 \quad \text{at} \quad x = 0, t > 0$$

2. Using a third-degree polynomial approximation, obtain temperature distribution in a semi-infinite solid $0 \leq x < \infty$ which is initially at temperature T_i and for times $t > 0$ the boundary surface at $x = 0$ is subjected to a prescribed heat flux which varies with time, i.e.,

$$k \frac{\partial T}{\partial x} = f(t) \quad \text{at} \quad x = 0, t > 0$$

3. A semi-infinite region $0 \leq x < \infty$ is initially at zero temperature, for times $t > 0$ the boundary surface at $x = 0$ is subjected to a constant temperature T_0. The thermal conductivity of the solid strongly depends on temperature and cannot be assumed constant. Using a fourth-degree polynomial approximation, determine a relation for temperature distribution in the solid with integral methods.

4. Using a second-degree polynomial approximation, determine the rate of ablation of a semi-infinite solid $0 \leq x < \infty$ subjected to aerodynamic heating at the boundary surface $x = 0$ at constant rate in the form

$$\text{Heat flux} = f \equiv \text{const.} \quad \text{at} \quad x = 0$$

Compare this solution with those given by Eqs. 7-134 and 7-137 for a fourth-degree polynomial approximation.

5. Consider the triangular region shown in Fig. 7-11b. Heat is generated within this region at a constant rate of g Btu/hr ft^3 while the boundaries are kept at zero temperature. Using Galerkin's method and taking only one-term approximation for the temperature profile, determine a relation for the approximate temperature distribution in the region—i.e., choose the first-term approximation in the form

$$T_1(x, y) = c_1(y - \alpha x)(y + \beta x)(L - x)$$

8

Nonlinear Boundary-value Problems of Heat Conduction

Nonlinearities actually exist in the physical problems; but because of the difficulties associated with the solution of nonlinear problems simplifying assumptions are usually made to linearize them. For example, constant thermal conductivity and linearized boundary conditions are generally assumed. When the temperature change is large or thermal conductivity varies greatly with temperature, the assumption of constant thermal conductivity may lead to significant error. When thermal radiation takes place at high temperatures, the fourth-power radiation law should be considered instead of linearized boundary conditions. Solution of nonlinear problems is not straightforward, and each problem should be treated individually because "there is no extant theory for nonlinear partial differential equations of any order of nonlinearity [1]." There are a number of ingeneous transformations that have been introduced into the literature of nonlinear partial differential equations with the purpose of linearizing them or reducing the order of the differential equation or achieving some reduction in complexity. An excellent treatment of several techniques for the solution of nonlinear boundary-value problems of engineering is given by Ames [1]. In this chapter we examine some of the techniques that are pertinent to the solution of nonlinear boundary-value problems of heat conduction. The solution of one-dimensional, time-dependent boundary-value problems of heat conduction with variable thermal conductivity; problems involving nonlinear boundary conditions and those involving change of phase will be examined. The use of transformation that will achieve a reduction in the number of independent variables of the differential equation of heat conduction will be discussed.

8-1. SEMI-INFINITE REGION WITH NONLINEAR BOUNDARY CONDITION

In this section we examine the solution of one-dimensional, time-dependent boundary-value problems of heat conduction for a semi-infinite region with nonlinear boundary condition [2]. For example, the natural convection boundary condition in the form

$$-k\frac{\partial T}{\partial x} = h(T_\infty - T)^{5/4} \quad \text{at the boundary surface} \tag{8-1}$$

or the fourth-power thermal radiation boundary condition

$$-k\frac{\partial T}{\partial x} = \sigma\epsilon(T_\infty^4 - T^4) \quad \text{at the boundary surface} \tag{8-2}$$

where σ = Stefan-Boltzmann constant
ϵ = emissivity of the surface
T_∞ = temperature of the surrounding

are nonlinear boundary conditions because a power of temperature is involved. In such problems a quantity of greatest physical interest is usually the temperature at the boundary surface because of its extremum character. For example, in the problem of heating of a semi-infinite solid the boundary surface will always attain a maximum value, and under cooling conditions it will attain a minimum value with respect to the interior points.

In the following example we consider the surrounding medium to be at a constant temperature T_∞. By introducing a dimensionless temperature

$$U = \frac{T}{T_\infty} \tag{8-3}$$

the above nonlinear boundary conditions may be represented with a general expression in the form

$$-\frac{\partial U}{\partial x} = f(U_s) \tag{8-4}$$

where U_s = dimensionless temperature of the boundary surface
$f(U_s)$ = a nonlinear analytic function in U_s.

For a semi-infinite solid $0 \le x < \infty$ which is initially at a constant temperature $U_i = T_i/T_\infty$, the boundary-value problem of heat conduction is given as

$$\frac{\partial^2 U}{\partial x^2} = \frac{\partial U}{\partial \tau} \quad \text{in} \quad 0 \le x < \infty,\ \tau > 0 \tag{8-5a}$$

$$-\frac{\partial U}{\partial x} = f(U_s) \quad \text{at} \quad x = 0,\ \tau > 0 \tag{8-5b}$$

$$U = U_i \quad \text{in} \quad 0 \le x < \infty,\ \tau = 0 \tag{8-5c}$$

where

$$U = \frac{T(x,\tau)}{T_\infty} = \text{dimensionless temperature} \tag{8-6a}$$

$$U_s = \frac{T_s(\tau)}{T_\infty} = \text{dimensionless temperature at the boundary surface } x = 0 \tag{8-6b}$$

$$f(U_s) = \text{a nonlinear analytic function in } U_s \tag{8-6c}$$

$$\tau = \alpha \cdot t \tag{8-6d}$$

Figure 8-1 shows the geometry and the boundary condition.

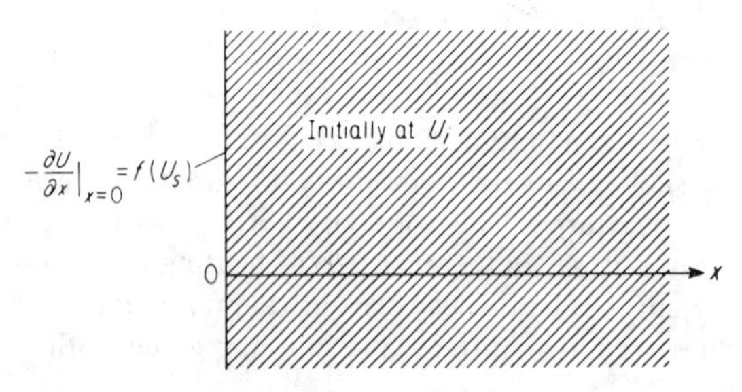

FIG. 8-1. Semi-infinite region $0 \leq x < \infty$ with nonlinear boundary condition at $x = 0$.

The nonlinear heat conduction problem described by the system Eq. 8-5 will now be solved for the boundary surface temperature, U_s, at the surface $x = 0$. One method of solving system 8-5 is to transform it into an integral equation in the form[1]

$$U(x, \tau) = U_i + \frac{1}{\sqrt{\pi}} \int_{\tau'=0}^{\tau} \frac{e^{-x^2/4(\tau-\tau')}}{(\tau - \tau')^{1/2}} \cdot f(U_s) \cdot d\tau' \tag{8-7}$$

Temperature at the boundary surface $x = 0$ is obtained by setting in Eq. 8-7, $x = 0$.

$$U_s(\tau) = U_i + \frac{1}{\sqrt{\pi}} \int_{\tau'=0}^{\tau} \frac{1}{(\tau - \tau')^{1/2}} f[U_s(\tau')] \cdot d\tau' \tag{8-8}$$

Equation 8-8 is a singular, nonlinear Volterra integral equation.

A Volterra integral equation may be solved by a number of analytical methods [3,4]; but in the present problem the singular character of the kernel

$$\frac{1}{(\tau - \tau')^{1/2}} \tag{8-9}$$

at the upper limit τ and the nonlinearity of function $f[U_s(\tau)]$ restricts the choice of solutions.

[1]This result was obtained by Chambre [2] by means of Laplace transform and by using convolution integral for inversion of the product of Laplace transform of two functions. The same result can be obtained with an integral-transform technique; this is left to the reader.

The *method of successive approximations* can be used to solve the integral Eq. 8-8. In this method one chooses an initial guess $U_{0s}(\tau)$ to the solution of the problem. If this approximation is substituted on the right-hand side of Eq. 8-8 and the integration is carried out, the resulting temperature $U_{1s}(\tau)$ is the first approximation. If $U_{1s}(\tau)$ is substituted on the right-hand side of Eq. 8-8 and integration is performed the resulting temperature $U_{2s}(\tau)$ is the second approximation. If this process is repeated, one obtains a set of functions $U_{ns}(\tau)$, with $n = 1, 2, 3, \ldots$, which can converge to the solution of the integral Eq. 8-8.

In practice, however, if a sufficiently good initial approximation to the solution of the problem is known, only two or three iterations are needed. Furthermore if $f(U_s)$ is a *monotone* function (i.e., either increasing or decreasing) as in the case of boundary conditions 8-1 or 8-2, the successive approximations lie above and below the actual solution; that is, the actual solution lies between two successive approximate solutions.

To remove the singularity $(\tau - \tau')^{-1/2}$ from the integral equation 8-8, Chambre introduced new variables $\bar{\tau}$, $\bar{\tau}'$ as

$$\tau = \bar{\tau}^2 \quad \text{and} \quad \tau' = \bar{\tau}'^2 \tag{8-10}$$

and a new variable of integration ξ as

$$\bar{\tau}^2 - \bar{\tau}'^2 = \xi^2 \tag{8-11}$$

Substituting the new variables 8-10 and 8-11 into Eq. 8-8, we obtain

$$U_s(\bar{\tau}^2) = U_i + \frac{2}{\sqrt{\pi}} \int_{\xi=0}^{\bar{\tau}} f[U_s(\bar{\tau}^2 - \xi^2)] \cdot d\xi \tag{8-12}$$

The integral equation is written in the form of a recursion relation:

$$U_{n+1,s}(\bar{\tau}^2) = U_i + \frac{2}{\sqrt{\pi}} \int_{\xi=0}^{\bar{\tau}} f[U_{n,s}(\bar{\tau}^2 - \xi^2)] \cdot d\xi \tag{8-13}$$

where $n = 0, 1, 2, 3, \ldots$.

To a limited extent the integration in Eq. 8-13 can be carried out by analytical means, but because of the complexity of the function $f[U_{n,s}]$ integrations are performed by numerical means. Chambre used the recursion relation Eq. 8-13 to solve the heat-conduction problems for a semi-infinite region $0 \leq x \leq \infty$ which is initially at zero temperature (i.e. $U_i = 0$) and for times $t > 0$, the boundary surface at $x = 0$ is heated by radiation according to the fourth-power radiation law from a surrounding at constant temperature T_∞. In this case the nonlinear boundary-condition function $f(U_s)$ is in the form

$$-\left.\frac{\partial T}{\partial x}\right|_{x=0} = \frac{\sigma\epsilon}{k}(T_\infty^4 - T^4)\,|_{x=0}$$

or

$$-\left.\frac{\partial U}{\partial x}\right|_{x=0} = \frac{\sigma\epsilon T_\infty^3}{k}(1 - U^4)\,|_{x=0} \equiv f(U_s) \tag{8-14}$$

An approximate solution of this problem was obtained in Chapter 7 by means of the *integral method* and the resulting solution was given by Eq. 7-84. This approximate solution is used as the initial guess in the recursion relation 8-13.

Figure 8-2 shows the dimensionless surface temperature U_s as a function of the time variable $\alpha(\sigma\epsilon/k)^2 t$ for the first and the improved second approximations. The two curves are very close. Since the actual solution lies between the first and the second approximations, a new curve representing the arithmetic mean of the two approximations is drawn between them.

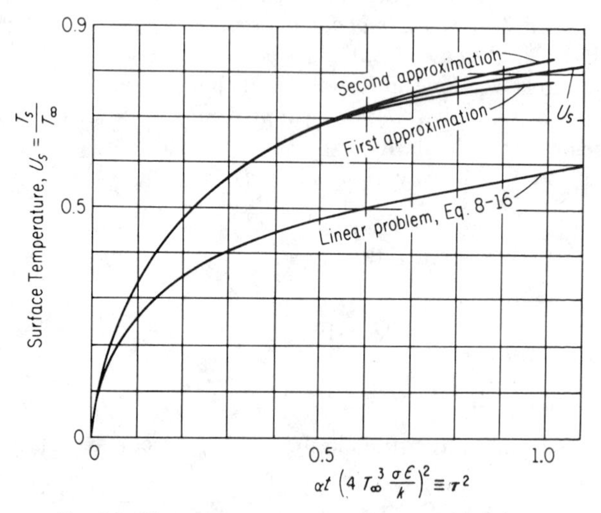

FIG. 8-2. The surface temperature for a semi-infinite solid initially at zero temperature for heating at $x = 0$ according to the fourth power radiation law and according to Newton's law of cooling. (Paul L. Chambre, *J. Appl. Phys.*, Vol. 30, November 1959, p. 1683.)

It is interesting to compare the above solution with the solution for a linearized boundary condition in the form

$$-\left.\frac{\partial U}{\partial x}\right|_{x=0} = 4\,T_\infty^3\,\frac{\sigma\epsilon}{k}\,(1 - U)\,|_{x=0} \tag{8-15}$$

The analytical solution for this problem is available and the relation for the boundary surface temperature at $x = 0$ is

$$U_s = 1 - e^{\bar{\tau}^2} \cdot \operatorname{erfc}(\bar{\tau}) \tag{8-16}$$

where

$$\bar{\tau}^2 = \alpha \left(\frac{\sigma\epsilon}{k}\right) t$$

Solution 8-16 is also included in Fig. 8-2 for comparison. It is apparent that the surface temperature will equalize with the surrounding temperature at a much faster rate with the fourth-power radiation law than with the linearized boundary condition as given by Eq. 8-16.

8-2. PROBLEMS INVOLVING TEMPERATURE-DEPENDENT THERMAL PROPERTIES— USE OF KIRCHHOFF TRANSFORMATION

When thermal properties of the solid vary with temperature but independent of position, the differential equation of heat conduction is in the form

$$\rho c_p \frac{\partial T}{\partial t} = \nabla \cdot k \nabla T + g \tag{8-17}$$

where it is assumed that ρ, c_p, k are temperature-dependent, but the heat generation term $g \equiv g(\bar{r}, t)$ is independent of temperature. In such cases a change of dependent variable by means of *Kirchhoff transformation* [1, 5, 6] will remove the thermal conductivity in Eq. 8-17 outside the differential operator. The procedure is as follows.

We define a new dependent variable in the form

$$U = \int_0^T \frac{k(T')}{k_0} \cdot dT' \tag{8-18}$$

where $U \equiv U(T)$

$T \equiv T(r, t)$

k_0 = value of thermal conductivity for T equal to zero

The transformation 8-18 is called the Kirchhoff transformation.

Since k is a function of temperature, Eq. 8-17 is written in the form

$$\rho c_p \frac{\partial T}{\partial t} = k \nabla^2 T + \nabla k \cdot \nabla T + g \tag{8-19}$$

Expressing ∇k in the form

$$\nabla k = \frac{dk}{dT} \nabla T \tag{8-20}$$

Equation 8-19 becomes

$$\rho c_p \frac{\partial T}{\partial t} = k \nabla^2 T + \frac{dk}{dT} (\nabla T)^2 + g \tag{8-21}$$

In Eq. 8-21 we transform the dependent variable T into the new variable U by means of the transformation 8-18.

From Eq. 8-18 we have the following relations:

$$\frac{\partial U}{\partial t} = \frac{k}{k_0} \frac{\partial T}{\partial t} \tag{8-22a}$$

$$\nabla U = \frac{dU}{dT} \nabla T = \frac{k}{k_0} \nabla T \tag{8-22b}$$

$$\nabla^2 U = \nabla\left[\frac{k}{k_0}\nabla T\right] = \frac{1}{k_0}[\nabla k \cdot \nabla T + k\nabla^2 T]$$

$$= \frac{1}{k_0}\left[\frac{dk}{dT}(\nabla T)^2 + k\nabla^2 T\right] \tag{8-22c}$$

Substituting 8-22a and c into Eq. 8-21 we obtain

$$\frac{1}{\alpha}\frac{\partial U}{\partial t} = \nabla^2 U + \frac{g}{k_0} \tag{8-23}$$

Since the thermal diffusivity α, in Eq. 8-23, is a function of temperature, Eq. 8-23 is still nonlinear but simpler in form. Assuming variation of thermal diffusivity with temperature is negligible, Eq. 8-23 becomes a linear differential equation of heat conduction which can be solved with the usual techniques provided that the boundary conditions for the problem are also transformed.

In the case of steady-state problems the left-hand side of Eq. 8-23 vanishes, hence there is no approximation associated with the choice of α as constant.

We now examine the transformation of a prescribed temperature and heat flux boundary conditions by means of the transformation 8-18.

Prescribed Temperature. Let the temperature at the boundary surface be prescribed in the form

$$T = f_i(\bar{r}, t) \quad \text{at the boundary surface } i \tag{8-24}$$

Transformation of this boundary condition by means of the transformation 8-18 is also a prescribed temperature at the boundary surface. To illustrate this, consider the thermal conductivity depends on temperature in the form,

$$k(T) = k_0(1 + \beta T) \tag{8-25}$$

then transformation 8-18 becomes

$$U = \int_0^T (1 + \beta T')\cdot dT' = T + \frac{\beta}{2}T^2 \tag{8-26}$$

and transformation of boundary condition 8-24 by means of Eq. 8-26 gives

$$U = f_i(\bar{r}, t) + \frac{\beta}{2}f_i^2(\bar{r}, t) \equiv f_i^*(\bar{r}, t) \quad \text{at boundary surface } i \tag{8-27}$$

which is again a prescribed temperature boundary condition.

Prescribed Heat Flux. Consider the heat flux at the boundary surface be prescribed in the form

$$k(T)\cdot\frac{\partial T}{\partial n_i} = f_i(\bar{r}, t) \quad \text{at the boundary surface } i \tag{8-28}$$

which is nonlinear because thermal conductivity depends on temperature. This nonlinear prescribed heat flux boundary condition is transformed into a linear prescribed heat-flux boundary condition by means of the transformation 8-18.

From the transformation 8-18 (See: Eq. 8-22b) we have

$$\frac{\partial U}{\partial n_i} = \frac{k(T)}{k_0}\frac{\partial T}{\partial n_i} \quad \text{at the boundary surface } i \tag{8-29}$$

Eliminating $\partial T/\partial n_i$ between Eq. 8-28 and 8-29 the transformed boundary condition becomes

$$k_0 \frac{\partial U}{\partial n_i} = f_i(\bar{r}, t) \quad \text{at the boundary surface } i \tag{8-30}$$

which is a linear, prescribed heat-flux boundary condition.

We illustrate the use of Kirchhoff transformation in the solution of heat-conduction problems with temperature-dependent thermal conductivity with the following example.

EXAMPLE 8-1. A slab $0 \le x \le L$ is initially at temperature $F(x)$. For times $t > 0$ there is heat generation within the solid at a rate of $g(x, t)$ Btu/hr ft^3 while the boundary condition at $x = 0$ is insulated and that at $x = L$ is kept at a temperature $f(t)$.

The thermal conductivity of the solid is assumed to depend on temperature in the form

$$k(T) = k_0(1 + \beta T) \tag{8-31}$$

but the thermal diffusivity of the solid can be taken as constant.

The boundary-value problem of heat conduction for the case of variable thermal conductivity is given as

$$\rho c_p \frac{\partial T}{\partial t} = \frac{\partial}{\partial x}\left(k \frac{\partial T}{\partial x}\right) + g(x, t) \quad \text{in} \quad 0 \le x \le L,\ t > 0 \tag{8-32}$$

$$\frac{\partial T}{\partial x} = 0 \quad \text{at} \quad x = 0,\ t > 0 \tag{8-33}$$

$$T = f(t) \quad \text{at} \quad x = L,\ t > 0 \tag{8-34}$$

$$T = F(x) \quad \text{in} \quad 0 \le x \le L,\ t = 0 \tag{8-35}$$

For thermal conductivity given in the form 8-31, the transformation 8-18 becomes

$$U = \frac{1}{k_0}\int_0^T k(T')\cdot dT' = \int_0^T (1 + \beta T')\,dT' = T + \frac{\beta}{2}T^2 \tag{8-36}$$

By means of the transformation 8-36 the differential equation of heat conduction, the boundary and initial conditions in the above system are transformed into the following system:

$$\frac{1}{\alpha}\frac{\partial U}{\partial t} = \frac{\partial^2 U}{\partial x^2} + \frac{g(x, t)}{k_0} \quad \text{in} \quad 0 \le x \le L,\ t > 0 \tag{8-37}$$

$$\frac{\partial U}{\partial x} = 0 \quad \text{at} \quad x = 0,\ t > 0 \tag{8-38}$$

$$U = \left[f(t) + \frac{\beta}{2} f^2(t) \right] \equiv \phi(t) \quad \text{at} \quad x = L,\, t = 0 \tag{8-39}$$

$$U = \left[F(x) + \frac{\beta}{2} F^2(x) \right] \equiv \psi(x) \quad \text{in} \quad 0 \leq x \leq L,\, t = 0 \tag{8-40}$$

Assuming the thermal diffusivity α can be taken as constant, the above system is a linear boundary value of heat conduction for a slab $0 \leq x \leq L$ for the temperature function $U(x, t)$, which is easily solved with the integral-transform technique discussed in Chapter 2.

When the function $U(x, t)$ is determined from the solution of the above system, transformation from $U(x, t)$ to $T(x, t)$ is made by means of the transformation 8-36—that is,

$$T(x, t) = \frac{1}{\beta} \left[\sqrt{1 + 2\beta U(x, t)} - 1 \right] \tag{8-36a}$$

8-3. TRANSFORMATION OF INDEPENDENT VARIABLE—USE OF BOLTZMANN TRANSFORMATION

A method in the solution of partial differential equations (linear or nonlinear) is the transformation of the independent variable in such a manner as to achieve a reduction in the number of independent variables. Later in this chapter a general method will be given for achieving such transformation; however, to illustrate the basic concepts involved and the restrictions to be imposed on the boundary and initial conditions during such transformation we first examine one simple case.

The nonlinear *partial differential* equation of heat conduction

$$\rho c_p \frac{\partial T}{\partial t} = \frac{\partial}{\partial x} \left(k \frac{\partial T}{\partial x} \right) \tag{8-41}$$

can be transformed into a nonlinear, second-order *ordinary differential* equation for the independent variable η, by means of the classical *Boltzmann transformation* [Ref. 1, p. 34]

$$\eta = \frac{x}{\sqrt{t}} \tag{8-42}$$

This transformation is useful provided that the boundary and initial conditions for the problem can also be transformed. The original differential Eq. 8-41 requires three auxiliary conditions (i.e., two boundary conditions and one initial condition) for its solution, but when it is transformed into a second order ordinary differential equation in η it requires only two auxiliary conditions on η. This reduction in the auxiliary conditions from three to two requires that two of the original auxiliary conditions should coalesce under the transformation. This condition imposes restriction on the choice of auxiliary conditions that can be used with the differential Eq. 8-41 if the transformation 8-42 be applicable.

We now examine the transformation of the differential equation 8-41 and the restrictions to be imposed on the auxiliary conditions under the transformation 8-42.

From Eq. 8-42 we write the following relations:

$$\frac{\partial \eta}{\partial x} = \frac{1}{t^{1/2}} \tag{8-43a}$$

$$\frac{\partial \eta}{\partial t} = -\frac{1}{2}\frac{x}{t^{3/2}} = -\frac{1}{2}\frac{\eta}{t} \tag{8-43b}$$

Various terms in the differential equation of heat conduction, Eq. 8-41, are written as

$$\frac{\partial T}{\partial x} = \frac{\partial \eta}{\partial x}\frac{dT}{d\eta} = \frac{1}{t^{1/2}}\frac{dT}{d\eta} \tag{8-44a}$$

$$\frac{\partial}{\partial x}\left(k\frac{\partial T}{\partial x}\right) = \frac{1}{t}\frac{d}{d\eta}\left[k\frac{dT}{d\eta}\right] \tag{8-44b}$$

$$\frac{\partial T}{\partial t} = \frac{\partial \eta}{\partial t}\frac{dT}{d\eta} = -\frac{1}{2}\frac{\eta}{t}\frac{dT}{d\eta} \tag{8-44c}$$

Substituting Eqs. 8-44b and 8-44c into the differential equation 8-41, we obtain the following nonlinear, second-order, ordinary differential equation:

$$\frac{d}{d\eta}\left(k\frac{dT}{d\eta}\right) + \frac{1}{2}\rho c_p \eta\frac{dT}{d\eta} = 0 \tag{8-45}$$

where $T \equiv T(\eta)$

$\eta \equiv \eta(x, t)$

The boundary and the initial conditions of the original problem should be such that under the transformation 8-42 two of these conditions should coalesce. These conditions will be satisfied if the range of the x variable extends to infinity, the boundary condition at $x = 0$ and the initial conditions are constants.

Consider the following nonlinear boundary-value problem of heat conduction:

$$\frac{\partial}{\partial x}\left(k\frac{\partial T}{\partial x}\right) = \rho c_p \frac{\partial T}{\partial t} \quad \text{in} \quad 0 \le x < \infty,\ t > 0 \tag{8-46a}$$

$$T = T_0 \quad \text{at} \quad x = 0,\ t > 0 \tag{8-46b}$$

$$T = T_i \quad \text{at} \quad x \to \infty,\ t \ge 0 \tag{8-46c}$$

$$T = T_i \quad \text{in} \quad 0 \le x < \infty,\ t = 0 \tag{8-46d}$$

where $T \equiv T(x, t)$.

Under the transformation 8-42 the system 8-46 transforms into

$$\frac{d}{d\eta}\left(k\frac{dT}{d\eta}\right) + \frac{1}{2}\rho c_p \eta\frac{dT}{d\eta} = 0 \quad \text{in} \quad 0 \le \eta < \infty \tag{8-47a}$$

$$T = T_0 \quad \text{at} \quad \eta = 0 \tag{8-47b}$$

$$T = T_i \quad \text{at} \quad \eta \to \infty \tag{8-47c}$$

where $T \equiv T(\eta)$.

It is apparent that the initial condition and the condition at infinity of the original problem 8-46 coalesced into one condition after the transformation. Several methods of solution of system 8-47 are discussed by Crank [7].

8-4. SEMI-INFINITE REGION WITH VARIABLE THERMAL CONDUCTIVITY

We now examine the solution of one-dimensional, time-dependent heat-conduction equation for a semi-infinite solid with variable thermal conductivity by combined application of Boltzmann transformation and iterative-solution technique. As in the previous chapter, we assume that initially the solid is at a constant temperature T_i and for times $t > 0$ the boundary surface at $x = 0$ is kept at a constant temperature T_0. The density and specific heat of the solid are taken as constant.

The boundary-value problem of heat conduction is

$$\rho c_p \frac{\partial T}{\partial t} = \frac{\partial}{\partial x}\left(k \frac{\partial T}{\partial x}\right) \quad \text{in} \quad 0 \le x < \infty,\ t > 0 \tag{8-48a}$$

$$T = T_0 \quad \text{at} \quad x = 0,\ t > 0 \tag{8-48b}$$

$$T = T_i \quad \text{at} \quad x \to \infty,\ t \ge 0 \tag{8-48c}$$

$$T = T_i \quad \text{at} \quad 0 \le x < \infty,\ t = 0 \tag{8-48d}$$

where $T \equiv T(x, t)$.

For convenience we define the following dimensionless quantities

$$\tilde{T} = \frac{T - T_i}{T_0 - T_i} \qquad \tilde{k} = \frac{k}{k_0} \tag{8-49}$$

where k_0 = value of thermal conductivity at a constant reference temperature. Then system 8-48 is written in the form

$$\frac{\partial \tilde{T}}{\partial \tau} = \frac{\partial}{\partial x}\left(\tilde{k} \frac{\partial \tilde{T}}{\partial x}\right) \quad \text{in} \quad 0 \le x < \infty,\ \tau > 0 \tag{8-50a}$$

$$\tilde{T} = 1 \quad \text{at} \quad x = 0,\ \tau > 0 \tag{8-50b}$$

$$\tilde{T} = 0 \quad \text{at} \quad x \to \infty,\ \tau \ge 0 \tag{8-50c}$$

$$\tilde{T} = 0 \quad \text{in} \quad 0 \le x < \infty,\ \tau = 0 \tag{8-50d}$$

where

$$\tau = \alpha_0 t \qquad \alpha_0 = \frac{k_0}{\rho c_p} \qquad \rho c_p = \text{const.}$$

Yang [8] examined the solution of system 8-50 by means of the method of successive approximations after reducing system 8-50 into a nonlinear, ordinary differential equation with Boltzmann transformation.

Consider the Boltzmann transformation

$$\eta = \frac{x}{\sqrt{\tau}} \tag{8-51}$$

Under the transformation 8-51, system 8-50 becomes

$$\tilde{k}\frac{d^2\tilde{T}}{d\eta^2} + \frac{d\tilde{k}}{d\tilde{T}}\left(\frac{d\tilde{T}}{d\eta}\right)^2 + \frac{1}{2}\eta\frac{d\tilde{T}}{d\eta} = 0 \quad \text{in} \quad 0 \le \eta < \infty \tag{8-52a}$$

$$\tilde{T} = 1 \quad \text{at} \quad \eta = 0 \tag{8-52b}$$

$$\tilde{T} = 0 \quad \text{at} \quad \eta \to \infty \tag{8-52c}$$

where $\tilde{T} \equiv \tilde{T}(\eta)$.

It is to be noted that system 8-52 is similar in form to system 8-47, but is in dimensionless form. System 8-52 can be solved numerically by the method of successive approximations.

We write differential equation 8-52a in the form

$$\frac{d^2\tilde{T}}{d\eta^2} = -f(\eta)\cdot\frac{d\tilde{T}}{d\eta} \tag{8-53}$$

where

$$f(\eta) = \left[\frac{1}{\tilde{k}}\frac{d\tilde{k}}{d\tilde{T}}\frac{d\tilde{T}}{d\eta} + \frac{1}{2k}\eta\right] \tag{8-54}$$

In order to integrate Eq. 8-53 we assume that $f(\eta)$ is a known function of η which is obtained from Eq. 8-54 by choosing a suitable initial guess for the temperature distribution, $\tilde{T}(\eta)$.

The first integration of Eq. 8-53 gives

$$\frac{d\tilde{T}}{d\eta} = c_0\cdot e^{-\int_0^{\eta} f(\eta')\cdot d\eta'} \tag{8-55}$$

where c_0 = the integration constant.

Integrating Eq. 8-55 from $\eta = 0$ to η, and by making use of the boundary condition $\tilde{T}|_{\eta=0} = 1$, we obtain

$$\tilde{T} - 1 = c_0\int_0^{\eta} e^{-\int_0^{\eta'} f(\eta'')\cdot d\eta''}\cdot d\eta' \tag{8-56}$$

The integration constant c_0 is evaluated by applying the second boundary condition $\tilde{T}|_{\eta=\infty} = 0$, and the result is

$$\tilde{T} = 1 - \frac{\int_0^{\eta} e^{-\int_0^{\eta'} f(\eta'')\cdot d\eta''}\cdot d\eta'}{\int_0^{\infty} e^{-\int_0^{\eta} f(\eta')\cdot d\eta'} d\eta} \tag{8-57a}$$

where

$$f(\eta) = \left[\frac{1}{\tilde{k}}\frac{d\tilde{k}}{d\tilde{T}}\frac{d\tilde{T}}{d\eta} + \frac{1}{2\tilde{k}}\eta\right] \tag{8-57b}$$

Equations 8-57a and 8-57b are written in the form of recursion formulas as

$$\tilde{T}_{n+1} = 1 - \frac{\displaystyle\int_0^{\eta} e^{-\int_0^{\eta'} f_n(\eta'')\cdot d\eta''} d\eta'}{\displaystyle\int_0^{\infty} e^{-\int_0^{\eta} f_n(\eta')\cdot d\eta'} d\eta} \tag{8-58a}$$

where

$$f_n(\eta) = \left[\frac{1}{\tilde{k}}\frac{d\tilde{k}}{d\tilde{T}}\frac{d\tilde{T}_n}{d\eta} + \frac{\eta}{2\tilde{k}}\right] \quad n = 0, 1, 2, 3, \ldots. \tag{8-58b}$$

Equations 8-58a and b are used to calculate the temperature distribution by the methods of successive approximations; the procedure is as follows. An initial guess, $\tilde{T}_0(\eta)$, is chosen for the temperature distribution and the function $f_0(\eta)$ is evaluated from Eq. 8-58b. Substituting this value of $f_0(\eta)$ into Eq. 8-58a, integrations are performed numerically for several mesh points in the range $0 \le \eta < \infty$ and a first approximation, $T_1(\eta)$, is obtained for the temperature distribution. Using this first approximation as the guess value the function $f_1(\eta)$ is evaluated from Eq. 8-58b and the second approximation for the temperature distribution, $T_2(\eta)$, is obtained from Eq. 8-58a. The second approximation is used to obtain a third approximation. The procedure is repeated until convergence is achieved.

To start the calculations an initial guess, $\tilde{T}_0(\eta)$, is needed for the temperature distribution and it can be taken as the solution of the problem 8-52 for the case of constant thermal conductivity. That is $\tilde{T}_0(\eta)$ function satisfies the following problem (i.e. $\tilde{k} = 1$)

$$\frac{d^2\tilde{T}_0}{d\eta^2} + \frac{1}{2}\eta\frac{d\tilde{T}_0}{d\eta} = 0 \quad \text{in} \quad 0 \le \eta < \infty \tag{8-59a}$$

$$\tilde{T}_0 = 1 \quad \text{at} \quad \eta = 0 \tag{8-59b}$$

$$\tilde{T}_0 = 0 \quad \text{at} \quad \eta \to \infty \tag{8-59c}$$

The solution of the system 8-59 is given as

$$\tilde{T}_0 = 1 - \operatorname{erf}\left(\frac{\eta}{2}\right) = 1 - \operatorname{erf}\left(\frac{x}{2\sqrt{\alpha_0 t}}\right) \tag{8-60}$$

Yang [8] calculated numerical examples by assuming a linear variation of thermal conductivity in the form

$$\tilde{k} = 1 + \beta\tilde{T} \tag{8-61}$$

Figure 8-3 shows the results of his calculations for $\beta = -0.5, 0, 0.5$. The solution for $\beta = 0$ corresponds to the case of constant thermal conductivity.

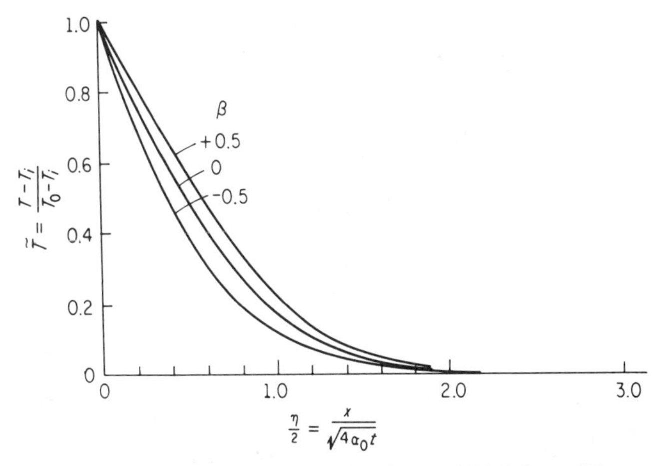

FIG. 8-3. Temperature distribution in a semi-infinite solid $0 \le x < \infty$ with variable thermal conductivity. (Kwang-Tzu Yang, *Trans. ASME, J. Appl. Mech.*, Vol. 79, 1958, p. 146.)

8-5. TRANSFORMATION OF INDEPENDENT VARIABLE USING SIMILARITY VIA ONE-PARAMETER GROUP-THEORY METHOD

Difficulties associated with the solution of nonlinear partial differential equations have led several authors to find suitable transformations that will reduce the number of independent variables in a partial differential equation. The Boltzmann transformation examined in the previous sections is one example of such transformations and one may find numerous other examples in the solution of boundary-layer problems in fluid mechanics. The method of solution of partial differential equations by the transformation of independent variables in such a manner as to achieve a reduction in the number of independent variables is called *similarity solution* of partial differential equations. The transformation variables that will achieve a reduction in the number of independent variables are called *similarity variables*.

There are several methods for developing similarity variables; they are discussed in detail in the texts by Hansen [9] and Ames [1]. The most fundamental and yet the most simple and straightforward of these methods is the similarity via "one-parameter group-theory" method which is based upon the general theory by Morgan [10]. Here we present the one-parameter group-theory method.

Let Σ be a system of partial differential equations given as

$$\Sigma : \phi_j(x_i, y_j) = 0 \qquad i = 1, 2, \ldots, m \quad j = 1, 2, \ldots, n \tag{8-62}$$

where x_i and y_j are respectively the independent and dependent variables.

Consider a one-parameter transformation group G_1 consisting of a set of transformations for the independent and dependent variables x_i and y_j, defined in the form

$$G_1 : \tilde{x}_i = a^{\beta_i} x_i \qquad i = 1, 2, \ldots, m$$
$$\tilde{y}_j = a^{\gamma_j} y_j \qquad j = 1, 2, \ldots, n \tag{8-63}$$

where the parameter a is a *nonzero* real number and the exponents β_i, γ_j are to be determined from the condition that the system of partial differential Eq. 8-62 remain *absolutely invariant* under the transformation G_1 given by Eq. 8-63. The system 8-62 is said *absolutely invariant* under the transformation G_1 if we have

$$\phi_j(x_i, y_j) = \phi_j(\tilde{x}_i, \tilde{y}_j) \tag{8-64}$$

When the transformation 8-63 is applied to the system of partial differential equations 8-62 and the condition is imposed that the system 8-62 should remain absolutely invariant under the transformation, one obtains a set of simultaneous algebraic equations for determining the unknown exponents β_i and γ_i.

Suppose the independent variable x_1 is to be eliminated from the system 8-62; then, the similarity variables that will achieve the removal of the independent variable x_1 when they are substituted in the system 8-62 are given as [1, p. 137].

$$\eta_i = \frac{x_i}{x_1^{\beta_i/\beta_1}} \qquad i = 2, 3, \ldots, m \tag{8-65a}$$

$$g_j(\eta_2, \eta_3, \ldots, \eta_m) = \frac{y_j(x_1, x_2, \ldots, x_m)}{x_1^{\gamma_j/\beta_1}} \qquad j = 1, 2, \ldots, n \tag{8-65b}$$

where $\beta_1 \neq 0$.

If the exponent β_1 is zero and yet we wish to eliminate the independent variable x_1, then we may choose the one-parameter group transformation for x_1 variable in the form

$$\tilde{x}_1 = x_1 + \ln a \quad \text{for} \quad \beta_1 = 0 \tag{8-66}$$

and transformations for the remaining variables are chosen the same as before, i.e.,

$$\tilde{x}_i = a^{\beta_i} x_i \qquad i = 2, 3, \ldots, m$$
$$\tilde{y}_j = a^{\gamma_j} y_j \qquad j = 1, 2, \ldots, n$$

Then the corresponding similarity variables are

$$\eta_i = \frac{x_i}{e^{\beta_i x_1}} \qquad i = 2, 3, \ldots, m \tag{8-67a}$$

$$g_j(\eta_2, \ldots, \eta_m) = \frac{y_j(x_1, \ldots, x_m)}{e^{\gamma_j x_1}} \qquad j = 1, 2, \ldots, n \tag{8-67b}$$

When the similarity variables given by Eq. 8-65 (or 8-67) are substituted into the system of partial differential equations 8-62 and differential operations associated with the system are performed, one obtains a new system of partial differential equations for the dependent variables $g_1, g_2, \ldots, g_n$ and the independent variables $\eta_2, \eta_3, \ldots, \eta_m$. In the new system, the number of independent variables is $(m - 1)$ which is one less than that in the original system 8-62.

This procedure can be repeated to achieve further reductions in the number of independent variables until a system of ordinary differential equations is obtained.

In the above analysis the transformation of the partial differential equation is performed with no concern about the transformation of the boundary and initial conditions associated with the problem. It is only after the similarity variables are established that one examines the transformation of the boundary and initial conditions; and it is usually where the difficulty starts. Any given boundary and initial condition will not always transform under the similarity variables established for the transformation of the differential equation. To achieve the transformation of the boundary and initial conditions, they should be such that some coalesce under the similarity transformation. Usually similarity in the independent variable which extends to infinity in one direction favors the transformation of the boundary and initial conditions.

In the next section we examine the application of similarity transformation via one-parameter group-transformation method to the solution of nonlinear partial differential equation of heat conduction.

8-6. SIMILARITY SOLUTION OF ONE-DIMENSIONAL, TIME-DEPENDENT HEAT CONDUCTION EQUATION WITH VARIABLE THERMAL CONDUCTIVITY AND SPECIFIC HEAT

Ames [11] examined the similarity solution via the one-parameter group-transformation of the one-dimensional, time-dependent differential equation of heat conduction with variable thermal conductivity and specific heat.

Consider the nonlinear, symmetrical differential equation of heat conduction in the form

$$\frac{1}{r^{m-1}} \frac{\partial}{\partial r} \left[r^{m-1} k \frac{\partial T}{\partial r} \right] = \rho c_p \frac{\partial T}{\partial t} \tag{8-68}$$

where $m = 1, 2, 3$ for the rectangular, cylindrical, and spherical coordinate systems respectively. The thermal conductivity and specific heat are functions of temperature, but variation of density with temperature is negligible and it can be taken as constant.

We assume that both thermal conductivity and specific heat are directly proportional to powers of temperature and taken in the form

$$k = k_0 \left(\frac{T}{T_0}\right)^s \tag{8-69a}$$

$$c_p = c_{p0} \left(\frac{T}{T_0}\right)^p \tag{8-69b}$$

where k_0 and c_{p0} are the thermal conductivity and specific heat at a reference temperature T_0 and the exponents s and p are constants.

Substituting Eq. 8-69 into Eq. 8-68 we obtain

$$\frac{1}{r^{m-1}} \frac{\partial}{\partial r}\left[r^{m-1}\left(\frac{T}{T_0}\right)^s \frac{\partial T}{\partial r}\right] = \frac{1}{\alpha_0}\left(\frac{T}{T_0}\right)^p \frac{\partial T}{\partial t} \tag{8-70}$$

where

$$\alpha_0 = \frac{k_0}{\rho c_{p0}} = \text{const.}$$

We define a new dependent variable U in the form

$$\frac{T}{T_0} = U^{\frac{1}{p+1}} \tag{8-71}$$

Then Eq. 8-70 becomes

$$\frac{1}{r^{m-1}} \frac{\partial}{\partial r}\left[r^{m-1} \cdot U^{\frac{s-p}{p+1}} \frac{\partial U}{\partial r}\right] = \frac{1}{\alpha_0} \frac{\partial U}{\partial t} \tag{8-72}$$

Defining,

$$\frac{s-p}{p+1} \equiv n > 0 \tag{8-73}$$

Equation 8-72 is written as

$$\frac{1}{r^{m-1}} \frac{\partial}{\partial r}\left[r^{m-1} \cdot U^n \frac{\partial U}{\partial r}\right] = \frac{1}{\alpha_0} \frac{\partial U}{\partial t} \tag{8-74}$$

We now examine the similarity solution of the nonlinear differential equation of heat conduction 8-74 in a semi-infinite region $0 \le r < \infty$ by following Ames [11]. We assume that the region is initially at zero temperature and for times $t > 0$ the boundary surface at $r = 0$ is kept at a prescribed temperature which may be a function of time.

The nonlinear boundary-value problem of heat conduction is given in the form

$$\frac{1}{r^{m-1}} \frac{\partial}{\partial r}\left[r^{m-1} U^n \frac{\partial U}{\partial r}\right] = \frac{\partial U}{\partial \tau} \quad \text{in} \quad 0 \le r < \infty,\ \tau > 0 \tag{8-75}$$

$$U = f(\tau) \quad \text{at} \quad r = 0,\ \tau > 0 \tag{8-76a}$$

$$U = 0 \quad \text{at} \quad r \to \infty,\ \tau \ge 0 \tag{8-76b}$$

$$U = 0 \quad \text{in} \quad 0 \le r < \infty,\ \tau = 0 \tag{8-76c}$$

where $\tau = \alpha_0 t$

$m = 1, 2, 3$ for rectangular, cylindrical, and spherical geometries

Consider the one-parameter group-transformation in the form

$$\tilde{r} = a^b r \qquad \tilde{\tau} = a^d \tau \qquad \tilde{U} = a^e U \tag{8-77}$$

where the parameter a is a nonzero, real number and the exponents b, d, e are real numbers to be determined.

We substitute the new variables 8-77 into the differential equation 8-75,

$$a^{2b-(n+1)e} \cdot \frac{1}{\tilde{r}^{m-1}} \frac{\partial}{\partial \tilde{r}} \left(\tilde{r}^{m-1} \tilde{U}^n \frac{\partial \tilde{U}}{\partial \tilde{r}} \right) = a^{d-e} \cdot \frac{\partial \tilde{U}}{\partial \tilde{\tau}} \tag{8-78}$$

The differential equation 8-75 remains *absolutely invariant* under the transformation 8-77 provided

$$2b - (n+1)e = d - e$$

or

$$2b - ne = d \tag{8-79}$$

The partial differential equation 8-75 involves two independent variables, τ and r, both of which extend to infinity; and one may seek *similarity* either in the τ variable (i.e., τ variable is to be eliminated) or in the r variable (i.e., r variable is to be eliminated). To illustrate the procedure we examine similarity first in the τ variable and then in the r variable.

Similarity in τ. Since the independent variable τ is to be eliminated we divide both sides of Eq. 8-79 by the exponent d and assume that $d \neq 0$:

$$2\left(\frac{b}{d}\right) - n\left(\frac{e}{d}\right) = 1 \tag{8-80}$$

For convenience we define

$$\frac{b}{d} \equiv A \tag{8-81a}$$

Then

$$\frac{e}{d} = \frac{2A - 1}{n} \tag{8-81b}$$

The similarity variables η_i and g_j are now determined according to Eq. 8-65:

$$\eta = \frac{r}{\tau^{b/d}} = \frac{r}{\tau^A} \tag{8-82a}$$

$$g(\eta) = \frac{U(r,\tau)}{\tau^{e/d}} = \frac{U(r,\tau)}{\tau^{\frac{2A-1}{n}}} \tag{8-82b}$$

where A is an arbitrary constant (i.e., degree of freedom) and will be selected later in the analysis.

When the similarity variables 8-82 are substituted into the differential equation of heat conduction 8-75 we obtain the following ordinary differential equation:

$$\frac{d}{d\eta}\left[\eta^{m-1} \cdot g^n \cdot \frac{dg}{d\eta}\right] = \frac{2A - 1}{n} g\eta^{m-1} - A\eta^m \frac{dg}{d\eta} \tag{8-83}$$

where $g \equiv g(\eta)$.

For the right-hand side of Eq. 8-83 to be of the form

$$-\frac{d}{d\eta}(B\eta^m \cdot g) \tag{8-84}$$

requires that we choose

$$\begin{aligned} B &= A \\ -Bm &= \frac{2A - 1}{n} \end{aligned} \tag{8-85}$$

From Eq. 8-85 we find

$$B = A = \frac{1}{2 + mn} \tag{8-86}$$

Then Eq. 8-83 is written in the form

$$\frac{d}{d\eta}\left[\eta^{m-1} \cdot g^n \cdot \frac{dg}{d\eta}\right] + \frac{1}{2 + mn}\frac{d}{d\eta}(\eta^m \cdot g) = 0 \tag{8-87}$$

Thus we reduced the partial differential equation 8-75 into a second-order ordinary differential equation.

We now examine the transformation of the boundary and initial conditions in Eq. 8-76 by means of the similarity variables in Eq. 8-82 for $A = 1/(2 + mn)$, that is

$$\eta = \frac{r}{\tau^{\frac{1}{2+mn}}} \tag{8-88a}$$

$$g(\eta) = U(r, \tau) \cdot \tau^{\frac{m}{2+mn}} \tag{8-88b}$$

and the boundary condition for Eq. 8-87 becomes

$$g(\eta) = \tau^{\frac{m}{2+mn}} \cdot f(\tau) \quad \text{for} \quad \eta = 0 \tag{8-89a}$$

$$g(\eta) = 0 \quad \text{for} \quad \eta = \infty \tag{8-89b}$$

It is to be noted that the boundary condition 8-76b (i.e., at $r \to \infty$) and the initial condition 8-76c are converged into a single boundary condition 8-89b. However, the transformed boundary condition 8-89a should be independent of the independent variable τ. To achieve this condition we choose the boundary-condition function $f(\tau)$ in the form

$$f(\tau) = \tau^{-\frac{m}{2+mn}} \tag{8-90}$$

With this restriction on the choice of the boundary condition function, $f(\tau)$, the boundary conditions 8-89 become

$$g(\eta) = 1 \quad \text{for} \quad \eta = 0 \tag{8-91a}$$

$$g(\eta) = 0 \quad \text{for} \quad \eta = \infty \tag{8-91b}$$

Summarizing the above procedure, the partial differential equation of heat conduction 8-75 and its boundary and initial conditions 8-76 are reduced to a second-order, ordinary differential equation 8-87 subject to the boundary conditions 8-91 by means of the similarity variables 8-88. In order to achieve the transformation of the boundary conditions we choose the boundary-condition function $f(\tau)$ must be in the form as given by Eq. 8-90.

It is to be noted that the special case of $A = \frac{1}{2}$ the similarity variables 8-82 reduces to

$$\eta = \frac{r}{\sqrt{\tau}} \tag{8-92a}$$

$$g(\eta) = U(r, \tau) \tag{8-92b}$$

which is the Boltzmann similarity variable considered previously.

Similarity in *r*. We now examine the similarity in the r variable. We divide both sides of Eq. 8-79 by the exponent b and assume that $b \neq 0$.

$$2 - n\frac{e}{b} = \frac{d}{b} \tag{8-93}$$

For convenience we define

$$\frac{d}{b} \equiv A \tag{8-94a}$$

then

$$\frac{e}{b} = \frac{2 - A}{n} \tag{8-94b}$$

The similarity variables η_i and g_j are determined by means of the relations 8-65, as

$$\eta = \frac{\tau}{r^{d/b}} = \frac{\tau}{r^A} \tag{8-95a}$$

$$g(\eta) = \frac{U(r, \tau)}{r^{e/b}} = \frac{U(r, \tau)}{r^{\frac{2-A}{n}}} \tag{8-95b}$$

Substituting these similarity variables into the differential equation of heat conduction 8-75, we obtain the following ordinary differential equation

$$g^n \left[A^2\eta^2 \frac{d^2g}{d\eta^2} + A\eta(1 + A - 2\beta)\frac{dg}{d\eta} + \beta(\beta - 1)g\right]$$
$$+ ng^{n-1}\left(\beta g - A\eta\frac{dg}{d\eta}\right)^2 + (m - 1)\left(\beta g - A\eta\frac{dg}{d\eta}\right) = \frac{dg}{d\eta} \tag{8-96}$$

where

$$g \equiv g(\eta)$$

$$\beta = \frac{2 - A}{n}$$

If we choose $A = 2$, we have $\beta = 0$ and Eq. 8-96 simplifies to

$$\frac{d}{d\eta}\left(4\eta^2 g^n \frac{dg}{d\eta}\right) - 2\eta g^n \frac{dg}{d\eta} - [2(m - 1)\eta + 1]\frac{dg}{d\eta} = 0 \tag{8-97}$$

Then the similarity variables 8-95 becomes

$$\eta = \frac{\tau}{r^2} \tag{8-98a}$$

$$g(\eta) = U(r, \tau) \tag{8-98b}$$

The boundary conditions 8-76 will transform by means of the similarity variables 8-98 if we choose $f(\tau) = 1$. The resulting transformed boundary conditions are

$$g(\eta) = 1 \quad \text{for} \quad \eta = \infty \tag{8-99a}$$

$$g(\eta) = 0 \quad \text{for} \quad \eta = 0 \tag{8-99b}$$

and in this case the boundary condition 8-76b and the initial condition 8-76c coalesced after the transformation.

The resulting ordinary differential equations obtained above are easier to solve with a digital computer than the original nonlinear partial differential equation.

8-7. SIMILARITY TRANSFORMATION OF TWO-DIMENSIONAL, TIME-DEPENDENT HEAT-CONDUCTION EQUATION WITH VARIABLE THERMAL CONDUCTIVITY AND SPECIFIC HEAT

Ames [1, 11] also examined similarity transformation via one-parameter group-transformation of two-dimensional, time-dependent differential equation of heat conduction with variable thermal conductivity and specific heat.

Consider a two-dimensional differential equation of heat conduction in the form

$$\frac{\partial}{\partial x}\left[k\frac{\partial T}{\partial x}\right] + \frac{\partial}{\partial y}\left[k\frac{\partial T}{\partial y}\right] = \rho c_p \frac{\partial T}{\partial t} \tag{8-100}$$

The thermal conductivity and specific heat are assumed to be functions of temperature, but density is taken as constant. As in the previous case we assume that both thermal conductivity and specific heat are taken in the form

$$k = k_0\left(\frac{T}{T_0}\right)^s \tag{8-101a}$$

$$c_p = c_{p0}\left(\frac{T}{T_0}\right)^p \tag{8-101b}$$

where k_0 and c_{p0} are measured at a reference temperature T_0; the exponents s and p are constants.

Substituting Eq. 8-101 into Eq. 8-100,

$$\frac{\partial}{\partial x}\left[\left(\frac{T}{T_0}\right)^s\frac{\partial T}{\partial x}\right] + \frac{\partial}{\partial y}\left[\left(\frac{T}{T_0}\right)^s\frac{\partial T}{\partial y}\right] = \frac{1}{\alpha_0}\left(\frac{T}{T_0}\right)^p\frac{\partial T}{\partial t} \tag{8-102}$$

where

$$\alpha_0 = \frac{k_0}{\rho c_{p0}} = \text{const.}$$

Defining new variables,

$$\frac{T}{T_0} = U^{1/p+1} \tag{8-103a}$$

$$\alpha_0 t = \tau \tag{8-103b}$$

$$\frac{s-p}{p+1} = n > 0 \tag{8-103c}$$

Equation 8-102 becomes

$$\frac{\partial}{\partial x}\left(U^n \frac{\partial U}{\partial x}\right) + \frac{\partial}{\partial y}\left(U^n \frac{\partial U}{\partial y}\right) = \frac{\partial U}{\partial \tau} \tag{8-104}$$

This partial differential equation may be reduced to an ordinary differential equation by successive application of the one-dimensional group-transformation. Transformation will be performed in two steps.

Step 1. We shall seek similarity in τ, that is we remove from the differential equation 8-104 the τ variable. We define the following one-parameter group-transformation for the τ, x, y and U variables.

$$\bar{x} = a^{c_1}x \qquad \bar{y} = a^{c_2}y \qquad \bar{\tau} = a^{c_3}\tau \qquad \overline{U} = a^{c_4}U \tag{8-105}$$

Substituting Eq. 8-105 into differential equation 8-104,

$$a^{2c_1-(n+1)c_4}\frac{\partial}{\partial \bar{x}}\left(\overline{U}^n \frac{\partial \overline{U}}{\partial \bar{x}}\right) + a^{2c_2-(n+1)c_4}\frac{\partial}{\partial \bar{y}}\left(\overline{U}^n \frac{\partial \overline{U}}{\partial \bar{y}}\right) = a^{c_3-c_4}\frac{\partial \overline{U}}{\partial \tau} \tag{8-106}$$

For Eq. 8-104 to remain absolutely invariant under the transformation, the exponents in Eq. 8-106 are related by

$$2c_1 - (n+1)c_4 = 2c_2 - (n+1)c_4 = c_3 - c_4 \tag{8-107a}$$

or in the form

$$2c_1 - nc_4 = 2c_2 - nc_4 = c_3 \tag{8-107b}$$

Since we seek similarity in τ, we divide both sides of Eq. 8-107b by the exponent c_3, and for $c_3 \neq 0$ we obtain

$$2\frac{c_1}{c_3} - n\frac{c_4}{c_3} = 2\frac{c_2}{c_3} - n\frac{c_4}{c_3} = 1 \tag{8-108}$$

For convenience, if we choose

$$\frac{c_1}{c_3} = \frac{c_2}{c_3} = A \tag{8-109a}$$

then we find

$$\frac{c_4}{c_3} = \frac{2A-1}{n} \tag{8-109b}$$

where A is an arbitrary constant.

The resulting similarity variables are determined by means of the relation 8-65, and are in the form

$$\eta_1 = \frac{x}{\tau^{c_1/c_3}} = \frac{x}{\tau^A} \qquad \text{(8-110a)}$$

$$\eta_2 = \frac{y}{\tau^{c_2/c_3}} = \frac{y}{\tau^A} \qquad \text{(8-110b)}$$

$$g(\eta_1, \eta_2) = \frac{U(x, y, \tau)}{\tau^{c_4/c_3}} = \frac{U(x, y, \tau)}{\tau^{2A-1/n}} \qquad \text{(8-110c)}$$

Substituting the similarity variables 8-110 into the partial differential equation of heat conduction 8-104, we obtain the following partial differential equation for $g(\eta_1, \eta_2)$.

$$\frac{\partial}{\partial \eta_1}\left[g^n \frac{\partial g}{\partial \eta_1}\right] + \frac{\partial}{\partial \eta_2}\left[g^n \frac{\partial g}{\partial \eta_2}\right] + A\left[\eta_1 \frac{\partial g}{\partial \eta_1} + \eta_2 \frac{\partial g}{\partial \eta_2}\right] - \frac{2A - 1}{n} g = 0 \qquad \text{(8-111)}$$

where $g \equiv g(\eta_1, \eta_2)$. Equation 8-111 is put into the symmetric form by choosing

$$A = \frac{1}{2n + 2} \qquad \text{(8-112)}$$

and we obtain

$$\frac{\partial}{\partial \eta_1}\left(\frac{\partial g^{n+1}}{\partial \eta_1} + \frac{g\eta_1}{2}\right) + \frac{\partial}{\partial \eta_2}\left(\frac{\partial g^{n+1}}{\partial \eta_2} + \frac{g\eta_2}{2}\right) = 0 \qquad \text{(8-113)}$$

where $g \equiv g(\eta_1, \eta_2)$.

Step 2. In Eq. 8-113 further reduction in the number of independent variables is possible by the repeated application of one-parameter group-transformation. We shall now seek similarity in η_1. We choose a new transformation in the form

$$\bar{\eta}_1 = b^{e_1}\eta_1 \qquad \bar{\eta}_2 = b^{e_2}\eta_2 \qquad \bar{g} = b^{e_3}g \qquad \text{(8-114)}$$

Applying the transformation 8-114 into the differential equation 8-113 and by imposing the condition that the differential equation remain absolutely invariant under the transformation, we obtain the following relations among the exponents:

$$2e_1 - (n + 1)e_3 = -e_3 = 2e_2 - (n + 1)e_3 \qquad \text{(8-115)}$$

As we seek similarity in η_1, both sides of Eq. 8-115 are divided by e_1 and for $e_1 \neq 0$ we obtain

$$2 - (n + 1)\frac{e_3}{e_1} = -\frac{e_3}{e_1} = 2\frac{e_2}{e_1} - (n + 1)\frac{e_3}{e_1} \qquad \text{(8-116)}$$

solution of which yields

$$\frac{e_3}{e_1} = \frac{2}{n} \qquad \frac{e_2}{e_1} = 1 \qquad \text{(8-117)}$$

The similarity variables are determined by means of Eq. 8-65, and are given as

$$\xi = \frac{\eta_2}{\eta_1^{e_2/e_1}} = \frac{\eta_2}{\eta_1} \tag{8-118a}$$

$$H(\xi) = \frac{g(\eta_1, \eta_2)}{\eta_1^{e_3/e_1}} = \frac{g(\eta_1, \eta_2)}{\eta_1^{2/n}} \tag{8-118b}$$

Substituting the similarity variables 8-118 into the partial differential equation 8-113 we obtain the following ordinary differential equation:

$$H^{n-1}\left[(1 + \xi^2)\frac{d^2H}{d\xi^2} + 2\xi\left(1 - \frac{2}{n}\right)\frac{dH}{d\xi} + \frac{2}{n}\left(\frac{2}{n} - 1\right)H\right]$$
$$+ nH^{n-2}\left[\left(\frac{dH}{d\xi}\right)^2 + \left(\frac{2H}{n} - \xi\frac{dH}{d\xi}\right)^2\right] + \frac{1}{n} = 0 \tag{8-119}$$

where $H \equiv H(\xi)$.

Summarizing, we removed from the partial differential equation of heat conduction 8-104 two of the independent variables and reduced it to the ordinary differential equation 8-119. Combining the transformation we obtain the following similarity variables:

$$\xi = \frac{\eta_2}{\eta_1} = \frac{y/\tau^A}{x/\tau^A} = \frac{y}{x} \tag{8-120a}$$

$$H(\xi) = \frac{g(\eta_1, \eta_2)}{\eta_1^{2/n}} = \frac{U(x, y, \tau)/\tau^{2A-1/n}}{\eta_1^{2/n}/\tau^{2A/n}} = \frac{U(x, y, \tau)}{x^{2/n}} \cdot \tau^{1/n} \tag{8-120b}$$

Equation 8-119 is a second-order ordinary differential equation which requires two boundary conditions for solution, therefore the auxiliary condition for the original differential equation 8-104 should be such that they should consolidate into two conditions under the transformation 8-120.

8-8. SIMILARITY SOLUTION OF MELTING OF A SLAB INITIALLY AT FUSION TEMPERATURE

Consider a solid slab $0 \leq x \leq 1$ which is initially at the melting temperature T_m. For times $t > 0$ the boundary surface $x = b$ is kept insulated while a constant temperature T_0 higher than the melting temperature is applied at the boundary $x = 0$. The solid will melt as a result of this applied temperature and the liquid–solid interface will move in the positive x direction. Figure 8-4 shows the location $s(t)$ of the liquid–solid interface at any instant of time t and the boundary conditions for the problem. There is no temperature gradient in the solid phase because it is assumed at the melting temperature. Assuming no natural convection in the liquid phase, heat transfer from the hot boundary surface

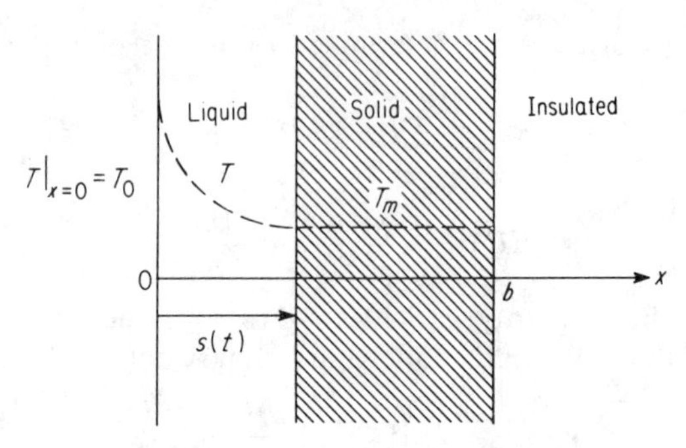

FIG. 8-4. The boundary conditions and the location of solid-liquid interface for melting of a slab $0 \le x \le b$, initially at fusion temperature.

$x = 0$ to the liquid–solid interface takes place through the liquid layer by pure conduction. The boundary-value problem of heat conduction for the liquid phase is given as

$$\frac{\partial^2 T}{\partial x^2} = \frac{1}{\alpha}\frac{\partial T}{\partial t} \quad \text{in} \quad 0 \le x \le s(t),\ t > 0 \tag{8-121}$$

with boundary conditions

$$T = T_0 \quad \text{at} \quad x = 0,\ t > 0 \tag{8-122a}$$

$$T = T_m \quad \text{at} \quad x = s(t),\ t > 0 \tag{8-122b}$$

$$-k\frac{\partial T}{\partial x} = \rho L \frac{ds(t)}{dt} \quad \text{at} \quad x = s(t),\ t > 0 \tag{8-122c}$$ [2]

where $s(t)$ = location of the liquid–solid interface
L = latent heat of fusion

Ruoff [12] obtained the solution of this melting problem by means of a simple similarity transformation. Defining the similarity variable as

$$\eta = \frac{x}{2(\alpha t)^{1/2}} \tag{8-123}$$

where

$$\eta = \eta(x, t)$$

the differential equation of heat conduction 8-121 is transformed as

$$\frac{d^2 T}{d\eta^2} + 2\eta\frac{dT}{d\eta} = 0 \quad \text{in} \quad 0 \le \eta \le \eta_s(t) \tag{8-124}$$

[2] This boundary condition is obtained from the general boundary condition 7-103

$$\left[k_s \frac{\partial T_s}{\partial x} - k_l \frac{\partial T_l}{\partial x}\right]_{x=s(t)} = \rho L \frac{ds(t)}{dt}$$

by setting $\partial T_s/\partial x = 0$.

and the boundary conditions are transformed as

$$T\,\big|_{\eta=0} = T_0 \tag{8-125a}$$

$$T\,\big|_{\eta=\eta_s(t)} = T_m \tag{8-125b}$$

$$-\left.\frac{dT}{d\eta}\right|_{\eta=\eta_s(t)} = 2A\left[\eta_s(t) + 2t\,\frac{d\eta_s(t)}{dt}\right] \tag{8-125c}$$

where

$$A = \frac{L}{c_p}$$

Differential Eq. 8-124 is integrated once to yield

$$\frac{dT}{d\eta} = \beta e^{-\eta^2} \tag{8-126}$$

where β = the first integration constant. The second integration of Eq. 8-126 from $\eta = 0$ to $\eta = \eta_s(t)$ gives

$$\int_0^{\eta_s(t)} dT = \beta \int_0^{\eta_s(t)} e^{-\eta^2} d\eta + c \tag{8-127}$$

Applying the boundary conditions 8-125a and 8-125b, Eq. 8-127 gives

$$T_m - T_0 = \beta \int_0^{\eta_s(t)} e^{-\eta^2}\, d\eta \tag{8-128}$$

The upper limit of integration should be a constant since the left-hand side of Eq. 8-128 is constant. Hence we take

$$\eta_s(t) = \gamma = \text{const.} \tag{8-129}$$

In view of the relation 8-129 the boundary condition at the liquid–solid interface, Eq. 8-125c becomes

$$-\left.\frac{dT}{d\eta}\right|_{\eta=\gamma} = 2A\gamma \tag{8-130}$$

and from Eq. 8-126 for $\eta_s(t) = \gamma$ we have

$$\left.\frac{dT}{d\eta}\right|_{\eta=\gamma} = \beta e^{-\gamma^2} \tag{8-131}$$

From Eqs. 8-130 and 8-131,

$$\beta e^{-\gamma^2} = -2A\gamma \tag{8-132}$$

For $\eta_s(t) = \gamma$ Eq. 8-128 becomes

$$T_m - T_0 = \beta \int_0^{\gamma} e^{-\eta^2}\, d\eta \tag{8-133}$$

Equations 8-132 and 8-133 are two relations for determining the two unknown coefficients β and γ. For example, eliminating β between Eqs. 8-132 and 8-133

$$T_m - T_0 = -2A\gamma\, e^{\gamma^2} \int_0^{\gamma} e^{-\eta^2}\, d\eta$$

or in the form

$$\gamma e^{\gamma^2} \operatorname{erf}(\gamma) = \frac{T_0 - T_m}{A\sqrt{\pi}} \tag{8-134}$$

when γ is evaluated from Eq. 8-134 the unknown coefficient β is determined from Eq. 8-132.

Knowing β, we can evaluate the temperature distribution in the liquid phase from

$$T - T_0 = \beta \int_0^{\eta} e^{-\eta'^2} \cdot d\eta' \quad \text{subject to} \quad 0 \le \eta \le \gamma \tag{8-135}$$

The location of the liquid–solid interface is obtained from Eq. 8-123 by setting $\eta = \gamma$:

$$s(t) = 2\gamma(\alpha t)^{1/2} \tag{8-136}$$

8-9. CHARTS FOR NONLINEAR, TRANSIENT HEAT-CONDUCTION PROBLEMS

Zerkle and Sunderland [13] prepared temperature-time charts for a slab of thickness L, insulated on one face, and subjected to the fourth-power thermal radiation at the other face. The slab is assumed to be homogeneous, isotropic, with constant thermal properties and initially at uniform temperature T_i. The environment temperature T_e is taken constant. The boundary-value problem of heat conduction for the problem considered is given in the form

$$\frac{\partial^2 T}{\partial x^2} = \frac{1}{\alpha}\frac{\partial T}{\partial t} \quad \text{in} \quad 0 \le x \le L, t > 0 \tag{8-137a}$$

$$\frac{\partial T}{\partial x} = 0 \quad \text{at} \quad x = 0, t > 0 \tag{8-137b}$$

$$-k\frac{\partial T}{\partial x} = \sigma \mathfrak{F}_{se}[T^4 - T_e^4] \quad \text{at} \quad x = L, t > 0 \tag{8-137c}$$

$$T = T_i \quad \text{in} \quad 0 \le x \le L, t = 0 \tag{8-137d}$$

where $\mathfrak{F}_{se}$ = radiation interchange factor between slab and environment

σ = Stefan-Boltzmann constant, 0.174×10^{-8} Btu/hr ft$^{2\circ}$R^4

The nonlinear heat-conduction problem described by system 8-137 is solved by Zerkle and Sunderland [13] by means of a thermal-electrical analog computor for both heating and cooling situations. Figures 8-5 and 8-6 show the charts by Zerkle and Sunderland for the radiant *heating* of the slab for the values of T_i/T_e equal to 0.0 and 0.5 respectively. In each of these charts dimensionless temperature $(T - T_i)/(T_e - T_i)$ is plotted as a function of dimensionless time $\alpha t/L^2$ for several different values of the parameter $k/\sigma \mathfrak{F}_{se} T_e^3 L$ at three different positions in the slab.

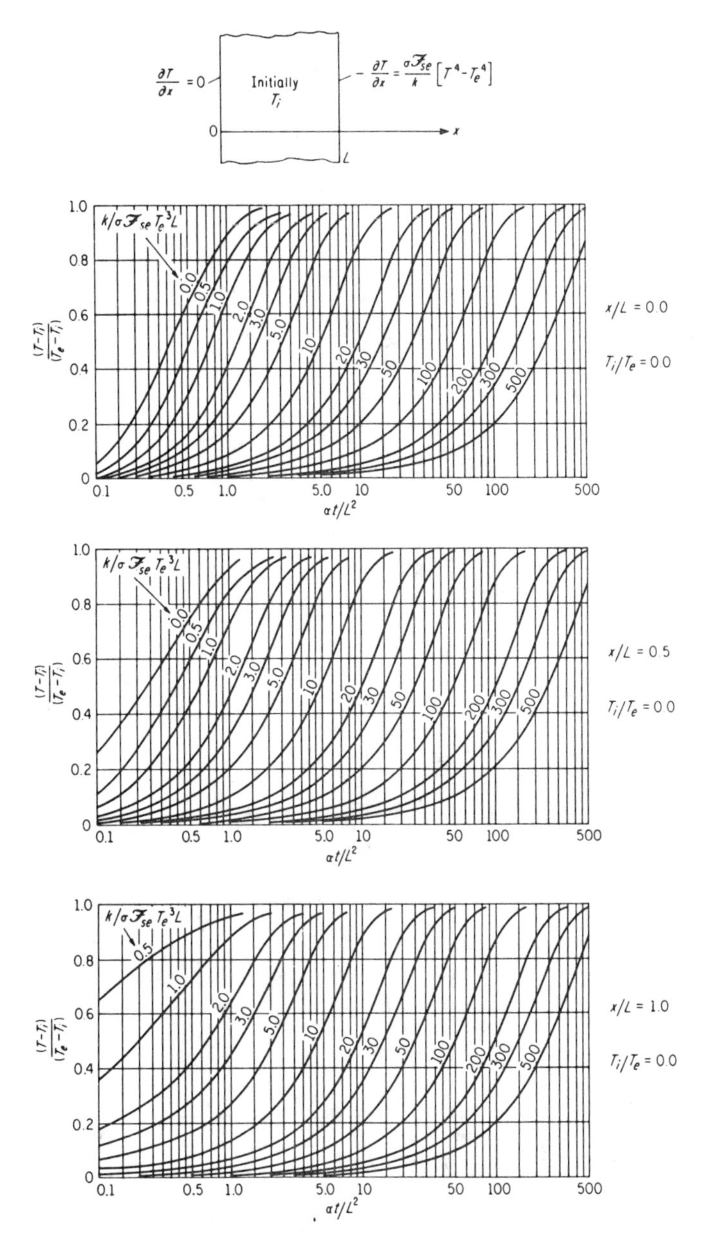

FIG. 8-5. Transient temperature distribution in a slab subjected to radiant heating when $\frac{T_i}{T_e} = 0.0$. (From Zerkle and Sunderland, Ref. 13.)

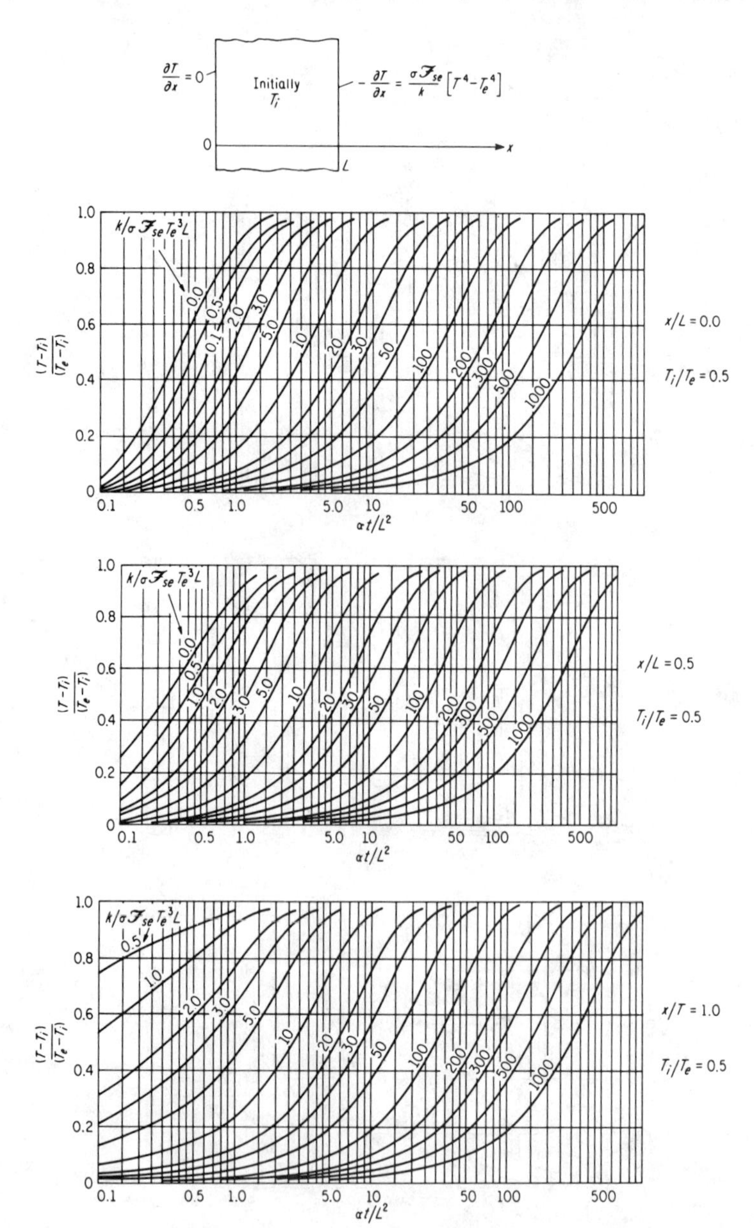

FIG. 8-6. Transient temperature distribution in a slab subjected to radiant heating when $\frac{T_i}{T_e} = 0.5$. (From Zerkle and Sunderland, Ref. 13.)

Figures 8-7 and 8-8 show similar charts for the radiant *cooling* of the slab for the values of T_e/T_i equal to 0.0 and 0.5 respectively. In these charts, the dimensionless temperature is plotted as a function of dimensionless time for several different values of the dimensionless parameter $(k/\sigma \cdot \mathfrak{F}_{se} \cdot T_i^3 \cdot L)$ at three different positions in the slab. Although it is not presented here because of space considerations, the original reference by Zerkle and Sunderland [13] includes several additional charts for several different values of T_i/T_e and T_e/T_i.

In recent years heat-conduction problems involving large temperature differences have found an increased number of engineering applications. In such problems if the thermal conductivity of the material varies with temperature, the assumption of constant thermal conductivity may introduce significant error in the heat-transfer analysis; on the other hand, if the thermal conductivity is allowed to vary with temperature in the analysis the problem becomes nonlinear. To illustrate the effect of variable thermal conductivity on the transient-temperature distribution, Dowty and Haworth [14] solved with numerical methods the heat-conduction problem in a slab by assuming thermal conductivity of the material varied linearly with temperature in the form

$$k = k_0(1 + \beta T) \tag{8-138}$$

where β = temperature coefficient of thermal conductivity, but volumetric specific heat of the material remained constant.

The heat-conduction problem solved by Dowty and Haworth was for a slab, initially at uniform temperature T_i, and for times $t > 0$ both boundary surfaces are kept at constant and equal temperature T_b. Choosing the origin of the x-axis at the center plane, the boundary-value problem of heat conduction is in the form[3]

$$\frac{\partial}{\partial x}\left(k \frac{\partial T}{\partial x}\right) = \rho c_p \frac{\partial T}{\partial t} \quad \text{in} \quad 0 \le x \le L, t > 0 \tag{8-139a}$$

$$\frac{\partial T}{\partial x} = 0 \quad \text{at} \quad x = 0, t > 0 \tag{8-139b}$$

$$T = T_b \quad \text{at} \quad x = L, t \ge 0 \tag{8-139c}$$

$$T = T_i \quad \text{in} \quad 0 \le x \le L, t = 0 \tag{8-139d}$$

where thermal conductivity is assumed to vary with temperature linearly according to Eq. 8-138 and volumetric specific heat of the material remained constant.

The transient temperature distribution in the slab obtained from the numerical solution of system 8-139 is presented in dimensionless form

[3] Dowty and Haworth consider a slab of thickness L with the origin of the x-axis at the boundary surface of the slab. For uniformity of presentation we consider a slab of thickness $2L$ with the origin of the x-axis at the center plane.

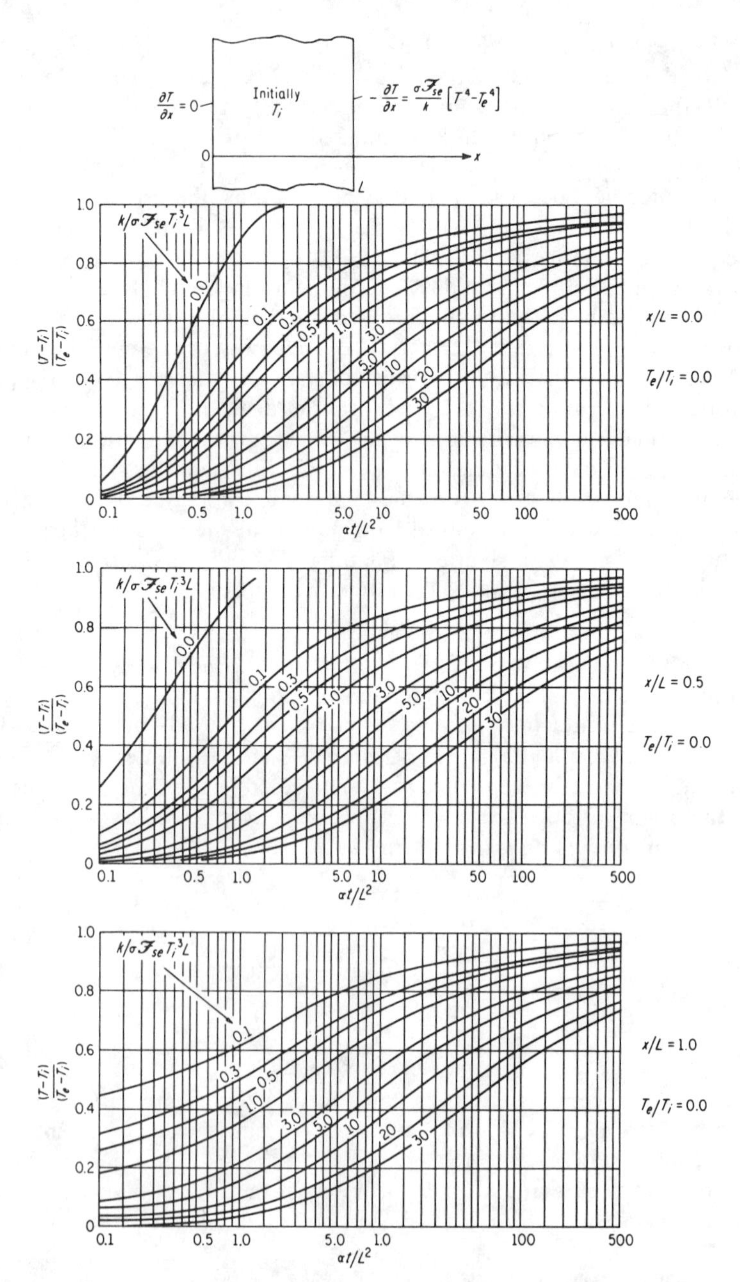

FIG. 8-7. Transient temperature distribution in a slab subjected to radiant cooling when $\frac{T_e}{T_i} = 0.0$. (From Zerkle and Sunderland, Ref. 13.)

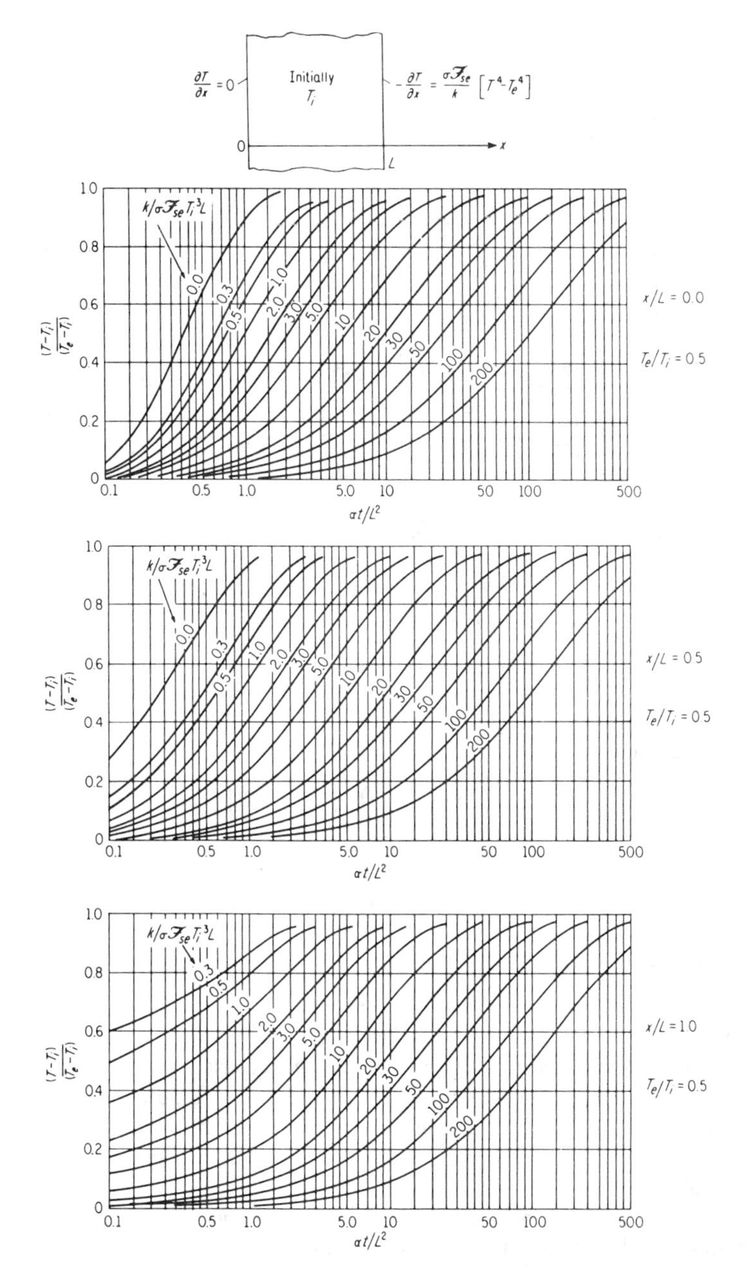

FIG. 8-8. Transient temperature distribution in a slab subjected to radiant cooling when $\frac{T_e}{T_i} = 0.5$. (From Zerkle and Sunderland, Ref. 13.)

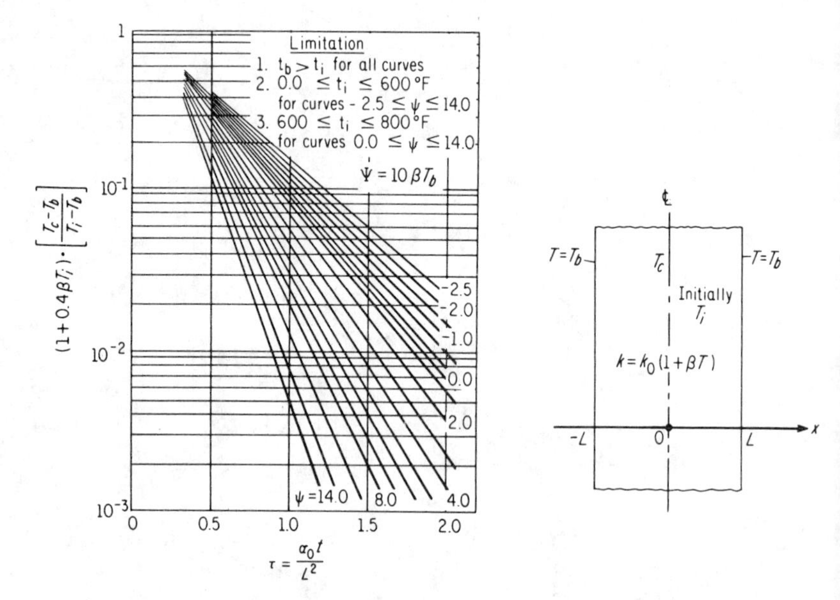

FIG. 8-9. Center-plane temperature T_c as a function of time for a slab with temperature dependent thermal conductivity and constant-temperature boundary condition. (From Dowty and Haworth, Ref. 14.)

similar to the Heisler charts. Figure 8-9 shows the dimensionless center-plane temperature

$$(1 + 0.4\beta T_i) \cdot \left[\frac{T_c - T_b}{T_i - T_b}\right]$$

as a function of the dimensionless time

$$\tau = \frac{\alpha_0 t}{L^2}$$

for various values of the parameter

$$\Psi = 10\,\beta T_b$$

The thermal diffusivity α_0 is calculated for each problem from the ratio of thermal conductivity at 0°F and the characteristic value of the volumetric specific heat for the particular temperature range considered. It is to be noted that in Fig. 8-9 the curve $\Psi = 0$ corresponds to situations where the material involved has constant thermal conductivity (i.e., $\beta = 0$).

Figure 8-10 is a position-correction chart which provides a means of evaluating the temperature at other positions in the slab. On this chart the dimensionless temperature

$$\frac{T - T_b}{T_c - T_b}$$

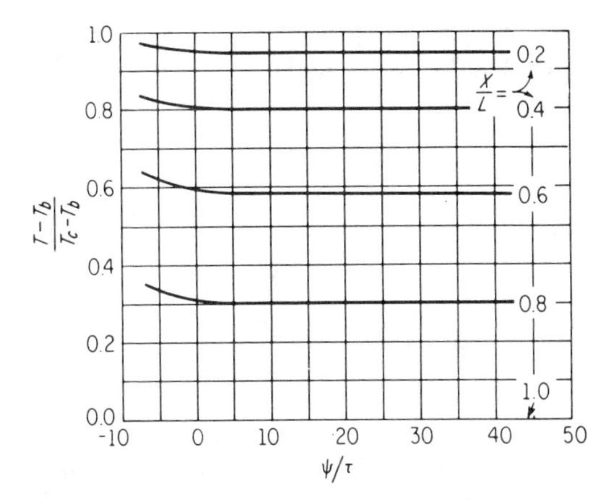

FIG. 8-10. Position-correction chart for use with Fig. 8-9. (From Dowty and Haworth, Ref. 14.)

is plotted as a function of the dimensionless group Ψ/τ for various values of the dimensionless position x/L. Temperature of a given position x/L for any given value of τ and Ψ can be determined by multiplying the value of

$$(1 + 0.4\beta T_i)\left[\frac{T_c - T_b}{T_i - T_b}\right]$$

from Fig. 8-9 by the value of $\dfrac{T - T_b}{T_c - T_b}$ at the corresponding value of x/L and Ψ/τ from Fig. 8-10.

Problems involving change of phase due to melting or solidification are becoming important. A quantity of interest in such problems is the position of solid–liquid interface as a function of time because it provides a means of determining the quantity of material melted or solidified. In the previous section we considered an analytical solution of one-dimensional melting problem for a solid initially at melting temperature and for times $t > 0$ where the boundary surface at $x = 0$ is kept at a constant temperature higher than the melting temperature. In a number of engineering applications involving melting or solidification it is desirable to have charts available for rapid estimation of the quantity of material melted or solidified. Kreith and Romie[15] prepared charts for the position of the solid–liquid interface as a function of time for the one-dimensional solidification problem in a semi-infinite (i.e., $0 \leq x \leq \infty$) region. It is assumed that the liquid region is initially at the melting temperature and for times $t > 0$ the boundary surface at $s = 0$ is subjected to convection into an environment at a constant temperature T_e which is below the melting temperature. Thermal properties of the material are assumed to be uniform and constant. The boundary-value problem of heat conduction for the solid phase is given as

$$\frac{\partial^2 T}{\partial x^2} = \frac{1}{\alpha}\frac{\partial T}{\partial t} \quad \text{in} \quad 0 \le x \le S, t > 0 \tag{8-140a}$$

The boundary conditions are

$$T = T_m \quad \text{at} \quad x = S, t > 0 \tag{8-140b}$$

$$k\frac{\partial T}{\partial x} = \rho L \frac{dS}{dt} \quad \text{at} \quad x = S, t > 0 \tag{8-140c}$$

$$k\frac{\partial T}{\partial x} = h(T - T_e) \quad \text{at} \quad x = 0, t > 0 \tag{8-140d}$$

The initial conditions are that no solidification has occurred and that the temperature of the liquid is uniformly at the fusion temperature. That is,

$$S = 0 \quad \text{and} \quad T = T_m \quad \text{for } t = 0$$

where S = location of the solid–liquid interface

L = latent heat of fusion

Kreith and Romie [15] solved system 8-140 by an analog computer and prepared a chart for the position of the solid–liquid interface as a function of time. Figure 8-11 shows the position of the solid–liquid interface $S \cdot h/k$ as a function of the dimensionless time $\alpha h^2 t/k^2$ for several different values of the dimensionless parameter $L/(T_m - T_e) \cdot c_p$. Temperature–time history of the surface $x = 0$ of the semi-infinite region is presented in Fig. 8-12.

It is of interest to compare the position of solid–liquid interface as obtained from Fig. 8-11 for the convection boundary condition at $x = 0$ with that predicted from Eq. 8-136 for a constant-temperature boundary condition at the boundary surface $x = 0$. (The solution to the problem in melting considered in the previous section is also a solution to the corresponding problem in solidification.) The depth of solid–liquid interface for the two problems differs by less than 10 percent, provided that the dimensionless parameter $S \cdot h/k$ is greater than unity [15]. The agreement between the two predicted depths of solidification improves as Sh/k becomes larger. The depth of solidification predicted by convection boundary condition will always be somewhat less than that predicted by the constant-temperature boundary condition at $x = 0$.

Figures 8-11 and 8-12 have been obtained for the process of solidification. However, the solutions are equally applicable to the corresponding process of melting if it is assumed that the heat transfer within the liquid is purely by conduction (i.e., convection effects are neglected). For melting problem, the thermal properties, k, ρ, c_p and α are evaluated for the

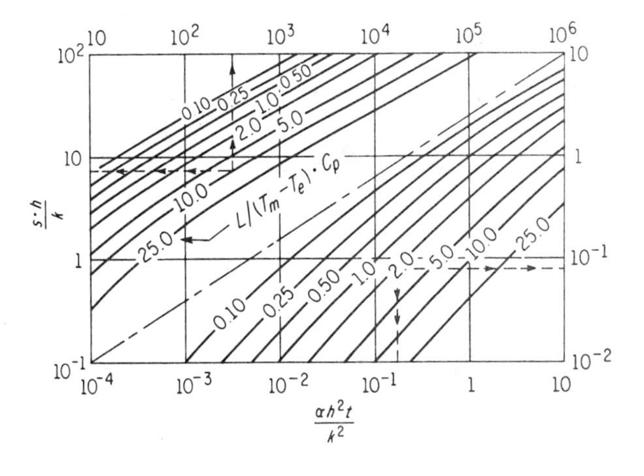

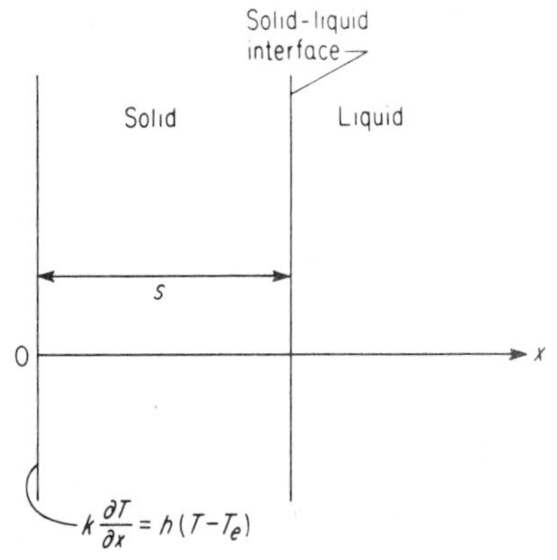

FIG. 8-11. Depth of solidification as a function of time for a semi-infinite region subjected to boundary condition at the surface $x = 0$. (From Krieth and Romie, Ref. 15.)

liquid phase. The latent heat of fusion is taken as a negative number for melting but a positive number for solidification.

In the problems of melting or solidification treated in the literature, the effect of natural convection in the liquid phase has been neglected to simplify the problem. Experimental investigations in this area are very few and most of the mathematical results presented in the literature have not been verified experimentally. Recently Boger and Westwater [16] experimentally investigated the effect of buoyancy on the melting and freezing process. Measurements made of interfacial velocities and transient and steady-state temperature profiles during the freezing and melting of water with heat flow up and down. With no natural convec-

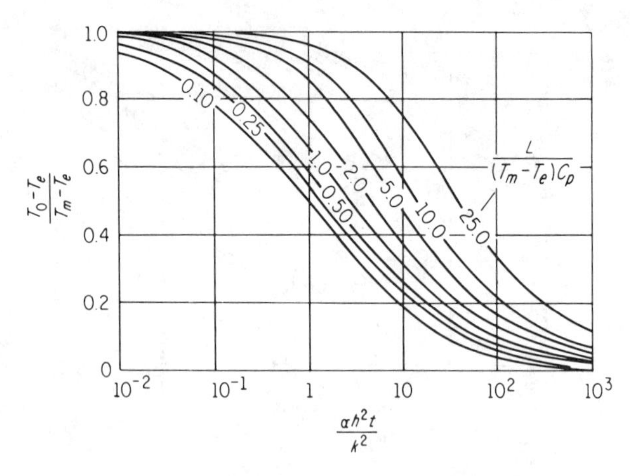

FIG. 8-12. Temperature–time history of the surface $x = 0$ during solidification of a semi-infinite region subjected to convection boundary condition at the surface $x = 0$. (From Krieth and Romie, Ref. 15.)

tion the experimental findings agreed with predictions found with the theory. The Rayleigh number, Ra, is used as the criterion for the onset of natural convection, i.e.

$$\text{Ra} = \frac{a^3 \beta \, \Delta T \, \rho_L^2 g c_{pL}}{\mu_L k_L}$$

where

β = coefficient of expansion, °F^{-1}
a = characteristic length, ft
ΔT = temperature drop in the liquid, °F
c_{pL} = specific heat of liquid, Btu/lb °F
g = acceleration of gravity, ft/hr^2
μ_L = viscosity of liquid, lb/ft hr
k_L = thermal conductivity of liquid, Btu/hr ft °F
ρ_L = density of liquid, lb/ft^3

It was found that pure conduction occurs for Rayleigh numbers below 1700; natural convection effects become important for Rayleigh numbers above 1700. For large Rayleigh numbers the theoretical predictions using the pure conduction model was found to be seriously in error. For example, in one of the melting experiments performed at Rayleigh numbers of the order of 10^7 it was found that as a result of natural convection over 400 percent more ice melted than would be predicted by pure conduction only.

For additional references on theoretical and experimental investigation of heat conduction with melting or solidification, the reader should refer to the literature review on this subject by Muehlbauer and Sunderland [17] in which a total of 146 references are cited.

REFERENCES

1. W. F. Ames, *Nonlinear Partial Differential Equations in Engineering*, Academic Press, New York, 1965.

2. Paul L. Chambre, "Nonlinear Heat Transfer Problem," *Journal of Applied Physics*, Vol. 30, No. 11 (1959), pp. 1683–1688.

3. W. V. Lovitt, "Linear Integral Equations," McGraw-Hill Book Company, Inc., (1924), p. 3, 7, 13.

4. R. Courant and H. Hilbert, *Methods of Mathematical Physics*, Vol. 1, Interscience Publishers, Inc., New York, 1953.

5. H. S. Carslaw and J. C. Jaeger, *Conduction of Heat in Solids*, Clarendon Press, Oxford, 1959, pp. 10–11.

6. H. H. Cobble, "Nonlinear Heat Transfer in Solids," Technical Report No. 8, January 1963. Mechanical Engineering Department, Eng. Expt. Station, New Mexico State University, University Park, N.M.

7. J. Crank, *Mathematics of Diffusion*, Oxford University Press, London, 1956, pp. 147–185.

8. Kwang-Tzu Yang, "Transient Conduction in a Semi-Infinite Solid with Variable Thermal Conductivity," *Trans. ASME, J. Applied Mechanics*, Vol. 79 (1958), pp. 146–147.

9. Arthur G. Hansen, *Similarity Analysis of Boundary Value Problems in Engineering*, Prentice-Hall, Inc., Englewood Cliffs, N. J., 1964.

10. A. J. A. Morgan, "The Reduction by One of the Number of Independent Variables in Some System of Partial Differential Equations," *Quart. J. Math.* (Oxford), Vol. 3 (1951), pp. 250–259.

11. W. F. Ames, "Similarity for the Nonlinear Diffusion Equation," *Ind. Eng. Chem. Fundamentals*, Vol. 4 (1965), pp. 72–76.

12. Arthur L. Ruoff, "An Alternate Solution of Stefan's Problem," *Quarterly of Applied Mathematics*, Vol. 16 (1958); pp. 197–201.

13. R. D. Zerkle and J. Edward Sunderland, "The Transient Temperature Distribution in a Slab Subject to Thermal Radiation," *J. Heat Transfer* (February 1965), pp. 117–130.

14. E. L. Dowty and D. R. Haworth, "Solution Charts for Heat Conduction in Materials with Variable Thermal Conductivity," *ASME paper* 65–WA/HT–29.

15. F. Kreith and F. E. Romie, "A Study of the Thermal Diffusion Equation with Boundary Conditions Corresponding to Solidification or Melting of Materials Initially at the Fusion Temperature," *Proceedings of the Physics Society*, Vol. 68 (1955), pp. 277–291.

16. D. V. Boger and J. W. Westwater, "Effects of Buoyancy on the Melting and Freezing Process," *ASME paper* 66–WA/HT–31.

17. J. C. Muehlbauer and J. Edward Sunderland, "Heat Conduction with Freezing or Melting," *Applied Mechanics Reviews*, Vol. 18, No. 12 (December 1965), pp. 951–959.

PROBLEMS

1. A slab $0 \le x \le L$ is initially at zero temperature, for times $t > 0$ the boundary surface at $x = 0$ is kept at zero temperature and that at $x = L$ at a constant temperature T_L and there is heat generation within the solid at a constant rate of g Btu/hr ft^3. Assuming thermal conductivity of the solid depends on temperature of the form

$$k = k_0(1 + \beta T)$$

(i) Find a relation for temperature distribution in the slab for a constant thermal diffusivity. (ii) Find a relation for the steady-state temperature distribution in the slab and discuss the effects of the parameter β on the steady-state temperature.

2. Determine the steady-state temperature distribution in a long solid cylinder of radius $r = b$ in which heat is generated at a constant rate of g Btu/hr ft^3, while the boundary surface at $r = b$ is kept at zero temperature. Assume that the thermal conductivity of the solid depends on temperature of the form

$$k = k_0(1 + \beta T)$$

3. Assuming thermal conductivity of the solid linearly dependent upon temperature as given in the above problems, determine the steady-state temperature distribution in a hollow cylinder $a \le r \le b$, with heat generation within the solid at a constant rate of g Btu/hr ft^3 and subjected to zero temperature at the boundary surfaces $r = a$ and $r = b$.

4. Determine the steady-state temperature distribution in a rectangular solid $0 \le x \le a, 0 \le y \le b$, in which heat is generated at a constant rate of g Btu/hr ft^3 while the boundary surfaces at $x = 0$ and $y = 0$ are kept insulated and those at $x = a$ and $y = b$ are subjected to zero temperature. Thermal conductivity of the solid linearly depends on temperature in the form

$$k = k_0(1 + \beta T)$$

5. When the thermal conductivity k, the mass specific heat (ρc_p) of a solid are slowly varying function of temperature in the form

$$k = k_0 \frac{T}{T_0} \qquad \rho c_p = (\rho c_p)_0 \cdot \frac{T}{T_0}$$

where the subscript zero refers to the reference condition, show that the heat-conduction equation

$$\rho c_p \frac{\partial T}{\partial t} = \frac{\partial}{\partial x}\left(k \frac{\partial T}{\partial x}\right)$$

can be reduced to a form

$$\frac{\partial U}{\partial \tau} = \frac{\partial^2 U}{\partial \bar{x}^2}$$

where

$$U = T^2 \qquad \bar{x} = \frac{x}{L} \qquad \tau = \frac{\alpha_0 t}{L^2} \qquad \alpha_0 = \frac{k_0}{(\rho c_p)_0}$$

L = a reference length

6. The one-dimensional, time-dependent heat-conduction problem for a semi-infinite region given in the form

$$\frac{\partial^2 T}{\partial x^2} = \frac{1}{\alpha}\frac{\partial T}{\partial t} \qquad 0 \le x \le \infty,\ t > 0$$

$$T = T_0 \quad \text{at} \quad x = 0,\ t > 0$$

$$T = 0 \quad \text{at} \quad x \to \infty,\ t > 0$$

$$T = 0 \quad 0 \le x \le \infty,\ t = 0$$

can be reduced to an ordinary differential equation in the form

$$\frac{d^2 F}{d\eta^2} + \frac{\eta}{2\alpha}\frac{dF}{d\eta} = 0$$

with

$$F = T_0 \quad \text{for} \quad \eta = 0$$

$$F = 0 \quad \text{for} \quad \eta = \infty$$

by means of similarity variables

$$F(\eta) = T(x, t) \qquad \eta = \frac{x}{t^{1/2}}$$

Derive these similarity variables by applying one-parameter group-theory.

Explain why the transformation of the boundary and initial conditions with the above similarity variables would not be satisfactory if the region were finite, i.e., $0 \le x \le L$, instead of semi-infinite.

9

Numerical Solution of Heat-Conduction Problems

In the previous chapters we examined analytical methods for solving the boundary-value problems of heat conduction. Digital and analog computers are of great value for solving the problems that cannot be handled analytically because they involve complicated geometries and boundary conditions, or because the numerical evaluation of the analytical solution becomes too laborious. Digital computers are frequently used because of their versatility, high-speed precision, and the fact that large-capacity, high-speed computers are now available to handle the most complicated problems.

In this chapter we examine the finite-difference approximation of the boundary-value problems of heat conduction for numerical calculations by means of digital computers. A basic concept in the finite-difference approximation of a differential equation is the use of Taylor series expansion of the function. There are several methods for approximating the derivative at a given point by finite differences; these will be discussed with particular emphasis on the solution of the heat-conduction equation with finite differences. There are the important problems of stability and convergence of finite-difference solutions. Since the analysis of stability and convergence is beyond the scope of the present work, only the results pertinent to the solution of heat-conduction equations will be given. For treatment of the numerical solution of partial differential equations of other types, and the problems of stability and convergence, the reader should refer to any one of the several references [1, 2, 3]; the literature on the subject of stability and convergence is tremendous and ever growing.

Once the finite-difference approximation of a boundary-value problem of heat conduction is complete, the problem reduces to the solution of a set of algebraic equations. There are several methods available for the solution of algebraic equations; some of these methods will be discussed later in this chapter. The computer languages such as FORTRAN [4] and ALGOL [5] are usually used in translating the set of numerical procedures into a set of commands that can be executed by a digital computer; but such languages will not be discussed in this book.

With the advent of high-speed, large-capacity digital computers, novel techniques have been developed for the solution of differential equations. One such technique is the *Monte Carlo method* which is in essence the method of statistical trials. The application of Monte Carlo method to the solution of heat conduction problems will also be discussed.

9-1. FINITE-DIFFERENCE APPROXIMATION OF DERIVATIVES

A frequently used and readily applicable method in the numerical solution of differential equations is the method of finite-difference approximation of the partial derivatives. The method is considered approximate in that the derivative at a given point is represented by a derivative taken over a finite interval across the point. However, the accuracy of such an approximation can be controlled by choosing the interval as small as possible at the expense of increased labor in solving the resulting increased number of algebraic equations.

We first consider the finite-difference approximation of the first derivative at a given point. There are several methods of approximating the first derivative of a function at a given point. Let $u(x)$ be a function which is finite, continuous, and single-valued. We assume that derivatives of this function are also finite, continuous, and single-valued. Figure 9-1 shows the values of function $u(x)$ at the point x and at the two neighboring points $(x - h)$ and $(x + h)$. We consider the following three different methods of finite-difference approximations of the first derivative of function $u(x)$ at the point x.

Central-Difference Approximation. Derivate of a function $u(x)$ at the point x by using a central-difference approximation is defined as

$$\left.\frac{du(x)}{dx}\right|_x = \frac{u(x + h) - u(x - h)}{2h} \tag{9-1}$$

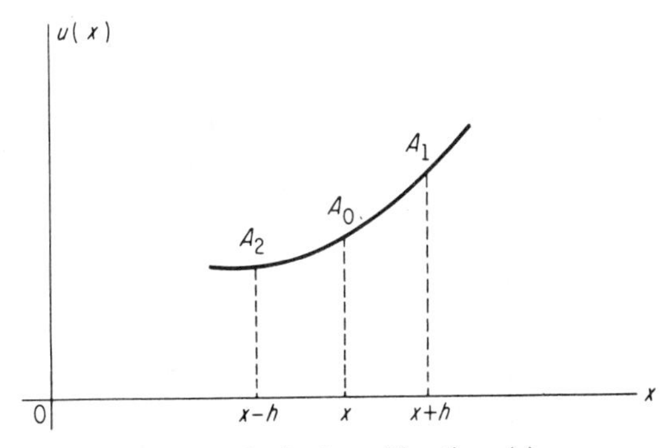

FIG. 9-1. Derivatives of function $u(x)$.

Forward-Difference Approximation. Derivative of function $u(x)$ at the point x using forward-difference approximation is defined as

$$\left.\frac{du(x)}{dx}\right|_x = \frac{u(x+h) - u(x)}{h} \tag{9-2}$$

Backward-Difference Approximation. Derivative of $u(x)$ at x using backward-difference approximation is defined as

$$\left.\frac{du(x)}{dx}\right|_x = \frac{u(x) - u(x-h)}{h} \tag{9-3}$$

The finite-difference approximation of the first derivative as defined above does not provide any information concerning the order of error involved in such approximations. The same results can be derived by Taylor series expansion of the function $u(x)$ about the point x; then, the results include an additional term representing the order of magnitude of the error involved for each of these approximations. To illustrate the procedure, we derive the above results using the Taylor series expansion. When a function $u(x)$ and its derivatives are finite, continuous, and single-valued, the function $u(x)$ can be expanded in the form of the Taylor series about the point x as

$$u(x+h) = u(x) + h\frac{du}{dx} + \frac{h^2}{2!}\frac{d^2u}{dx^2} + \frac{h^3}{3!}\frac{d^3u}{dx^3} + \frac{h^4}{4!}\frac{d^4u}{dx^4} + \cdots \tag{9-4}$$

$$u(x-h) = u(x) - h\frac{du}{dx} + \frac{h^2}{2!}\frac{d^2u}{dx^2} - \frac{h^3}{3!}\frac{d^3u}{dx^3} + \frac{h^4}{4!}\frac{d^4u}{dx^4} - \cdots \tag{9-5}$$

The central-difference approximation is immediately obtained by subtracting Eq. 9-5 from Eq. 9-4.

$$\left.\frac{du(x)}{dx}\right|_x = \frac{u(x+h) - u(x-h)}{2h} + 0(h^2) \tag{9-6}$$

The term $0(h^2)$ on the right-hand side indicates that the error involved as a result of cutting off the infinite series is of the order of h^2.

The forward-difference approximation can be obtained from Eq. 9-4 as

$$\left.\frac{du(x)}{dx}\right|_x = \frac{u(x+h) - u(x)}{h} + 0(h) \tag{9-7}$$

and the backward-difference approximation can be obtained from Eq. 9-5 as

$$\left.\frac{du(x)}{dx}\right|_x = \frac{u(x) - u(x+h)}{h} + 0(h) \tag{9-8}$$

It is to be noted that the error involved in the forward and backward finite-difference approximations is of the order of h, whereas in the central-difference approximation is of the order of h^2. Therefore, the central-difference approximation of the first derivative is more accurate than forward or backward-difference approximations.

A similar procedure can be extended to finite-difference approximation of the second derivative of a given function. By adding the Taylor series expansions 9-4 and 9-5 we obtain

$$\left.\frac{d^2u(x)}{dx^2}\right|_x = \frac{u(x-h) + u(x+h) - 2u(x)}{h^2} + 0(h^2) \tag{9-9}$$

The error involved is of the order of h^2.

The foregoing concepts on finite-difference approximation of a function of one independent variable can easily be extended to finite-difference approximation of functions involving more than one independent variable. We illustrate this by the finite-difference approximation of two-dimensional Laplacian of function $u(x, y)$:

$$\nabla^2 u(x,y) \equiv \frac{\partial^2 u(x,y)}{\partial x^2} + \frac{\partial^2 u(x,y)}{\partial y^2} \tag{9-10}$$

We interlace the region x, y with a rectangular net of sides $(\Delta x, \Delta y)$ as shown in Fig. 9-2, and for convenience denote the mesh sizes by

$$\Delta x \equiv h \qquad \Delta y \equiv l$$

The function $u(x, y)$ can be expanded by the Taylor series expansion about the point (x, y) with respect to the x and y variables in a similar manner as those given by Eqs. 9-4 and 9-5. By combining the resulting expansions it can be shown that the finite-difference approximation of the two-dimensional Laplacian, Eq. 9-10, is given as

$$\nabla^2 u\,|_{x,y} = \frac{u(x-h,y) + u(x+h,y) - 2u(x,y)}{h^2} + \frac{u(x,y-l) + u(x,y+l) - 2u(x,y)}{l^2} + 0(h^2 + l^2) \tag{9-11a}$$

If a square net of mesh size h were drawn over the region this equation reduces to

$$\nabla^2 u\,|_{x,y} = \frac{u(x-h,y) + u(x+h,y) + u(x,y-l) + u(x,y+l) - 4u(x,y)}{h^2} + 0(h^2) \tag{9-11b}$$

This topic will be further discussed in later sections on finite-difference representation of Laplacian in the cylindrical and spherical coordinate systems.

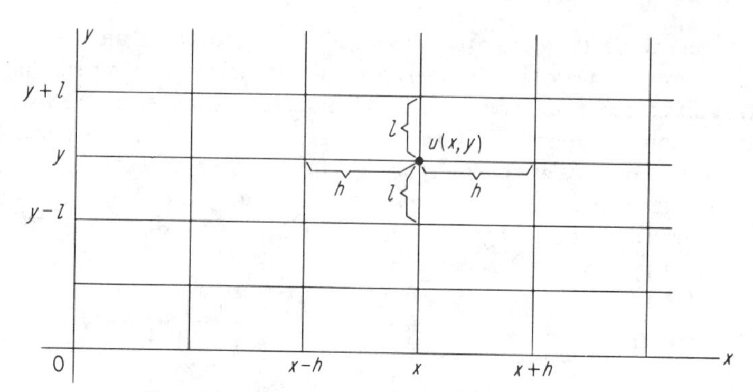

FIG. 9-2. A rectangular network of mesh size (h, l).

9-2. ERRORS INVOLVED IN FINITE DIFFERENCES

In the process of numerical solution of differential equations the derivatives are approximated with finite-difference expressions at each nodal point, and the solution of differential equation is reduced to the solution of a set of algebraic equations. On the other hand, error is introduced at each step of calculation due to approximations involved in finite-differencing and numerical calculations. Cumulative effect of such errors on the final solution and stability of difference equations is very important and deserves special consideration.

Round-off Error. Numerical calculations are carried out only to a finite number of decimal places. At each step the error involved from the rounding off of the numerical calculations is called the *round-off error.* In linear problems the effects of these errors superimpose themselves in the solution. The use of small size mesh, although desirable for better approximation of the differential equation, increases the cumulative effect of round-off error. Therefore, one cannot always say that decreasing the mesh size always increases the accuracy in finite-difference calculations. On the other hand, carrying out the numerical calculations at the intermediate stages to two or more decimal places helps to reduce the cumulative effect of round-off error. The distribution of round-off error has many features of a random process, and it is likely that the effects of these errors will generally cancel each other in part. Therefore, it is very difficult to determine exactly the order of magnitude of cumulative departure of the solution due to round-off errors.

Stability of Finite-Difference Solutions. At each stage of calculations, no matter how small, some round-off errors will be introduced. Let δ be the maximum absolute value of the error introduced to the calculations at each mesh point and u^* the resulting numerical solution of the differential equation with finite differences. If u is the solution of the differential equation with finite differences, assuming no round-off errors are involved in the calculations, the difference $u^* - u$ is the departure of the numerical solution from the solution u resulting from the round-off errors. The solution of finite-difference equations is called *stable* if the value of $u^* - u$ tends to zero, as the error δ tends to zero and does not increase exponentially as the mesh size tends to zero.

Stability consideration is important in the solution of the differential equation with finite differences. Several methods have been used in the literature to establish the conditions for the stability of difference equations. Richtmeyer [3] uses von Neumann's Fourier series expansion method in analyzing the stability of time-dependent heat-conduction equations. Consider one-dimensional boundary-value problems of heat conduction given as

$$\frac{\partial T}{\partial t} = \alpha \frac{\partial^2 T}{\partial x^2} \quad \text{in} \quad 0 \le x \le L, t > 0 \tag{9-12a}$$

$$T = 0 \quad \text{at} \quad x = 0 \text{ and } x = L,\ t > 0 \tag{9-12b}$$

$$T = F(x) \quad \text{in} \quad 0 \le x \le L,\ t = 0 \tag{9-12c}$$

The finite-difference expression for the differential Eq. 9-12a may be written in the form

$$\frac{T_j^{n+1} - T_j^n}{\Delta t} = \alpha \frac{T_{j-1}^n + T_{j+1}^n - 2T_j^n}{(\Delta x)^2} \tag{9-13}$$

where Δx and Δt are the space and time increments such that

$$x = j \cdot \Delta x \qquad j = 0, 1, 2, 3, \ldots, N$$

$$t = n \cdot \Delta t \qquad n = 0, 1, 2, 3, \ldots$$

Assuming the solution of the difference Eqs. 9-13 can be represented by a Fourier series in which the time and space variables are separated, a general term of the expansion is written in the form

$$e^{\gamma t} \cdot e^{i\beta x} \equiv e^{\gamma(n\Delta t)} \cdot e^{i\beta(j\Delta x)} \equiv \xi^{(n)} \cdot e^{i\beta(j\Delta x)}$$

We substitute this in the finite-difference Eq. 9-13 and rearrange the terms

$$\frac{\xi^{(n+1)}}{\xi^{(n)}} = 1 - \frac{2\alpha\Delta t}{(\Delta x)^2}\left[1 - \frac{e^{i\beta\Delta x} + e^{-i\beta\Delta x}}{2}\right]$$

or

$$\frac{\xi^{(n+1)}}{\xi^{(n)}} = 1 - \frac{2\alpha\Delta t}{(\Delta x)^2}[1 - \cos\beta\Delta x] \tag{9-14}$$

Any error, however small, when introduced into the calculations, will not be amplified beyond a limit during the numerical procedure, if the following condition is satisfied:

$$\text{Max}\left|\frac{\xi^{(n+1)}}{\xi^{(n)}}\right| \le 1$$

This ratio is the left-hand side of Eq. 9-14. By examining the right-hand side of Eq. 9-14 we note that this condition is satisfied if

$$\frac{2\alpha\Delta t}{(\Delta x)^2} \leq 1$$

This relationship gives the stability criteria for stable solution of the finite-difference equations 9-13. This stability condition introduces a limit to the maximum size of time steps that can be chosen for a fixed Δx. Figure 9-3 illustrates what happens to the numerical solutions when the

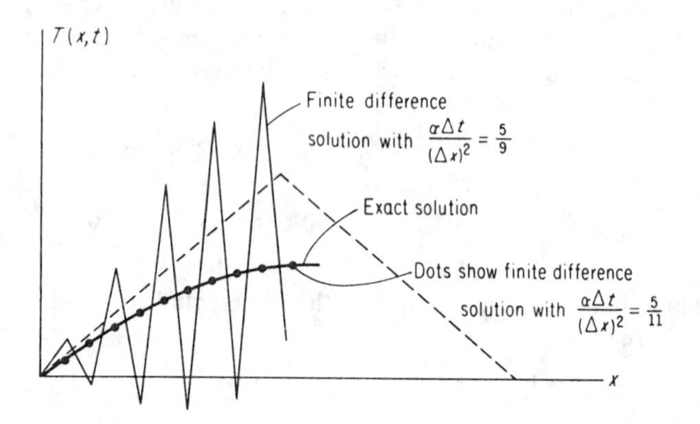

FIG. 9-3. Effect of parameter $\frac{\alpha\Delta t}{(\Delta x)^2}$ on the stability of finite-difference solution of one-dimensional heat-flow problem.

stability criteria is violated. This figure shows the numerical solution of transient heat-conduction problem for a slab which is initially at a prescribed temperature and for times $t > 0$ the boundaries of which are kept at zero temperature. Numerical calculations performed with a time and space interval satisfying a condition

$$\frac{\alpha \cdot \Delta t}{(\Delta x)^2} = \frac{5}{11} < \frac{1}{2}$$

are in good agreement with the exact analytical solution. Numerical solution of the same problem with slightly larger time steps which violated the stability criteria, i.e.,

$$\frac{\alpha\Delta t}{(\Delta x)^2} = \frac{5}{9} > \frac{1}{2}$$

results in an unstable solution.

The stability analysis via Fourier series technique as discussed above does not automatically include the effects of boundary conditions on the stability of difference system. A *matrix method* of stability analysis which automatically includes the effects of boundary conditions is described in Ref. 1, Chapter 18 and in Ref. 2.

It is to be pointed out that the form of difference equations depends on the type of differencing scheme used, the type of the differential equation, and the boundary conditions. Accordingly, the stability criteria given above cannot be generalized for other systems. Each system should be examined individually for its stability. Further, when the problems are nonlinear there is no general method that can be used effectively to determine the stability of the resulting finite-difference equations.

The stability analysis of complex systems is beyond the scope of the present work and the literature on stability is ever growing. For more detailed treatment of the subject of stability of difference equations the reader should refer to any one of the references [1,2,3].

Truncation Error. When expressing a partial differential equation in finite differences using Taylor series expansion, the series is truncated after a prescribed number of terms. The error involved in each step of calculation resulting from the truncation of the series is called the *truncation error*.

Consider the differential equation of heat conduction

$$\frac{\partial T}{\partial t} = \alpha \frac{\partial^2 T}{\partial x^2} \qquad 0 \leq x \leq L, t > 0 \tag{9-16}$$

The partial derivatives with respect to the space and time variables can be expressed in finite differences by subdividing the x,t domain into small intervals Δx and Δt, such that

$$x = i \cdot \Delta x \qquad i = 0,1,2, \ldots, N \qquad (\text{i.e., } N \cdot \Delta x = L)$$

$$t = n \cdot \Delta t \qquad n = 0,1,2, \ldots$$

Using forward differences, the time derivative is expressed as

$$\frac{T_i^{n+1} - T_i^n}{\Delta t} = \left[\frac{\partial T}{\partial t} + \frac{1}{2} \Delta t \frac{\partial^2 T}{\partial t^2}\right]_i^n \tag{9-17a}$$

By combining Taylor series expansions (i.e., see Eqs. 9-4 and 9-5) the space derivative is expressed as

$$\frac{T_{i-1}^n + T_{i+1}^n - 2T_i^n}{(\Delta x)^2} = \left[\frac{\partial^2 T}{\partial x^2} + \frac{1}{12} (\Delta x)^2 \frac{\partial^4 T}{\partial x^4}\right]_i^n \tag{9-17b}$$

Substituting 9-17a and b into 9-16, we obtain a finite-difference expression for the differential Eq. 9-16 with a truncation error

$$\left[\frac{1}{2} \Delta t \cdot \frac{\partial^2 T}{\partial t^2} - \alpha \frac{(\Delta x)^2}{12} \frac{\partial^2 T}{\partial x^4}\right] \tag{9-18}$$

at each nodal point i,n. The truncation error involves terms of order Δt and of order $(\Delta x)^2$. As the subdivisions are chosen smaller and smaller the truncation error is expected to become smaller; the smaller the truncation error the faster the convergence of the numerical solution to the true solution. On the other hand, with smaller-size network the number of nodal points hence the amount of computation involved increases. In practice, one may start with a larger-size network, reduce the mesh size gradually and compare the change in the solutions. A question may arise as to the manner one reduces the size of Δx and Δt steps in order to keep the magnitude of the truncation error at minimum. For the particular simple example under consideration this question is answered with the following consideration.

Since the temperature function T satisfies the differential equation of heat conduction (9-16), it also satisfies

$$\frac{\partial^2 T}{\partial t^2} = \alpha^2 \frac{\partial^4 T}{\partial x^4}$$

In view of this relation, the truncation error given by Eq. 9-18 becomes

$$\left[\frac{1}{2}\alpha^2 \Delta t - \alpha \frac{(\Delta x)^2}{12}\right] \frac{\partial^4 T}{\partial x^4}$$

The term in the bracket vanishes if Δx and Δt steps are made smaller in such a way as to satisfy

$$\frac{\alpha \Delta t}{(\Delta x)^2} = \frac{1}{6}$$

When the mesh sizes are chosen accordingly to this criteria finite-difference solution approaches the true solution of the differential equation at a much faster rate; but this result cannot be generalized for all heat-conduction problems.

There are several schemes available that express the differential equation in finite differences. The difference equation 9-13 considered above is called the *explicit form* because it gives an explicit relation for temperature at the nodal point j at time $(n + 1)\,\Delta t$ in terms of temperatures at the previous time $n\Delta t$. Richtmeyer [3] lists 13 different schemes, ranging from the *explicit form* to *fully implicit form* for finite differencing of one-dimensional time-dependent heat-conduction equation. When the boundary conditions involve derivatives one may use a forward, backward, or central-differencing scheme in expressing the boundary condition in finite differences. Each of these differencing schemes has its advantages and limitations. In the next two sections the use of *explicit* and *implicit* forms for the differential equation and the use of backward, forward, and central differencing schemes for the boundary conditions, their advantages and disadvantages will be illustrated with particular emphasis to the finite-difference approximation of the one-dimensional, time-dependent boundary value problem of heat conduction.

9-3. AN EXPLICIT METHOD OF FINITE DIFFERENCES FOR ONE-DIMENSIONAL HEAT-CONDUCTION PROBLEMS

Consider one-dimensional, time-dependent boundary-value problem of heat conduction in a finite region $0 \leq x \leq L$ subject to the boundary condition of the third kind at both surfaces.

$$\frac{\partial T}{\partial t} = \alpha \frac{\partial^2 T}{\partial x^2} \quad \text{in} \quad 0 \leq x \leq L, t > 0 \tag{9-19a}$$

$$-k_1 \frac{\partial T}{\partial x} + h_1 T = f_1 \quad \text{at} \quad x = 0, t > 0 \tag{9-19b}$$

$$k_2 \frac{\partial T}{\partial x} + h_2 T = f_2 \quad \text{at} \quad x = L, t > 0 \tag{9-19c}$$

$$T = F(x) \quad \text{in} \quad 0 \leq x \leq L, t = 0 \tag{9-19d}$$

We divide the x, t domain into small intervals Δx, Δt such that

$$x = i \cdot \Delta x \qquad i = 0,1,2,\ldots,N \qquad (\text{i.e., } N \cdot \Delta x = L)$$

$$t = n \cdot \Delta t \qquad n = 0, 1, 2, \ldots.$$

The temperature at the nodal point $i \cdot \Delta x$ at the time $n \cdot \Delta t$ is denoted by

$$T(i \cdot \Delta x, n \cdot \Delta t) \equiv T_i^n$$

Using forward differences in the time domain, time derivative of temperature is given as

$$\frac{\partial T}{\partial t} = \frac{T_i^{n+1} - T_i^n}{\Delta t} + 0(\Delta t)$$

Finite-difference expression for the partial derivative with respect to the space variable is given as

$$\frac{\partial^2 T}{\partial x^2} = \frac{T_{i-1}^n + T_{i+1}^n - 2T_i^n}{(\Delta x)^2} + 0(\Delta x)^2$$

Substituting these expressions in 9-19a, the explicit finite-difference form becomes

$$\frac{T_i^{n+1} - T_i^n}{\Delta t} = \alpha \frac{T_{i-1}^n + T_{i+1}^n - 2T_i^n}{(\Delta x)^2} \tag{9-20}$$

with a truncation error of order $0(\Delta t) + 0(\Delta x)^2$. Solving Eq. 9-20 for T_i^{n+1},

$$T_i^{n+1} = rT_{i-1}^n + (1 - 2r) T_i^n + rT_{i+1}^n \tag{9-21}$$

$$i = 1,2,3,\ldots,N - 1$$

$$n = 0,1,2,3,\ldots$$

where

$$r = \frac{\alpha \cdot \Delta t}{(\Delta x)^2} \tag{9-22}$$

Equation 9-21 is the explicit finite differences for the differential equation 9-19a. The boundary conditions in system 9-19 involve derivatives which should also be expressed in finite differences; in the followings we examine finite differences for the boundary conditions.

Explicit Scheme for Differential Equation, Backward and Forward Differences for Boundary Conditions. Using forward differences for the boundary condition at $x = 0$ and a backward difference for that at $x = L$, the finite-difference expressions for the boundary conditions 9-19b and 9-19c are

$$-k_1 \frac{T_1^{n+1} - T_0^{n+1}}{\Delta x} + h_1 T_0^{n+1} = f_1 \quad n = 0,1,2,\ldots \tag{9-23}$$

$$k_2 \frac{T_N^{n+1} - T_{N-1}^{n+1}}{\Delta x} + h_2 T_N^{n+1} = f_2 \quad n = 0,1,2,3,\ldots \tag{9-24}$$

with truncation error of the order of $0(\Delta x)$.

Solving Eqs. 9-23 and 9-24 for temperature at the boundary surfaces, we obtain

$$T_0^{n+1} = \frac{1}{1 + \dfrac{h_1 \Delta x}{k_1}} \left[T_1^{n+1} + \frac{f_1 \cdot \Delta x}{k_1} \right] \quad n = 0,1,2,3,\ldots \tag{9-25}$$

$$T_N^{n+1} = \frac{1}{1 + \dfrac{h_2 \Delta x}{k_2}} \left[T_{N-1}^{n+1} + \frac{f_2 \cdot \Delta x}{k_2} \right] \quad n = 0,1,2,3,\ldots \tag{9-26}$$

The finite-difference representation of differential equation of heat conduction 9-19a, by using an explicit difference, is given as

$$T_i^{n+1} = rT_{i-1}^n + (1 - 2r)T_i^n + rT_{i+1}^n \qquad i = 2,3,\ldots,N-1 \quad n = 0,1,2,3,\ldots \tag{9-27}$$

subject to the initial condition

$$T_i^0 = F(i \cdot \Delta x) \qquad i = 0,1,2,\ldots,N \tag{9-28}$$

For $i = 1$ and $i = N - 1$, Eq. 9-27 includes on the right-hand side the boundary-surface temperatures T_0^n and T_N^n respectively, which are given by Eqs. 9-25 and 9-26. Hence Eq. 9-27 provides $(N - 1)$ algebraic equations for the $(N - 1)$ unknown temperatures at the internal nodal points $i = 1,2,3,\ldots,N-1$. For the starting case of $n = 0$ the right-hand side of system 9-27 include the initial conditions which are prescribed by Eq. 9-28.

Summarizing, the $(N - 1)$ algebraic equations described by Eq. 9-27 for the $(N - 1)$ unknown temperatures $T_1^{n+1}, T_2^{n+1}, \ldots, T_{N-1}^{n+1}$, are given in the matrix form as

$$\begin{bmatrix} T_1^{n+1} \\ T_2^{n+1} \\ T_3^{n+1} \\ \vdots \\ T_{N-2}^{n+1} \\ T_{N-1}^{n+1} \end{bmatrix} = \begin{bmatrix} R_1 & r & 0 & 0 & \cdots & 0 & 0 & 0 \\ r & (1-2r) & r & 0 & \cdots & 0 & 0 & 0 \\ 0 & r & (1-2r) & r & \cdots & 0 & 0 & 0 \\ \vdots & & & & & & & \vdots \\ 0 & 0 & 0 & 0 & \cdots & r & (1-2r) & r \\ 0 & 0 & 0 & 0 & \cdots & 0 & r & R_2 \end{bmatrix} \begin{bmatrix} T_1^n \\ T_2^n \\ T_3^n \\ \vdots \\ T_{N-2}^n \\ T_{N-1}^n \end{bmatrix} + \begin{bmatrix} rT_0^0\delta_{0,n} + \dfrac{r}{\beta_1}\dfrac{f_1\Delta x}{k_1}(1-\delta_{0,n}) \\ 0 \\ 0 \\ \vdots \\ 0 \\ rT_N^0\delta_{0,n} + \dfrac{r}{\beta_2}\dfrac{f_2\Delta x}{k_2}(1-\delta_{0,n}) \end{bmatrix} \tag{9-29}$$

where

$$R_1 = (1-2r) + \frac{r}{\beta_1}(1-\delta_{0,n}) \tag{9-30}$$

$$R_2 = (1-2r) + \frac{r}{\beta_2}(1-\delta_{0,n}) \tag{9-31}$$

$$\beta_1 = 1 + \frac{h_1\Delta x}{k_1} \tag{9-32}$$

$$\beta_2 = 1 + \frac{h_2\Delta x}{k_2} \tag{9-33}$$

$$\delta_{0,n} = \text{Kroneker delta} = \begin{cases} 1 & \text{for } n = 0 \\ 0 & \text{for } n \neq 0 \end{cases}$$

$$n = 0,1,2,3,\ldots$$

T_i^0 for $i = 0,1,2,\ldots,N$ are as given by Eq. 9-28

$$r = \frac{\alpha\cdot\Delta t}{(\Delta x)^2}$$

In system 9-29 the equations are not coupled because an explicit scheme of finite differencing it used for the differential equation. Therefore, each equation is solved individually. Calculation is started with $n = 0$ and temperatures $T_1^1, T_2^1, T_3^1, \ldots, T_{N-1}^1$ are evaluated since temperatures on the right-hand side of Eqs. 9-29 are the initial conditions and known.

Knowing $T_1^1, T_2^1, \ldots, T_{N-1}^1$, the temperatures at the end of the second time step $T_1^2, T_2^2, T_3^2, \ldots, T_{N-1}^2$ are calculated from Eq. 9-29 by setting $n = 1$. The procedure is repeated to calculate temperatures at the following steps.

The boundary-surface temperatures T_0^{n+1} and T_N^{n+1} are calculated from Eqs. 9-25 and 9-26.

The entire set of equations are stable for [Ref. 2, p. 93]

$$0 < r \equiv \frac{\alpha \Delta t}{(\Delta x)^2} \leq \frac{1}{2}$$

The system 9-29 is for boundary conditions of the third kind at both boundaries. For other combinations of boundary conditions the terms involving the boundary-conditions functions should be adjusted accordingly.

Explicit Scheme for Differential Equation, Central Differences for Boundary Conditions. The truncation error associated with the finite difference approximation of the boundary conditions can be improved to $0(\Delta x)^2$ if the central-difference approximation is used for the boundary conditions. Then the finite difference approximation of the boundary conditions 9-19b and 9-19c becomes:

$$-k_1 \frac{T_1^{n+1} - T_{-1}^{n+1}}{2\Delta x} + h_1 T_0^{n+1} = f_1 \tag{9-34}$$

$$k_2 \frac{T_{N+1}^{n+1} - T_{N-1}^{n+1}}{2\Delta x} + h_2 T_N^{n+1} = f_2 \tag{9-35}$$

$$n = 0,1,2,\ldots$$

where the nodal points $i = -1$ and $i = N + 1$ are the images of $i = 1$ and $i = N - 1$ respectively as shown in Fig. 9-4.

Solving Eqs. 9-34 and 9-35 for the temperatures at the image points we obtain

$$T_{-1}^n = T_1^n + \frac{2\Delta x}{k_1}(f_1 - h_1 T_0^n) \tag{9-36}$$

$$T_{N+1}^n = T_{N-1}^n + \frac{2\Delta x}{k_2}(f_2 - h_2 T_N^n) \tag{9-37}$$

The finite difference expression for the differential equation of heat conduction 9-19a is

$$T_i^{n+1} = rT_{i-1}^n + (1 - 2r)\, T_i^n + rT_{i+1}^n \tag{9-38}$$

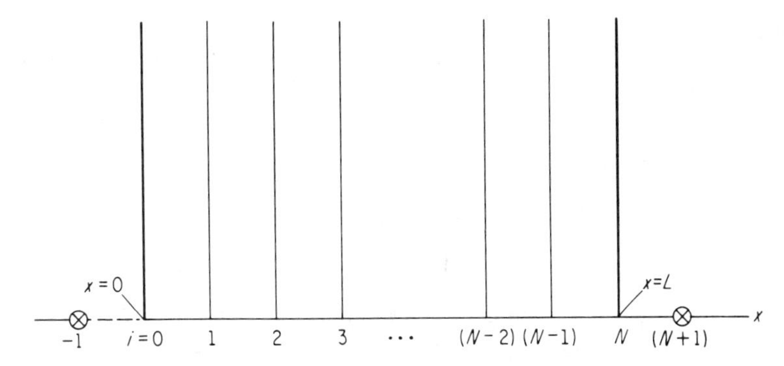

FIG. 9-4. The image nodal points $i = -1$ and $i = (N + 1)$.

where

$$i = 0,1,2,\ldots,N$$
$$n = 0,1,2,3,\ldots$$

For $i = 0$ and $i = N$ the right-hand side of system 9-38 includes the temperatures T^n_{-1} and T^n_{N+1} at the image points $i = -1$ and $i = N + 1$; these temperatures T^n_{-1} and T^n_{N+1} can be eliminated by means of relations 9-36 and 9-37. Then the system 9-38 gives $(N + 1)$ algebraic equations for the $(N + 1)$ unknown temperatures, and in the *matrix* form they are given as

$$\begin{bmatrix} T_0^{n+1} \\ T_1^{n+1} \\ T_2^{n+1} \\ \vdots \\ \vdots \\ T_{N-1}^{n+1} \\ T_N^{n+1} \end{bmatrix} = \begin{bmatrix} (1-2r\beta_1) & 2r & 0 & 0\ldots.0 & 0 & 0 \\ r & (1-2r) & r & 0\ldots.0 & 0 & 0 \\ 0 & r & (1-2r) & r\ldots.0 & 0 & 0 \\ \vdots & & & & & \vdots \\ 0 & 0 & 0 & 0\ldots.r & (1-2r) & r \\ 0 & 0 & 0 & 0\ldots.0 & 2r & (1-2r\beta_2) \end{bmatrix} \begin{bmatrix} T_0^n \\ T_1^n \\ T_2^n \\ \vdots \\ \vdots \\ T_{N-1}^n \\ T_N^n \end{bmatrix} + \begin{bmatrix} \dfrac{2r\cdot\Delta x f_1}{k_1} \\ 0 \\ 0 \\ \vdots \\ \vdots \\ 0 \\ \dfrac{2r\cdot\Delta x f_2}{k_2} \end{bmatrix} \tag{9-39}$$

where

$$\beta_1 = \left(1 + \frac{h_1 \Delta x}{k_1}\right), \quad \beta_2 = \left(1 + \frac{h_2 \Delta x}{k_2}\right)$$

The solution of system 9-39 is stable if [Ref. 2, p. 69]

$$r \leq \text{Minimum of} \left[\frac{1}{2 + \dfrac{h_1 \Delta x}{k_1}} \quad \text{or} \quad \frac{1}{2 + \dfrac{h_2 \Delta x}{k_2}} \right] \tag{9-40}$$

It is to be noted that each equation in system 9-39 is solved individually. The calculation is started with $n = 0$ and temperatures $T_0^1, T_1^1, T_2^1, \ldots, T_N^1$ are evaluated because the temperatures $T_0^0, T_1^0, T_2^0, \ldots, T_N^0$ on the right-hand side are the initial conditions. Knowing $T_0^1, T_1^1, \ldots, T_N^1$, the temperatures at the end of the second time step $T_0^2, T_1^2, \ldots, T_N^2$ are evaluated by setting $n = 1$; and so forth.

For other combinations of boundary conditions the system 9-39 should be modified accordingly. For example, for a boundary condition of the first kind at $x = 0$, the temperature T_0^{n+1} is a known quantity and the first equation is no longer needed. The right-hand side of the second equation should be modified to suit the boundary condition of the first kind at $x = 0$.

9-4. AN IMPLICIT METHOD OF FINITE DIFFERENCE (CRANK-NICOLSON METHOD)

In the previous section we have seen that an explicit method of finite-difference representation of the differential equation of heat conduction resulted in a set of simple algebraic relations which could easily be solved with a digital computer. On the other hand, the stability considerations have restricted the size of time step for a given value of Δx. If the space steps Δx are to be chosen rather small to improve accuracy and the calculations performed over a large period of time, computational problems become enormous. In such cases an *implicit method* is usually employed.

Consider the differential equation of heat conduction

$$\frac{\partial T}{\partial t} = \alpha \frac{\partial^2 T}{\partial x^2} \quad \text{in} \quad 0 \leq x \leq L, \ t > 0 \tag{9-41}$$

The finite-difference expression for Eq. 9-41 using a fully implicit form is given as

$$\frac{T_i^{n+1} - T_i^n}{\Delta t} = \alpha \cdot \frac{T_{i-1}^{n+1} + T_{i+1}^{n+1} - 2T_i^{n+1}}{(\Delta x)^2} \tag{9-42}$$

It is to be noted that the only difference between the explicit form given by Eq. 9-20 and the fully implicit form given above is the use of superscript $(n + 1)$ for the time steps on the right-hand side of Eq. 9-42. The fully implicit system 9-42 is stable for all values of the parameter [Ref. 2, p. 72]:

$$r \equiv \frac{\alpha \cdot \Delta t}{(\Delta x)^2}$$

On the other hand, equations is system 9-42 should be solved simultaneously, which is more involved computationally than solving each equation individually as in the case of the explicit method. The truncation error for system 9-42 is of the order of $0(\Delta t) + 0(\Delta x)^2$.

Crank and Nicolson [6] suggested a modified implicit method of finite differences in which an arithmetic average is taken of the right-hand side of the explicit form 9-20 and the fully implicit form 9-41. That is, Crank-Nicolson method of finite differences of differential equation 9-41 is given as

$$\frac{T_i^{n+1} - T_i^n}{\Delta t} = \frac{1}{2}\alpha\left[\frac{T_{i-1}^{n+1} + T_{i+1}^{n+1} - 2T_i^{n+1}}{(\Delta x)^2} + \frac{T_{i-1}^n + T_{i+1}^n - 2T_i^n}{(\Delta x)^2}\right] \tag{9-43}$$

The Crank-Nicolson method of finite differences requires a simultaneous solution of all equations for each time step; that is, if there are N internal mesh points, Eq. 9-43 gives N simultaneous algebraic equations for the N unknown temperatures for each time step. It is *stable* [Ref. 2, p. 65; Ref. 3, p. 93] for all values of r and the truncation error is the smallest, i.e., $0(\Delta t)^2 + 0(\Delta x)^2$.

We now examine the finite-difference representation of the heat-conduction problem considered previously.

$$\frac{\partial T}{\partial t} = \alpha \frac{\partial^2 T}{\partial x^2} \quad \text{in} \quad 0 \le x \le L,\ t > 0 \tag{9-19a}$$

$$-k_1 \frac{\partial T}{\partial x} + h_1 T = f_1 \quad \text{at} \quad x = 0,\ t > 0 \tag{9-19b}$$

$$k_2 \frac{\partial T}{\partial x} + h_2 T = f_2 \quad \text{at} \quad x = L,\ t > 0 \tag{9-19c}$$

$$T = F(x) \quad \text{in} \quad 0 \le x \le L,\ t = 0 \tag{9-19d}$$

Using Crank-Nicolson method for the differential equation and central differences for the boundary conditions, the finite-difference form of system 9-19 is

$$\frac{T_i^{n+1} - T_i^n}{\Delta t} = \frac{\alpha}{2}\left[\frac{T_{i-1}^{n+1} + T_{i+1}^{n+1} - 2T_i^{n+1}}{(\Delta x)^2} + \frac{T_{i-1}^n + T_{i+1}^n - 2T_i^n}{(\Delta x)^2}\right] \quad \text{for } i = 0, 1, 2, \ldots, N \tag{9-44a}$$

$$-k_1 \frac{T_1^n - T_{-1}^n}{2\Delta x} + h_1 T_0^n = f_1 \tag{9-44b}$$

$$k_2 \frac{T_{N+1}^n - T_{N-1}^n}{2\Delta x} + h_2 T_N^n = f_2 \tag{9-44c}$$

$$T_i^0 = F(i \cdot \Delta x) \quad \text{for} \quad i = 0, 1, 2, \ldots, N \tag{9-44d}$$

where the nodal points $i = -1$ and $i = N + 1$ are the images of the nodal points $i = 1$ and $i = N - 1$ respectively as shown in Fig. 9-4.

The above finite-difference equations are rearranged in the following form:

$$-rT_{i-1}^{n+1} + (2 + 2r)T_i^{n+1} - rT_{i+1}^{n+1} = rT_{i-1}^n + (2 - 2r)T_i^n + rT_{i+1}^n \qquad \text{for } i = 0, 1, 2, \ldots, N \tag{9-45a}$$

$$T_{-1}^n = T_1^n + \frac{2 \cdot \Delta x}{k_1}(f_1 - h_1 T_0^n) \tag{9-45b}$$

$$T_{N+1}^n = T_{N-1}^n + \frac{2 \cdot \Delta x}{k_2}(f_2 - h_2 T_N^n) \tag{9-45c}$$

$$T_i^0 = F(i \cdot \Delta x) \quad \text{for} \quad i = 0, 1, 2, \ldots, N \tag{9-45d}$$

Equation 9-45a gives $(N + 1)$ simultaneous algebraic equations for the $(N + 1)$ unknown temperatures at the time interval $(n + 1)$ in terms of the known temperatures at the previous time interval n. For $i = 0$ and $i = N$ system 9-45a includes two image temperatures at the image points $i = -1$ and $i = N + 1$, which can be eliminated by means of the boundary conditions 9-45b and 9-45c respectively. Then the resulting system of $(N + 1)$ simultaneous equations 9-45a in the matrix form are given as

$$\begin{bmatrix} (2 + 2r\beta_1) & -2r & 0 & 0 \cdots\cdots & 0 & 0 & 0 \\ -r & (2 + 2r) & -r & 0 \cdots\cdots & 0 & 0 & 0 \\ 0 & -r & (2 + 2r) & -r & 0 & 0 & 0 \\ \vdots & & & & & & \vdots \\ 0 & 0 & 0 & 0 \cdots\cdots & -r & (2 + 2r) & -r \\ 0 & 0 & 0 & 0 \cdots\cdots & 0 & -2r & (2 + 2r\beta_2) \end{bmatrix} \begin{bmatrix} T_0^{n+1} \\ T_1^{n+1} \\ T_2^{n+1} \\ \vdots \\ T_{N-1}^{n+1} \\ T_N^{n+1} \end{bmatrix}$$

$$= \begin{bmatrix} (2 - 2r\beta_1) & 2r & 0 & 0 \cdots\cdots & 0 & 0 & 0 \\ r & (2 - 2r) & r & 0 & 0 & 0 & 0 \\ 0 & r & (2 - 2r) & r & 0 & 0 & 0 \\ \vdots & & & & & & \vdots \\ 0 & 0 & 0 & 0 \cdots\cdots & r & (2 - 2r) & r \\ 0 & 0 & 0 & 0 \cdots\cdots & 0 & 2r & (2 - 2r\beta_2) \end{bmatrix}$$

$$\cdot \begin{bmatrix} T_0^n \\ T_1^n \\ T_2^n \\ \vdots \\ T_{N-1}^n \\ T_N^n \end{bmatrix} + \begin{bmatrix} \dfrac{4r\Delta x f_1}{k_1} \\ 0 \\ 0 \\ \vdots \\ 0 \\ \dfrac{4r\Delta x f_2}{k_2} \end{bmatrix} \tag{9-46}$$

$$n = 0, 1, 2, 3, \ldots$$

where

$$\beta_1 = \left(1 + \frac{h_1 \Delta x}{k_1}\right)$$

$$\beta_2 = \left(1 + \frac{h_2 \Delta x}{k_2}\right)$$

For generality the system 9-19 is considered subject to boundary condition of the third kind at both boundaries. For other combinations of boundary conditions the finite-difference system 9-46 should be modified accordingly.

The solution of system 9-46 is performed starting with $n = 0$; the system 9-46 becomes $(N + 1)$ algebraic simultaneous equations for the $(N + 1)$ unknown temperatures $T_0^1, T_1^1, T_2^1, \ldots, T_N^1$, since the temperature vector $T_0^0, T_1^0, T_2^0, \ldots, T_N^0$ on the right-hand side is known from the initial condition, i.e.,

$$T_i^0 = F(i \cdot \Delta x) \qquad i = 0, 1, 2, \ldots, N \tag{9-47}$$

Knowing the temperatures $T_0^1, T_1^1, \ldots, T_N^1$ we set $n = 1$ and evaluate temperatures $T_0^2, T_1^2, T_2^2, \ldots, T_N^2$.

The procedure is repeated for the following time steps.

It is apparent from the above discussion of the use of Crank-Nicolson implicit method that the differential equation of heat conduction is transformed into a set of algebraic equations which have to be solved simultaneously for each time step. In the following two sections the *direct* and the *iterative* methods of solution of simultaneous algebraic equations are examined.

9-5. A DIRECT METHOD OF SOLUTION OF SIMULTANEOUS ALGEBRAIC EQUATIONS

The implicit method of finite-difference approximation of the boundary-value problem of heat conduction has resulted in a set of simultaneous algebraic equations, 9-46, which may be written in the form

$$
\begin{aligned}
&b_0 T_0 + c_0 T_1 + 0 \quad + 0 \quad + 0 \ldots\ldots\ldots\ldots 0 = d_0 \\
&a_1 T_0 + b_1 T_1 + c_1 T_2 + 0 \quad + 0 \ldots\ldots\ldots\ldots 0 = d_1 \\
&0 \quad + a_2 T_1 + b_2 T_2 + c_2 T_3 + 0 \ldots\ldots\ldots\ldots 0 = d_2 \\
&\vdots \qquad\qquad\qquad\qquad\qquad\qquad \vdots \quad \vdots \\
&0 \ldots 0 + a_{N-2} T_{N-3} + b_{N-2} T_{N-2} + c_{N-2} T_{N-1} + 0 = d_{N-2} \\
&0 \ldots 0 + 0 \quad + a_{N-1} T_{N-2} + b_{N-1} T_{N-1} + c_{N-1} T_N = d_{N-1} \\
&0 \ldots\ldots\ldots\ldots 0 + 0 + 0 \quad + a_N T_{N-1} \quad + b_N T_N = d_N
\end{aligned}
\tag{9-48}
$$

The coefficients a, b, c, d are assumed to be known and the superscripts are removed from the temperatures because they are no longer needed.

System 9-48 includes $(N + 1)$ simultaneous equations for the $(N + 1)$ unknown temperatures $T_0, T_1, \ldots, T_N$ and is in the *tridiagonal* form; that is, when system 9-48 is written in the *matrix* form the coefficients of the matrix on the left-hand side will be zero everywhere except on the main diagonal and on the two diagonals parallel to it on either side.

A frequently used *direct* method for the solution of the linear system of algebraic equations is the *Gaussian elimination* method, which is based on successive subtraction of a suitable multiple of each equation from the following equation in such a manner as to eliminate one of the unknowns at a time. The Gaussian elimination procedure as applied to the tridiagonal system 9-48 is summarized as follows.

(1) A suitable multiple of the first equation in system 9-48 is subtracted from the second equation to eliminate T_0.

(2) A suitable multiple of the new second equation is subtracted from the third equation to eliminate T_1.

(3) A suitable multiple of the new equation before the last one is subtracted from the last equation to eliminate T_{N-1}.

(4) The unknown temperature T_N is immediately evaluated from the new last equation.

(5) The remaining unknowns $T_{N-1}, T_{N-2}, \ldots, T_1, T_0$ are evaluated by substitution in the reverse order.

The tridiagonal system 9-48 is transformed into a simpler one of *upper bidiagonal* form by applying the above elimination process and a systematic computing algorithm is obtained for determining the unknown coefficients as described below.

$$
\begin{aligned}
&T_0 + c_0^* T_1 + 0 \quad + 0 \quad + 0 \ldots + 0 = d_0^* \\
&0 + \quad T_1 + c_1^* T_2 + 0 \quad + 0 \quad + 0 = d_1^* \\
&0 + \quad 0 + T_2 \quad + c_2^* T_3 + 0 \quad + 0 = d_2^* \\
&\qquad\qquad\qquad\qquad\qquad \vdots \quad \vdots
\end{aligned}
\tag{9-49}
$$

$$
\begin{aligned}
\ldots .0 + T_{N-2} + c_{N-2} \cdot T_{N-1} + 0 \qquad &= d_{N-2}^* \\
\ldots .0 + 0 \quad + T_{N-1} \qquad + c_{N-1}^* \cdot T_N &= d_{N-1}^* \\
\ldots .0 + 0 \quad + 0 \qquad + T_N &= d_N^*
\end{aligned}
$$

where the coefficients c_k^* and d_k^* are calculated successively from the following recursion formulas:

$$
\begin{aligned}
c_0^* &= \frac{c_0}{b_0} \\
c_{k+1}^* &= \frac{c_{k+1}}{b_{k+1} - a_{k+1} \cdot c_k^*} \qquad \text{for} \qquad k = 0,1,2,\ldots, N-2 \qquad \text{(9-50)} \\
d_0^* &= \frac{d_0}{b_0} \\
d_{k+1}^* &= \frac{d_{k+1} - a_{k-1} \cdot d_k^*}{b_{k+1} - a_{k+1} \cdot c_k} \qquad \text{for} \qquad k = 0,1,2,\ldots, N-1
\end{aligned}
$$

Once the coefficients c_k^* and d_k^* are evaluated according to the recursion formulas 9-50, it is apparent from system 9-49 that the value of T_N is immediately evaluated from the last equation as

$$T_N = d_N^* \qquad \text{(9-51a)}$$

Starting with this known value of T_N, we evaluate successively the values of T_k from the following relation:

$$T_k = d_k^* - c_k^* \cdot T_{k+1}^* \quad \text{for} \quad k = N-1, N-2, \ldots, 2,1,0 \qquad \text{(9-51b)}$$

Computer subroutines are available for solving simultaneous algebraic equations by means of an elimination process as described above or by methods which are variations of the Gauss elimination process.

9-6. AN ITERATIVE METHOD OF SOLUTION OF SIMULTANEOUS ALGEBRAIC EQUATIONS

The *iterative* method is best suited to the solution of the systems whose matrix is sparse but not tridiagonal, i.e., the systems of two- and three-dimensional problems.

In the iterative method the calculation is started with an initial approximation which is improved by successive back substitutions. That is, a first approximation is used to calculate a second approximation, the second approximation is used to calculate the third, and so on. It is important that the iteration should converge and the convergence should be rapid. A discussion of several types of iterative methods is given in Ref. 7, pp. 373–389. One such method is the *extrapolated Liebmann method* [8] (sometimes called the method of *successive overrelaxation*) which is based on the technique of point iteration with successive overrelaxation.

We illustrate the *extrapolated Liebmann* method for the solution of simple heat-conduction equation:

$$\frac{\partial T}{\partial t} = \alpha \frac{\partial^2 T}{\partial x^2}$$

Using the Crank-Nicolson implicit method, we have

$$\frac{T_i^{n+1} - T_i^n}{\Delta t} = \frac{\alpha}{2}\left[\frac{T_{i-1}^{n+1} + T_{i+1}^{n+1} - 2T_i^{n+1}}{(\Delta x)^2} + \frac{T_{i-1}^n + T_{i+1}^n - 2T_i^n}{(\Delta x)^2}\right] \tag{9-52}$$

Solving Eq. 9-52 for T_i^{n+1},

$$(1 + r)\, T_i^{n+1} = \frac{r}{2}(T_{i-1}^{n+1} + T_{i+1}^{n+1}) + T_i^n + \frac{r}{2}(T_{i-1}^n + T_{i+1}^n - 2T_i^n) \tag{9-53}$$

In Eq. 9-53 we assume that the temperatures T^n's are known. Then for convenience we define

$$d_i \equiv T_i^n + \frac{r}{2}(T_{i-1}^n + T_{i+1}^n - 2T_i^n) \tag{9-54}$$

That is, d_i is a known quantity. Then Eq. 9-53 is written in the form

$$T_i = \frac{1}{2}\frac{r}{1 + r}(T_{i-1} + T_{i+1}) + \frac{b_i}{1 + r} \tag{9-55a}$$

where the superscript $(n + 1)$ is removed because it is no longer needed.

Equation 9-55a is an *iteration* formula for determining the value of T_i at the nodal point i when the values of T_{i-1} and T_{i+1} are taken from the previous iteration as a first approximation. The resulting values of T_i's are then used to evaluate an improved approximation, and so on. Then Eq. 9-55a is written in the form

$$T_i^{(s+1)} = \frac{1}{2}\frac{r}{1 + r}[T_{i-1}^{(s)} + T_{i+1}^{(s)}] + \frac{b_i}{1 + r} \tag{9-55b}$$

where the superscript (s) denotes the sth iteration.

Equation 9-55b is rearranged by adding and subtracting $T_i^{(s)}$ to the right-hand side,

$$T_i^{(s+1)} = T_i^{(s)} + \left[\frac{r}{2(1 + r)}(T_{i-1}^{(s)} + T_{i+1}^{(s)}) + \frac{b_i}{1 + r} - T_i^{(s)}\right] \tag{9-56}$$

To improve convergence in the process of iterations the terms in the bracket on the right-hand side of Eq. 9-56 are multiplied by a factor ω, which is called a *relaxation factor*, and Eq. 9-56 becomes

$$T_i^{(s+1)} = T_i^{(s)} + \omega\left[\frac{r}{2(1 + r)}(T_{i-1}^{(s)} + T_{i+1}^{(s)}) + \frac{b_i}{1 + r} - T_i^{(s)}\right]$$

or

$$T_i^{(s+1)} = (1 - \omega) T_i^{(s)} + \omega \left[\frac{r}{2(1 + r)} (T_{i-1}^{(s)} + T_{i+1}^{(s)}) + \frac{b_i}{1 + r} \right] \tag{9-57}$$

The value of the factor ω usually lies between 1 and 2. The iterative process so defined by Eq. 9-57 is called the *extrapolated Liebmann* method in which the calculations for each step are overrelaxed with the relaxation factor ω. A question arises as to the optimum value of ω_{opt} that will produce a maximum rate of convergence. The problem of determining the optimum value of ω_{opt} has been investigated [9, 10]. It is difficult to obtain the value of ω_{opt} analytically for simple geometries and it becomes impracticable for complex geometries.

For the one-dimensional, time-dependent heat-conduction equation considered above it is given as [Ref. 2, p. 29; Ref. 7, p. 380]

$$\omega_{\text{opt}} = \frac{2}{1 + \sqrt{1 - \mu^2}} \tag{9-58}$$

$$\mu = \frac{r}{1 + r} \cos \frac{\pi}{N}$$

with $(N - 1)$ being the total number of internal nodal points along a time row.

For the steady-state heat-conduction equation over a rectangular region with sides (a, b) subdivided into a square network of p equal divisions in the x-direction and q equal divisions in the y-direction, and for large values of p and q it is given as

$$\omega_{\text{opt}} \cong 2 - \sqrt{2}\,\pi(p^{-2} + q^{-2})^{1/2} \tag{9-59}$$

In many similar problems a considerable improvement in the convergence rate can be achieved by a suitable choice of ω_{opt} or an approximate value can be found empirically. Fowler and Volk [11] by combining analytical and empirical results have developed a method of computing ω_{opt} for use in their computer program for the solution of the steady heat-conduction equation.

9-7. ALTERNATING-DIRECTION IMPLICIT METHOD

The Crank-Nicolson implicit method discussed previously has the advantage that it is unconditionally stable for all values of the time step Δt. On the other hand, the computational problems become enormous when two- or three-dimensional heat-conduction equations are to be solved over a region requiring a large number of subdivisions. For example, for a three-dimensional problem with N interior points in each direction there are a total of N^3 interior points, hence $N^3 \times N^3$ matrix must be solved for each time increment. The procedure becomes obviously impractical if N exceeds, say, about 10.

Peaceman and Rachford [12] introduced an *alternating-direction implicit* method for use in problems involving a large number of internal nodal points. In this method the size of the matrix to be solved at each time step is reduced at the expense of solving the reduced matrix many times for each time step. For example, referring to the above illustration for a three-dimensional region with N interior points in each direction the use of alternating direction implicit method transforms the problem to the solving of an $N \times N$ matrix N^2 times for each time step, which is much easier than solving an $N^3 \times N^3$ matrix.

We illustrate the Peaceman and Rachford's alternating-direction implicit method for a two-dimensional, time-dependent heat conduction problem for a rectangular region (a, b) given as

$$\frac{\partial^2 T}{\partial x^2} + \frac{\partial^2 T}{\partial y^2} = \frac{1}{\alpha}\frac{\partial T}{\partial t} \quad \text{in} \quad 0 \le x \le a, 0 \le y \le b, t > 0 \tag{9-60a}$$

$$T = \text{prescribed at the boundaries for } t > 0 \tag{9-60b}$$

$$T = F(x, y) \quad \text{in} \quad 0 \le x \le a, 0 \le y \le b, t = 0 \tag{9-60c}$$

We construct a rectangular network of mesh size Δx, Δy over the rectangular region (a, b) as shown in Fig. 9-5; and divide the time domain into small intervals Δt. The temperature at the point x, y at time t is denoted by

$$T(x, y, t) = T(i\Delta x, \quad j\Delta y, \quad n\Delta t) \equiv T_{i,j}^{n} \tag{9-61}$$

The finite-difference approximation of the time derivative of the temperature is given as

$$\frac{\partial T}{\partial t} = \frac{T_{i,j}^{n+1} - T_{i,j}^{n}}{\Delta t} \tag{9-62}$$

In writing the finite-difference approximation for the second partial derivatives of temperature with respect to the space variable an *implicit* and

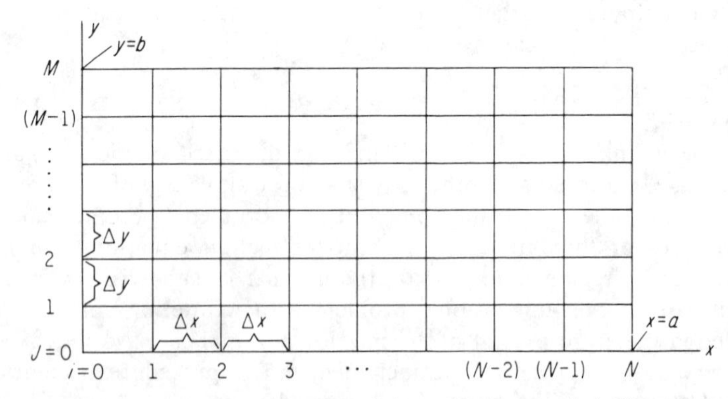

FIG. 9-5. A rectangular network drawn over a rectangle (a, b).

an *explicit* form is used alternately for $\partial^2 T/\partial x^2$ and $\partial^2 T/\partial y^2$ in successive time steps. For example, for the $(n+1)$th time step we choose an *implicit* form for $\partial^2 T/\partial x^2$ and an *explicit* form for $\partial^2 T/\partial y^2$;

$$\frac{\partial^2 T}{\partial x^2} = \frac{T_{i-1,j}^{n+1} + T_{i+1,j}^{n+1} - 2T_{i,j}^{n+1}}{(\Delta x)^2} \tag{9-63}$$

$$\frac{\partial^2 T}{\partial y^2} = \frac{T_{i,j-1}^{n} + T_{i,j+1}^{n} - 2T_{i,j}^{n}}{(\Delta y)^2} \tag{9-64}$$

Substituting Eqs. 9-62, 9-63, and 9-64 into the differential equation 9-60a we obtain

$$\frac{1}{\alpha}\frac{T_{i,j}^{n+1} - T_{i,j}^{n}}{\Delta t} = \frac{T_{i-1,j}^{n+1} + T_{i+1,j}^{n+1} - 2T_{i,j}^{n+1}}{(\Delta x)^2} + \frac{T_{i,j-1}^{n} + T_{i,j-1}^{n} - 2T_{i,j}^{n}}{(\Delta y)^2} \tag{9-65}$$

which is used for evaluating the temperatures at the $(n+1)$th time step.

For evaluating the temperatures at the $(n+2)$th time step we use an *implicit* form for $\partial^2 T/\partial y^2$ and *explicit* form for $\partial^2 T/\partial x^2$. The resulting finite-difference approximation of the differential Eq. 9-60a is

$$\frac{1}{\alpha}\frac{T_{i,j}^{n+2} - T_{i,j}^{n+1}}{\Delta t} = \frac{T_{i-1,j}^{n+1} + T_{i+1,j}^{n+1} - 2T_{i,j}^{n+1}}{(\Delta x)^2} + \frac{T_{i,j-1}^{n+2} + T_{i,j+1}^{n+2} - 2T_{i,j}^{n+2}}{(\Delta y)^2} \tag{9-66}$$

which is used for evaluating the temperature at the $(n+2)$th time step. The procedure is repeated alternately for the successive time steps.

The advantage of the alternating-direction implicit method over the Crank-Nicolson implicit method discussed above results from the fact that it simplifies the computations. Consider, for example, the solution of Eq. 9-65. If there are $(N-1)$ internal nodal points along the ox-axis, Eq. 9-65 is a system of $(N-1)$ algebraic equations for the $(N-1)$ unknown temperatures, which should be solved $(M-1)$ times if there are $(M-1)$ internal nodal points along the oy-axis. Therefore, when the alternating-direction implicit method is used for the above problem, $(N-1)$ simultaneous equations are solved $(M-1)$ times to evaluate temperatures at the nth time step, which is easier to do than to solve $(N-1)(M-1)$ simultaneous equations at once in the Crank-Nicolson method.

In the case of Eq. 9-66, a system of $(M-1)$ simultaneous algebraic equations are solved $(N-1)$ times to determine the temperatures at the $(n+2)$th time step.

For two-dimensional, time-dependent problems the alternating-direction implicit method has been shown to be stable for all values of time step [12]. However, little is known on the behavior and stability of this method for three-dimensional problems.

Smith and Spanier [13] used the alternating-direction implicit method for a digital computer program to solve the two-dimensional plane (x, y) and axially symmetric (r, z) steady-state heat-conduction problems. Mesh spacing is arranged to be completely variable and the program allows up to 5,000 points and as many as 99 regions to describe spacial variations in the material properties, heat-generation rates and boundary conditions.

Allada and Quon [14], in comparing the effectiveness of several numerical methods used the alternating-direction implicit method for the solution of the time-dependent heat-conduction problem for a cubical geometry. It was found that the solutions became unstable for certain values of the time step, Δt. The numerical solutions obtained with several methods were compared with the exact solution for the problem; it was found that the errors involved in all the numerical solutions examined, particularly in the three-dimensional cases, were not negligible.

Barakat and Clark [15] described an explicit-difference scheme which is unconditionally stable for the solution of multi-dimensional, time-dependent heat-conduction equation (see Prob. 6). The method possesses the advantages of the implicit scheme (i.e., there is no severe limitation on the size of time increment), and the simplicity of the explicit scheme.

9-8. FINITE DIFFERENCES IN THE CYLINDRICAL AND SPHERICAL COORDINATE SYSTEMS

In the previous sections we restricted our discussion to finite differencing of the heat conduction equation in the Cartesian Coordinate system. In the problems of heat conduction involving cylindrical and spherical geometries, it is desirable to write the finite difference equations in the cylindrical and spherical coordinates respectively; and for such cases the finite difference approximation of the Laplacian requires further considerations. A discussion of finite difference representation of the Laplacian in the cylindrical coordinate system is given by Schmidt [16]. In this section we examine the finite difference approximation of the Laplacian in the cylindrical and spherical coordinate systems by using Taylor series expansion.

Cylindrical Coordinate System. Consider the part of the Laplacian of temperature in the form

$$\nabla^2 T(r, \phi) = \frac{\partial^2 T}{\partial r^2} + \frac{1}{r}\frac{\partial T}{\partial r} + \frac{1}{r^2}\frac{\partial^2 T}{\partial \phi^2} \tag{9-67}$$

We interlace the region with a cylindrical (r, ϕ) network as shown in Fig. 9-6a, such that the coordinates of a nodal point (i, j) is given as

$$r = i \cdot \Delta r$$

$$\phi = j \cdot \Delta \phi$$

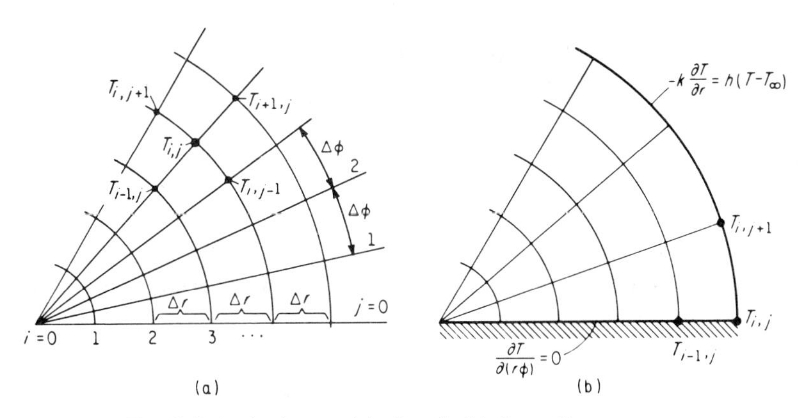

FIG. 9-6. An (r,ϕ) network in the cylindrical coordinate system.

The temperature at the nodal point (i,j) is denoted by

$$T(r,\phi)\,|_{i,j} = T(i\,\Delta r, j\,\Delta\phi) \equiv T_{i,j}$$

The first step in finite-difference procedure is to express the temperatures at the four neighboring nodal points surrounding the nodal point (i,j) in terms of the temperature at (i,j) and its derivatives at (i,j) by using Taylor series expansion. By neglecting the terms containing third- or higher-order derivatives the expansions become

$$T_{i+1,j} = T_{i,j} + \Delta r \left.\frac{\partial T}{\partial r}\right|_{i,j} + \frac{(\Delta r)^2}{2!}\left.\frac{\partial^2 T}{\partial r^2}\right|_{i,j} \tag{9-68a}$$

$$T_{i-1,j} = T_{i,j} - \Delta r \left.\frac{\partial T}{\partial r}\right|_{i,j} + \frac{(\Delta r)^2}{2!}\left.\frac{\partial^2 T}{\partial r^2}\right|_{i,j} \tag{9-68b}$$

$$T_{i,j+1} = T_{i,j} + (r\Delta\phi) \left.\frac{\partial T}{\partial(r\Delta\phi)}\right|_{i,j} + \frac{(r\Delta\phi)^2}{2!}\left.\frac{\partial T^2}{\partial(r\Delta\phi)^2}\right|_{i,j} \tag{9-68c}$$

$$T_{i,j-1} = T_{i,j} - (r\Delta\phi) \left.\frac{\partial T}{\partial(r\Delta\phi)}\right|_{i,j} + \frac{(r\Delta\phi)^2}{2!}\left.\frac{\partial^2 T}{\partial(r\Delta\phi)^2}\right|_{i,j} \tag{9-68d}$$

The finite-difference approximation of the Laplacian 9-67 at the mesh point (i,j) can be obtained from the Taylor series expansions given by Eq. 9-68 by the method of undetermined coefficients. Assuming the mesh point (i,j) is away from the boundaries the procedure is as follows:

We multiply both sides of Eq. 9-68a by A, 9-68b by B, 9-68c by C and 9-68d by D, add the resulting expressions, and collect the terms involving derivatives on one side.

$$(A - B)\Delta r \left.\frac{\partial T}{\partial r}\right|_{i,j} + (C - D)(r\Delta\phi) \left.\frac{\partial T}{\partial (r\Delta\phi)}\right|_{i,j}$$

$$+ (A + B)\frac{(\Delta r)^2}{2!}\left.\frac{\partial^2 T}{\partial r^2}\right|_{i,j} + (C + D)\frac{(r\Delta\phi)^2}{2!}\left.\frac{\partial^2 T}{\partial (r\phi)^2}\right|_{i,j}$$

$$= A(T_{i+1,j} - T_{i,j}) + B(T_{i-1,j} - T_{i,j}) + C(T_{i,j+1} - T_{i,j})$$

$$+ D(T_{i,j-1} - T_{i,j}) \qquad (9\text{-}69)$$

If the left-hand side of Eq. 9-69 is to represent the Laplacian 9-67, we obtain the following four algebraic equations for determining the four unknown coefficients A, B, C, D.

$$(A - B)\,\Delta r = \frac{1}{r}$$

$$(C - D)\,r\Delta\phi = 0$$

$$(A + B)\frac{(\Delta r)^2}{2!} = 1$$

$$(C + D)\frac{(r\Delta\phi)^2}{2!} = 1$$

Simultaneous solution of these four equations gives

$$A = \frac{1}{(\Delta r)^2} + \frac{1}{2r\Delta r} = \frac{1}{(\Delta r)^2}\left(1 + \frac{1}{2i}\right)$$

$$B = \frac{1}{(\Delta r)^2} - \frac{1}{2r\Delta r} = \frac{1}{(\Delta r)^2}\left(1 - \frac{1}{2i}\right)$$

$$C = D = \frac{1}{(r\Delta\phi)^2} = \frac{1}{i^2}\frac{1}{(\Delta r)^2(\Delta\phi)^2}$$

Substituting these coefficients on the right-hand side of Eq. 9-69, the finite-difference representation of the Laplacian 9-67 becomes

$$\left[\frac{\partial^2 T}{\partial r^2} + \frac{1}{r}\frac{\partial T}{\partial r} + \frac{1}{r^2}\frac{\partial^2 T}{\partial \phi^2}\right]_{i,j} = \frac{1}{(\Delta r)^2}$$

$$\cdot\left[\left(1 - \frac{1}{2i}\right)T_{i-1,j} + \left(1 + \frac{1}{2i}\right)T_{i+1,j} - 2T_{i,j}\right]$$

$$+ \frac{1}{i^2(\Delta r\cdot\Delta\phi)^2}\left[T_{i,j-1} + T_{i,j+1} - 2T_{i,j}\right] \qquad (9\text{-}70)$$

$$i = 1, 2, 3, 4, \ldots$$

$$j = 1, 2, 3, 4, \ldots$$

When the nodal point (i, j) lies on a boundary surface which is subjected to convective boundary condition, a similar procedure as described above is applied to obtain a finite-difference expression for the Laplacian at the point (i, j). For example, Fig. 9-6b shows a nodal point (i, j) at the intersection of two boundary surfaces one of which is insulated and the other is subjected to convection to a medium at constant temperature T_∞. The boundary conditions are given as

$$\frac{\partial T}{\partial(r\phi)} = 0 \qquad \text{at surface } \phi = j\Delta\phi$$

$$-k\frac{\partial T}{\partial r} = h(T - T_\infty) \quad \text{at surface } r = i\cdot\Delta r$$

Finite-difference representation of the Laplacian 9-67 at the nodal point (i, j) shown in Fig. 9-6b is obtained by making use of the following relations:

From Taylor series expansion,

$$T_{i-1,j} = T_{i,j} - \Delta r \left.\frac{\partial T}{\partial r}\right|_{i,j} + \frac{(\Delta r)^2}{2!}\left.\frac{\partial^2 T}{\partial r^2}\right|_{i,j} \tag{9-71a}$$

$$T_{i,j+1} = T_{i,j} + (r\Delta\phi)\left.\frac{\partial T}{\partial(r\phi)}\right|_{i,j} + \frac{(r\Delta\phi)^2}{2!}\left.\frac{\partial^2 T}{\partial(r\phi)^2}\right|_{i,j} \tag{9-71b}$$

From the boundary conditions,

$$\left.\frac{\partial T}{\partial(r\phi)}\right|_{i,j} = 0 \tag{9-71c}$$

$$\left.\frac{\partial T}{\partial r}\right|_{i,j} = \frac{h}{k}(T_\infty - T_{i,j}) \tag{9-71d}$$

We multiply both sides of Eq. 9-71a by A, Eq. 9-71b by B, Eq. 9-71c by C, and Eq. 9-71d by D, add the resulting expressions and collect the terms involving derivatives on one side,

$$(-A\,\Delta r + D)\left.\frac{\partial T}{\partial r}\right|_{i,j} + (Br\Delta\phi + C)\left.\frac{\partial T}{\partial(r\phi)}\right|_{i,j} + A\frac{(\Delta r)^2}{2!}\left.\frac{\partial^2 T}{\partial r^2}\right|_{i,j}$$

$$+ B\frac{(r\Delta\phi)^2}{2!}\left.\frac{\partial^2 T}{\partial(r\phi)^2}\right|_{i,j} = A(T_{i-1,j} - T_{i,j}) + B(T_{i,j+1} - T_{i,j})$$

$$+ D\frac{h}{k}(T_\infty - T_{i,j}) \tag{9-72}$$

If the left-hand side of Eq. 9-72 is to represent the Laplacian 9-67, we obtain

$$-A\,\Delta r + D = \frac{1}{r}$$

$$Br\Delta\phi + C = 0$$

$$A\frac{(\Delta r)^2}{2!} = 1$$

$$B\frac{(r\Delta\phi)^2}{2!} = 1$$

Simultaneous solution of these four algebraic equations gives the values of the four unknown coefficients as follows.

$$\begin{aligned} A &= \frac{2}{(\Delta r)^2} \\ B &= \frac{2}{(r\Delta\phi)^2} + \frac{2}{(i\Delta r\Delta\phi)^2} \\ C &= -\frac{2}{r\Delta\phi} = -\frac{2}{i\Delta r\Delta\phi} \\ D &= \frac{1}{r} + \frac{2}{\Delta r} = \frac{1}{\Delta r}\left(\frac{1}{i} + 2\right) \end{aligned} \tag{9-73}$$

Substituting these coefficients on the right-hand side of Eq. 9-72 we obtain the finite-difference expression for Laplacian at the point (i,j) shown in Fig. 9-6b.

$$\left[\frac{\partial^2 T}{\partial r^2} + \frac{1}{r}\frac{\partial T}{\partial r} + \frac{1}{r^2}\frac{\partial^2 T}{\partial \phi^2}\right]_{i,j} = \frac{2}{\Delta r^2}(T_{i-1,j} - T_{i,j}) + \frac{2}{i^2(\Delta r\Delta\phi)^2} \cdot (T_{i,j+1} - T_{i,j}) + \frac{2}{\Delta r}\left(1 + \frac{1}{2i}\right)\frac{h}{k}(T_\infty - T_{i,j}) \tag{9-74}$$

When the nodal point (i,j) is at the origin, $r = 0$, (i.e., $i = 0$), the finite-difference expression for the Laplacian 9-67 is given as

$$\left[\frac{\partial^2 T}{\partial r^2} + \frac{1}{r}\frac{\partial T}{\partial r} + \frac{1}{r^2}\frac{\partial^2 T}{\partial \phi^2}\right]_{r=0} = 4\,\frac{\bar{T}_1 - T_0}{(\Delta r)^2} \tag{9-75}$$

where $\bar{T}_1$ = arithmetic mean of the values of $T_{1,j}$ on a circle of radius Δr with center at the origin, $r = 0$

T_0 = value of temperature at the origin, $r = 0$

We now examine finite-difference representation of the Laplacian involving r and z variables, i.e.,

$$\nabla^2 T(r,z) = \frac{\partial^2 T}{\partial r^2} + \frac{1}{r}\frac{\partial T}{\partial r} + \frac{\partial^2 T}{\partial z^2} \tag{9-76}$$

In this case we interlace the region with (r,z) network as shown in Fig. 9-7, such that the coordinates of a nodal point (i,k) is given as

$$\begin{aligned} r &= i\Delta r \\ z &= k\Delta z \end{aligned}$$

The temperature at the nodal point (i,k) is denoted by

$$T(r,z)\,|_{i,k} = T(i\Delta r, k\Delta z) \equiv T_{i,k}$$

We express the temperatures at the four neighboring points surrounding the nodal point (i,j) by using Taylor series expansion and by neglecting the terms containing the third- and higher-order derivatives.

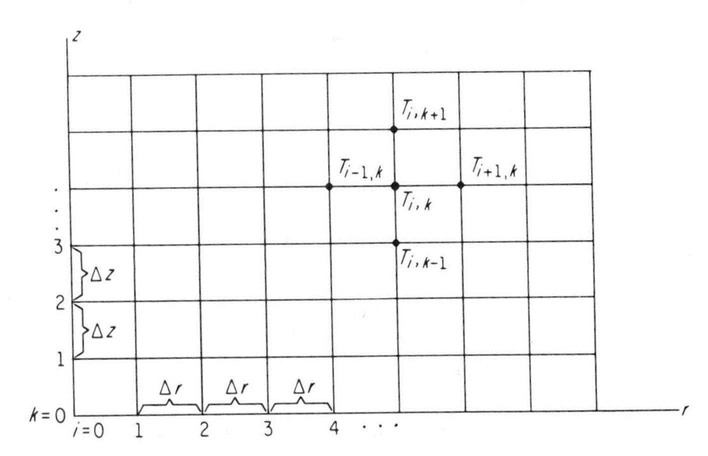

FIG. 9-7. An (r, z) network in the cylindrical coordinate system.

$$T_{i+1,k} = T_{i,k} + \Delta r \left.\frac{\partial T}{\partial r}\right|_{i,k} + \frac{(\Delta r)^2}{2!} \left.\frac{\partial^2 T}{\partial r^2}\right|_{i,k} \tag{9-77a}$$

$$T_{i-1,k} = T_{i,k} - \Delta r \left.\frac{\partial T}{\partial r}\right|_{i,k} + \frac{(\Delta r)^2}{2!} \left.\frac{\partial^2 T}{\partial r^2}\right|_{i,k} \tag{9-77b}$$

$$T_{i,k+1} = T_{i,k} + \Delta z \left.\frac{\partial T}{\partial z}\right|_{i,k} + \frac{(\Delta z)^2}{2!} \left.\frac{\partial^2 T}{\partial z^2}\right|_{i,k} \tag{9-77c}$$

$$T_{i,k-1} = T_{i,k} - \Delta z \left.\frac{\partial T}{\partial z}\right|_{i,k} + \frac{(\Delta z)^2}{2!} \left.\frac{\partial^2 T}{\partial a^2}\right|_{i,k} \tag{9-77d}$$

We multiply Eq. 9-77a by A, Eq. 9-77b by B, Eq. 9-77c by C, Eq. 9-77d by D, and add the resulting expressions:

$$\begin{aligned}(A - B)\Delta r \left.\frac{\partial T}{\partial r}\right|_{i,k} &+ (C - D)\Delta z \left.\frac{\partial T}{\partial z}\right|_{i,k} + (A + B)\frac{(\Delta r)^2}{2!} \left.\frac{\partial^2 T}{\partial r^2}\right|_{i,k} \\ &+ (C + D)\frac{(\Delta z)^2}{2!} \left.\frac{\partial^2 T}{\partial z^2}\right|_{i,k} = A(T_{i+1,k} - T_{i,k}) + B(T_{i-1,k} - T_{i,k}) \\ &+ C(T_{i,k+1} - T_{i,k}) + D(T_{i,k+1} - T_{i,k})\end{aligned} \tag{9-78}$$

Hence we obtain the following four algebraic relations for determining the four unknown coefficients:

$$(A - B)\Delta r = \frac{1}{r}$$

$$(C - D)\Delta z = 0$$

$$(A + B)\frac{(\Delta r)^2}{2!} = 1$$

$$(C + D)\frac{(\Delta z)^2}{2!} = 1$$

solution of which gives

$$A = \frac{1}{(\Delta r)^2} + \frac{1}{2r\Delta r} = \frac{1}{(\Delta r)^2}\left(1 + \frac{1}{2i}\right)$$

$$B = \frac{1}{(\Delta r)^2} - \frac{1}{2r\Delta r} = \frac{1}{(\Delta r)^2}\left(1 - \frac{1}{2i}\right)$$

$$C = D = \frac{1}{(\Delta z)^2}$$

Substituting these coefficients on the right-hand side of Eq. 9-78, the finite-difference expression for the Laplacian 9-76 becomes

$$\left(\frac{\partial^2 T}{\partial r^2} + \frac{1}{r}\frac{\partial T}{\partial r} + \frac{\partial^2 T}{\partial z^2}\right)_{i,k} = \frac{1}{(\Delta r)^2} \cdot \left[\left(1 - \frac{1}{2i}\right)T_{i-1,k} + \left(1 + \frac{1}{2i}\right)T_{i+1,k} - 2T_{i,k}\right] + \frac{1}{(\Delta z)^2}\left[T_{i,k-1} + T_{i,k+1} - 2T_{i,k}\right] \tag{9-79}$$

When the nodal point (i, k) is at the origin, $r = 0$, we have[1]

$$\lim_{r\to 0}\left(\frac{1}{r}\frac{\partial T}{\partial r}\right) = \frac{\partial^2 T}{\partial r^2}$$

Hence finite-difference expression for the Laplacian 9-76 for $r = 0$ becomes

$$\lim_{r\to 0}\left[\frac{\partial^2 T}{\partial r^2} + \frac{1}{r}\frac{\partial T}{\partial r} + \frac{\partial^2 T}{\partial z^2}\right] = \left[2\frac{\partial^2 T}{\partial r^2} + \frac{\partial^2 T}{\partial z^2}\right]_{0,k} = 4\,\frac{T_{1,k} - T_{0,k}}{(\Delta r)^2} + \frac{T_{0,k-1} + T_{0,k+1} - 2T_{0,k}}{(\Delta z)^2} \tag{9-80}$$

In the case of three-dimensional Laplacian of temperature, i.e.,

$$\nabla^2 T(r, \phi, z) = \frac{\partial^2 T}{\partial r^2} + \frac{1}{r}\frac{\partial T}{\partial r} + \frac{1}{r^2}\frac{\partial^2 T}{\phi^2} + \frac{\partial^2 T}{\partial z^2} \tag{9-81}$$

[1] By L'Hôpital's rule

$$\left.\frac{1}{r}\frac{\partial T}{\partial r}\right|_{r=0} = \frac{\dfrac{\partial}{\partial r}\left(\dfrac{\partial T}{\partial r}\right)}{\dfrac{\partial}{\partial r}(r)} = \left.\frac{\partial^2 T}{\partial r^2}\right|_{r=0}$$

we interlace the region with (r, ϕ, z) network as described above, such that the coordinate of a nodal point (i, j, k) is given by

$$r = i\Delta r \qquad \Phi = j\Delta\phi \qquad z = k\Delta z$$

and temperature at a nodal point (i, j, k) is denoted by

$$T(r, \phi, z)\,|_{i,j,k} = T(i\Delta r, j\Delta\phi, k\Delta z) \equiv T_{i,j,k}$$

By following a procedure similar to that described above it can be shown that the finite-difference approximation for the three-dimensional Laplacian 9-81 is given as

$$\nabla^2 T\,|_{i,j,k} = \frac{1}{(\Delta r)^2}\left[\left(1 - \frac{1}{2i}\right)T_{i-1,j,k} + \left[\left(1 + \frac{1}{2i}\right)T_{i+1,j,k} - 2T_{i,j,k}\right]\right.$$
$$+ \frac{1}{i^2(\Delta r\Delta\phi)^2}\,[T_{i,j-1,k} + T_{i,j+1,k} - 2T_{i,j,k}]$$
$$+ \frac{1}{\Delta z^2}\,[T_{i,j,k-1} + T_{i,j,k+1} - 2T_{i,j,k}] \qquad (9\text{-}82)$$

Spherical Coordinate System. The Laplacian of temperature in the spherical coordinate system can be expressed in finite differences by following a procedure similar to that described above for the cylindrical coordinate system. The region is interlaced with (r, ϕ, ψ) network and temperatures at the nodal points surrounding the nodal point under consideration are expressed with Taylor series expansion.

Consider, for example, the case involving spherical symmetry, i.e.,

$$\nabla^2 T(r) = \frac{\partial^2 T}{\partial r^2} + \frac{2}{r}\frac{\partial T}{\partial r} \qquad (9\text{-}83)$$

The finite-difference expression for Eq. 9-83 is

$$\left[\frac{\partial^2 T}{\partial r^2} + \frac{2}{r}\frac{\partial T}{\partial r}\right]_i = \frac{1}{(\Delta r)^2}\left[\left(1 - \frac{1}{i}\right)T_{i-1} + \left(1 + \frac{1}{i}\right)T_{i+1} - 2T_i\right] \qquad (9\text{-}84a)$$
$$\text{for } i = 1, 2, 3, 4, \ldots$$

If the region includes the origin $r = 0$ (i.e., $i = 0$), the finite difference expression 9-84a becomes[2]

$$\left[\frac{\partial^2 T}{\partial r^2} + \frac{2}{r}\frac{\partial T}{\partial r}\right]_{r=0} = 3\left.\frac{\partial^2 T}{\partial r^2}\right|_{r=0} = 6\,\frac{T_1 - T_0}{(\Delta r)^2} \qquad (9\text{-}84b)$$

[2] By L'Hôpital's rule

$$\left.\frac{2}{r}\frac{\partial T}{\partial r}\right|_{r=0} = 2\,\frac{\dfrac{\partial}{\partial r}\left(\dfrac{\partial T}{\partial r}\right)}{\dfrac{\partial}{\partial r}(r)} = 2\,\frac{\partial^2 T}{\partial r^2}$$

Then, at $r = 0$ we have

$$\left[\frac{\partial^2 T}{\partial r^2} + \frac{2}{r}\frac{\partial T}{\partial r}\right]_{r=0} = 3\left.\frac{\partial^2 T}{\partial r^2}\right|_{r=0} = 3\,\frac{T_{-1} - 2T_0 + T_1}{(\Delta r)^2} = 6\,\frac{T_1 - T_0}{(\Delta r)^2}$$

where T_0 and T_1 are the temperatures at the origin and at the nodal point next to the origin respectively.

9-9. AN IMPLICIT FINITE DIFFERENCE FOR THREE-DIMENSIONAL, TIME-DEPENDENT HEAT-CONDUCTION EQUATION

In this section we examine an implicit method of finite difference representation of a three-dimensional, time-dependent heat-conduction equation for finite regions with variable thermal conductivity and with heat generation within the solid.

Consider the differential equation of heat conduction in the Cartesian coordinate system in the form

$$\rho c \frac{\partial T}{\partial t} = \frac{\partial}{\partial x}\left(k\frac{\partial T}{\partial x}\right) + \frac{\partial}{\partial y}\left(k\frac{\partial T}{\partial y}\right) + \frac{\partial}{\partial z}\left(k\frac{\partial T}{\partial z}\right) + g(x,y,z,t) \quad \text{in a region } R \qquad (9\text{-}85)$$

We interlace the region R with a rectangular network (x, y, z) such that

$$x = i\,\Delta x$$
$$y = j\,\Delta y$$
$$z = k\Delta z$$

and subdivide the time domain into small time steps Δt such that

$$t = n\Delta t$$

The temperature at a nodal point $(i\Delta z,\ j\Delta y,\ k\Delta z)$ at a time $(n\Delta t)$ is denoted by

$$T(i\Delta x, j\Delta y, k\Delta z, n\Delta t) \equiv T^n_{i,j,k}$$

Using a "fully implicit" scheme, the finite difference expression for the differential equation 9-85 is given as

$$(\rho c)_{i,j,k} \frac{T^{n+1}_{i,j,k} - T^n_{i,j,k}}{\Delta t}$$
$$= \frac{k_{i-1/2,j,k}[T^{n+1}_{i-1,j,k} - T^{n+1}_{i,j,k}] - k_{i+1/2,j,k}[T^{n+1}_{i,j,k} - T^{n+1}_{i+1,j,k}]}{(\Delta x)^2}$$
$$+ \frac{k_{i,j-1/2,k}[T^{n+1}_{i,j-1,k} - T^{n+1}_{i,j,k}] - k_{i,j+1/2,k}[T^{n+1}_{i,j,k} - T^{n+1}_{i,j+1,k}]}{(\Delta y)^2}$$
$$+ \frac{k_{i,j,k-1/2}[T^{n+1}_{i,j,k-1} - T^{n+1}_{i,j,k}] - k_{i,j,k+1/2}[T^{n+1}_{i,j,k} - T^{n+1}_{i,j,k+1}]}{(\Delta z)^2}$$
$$+ g^{n+1/2}_{i,j,k} \qquad (9\text{-}86)$$

The subscript $i - \frac{1}{2}$ for the thermal conductivity denotes that a mean value of themal conductivity between the nodal points $(i - 1)$ and i is taken, similarly the subscript $i + \frac{1}{2}$ refers to the mean value of thermal conductivity between the nodal points i and $i + 1$, and so forth. For the heat generation term the superscript $n + \frac{1}{2}$ denotes that the heat generation is evaluated at a time $t + \Delta t/2$ in order to get a reasonably good average value of heat generation during the time step Δt.

Thomas and McRoberts [17] prepared a generalized computer program for solving the three-dimensional heat-conduction equation 9-81 with a digital computer with provision for convection, contact resistance, and thermal-radiation boundary conditions. They used fully implicit finite-difference scheme as given by Eq. 9-85, but expressed their finite-difference equations in a more compact form which is obtainable from Eq. 9-86 by multiplying both sides by $\Delta x \cdot \Delta y \cdot \Delta z$ and defining new symbols.

$$C_p \frac{T_p(t + \Delta t) - T_p(t)}{\Delta t} = G_p\left(t + \frac{\Delta t}{2}\right) + \sum_m K_{pm}[T_m(t + \Delta t) - T_p(t + \Delta t)] \tag{9-87a}$$

where

p = nodal point i, j, k
m = neighboring nodal points to the nodal point p
$T_p(t + \Delta t)$ = temperature at the nodal point p at time $(t + \Delta t)$
$C_p = (\rho c\, \Delta x\, \Delta y\, \Delta z)_p$ = heat capacity of the cell at the nodal point p
$G_p = (g_p \cdot \Delta x \cdot \Delta y \cdot \Delta z)$ = rate of heat generation in the cell at the nodal point p at time $(t + \Delta t/2)$
$K_{pm} = \dfrac{A_{pm} k_{pm}}{\Delta L_{pm}}$ = mean conductance of the heat flow path pm
A_{pm} = Interface area between the cells p and m
ΔL_{pm} = distance between the nodal points p and m

Solving Eq. 9-87a for $T_p(t + \Delta t)$, we obtain

$$T_p(t + \Delta t) = \frac{C_p T_p(t) + \Delta t\left[G_p\left(t + \frac{\Delta t}{2}\right) + \sum_m K_{pm} T_m(t + \Delta t)\right]}{C_p + \Delta t \sum_m K_{pm}} \tag{9-87b}$$

where $p = 1,2,3, \ldots,$ number of nodal points in the region R.

In Eq. 9-87b temperatures $T_p(t + \Delta t)$ are unknowns. If there are $p = 1,2,3, \ldots, N$ nodal points in the region, Eq. 9-87b provides N simultaneous algebraic equations for the N unknowns for each time step Δt.

Thomas and Roberts used "extrapolated Liebmann" method of iteration to solve the system 9-87b; that is, successive values of the temperature $T_p(t + \Delta t)$ are evaluated by means of the following iteration formula:

$$T_p^{(s+1)}(t + \Delta t) = (1 - \omega) T_p^s(t + \Delta t) + \omega \frac{C_p T_p^{(s)}(t) + \Delta t \left[G_p + \sum_m K_{pm} T_m^{(s)}(t + \Delta t) \right]}{C_p + \Delta t \sum_m K_{pm}} \tag{9-89}$$

where superscript s = number of iterations
ω = the relaxation factor

The steady-state formula is obtained from Eq. 9-87a by setting $T_p(t) = T_p(t + \Delta t)$, solving for $T_p(t)$ applying the extrapolated Liebmann method of iteration. The resulting iteration formula for the steady-state case is

$$T_p^{(s+1)} = (1 - \omega) T_p^{(s)} + \omega \frac{G_p + \sum_m K_{pm} T_m^{(s)}}{\sum_m K_{pm}} \tag{9-90}$$

When the nodal point p is situated next to a surface involving convective or radiation boundary condition or contact-resistance, evaluation of K_{pm} requires different consideration. We examine evaluation of K_{pm} for some of these cases.

Convection Boundary Condition. When the nodal point p is situated next to a boundary surface, A_s, from which heat is transferred to the neighboring nodal point m by convection, K_{pm} is evaluated from the relation

$$\frac{1}{K_{pm}} = \frac{1}{hA_s} + \frac{1}{K_{ps}} \tag{9-91}$$

where A_s = area of boundary surface
h_s = convection heat-transfer coefficient at boundary surface A_s
K_{ps} = conductance of heat-flow path between the nodal point p and boundary surface A_s

Contact Resistance. When the boundary surface between the cells p and m involves contact resistance, the value of K_{pm} is evaluated from

$$\frac{1}{K_{pm}} = \frac{1}{K_{ps}} + \frac{1}{A_c h_c} + \frac{1}{K_{sm}} \tag{9-92}$$

where A_c = surface area of contact between cells p and m
h_c = contact conductance
K_{ps}, K_{sm} = conductance of heat-flow paths between nodal point p and boundary surface s; and between boundary surface s and nodal point m respectively

9-10. CURVED BOUNDARIES

When a boundary is not exactly at a mesh distance away from the neighboring nodal points, as is the case with regions having irregular boundaries, it intersects the network strings. The finite-difference approximation of space derivatives at a nodal point having strings intersected by an irregular boundary may require special considerations to improve accuracy. To illustrate the finite-differencc approximation of derivatives at such nodal points we examine the Laplacian of temperature

$$\nabla^2 T(x, y) \equiv \frac{\partial^2 T}{\partial x^2} + \frac{\partial^2 T}{\partial y^2} \tag{9-93}$$

at two different nodal points having one and two strings intersected by the boundary.

One String Intersected by the Boundary. Figure 9-8 shows a curved boundary intersecting a string from a nodal point (N, j). We first evaluate the finite-difference approximation of $\frac{\partial^2 T}{\partial x^2}$ at the nodal point (N, j).

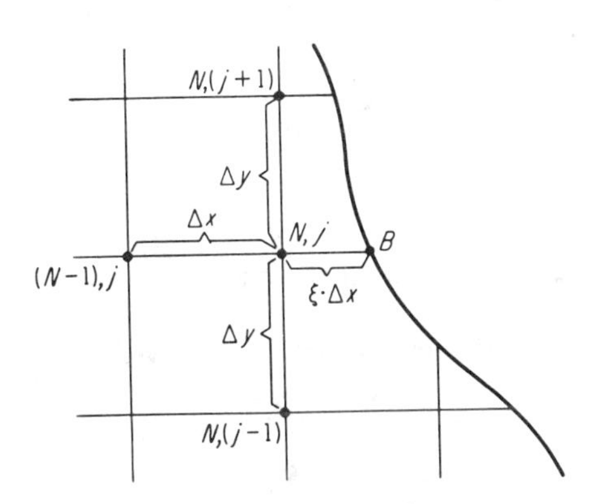

FIG. 9-8. A curved boundary intersecting x-direction string from the nodal point (N, j).

A Taylor-series expansion of temperature at the point B, T_B in powers of $(\xi \Delta x)$ gives

$$T_B = T_{N,j} + (\xi \Delta x) \left.\frac{\partial T}{\partial x}\right|_{N,j} + \frac{(\xi \Delta x)^2}{2!} \left.\frac{\partial^2 T}{\partial x^2}\right|_{N,j} + 0(\Delta x)^3 \tag{9-94}$$

where $\xi \Delta x$ = distance between the nodal point (N, j) and the point B on the boundary;

$$0 \leq \xi \leq 1$$

A Taylor series expansion of $T_{N-1,j}$, in powers of $(-\Delta x)$ gives

$$T_{N-1,j} = T_{N,j} - \Delta x \cdot \left.\frac{\partial T}{\partial x}\right|_{N,j} + \frac{(\Delta x)^2}{2!} \left.\frac{\partial^2 T}{\partial x^2}\right|_{N,j} + 0(\Delta x)^3 \tag{9-95}$$

Eliminating $\partial T/\partial x_{N,j}$ between these two last relations, we obtain

$$\left.\frac{\partial^2 T}{\partial x^2}\right|_{N,j} = \frac{2}{(\Delta x)^2}\left[\frac{T_B}{\xi(1+\xi)} + \frac{T_{N-1,j}}{1+\xi} - \frac{T_{N,j}}{\xi}\right] \tag{9-96}$$

with an error of the order of (Δx).

The finite-difference representation of

$$\left.\frac{\partial^2 T}{\partial y^2}\right|_{N,j}$$

is given as

$$\left.\frac{\partial^2 T}{\partial y^2}\right|_{N,j} = \frac{T_{N,j-1} + T_{N,j+1} - 2T_{N,j}}{(\Delta y)^2} \tag{9-97}$$

Substituting Eqs. 9-96 and 9-97 into Eq. 9-93, we obtain

$$\left[\frac{\partial^2 T}{\partial x^2} + \frac{\partial^2 T}{\partial y^2}\right]_{N,j} = \frac{2}{(\Delta x)^2}\left[\frac{T_B}{\xi(1+\xi)} + \frac{T_{N-1,j}}{1+\xi} - \frac{T_{N,j}}{\xi}\right] + \frac{1}{(\Delta y)^2}\left[T_{N,j-1} + T_{N,j+1} - 2T_{N,j}\right] \tag{9-98}$$

It is to be noted that for $\xi = 1$, Eq. 9-98 reduces to the usual finite-difference approximation in the (x, y) coordinate system discussed previously.

It is also of interest to evaluate the finite-difference approximation of the first partial derivative with respect to the x variable at the nodal point (N,j). By eliminating $\partial^2 T/\partial x^2$ between Eqs. 9-94 and 9-95, we obtain

$$\left.\frac{\partial T}{\partial x}\right|_{N,j} = \frac{1}{\Delta x}\left[\frac{T_B}{\xi(1+\xi)} - \frac{\xi}{1+\xi}T_{N-1,j} - \frac{1-\xi}{\xi}T_{N,j}\right] \tag{9-99}$$

with an error of the order of $(\Delta x)^2$.

Two Strings Intersected by the Boundary. Figure 9-9 shows a situation where a curved boundary intersects both strings from a nodal point (N, M). By the Taylor series expansion of temperatures in powers of $(\xi\Delta x)$ and $(\eta\Delta y)$ we have

$$T_B = T_{N,M} + (\xi\Delta x)\left.\frac{\partial T}{\partial x}\right|_{N,M} + \frac{(\xi\Delta x)^2}{2!}\left.\frac{\partial^2 T}{\partial x^2}\right|_{N,M} + 0(\Delta x)^3 \tag{9-100}$$

$$T_C = T_{N,M} + (\eta\Delta y)\left.\frac{\partial T}{\partial y}\right|_{N,M} + \left(\frac{\eta\Delta y}{2!}\right)^2\left.\frac{\partial^2 T}{\partial y^2}\right|_{N,M} + 0(\Delta y)^3 \tag{9-101}$$

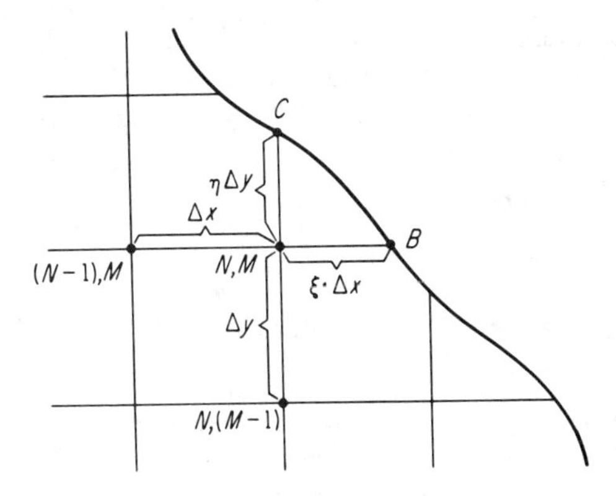

FIG. 9-9. A curved boundary intersecting both x- and y-direction strings from the nodal point (N, M).

where

$$0 \le \xi \le 1$$
$$0 \le \eta \le 1$$

By the Taylor series expansion in powers of $(-\Delta x)$ and $(-\Delta y)$ we have

$$T_{N-1,M} = T_{N,M} - \Delta x \left.\frac{\partial T}{\partial x}\right|_{N,M} + \frac{(\Delta x)^2}{2!}\left.\frac{\partial^2 T}{\partial x^2}\right|_{N,M} + 0(\Delta x)^3 \tag{9-102}$$

$$T_{N,M-1} = T_{N,M} - \Delta y \left.\frac{\partial T}{\partial y}\right|_{N,M} + \frac{(\Delta y)^2}{2!}\left.\frac{\partial^2 T}{\partial y^2}\right|_{N,M} + 0(\Delta y)^3 \tag{9-103}$$

By combining the above results $\left.\frac{\partial T}{\partial x}\right|_{N,M}$ and $\left.\frac{\partial T}{\partial y}\right|_{N,M}$ are eliminated, and we obtain the finite-difference representation of the Laplacian 9-93 at the nodal point (N, M) in the form

$$\left[\frac{\partial^2 T}{\partial x^2} + \frac{\partial^2 T}{\partial y^2}\right]_{N,M} = \frac{2}{(\Delta x)^2}\left[\frac{T_B}{\xi(1+\xi)} + \frac{T_{N-1,M}}{1+\xi} - \frac{T_{N,M}}{\xi}\right] + \frac{2}{(\Delta y)^2}\left[\frac{T_c}{\eta(1+\eta)} + \frac{T_{N,M-1}}{1+\eta} - \frac{T_{N,M}}{\eta}\right] \tag{9-104}$$

For a square network, $\Delta x = \Delta y \equiv h$, Eq. 9-104 reduces to

$$\left[\frac{\partial^2 T}{\partial x^2} + \frac{\partial^2 T}{\partial y^2}\right]_{N,M} = \frac{2}{h^2}\left[\frac{T_B}{\xi(1+\xi)} + \frac{T_C}{\eta(1+\eta)} + \frac{T_{N-1,M}}{1+\xi} + \frac{T_{N,M-1}}{1+\eta} - \left(\frac{1}{\xi} + \frac{1}{\eta}\right)T_{N,M}\right] \tag{9-105}$$

with an error of the order of h.

The finite-difference representation of the first partial derivative with respect to the y variable at the nodal point (N, M) can be evaluated from Eqs. 9-101 and 9-103 by eliminating $\partial^2 T/\partial y^2$ between them. We obtain

$$\left.\frac{\partial T}{\partial y}\right|_{N,M} = \frac{1}{\Delta y}\left[\frac{T_C}{\eta(1+\eta)} - \frac{\eta}{1+\eta} T_{N,M-1} - \frac{1-\eta}{\eta} T_{N,M}\right] \tag{9-106}$$

with an error of the order of $(\Delta y)^2$.

9-11. VARIOUS FORMS OF APPROXIMATIONS FOR DERIVATIVES

There are several other formulas for representing derivatives at any given point in terms of the data at the neighboring points; here we present a list of formulas that may be used in finite differencing.

Let $T_0, T_1, T_2, \ldots,$ be temperatures at the nodal points $0, 1, 2, \ldots,$ which are situated along the x-axis with equal spacing Δx. A Taylor series expansion of the derivative of temperature at the point 0 in terms of the derivatives at the point 1 is given as [See: Eq. 9-5]

$$\left(\frac{dT}{dx}\right)_0 = \left(\frac{dT}{dx}\right)_1 - \Delta x\left(\frac{d^2T}{dx^2}\right)_1 + \frac{(\Delta x)^2}{2!}\left(\frac{d^3T}{dx^3}\right)_1 - \cdots$$

where, $\left(\frac{dT}{dx}\right)_1$ and $\left(\frac{d^2T}{dx^2}\right)_1$ are evaluated according to Eqs. 9-6 and 9-9 respectively,

$$\left(\frac{dT}{dx}\right)_0 = \frac{T_2 - T_0}{2\Delta x} - \Delta x \cdot \frac{T_2 + T_0 - 2T_1}{\Delta x^2} + 0(\Delta x)^2$$

or

$$\left(\frac{dT}{dx}\right)_0 = \frac{1}{2\Delta x}[-3T_0 + 4T_1 - T_2] + 0(\Delta x)^2 \tag{9-107}$$

Equation 9-107 gives the first derivative of temperature at the nodal point 0 in terms of the temperatures at the nodal points 0, 1 and 2 with a truncation error of the order of $0(\Delta x)^2$. Following a similar procedure several other finite difference expressions can be derived for the first and second space derivatives of temperature. Berezin and Zhidkov [18] use the Lagrange interpolation to obtain relations for the first and second derivatives at any given point in terms of the values at the neighboring points. We present below some of their results.

First derivatives

Three-point formulas:

$$\left.\frac{dT}{dx}\right|_0 = \frac{1}{2\Delta x}[-3T_0 + 4T_1 - T_2] + 0(\Delta x)^2 \tag{9-107}$$

$$\left.\frac{dT}{dx}\right|_1 = \frac{1}{2\Delta x}[T_2 - T_0] + 0(\Delta x)^2 \tag{9-108}$$

$$\left.\frac{dT}{dx}\right|_2 = \frac{1}{2\Delta x}[T_0 - 4T_1 + 3T_2] + 0(\Delta x)^2 \tag{9-109}$$

Four-point formulas:

$$\left.\frac{dT}{dx}\right|_0 = \frac{1}{6\Delta x}[-11T_0 + 18T_1 - 9T_2 + 2T_3] + 0(\Delta x)^3 \tag{9-110}$$

$$\left.\frac{dT}{dx}\right|_1 = \frac{1}{6\Delta x}[-2T_0 - 3T_1 + 6T_2 - T_3] + 0(\Delta x)^3 \tag{9-111}$$

$$\left.\frac{dT}{dx}\right|_2 = \frac{1}{6\Delta x}[T_0 - 6T_1 + 3T_2 + 2T_3] + 0(\Delta x)^3 \tag{9-112}$$

$$\left.\frac{dT}{dx}\right|_3 = \frac{1}{6\Delta x}[-2T_0 + 9T_1 - 18T_2 + 11T_3] + 0(\Delta x)^3 \tag{9-113}$$

Second derivatives

Three-point formulas:

$$\left.\frac{d^2T}{dx^2}\right|_0 = \frac{1}{(\Delta x)^2}[T_0 - 2T_1 + T_2] + 0(\Delta x)^2 \tag{9-114}$$

$$\left.\frac{d^2T}{dx^2}\right|_1 = \frac{1}{(\Delta x)^2}[T_0 - 2T_1 + T_2] + 0(\Delta x)^2 \tag{9-115}$$

$$\left.\frac{d^2T}{dx^2}\right|_2 = \frac{1}{(\Delta x)^2}[T_0 - 2T_1 + T_2] + 0(\Delta x)^2 \tag{9-116}$$

Four-point formulas:

$$\left.\frac{d^2T}{dx^2}\right|_0 = \frac{1}{6(\Delta x)^2}[12T_0 - 30T_1 + 24T_2 - 6T_3] + 0(\Delta x)^2 \tag{9-117}$$

$$\left.\frac{d^2T}{dx^2}\right|_1 = \frac{1}{6(\Delta x)^2}[6T_0 - 12T_1 + 6T_2] + 0(\Delta x)^2 \tag{9-118}$$

$$\left.\frac{d^2T}{dx^2}\right|_2 = \frac{1}{6(\Delta x)^2}[6T_1 - 12T_2 + 6T_3] + 0(\Delta x)^2 \tag{9-119}$$

$$\left.\frac{d^2T}{dx^2}\right|_3 = \frac{1}{6(\Delta x)^2}[-6T_0 + 24T_1 - 30T_2 + 12T_3] + 0(\Delta x)^2 \tag{9-120}$$

9-12. FINITE DIFFERENCES FOR PROBLEMS INVOLVING CHANGE OF PHASE

In Chapters 7 and 8 we examined analytical methods for solving the heat-conduction problem involving change of phase by assuming natural convection effects are negligible and that heat transfer is by pure conduction. Several investigators applied finite differences to solve heat-conduction problems involving change of phase.

Foster [19] investigated the solidification (or melting) of materials initially at fusion temperature. Citron [20] obtained a numerical solution for the melting of a slab insulated on one side and exposed to a varying temperature on the other side. Sunderland and Grosh [21] studied the temperature distribution in a semi-infinite solid initially at a constant temperature, then suddenly heated by convection at the boundary surface; the molten material was assumed to be removed from the surface as soon as it was formed. Murray and Landis [22] proposed an improved numerical method for the solution of one-dimensional solidification (or melting) problem that yielded greater accuracy than the conventional finite differences especially in the region near the fusion front. Springer and Olson [23] extended the method used by Murray and Landis [22] to the problem of solidification for a two-dimensional (r, z), multiregion (in the r-direction) hollow cylinder with variable thermal conductivity.

In the problems of heat conduction involving change of phase the finite-difference representation of derivatives of temperature with respect to the space variable at the nodal points near the solid–liquid interface require special consideration because the solid–liquid interface is moving. To illustrate the procedure we examine finite-difference representation of space derivatives for a solidification problem. For this purpose we choose a simplified version of the solidification problem solved by Springer and Olson [23] who used a finite-differencing scheme which is similar to one of the methods proposed by Murray and Landis [22].

Consider solidification in the r direction of a liquid contained between the walls of a long, hollow cylinder of inner and outer radii r_i and r_0 respectively, as a result of removal of heat uniformly at the inner surface of the cylinder. Prior to the start of solidification, temperature distribution within the liquid can be determined by the conventional heat transfer analysis. After the start of solidification the analysis becomes more involved because the problem involves a solid–liquid boundary which moves in the positive r direction. Figure 9-10 shows the geometry under consideration and the position of the solid–liquid interface, ϵ, at any instant of time t after the start of solidification. The mathematical formulation of the problem of solidification is as follows.

For the solid region:

$$\frac{1}{\alpha_s}\frac{\partial \theta_s}{\partial t} = \frac{\partial^2 \theta_s}{\partial r^2} + \frac{1}{r}\frac{\partial \theta_s}{\partial r} \tag{9-121}$$

For the liquid region:

$$\frac{1}{\alpha_L}\frac{\partial \theta_L}{\partial t} = \frac{\partial^2 \theta_L}{\partial r^2} + \frac{1}{r}\frac{\partial \theta_L}{\partial r} \tag{9-122}$$

where $\theta \equiv \dfrac{T - T_F}{T_F}$

T_F = fusion temperature

and subscripts s and L refer to solid and liquid phases respectively.

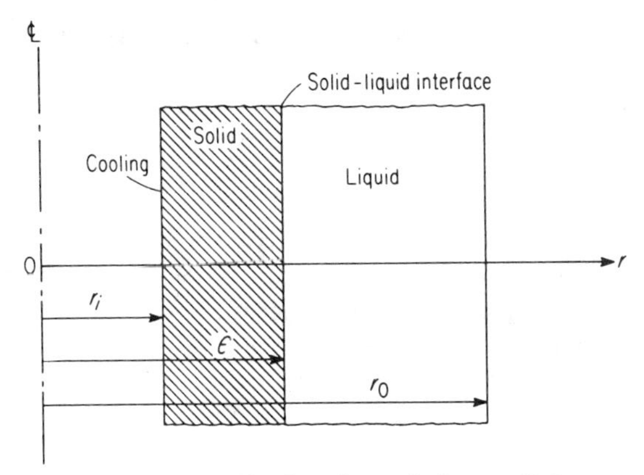

FIG. 9-10. Solidification in a hollow cylinder (r_i, r_0).

The above two equations are coupled at the solid–liquid interface with the following relation (see Eq. 7-103).

$$\frac{d\epsilon}{dt} = \frac{T_F}{\rho L}\left[k_s \frac{\partial \theta_s}{\partial r}\bigg|_\epsilon - k_L \frac{\partial \theta_L}{\partial r}\bigg|_\epsilon\right] \tag{9-123}$$

where $\rho_s = \rho_L = \rho$ is assumed
L = latent heat of fusion

Equations 9-121, 9-122, and 9-123 are expressed in dimensionless form as

$$\frac{\partial \theta_s}{\partial \bar{t}} = \beta_s \left[\frac{\partial^2 \theta_s}{\partial \bar{r}^2} + \frac{1}{1+\bar{r}}\frac{\partial \theta_s}{\partial \bar{r}}\right] \tag{9-124}$$

$$\frac{\partial \theta_L}{\partial \bar{t}} = \beta_L \left[\frac{\partial^2 \theta_L}{\partial \bar{r}^2} + \frac{1}{1+\bar{r}}\frac{\partial \theta_L}{\partial \bar{r}}\right] \tag{9-125}$$

$$\frac{d\bar{\epsilon}}{d\bar{t}} = \frac{\partial \theta_s}{\partial \bar{r}} - K\frac{\partial \theta_L}{\partial \bar{r}} \tag{9-126}$$

where the dimensionless quantities are defined as

$$\begin{aligned} \bar{r} &= \frac{r - r_i}{r_i} \\ \bar{t} &= \frac{T_F k_s}{\rho L r_i^2} t \\ K &= \frac{k_L}{k_s} \\ \bar{\epsilon} &= \frac{\epsilon - r_i}{r_i} \\ \beta_L &= k_L \left(\frac{L}{k_s c_p T_F}\right) \qquad \beta_s = k_s \left(\frac{L}{k_s c_p T_F}\right) \end{aligned} \tag{9-127}$$

We now examine the finite difference representation of various derivatives in Eqs. 9-124, 9-125, and 9-126.

We subdivide the region into small intervals $\Delta\bar{r}$ such that

$$\bar{r} = j \cdot \Delta\bar{r} \qquad j = 0,1,2,\ldots,N \tag{9-128}$$

and the time domain into small time steps $\Delta\bar{t}$ such that

$$\bar{t} = n \cdot \Delta\bar{t} \qquad n = 0,1,2,3,4,\ldots \tag{9-129}$$

Figure 9-11 shows the $\bar{r}$ network drawn over the region and the position of the solid–liquid interface. Since the solid–liquid interface does not necessarily coincide with the nodal points, its space coordinate is denoted by

$$\bar{\epsilon} = q \cdot \Delta\bar{r} + \xi \cdot \Delta\bar{r} \tag{9-130}$$

where q = coordinate of the nodal point near the solid–liquid interface (i.e., $j = q$), as shown in Fig. 9-11

$\xi\Delta\bar{r}$ = distance of the solid–liquid interface from the nodal point q, and $-0.5 \le \xi \le 0.5$

The finite-difference representation of time derivatives of functions θ and $\bar{\epsilon}$ are given as

$$\frac{\partial\theta}{\partial t} = \frac{\theta_j^{n+1} - \theta_j^n}{\Delta\bar{t}} \qquad \begin{aligned} j &= 0, 1, 2, \ldots, N \\ n &= 0, 1, 2, 3, \ldots \end{aligned} \tag{9-131}$$

$$\frac{d\bar{\epsilon}}{d\bar{t}} = \frac{\epsilon_j^{n+1} - \bar{\epsilon}_j^{\,n}}{\Delta\bar{t}} \qquad \begin{aligned} j &= 0, 1, 2, \ldots, N \\ n &= 0, 1, 2, 3, \ldots \end{aligned} \tag{9-132}$$

The finite-difference equations 9-131 and 9-132 for time derivatives are applicable at all nodal points in the region. On the other hand, the finite differences for the space derivatives depend on the position of the nodal points relative to the solid-liquid interface and the boundaries. Therefore, we examine the finite-difference representation of the space derivatives for each such location.

Nodal Points Away from the Solid–Liquid Interface and the Boundaries. Consider the nodal points which are remote from the solid–liquid inter-

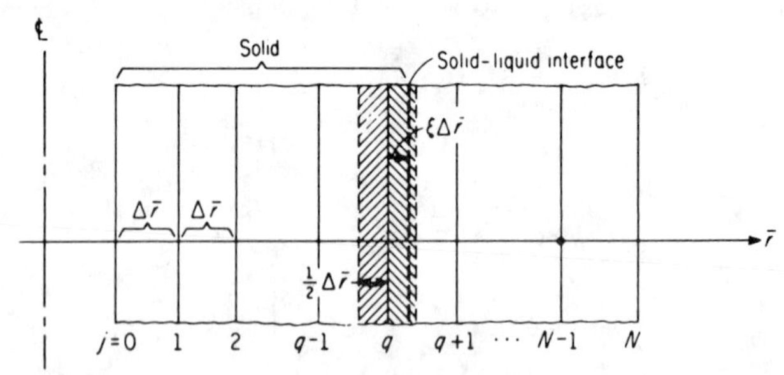

FIG. 9-11. The $\bar{r}$ network subdividing a hollow cylinder into equal space intervals $\Delta\bar{r}$.

face and the boundaries. The finite-difference representation of space derivatives of the temperature at such a nodal point j, are given as

$$\left.\frac{\partial \theta}{\partial \bar{r}}\right|_j = \frac{\theta_{j+1}^n - \theta_{j-1}^n}{2\Delta \bar{r}} \qquad \text{all nodal points except } j = 0, N, q, (q+1), (q-1) \tag{9-133}$$

$$\left.\frac{\partial^2 \theta}{\partial \bar{r}^2}\right|_j = \frac{\theta_{j-1}^n + \theta_{j+1}^n - 2\theta_j^n}{(\Delta \bar{r})^2} \qquad \text{all nodal points except } j = 0, N, q, (q+1), (q-1) \tag{9-134}$$

which are similar in form to the finite-difference approximation of the space derivatives discussed in the previous sections.

Nodal Points Near the Solid–Liquid Interface. Let $j = q$ be the nodal point at a distance of $\xi \Delta r$ from the solid–liquid interface as shown in Fig. 9-12. The finite-difference approximation of the space derivatives of

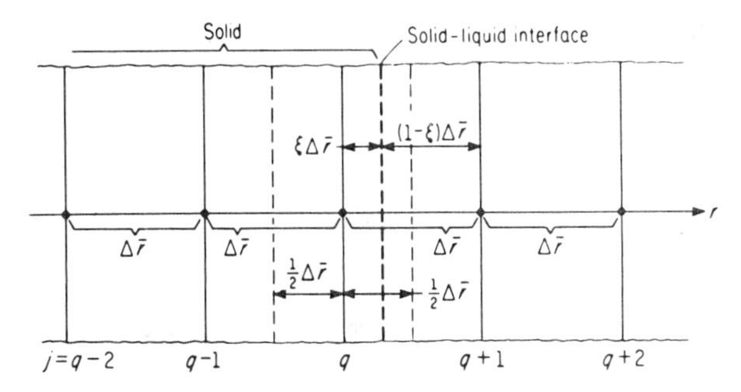

FIG. 9-12. The nodal points in the neighborhood of the solid–liquid interface.

temperature at the nodal points $(q - 1)$, q and $(q + 1)$ may be determined as described below.

Nodal Point $(q + 1)$. Finite-difference representation of space derivatives $\partial T/\partial r$ and $\partial^2 T/\partial r^2$ at the nodal point $(q + 1)$ is obtainable by Taylor series expansion about the nodal point $(q + 1)$ of the temperatures at the solid–liquid interface and at the nodal point $(q + 2)$.

Let T_F = temperature at the solid–liquid interface, i.e., the fusion temperature, and

$(1 - \xi)\Delta \bar{r}$ = distance between the nodal point $(q + 1)$ and the solid-liquid interface.

The Taylor series expansion of T_F in powers of $-(1 - \xi)\Delta \bar{r}$ about the nodal point $(q + 1)$ gives

$$T_F = T_{q+1} - (1 - \xi) \cdot \Delta \bar{r} \left.\frac{\partial T}{\partial \bar{r}}\right|_{q+1} + \frac{(1-\xi)^2(\Delta \bar{r})^2}{2!} \left.\frac{\partial^2 T}{\partial \bar{r}^2}\right|_{q+1} + 0\,(\Delta \bar{r})^3 \tag{9-135}$$

and the expansion of T_{q+2} in powers of $\Delta\bar{r}$ about the nodal point $(q + 1)$ gives

$$T_{q+2} = T_{q+1} + \Delta\bar{r}\left.\frac{\partial T}{\partial \bar{r}}\right|_{q+1} + \frac{(\Delta\bar{r})^2}{2!}\left.\frac{\partial^2 T}{\partial \bar{r}^2}\right|_{q+1} + 0(\Delta\bar{r})^3 \qquad (9\text{-}136)$$

Eliminating $\left.\frac{\partial^2 T}{\partial \bar{r}^2}\right|_{q+1}$ between Eqs. 9-135 and 9-136 we obtain the finite-difference expression for the first-order space derivative of temperature at the nodal point $(q + 1)$ as

$$\left.\frac{\partial T}{\partial \bar{r}}\right|_{q+1} = \frac{1}{\Delta\bar{r}}\left[-\frac{1}{(1-\xi)(2-\xi)}T_F + \frac{\xi}{1-\xi}T_{q+1} + \frac{1-\xi}{2-\xi}T_{q+2}\right] \qquad (9\text{-}137)$$

with a truncation error of the order of $(\Delta\bar{r})^2$.

Defining a dimensionless temperature at the time step (n) as

$$\theta^n = \frac{T - T_F}{T_F} \qquad (9\text{-}138)$$

Eq. 9-137 reduces to

$$\left.\frac{\partial \theta^n}{\partial \bar{r}}\right|_{q+1} = \frac{1}{\Delta\bar{r}}\left[\frac{\xi}{1-\xi}\theta^n_{q+1} + \frac{1-\xi}{2-\xi}\theta^n_{q+2}\right] \quad \text{for } q \le N - 2 \qquad (9\text{-}139)$$

Similarly, by eliminating $\left.\frac{\partial T}{\partial \bar{r}}\right|_{q+1}$ between Eqs. 9-135 and 9-136 it can be shown that the finite-difference representation of the second space derivative of temperature at the nodal point $(q + 1)$ is given as

$$\left.\frac{\partial^2 \theta^n}{\partial \bar{r}^2}\right|_{q+1} = \frac{2}{(\Delta\bar{r})^2}\left[\frac{1}{2-\xi}\theta^n_{q+2} - \frac{1}{1-\xi}\theta^n_{q+1}\right] \quad \text{for } q \le N - 2 \qquad (9\text{-}140)$$

with a truncation error of the order of $(\Delta\bar{r})$.

Nodal Point $(q - 1)$. The finite-difference representation of space derivatives of temperature at the nodal point $j = (q - 1)$ is obtained by the Taylor series expansion in a similar manner.

The Taylor series expansion of T_F in powers of $(1 + \xi)\cdot\Delta\bar{r}$ about the nodal point $(q - 1)$ gives

$$T_F = T_{q-1} + (1+\xi)\cdot\Delta\bar{r}\left.\frac{\partial T}{\partial \bar{r}}\right|_{q-1} + \frac{(1+\xi)^2\cdot(\Delta\bar{r})^2}{2!}\left.\frac{\partial^2 T}{\partial \bar{r}^2}\right|_{q-1} + 0(\Delta\bar{r})^3 \qquad (9\text{-}141)$$

A similar expansion of T_{q-2} in powers of $-\Delta\bar{r}$ about $(q - 1)$ gives

$$T_{q-2} = T_{q-1} - \Delta\bar{r}\left.\frac{\partial T}{\partial \bar{r}}\right|_{q-1} + \frac{(\Delta\bar{r})^2}{2!}\left.\frac{\partial^2 T}{\partial \bar{r}^2}\right|_{q-1} + 0(\Delta\bar{r})^3 \qquad (9\text{-}142)$$

From Eqs. 9-141 and 9-142 we obtain

$$\left.\frac{\partial \theta^n}{\partial \bar{r}}\right|_{(q-1)} = \frac{1}{\Delta \bar{r}}\left[\frac{\xi}{1+\xi}\,\theta^n_{q-1} - \frac{1+\xi}{2+\xi}\,\theta^n_{q+2}\right] \quad \text{for } q \geq 2 \qquad (9\text{-}143)$$

with truncation error of the order of $(\Delta \bar{r})^2$; and

$$\left.\frac{\partial^2 \theta^n}{\partial \bar{r}^2}\right|_{(q-1)} = \frac{2}{(\Delta \bar{r})^2}\left[\frac{1}{2+\xi}\,\theta^n_{q-2} - \frac{1}{1+\xi}\,\theta^n_{q-1}\right] \quad \text{for } q \geq 2 \qquad (9\text{-}144)$$

with a truncation error of the order of $\Delta \bar{r}$.

Nodal Point q. When the nodal point ($j = q$) lies between the solid–liquid interface and the inner radius as shown in Fig. 9-12, a backward-difference scheme is used to express the space derivatives of the temperature at the nodal point q in terms of the values of the temperature at the nodal points $(q - 1)$ and $(q - 2)$. Such a finite-difference scheme is included among the relations given in the previous section [i.e. Eqs. 9-109 and 9-116] and we have

$$\left.\frac{\partial \theta}{\partial \bar{r}}\right|_{q} = \frac{1}{2\Delta \bar{r}}\,[\theta^n_{q-2} - 4\theta^n_{q-1} + 3\theta^n_{q}] \quad \text{for } 2 \leq q \leq N-1 \qquad (9\text{-}145)$$

$$\left.\frac{\partial^2 \theta}{\partial r^2}\right|_{q} = \frac{1}{(\Delta \bar{r})^2}\,[\theta^n_{q-2} - 2\theta^n_{q-1} + \theta^n_{q}] \quad \text{for } 2 \leq q \leq N-1 \qquad (9\text{-}146)$$

When the nodal point q lies between the outer radius and the solid–liquid interface, the space derivatives at q are expressed using the values of temperature at $(q + 1)$ and $(q + 2)$. Such finite-difference representations can be obtained from the relations given in the previous section [i.e. Eqs. 9-107 and 9-114] and given as

$$\left.\frac{\partial \theta}{\partial \bar{r}}\right|_{q} = \frac{1}{2\Delta \bar{r}}\,[-3\theta^n_{q} + 4\theta^n_{q+1} - \theta^n_{q+2}] \quad \text{for } 2 \leq q \leq N-2 \qquad (9\text{-}147)$$

$$\left.\frac{\partial^2 \theta}{\partial \bar{r}^2}\right|_{q} = \frac{1}{(\Delta \bar{r})^2}\,[\theta^q_{n} - 2\theta^n_{q+1} + \theta^n_{q+2}] \quad \text{for } 2 \leq q \leq N-2 \qquad (9\text{-}148)$$

Coupling Condition at the Solid-Liquid Interface. The coupling condition at the solid-liquid interface was given as

$$\frac{d\bar{\epsilon}}{dt} = \left.\frac{\partial \theta_s}{\partial \bar{r}}\right|_{\bar{\epsilon}} - K\left.\frac{\partial \theta_L}{\partial \bar{r}}\right|_{\bar{\epsilon}} \qquad (9\text{-}149)$$

The finite-difference representation of the time derivative on the left-hand side of this equation was given previously (i.e., Eq. 9-132). The finite-difference approximation of the space derivatives are obtained by taking the first two terms in the Taylor series expansion and by applying backward and forward schemes. The results are

$$\left.\frac{\partial \theta_s}{\partial \bar{r}}\right|_{\epsilon} = \frac{1}{\Delta \bar{r}}\left[\frac{1+\xi}{2+\xi}\,\theta^n_{q-2} - \frac{2+\xi}{1+\xi}\,\theta^n_{q-1}\right] \quad \text{for } q \geq 2 \qquad (9\text{-}150)$$

$$\left.\frac{\partial \theta_L}{\partial \bar{r}}\right|_{\xi} = \frac{1}{\Delta \bar{r}}\left[\frac{2-\xi}{1-\xi}\,\theta_{q+1}^n - \frac{1-\xi}{2-\xi}\,\theta_{q+2}^n\right] \qquad \text{for } q \le N-2 \tag{9-151}$$

Boundary Regions. The simplest of the boundary conditions is the case when the surface temperature is prescribed (i.e., boundary condition of the first kind). When the boundary condition is of the second kind or of the third kind, it includes the space derivative of temperature normal to the boundary surface. In such cases the finite-difference representation is needed for the space derivative of temperature at the boundary surfaces.

Consider, for example, the case of prescribed heat flux at the inner boundary surface in the form

$$k_s \left.\frac{\partial T}{\partial r}\right|_{j=0} = -f_0 \tag{9-152}$$

or in the dimensionless form as

$$\left.\frac{\partial \theta}{\partial \bar{r}}\right|_{j=0} = -\frac{f_0 r_i}{k_s T_F} \tag{9-153}$$

Assuming the solid-liquid interface is not adjacent to the inner boundary surface, the finite differences for the second partial derivative of temperature with respect to the space variable at the inner boundary surface is given as

$$\left.\frac{\partial^2 \theta}{\partial \bar{r}^2}\right|_{j=0} = \frac{2}{\Delta \bar{r}}\left[\frac{\theta_1^n - \theta_0^n}{\Delta \bar{r}} + \frac{f_0 \cdot r_i}{k_s T_F}\right] \qquad q \neq 0,1 \tag{9-154}$$

When the solid–liquid interface separates the boundary from the adjacent nodal point, the first term in Eq. 9-154 should be modified and it becomes

$$\left.\frac{\partial^2 \theta}{\partial \bar{r}^2}\right|_{j=0} = \frac{2}{\bar{\epsilon}^n}\left[-\frac{\theta_1^n}{\bar{\epsilon}^n} + \frac{f_0 r_i}{k_s T_F}\right] \qquad \text{for } q = 0,1 \tag{9-155}$$

It is to be noted that when the solid–liquid interface is adjacent to the boundary surface, the finite-difference equation 9-155 becomes infinite as the solid–liquid interface approaches to the inner boundary surface. This difficulty, which occurs at the start of solidification, is avoided by starting the solution at a small distance away from the boundary.

Similar consideration applies for finite differences at the outer boundary.

With the method described above differential equations of heat conduction for the solidification problem can be reduced to a set of algebraic equations, solution of which for each time step yields the temperatures at the nodal points.

The stability of the solution of finite-difference equations is an important matter. Springer and Olson [23] found that the simple stability criterion for the uncoupled equations was sufficiently restrictive to insure stability for all cases investigated.

9-13. MONTE CARLO METHODS IN THE SOLUTION OF HEAT-CONDUCTION PROBLEMS

Monte Carlo methods are recent innovations which are used to obtain approximate solutions to mathematical and physical problems. The method derives its name from the fact that it is based on probability sampling techniques. A large scale application of Monte Carlo methods is associated with the problems of radiation shielding and nuclear reactor criticality. Its application in the solutions of problems of thermal radiation can be found in a series of recent papers by Howell and Perlmutter [24, 25, 26, 27]. Its application to other branches of science and engineering is summarized in a book by Hammersley and Handscomb [28]. Use of Monte Carlo methods in the solution of partial differential equation in the form

$$\frac{\partial^2 U}{\partial x^2} + \frac{\partial^2 U}{\partial y^2} + \frac{k}{y}\frac{\partial U}{\partial y} = 0 \tag{9-156}$$

in rectangular regions subject to prescribed values of U at the boundaries is examined by Ehrlich [29]. Application of Monte Carlo methods to the solution of multidimensional steady and time dependent boundary value problems of heat conduction in finite regions of arbitrary shape, with or without internal heat generation within the solid, may be found in the book by Buslenko, et. al. [30], in the paper by Haji-Sheik and Sparrow [31] and in the Ph.D. dissertation by Haji-Sheik [32]. In the following sections we present briefly Monte Carlo methods of solution of steady-heat conduction problems; for more detailed treatment and for the solution of time-dependent problems the reader should refer to references [30, 31 and 32].

Before presenting the application of Monte Carlo methods we give a brief account of those statistical terms which arise in the analyses.

Definition of Some Statistical Terms [28]. *Probability* is a number that lies between 0 and 1, both inclusive. An event with zero probability never occurs and that with probability of unity surely occurs. A *random event* is an event which has a chance of happening and probability is a numerical measure of that chance.

Consider a set of events each characterized by a number η. A *probability distribution function* $F(\xi)$ is defined as the probability that the event which occurs has a value of η not exceeding ξ. If $T(\eta)$ is a function of η, the *expected value* (or the *mean value*) of $T(\eta)$ is defined as

$$T_m = \int T(\xi) \cdot dF(\xi) \tag{9-157}$$

where the integral is taken over all values of ξ.

If the probability distribution function $F(\xi)$ has a derivative, i.e.,

$$\frac{dF(\xi)}{d\xi} = f(\xi) \tag{9-158}$$

Equation 9-157 is written in the form

$$T_m = \int T(\xi)\cdot f(\xi)\cdot d\xi \tag{9-159}$$

If $F(\xi)$ is a step function with steps of height f_j at the points ξ_j, Eq. 9-157 is written in the form

$$T_m = \sum_j T(\xi_j)\cdot f_j \tag{9-160}$$

The functions $f(\xi)$ and f_j in Eqs. 9-158 and 9-160 called the *probability density functions* of the random variable η.

Preliminary Considerations on the Monte Carlo Methods. To introduce the reader to the basic concepts associated with the Monte Carlo method we present a brief qualitative account of the Monte Carlo method as applied to the solution of steady heat-conduction problems in finite regions.

Consider an arbitrarily shaped, homogeneous, finite region subjected to a prescribed linear boundary condition and with heat generation within the solid at a uniform rate. Suppose it is desired to determine the steady temperature at any point P in the region. A Monte Carlo method of solving for the temperature at the point P may be summarized as follows.

A random-walking *first particle*[3] is set into motion from the point P for *random-walk* in the region; the rules governing the steps to be taken during the random-walk process will be described later in the analysis. The particle wanders from point to point in the region and eventually strikes the boundary; then the fate of the particle for subsequent motion is determined according to the type of the boundary condition at the point where the particle strikes the boundary. In this chapter, for heat conduction problems, we shall be concerned with three different types of boundary conditions; prescribed temperature, prescribed heat flux, and convection boundary conditions. When the temperature is prescribed the boundary is called an *absorbing barrier* and the walk of the particle is terminated upon striking the boundary. When the heat flux is prescribed the boundary is called a *reflecting barrier* and the particle, upon striking such a boundary, is reflected back into the region in the direction normal to the boundary and the particle continues the random-walk until it strikes another portion of the boundary where the walk may be terminated. When the boundary is subjected to convection it is called a *partially absorbing* and *partially reflecting* barrier. The particle striking such a boundary is either reflected back into the region where it continues the random-walk, or it is absorbed by the boundary and the walk is terminated. The probability for the particle being absorbed or reflected back into the region at the boundary subjected to convection depends on the

[3]The usage of "particle" has no physical significance except for convenience in describing the procedure.

relative magnitudes of the heat transfer coefficient for convection and the thermal conductivity of the solid. (These conditions will be discussed in detail later in the analysis.) It is apparent from the above description of the random-walk process that the particle which is set into motion from the point P will be eventually absorbed at the boundary and the walk will be terminated provided that all the boundaries of the region are not subjected to a boundary condition of the second kind. As soon as the walk of the first particle is terminated the *effective random-walk temperature* $T_E(1)$ for the random-walk of the first particle is recorded. The rules for evaluating the effective random-walk temperature will be discussed later in the analysis.

Second, third, ... , and Nth particles are successively released from the point P for random-walk in the region. For each of these particles, upon their absorption at the boundary, the effective random-walk temperatures $T_E(2)$, $T_E(3)$, ..., and $T_E(N)$ are recorded. For sufficiently large numbers of particles performing random-walk in the region, the sum of the effective random-walk temperatures divided by the number of particles released (N) will give the mathematical expectation of the temperature at the point P (approximate temperature at the point P).

It is to be noted that, since the Monte Carlo method is based on the laws of probability, a large number of walks should be considered if the result is to closely approximate the temperature at the point P. A vast number of calculations are needed to obtain accurate Monte Carlo solutions, but with the availability of large-capacity, high-speed digital computers such calculations can be performed with no difficulty.

Two different random-walk procedures, called the *fixed-random-walk* and the *floating random-walk*, have been considered by Haji-Sheikh and Sparrow for the solution of heat-conduction problems. In the following sections we examine the fixed random-walk and the floating random-walk procedures in the solution of steady heat-conduction problems.

9-14. THE FIXED-RANDOM-WALK MONTE CARLO IN THE SOLUTION OF STEADY-STATE PROBLEMS

To introduce the procedure in the fixed-random-walk Monte Carlo we examine the following two-dimensional, steady-state heat-conduction problem.

Consider an arbitrary-shaped, homogeneous, two-dimensional finite region subjected to prescribed temperature at the boundaries (i.e., boundary condition of the first kind), with heat generation within the solid at a uniform rate of G Btu/hr ft^3. Suppose we desire to determine the steady temperature at an arbitrary point o in the region.

The region is interlaced with a square network of mesh size l, so that the point o is a mesh point, as shown in Fig. 9-13. The steady-state tem-

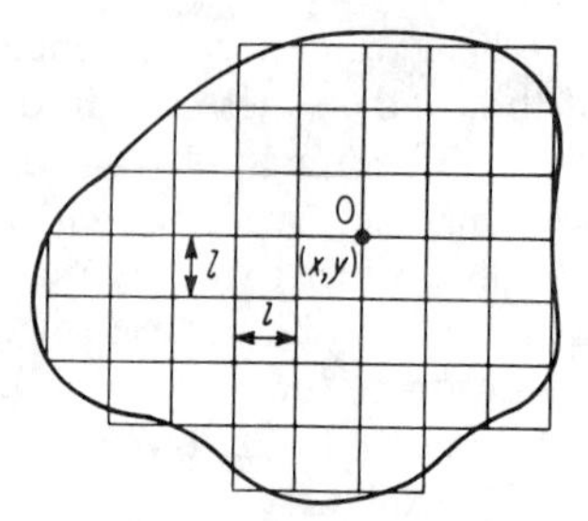

FIG. 9-13. An arbitrary-shaped two-dimensional finite region interfaced with a square network of mesh size l.

perature in the region satisfies Poisson's equation, and the corresponding finite-difference representation of this equation for the temperature $T(x, y)$ at the nodal point $0(x, y)$ is given in the form

$$T(x,y) = p_{x^+} \cdot T(x + l, y) + p_{x^-} \cdot T(x - l, y) + p_{y^+} \cdot T(x, y + l) + p_{y^-} \cdot T(x, y - l) + \frac{Gl^2}{4} \tag{9-161a}$$

where $p_{x^+} = p_{x^-} = p_{y^+} = p_{y^-} = \frac{1}{4}$ (9-161b)

G = heat-generation rate per unit volume
l = mesh size
k = thermal conductivity

A probabilistic interpretation of the finite-difference equation 9-161 is as follows. Consider a random-walking particle instantaneously situated at the nodal point $0(x, y)$ and prepared to step to one of the four neighboring nodal points $(x + l, y)$, $(x - l, y)$, $(x, y + l)$ or $(x, y - l)$. The particle has equal probability to step to any one of these four neighboring points (i.e., 0.25). The direction in which the particle takes the first step is determined by the following considerations.

Suppose a digital-computer program is available to supply upon request a random number of magnitude between zero and unity.[4] The magnitude of this random number M is related to the direction in which the particle at $0(x, y)$ takes the first step, by the following consideration.

$$\begin{aligned}
&\text{If } 0 \le M < 0.25, \text{ particle steps from } (x, y) \text{ to } (x + l, y)\\
&\text{if } 0.25 \le M < 0.50, \text{ particle steps from } (x, y) \text{ to } (x - l, y)\\
&\text{if } 0.50 \le M < 0.75, \text{ particle steps from } (x, y) \text{ to } (x, y + l)\\
&\text{if } 0.75 \le M < 1, \text{ particle steps from } (x, y) \text{ to } (x, y - l)
\end{aligned} \tag{9-162}$$

[4] The standard procedure is available to produce random numbers uniformly distributed between two given limits [34]. Computer subroutines are available for generating random numbers [35]. A computer program for generating uniformly distributed random numbers between 0 and 1 is given in Appendix A of Ref. 31.

A *first* particle situated at the point $0(x, y)$ takes the first step to one of the four neighboring mesh points according to the rules set up by Eq. 9-162 as soon as the digital computer supplies a random number of magnitude M between 0 and 1. The second, third, ..., steps of the *first* particle are accomplished by similar considerations. The particle wanders from point to point in the region and eventually reaches to a nodal point situated on the boundary (i.e., the particle strikes the boundary). Since the boundary condition for the present problem is assumed to be of the first kind, the boundary is an *absorbing barrier* and the walk is terminated at the boundary. (The probability for an infinitely long walk in the region is zero [32].) The effective random-walk temperature for the random-walk of the first particle, $T_E(1)$ is given as

$$T_E(1) = T_b(1) + m_1 \cdot \left(\frac{Gl^2}{4k}\right) \tag{9-163}$$

where $T_b(1)$ = temperature of the boundary at the point where the random walk of the first particle is terminated

m_1 = number of steps executed by the first random-walking particle to reach the boundary

G = heat-generation rate per unit volume

l = mesh size

k = thermal conductivity

It is to be noted that the quantity $Gl^2/4k$ which is scored following each step is the contribution to the temperature of each nodal point of heat generation within the square mesh. Therefore, the effective random-walk temperature given by Eq. 9-163 is equal to the boundary temperature $T_b(1)$ plus the temperature rise due to heat generation at each nodal point the particle has traveled.

A second, third, fourth, ..., and Nth random-walking particles are successively released from the point $0(x, y)$ for random-walk in the region as described above; upon the termination of each walk at the boundary the corresponding effective random-walk temperatures $T_E(2)$, $T_E(3)$, $T_E(4), \ldots$, and $T_E(N)$ are recorded.

The Monte Carlo solution for the *expected temperature* $T_m(x, y)$ at the point (x, y) is related to the effective random-walk temperatures by the following relation.

$$\begin{aligned} T_m(x, y) &= \frac{1}{N}\sum_{j=1}^{N} T_E(j) \\ &= \frac{1}{N}\sum_{j=1}^{N}\left[T_b(j) + m_j \cdot \left(\frac{Gl^2}{4k}\right)\right] \\ &= \frac{1}{N}\sum_{j=1}^{N} T_b(j) + \left(\frac{Gl^2}{4k}\right)\cdot\frac{1}{N}\sum_{j=1}^{N} m_j \end{aligned} \tag{9-164}$$

In the limit as $N \to \infty$ the expected temperature $T_m(x, y)$ as given by Eq. 9-164 is identical with the finite-difference solution of the same problem by using a square network of mesh size l. The error involved from the finiteness of N may be estimated by making use of the central-limit theorem of probability theory [33].

It is to be noted that the Monte Carlo method of solution as described above is capable of solving for the temperature of any isolated point in the region without solving for the temperature of the other mesh points. With the finite differences, on the other hand, simultaneous solution of the temperatures at all the nodal points in the region is required. The Monte Carlo method of solution is effective if the temperature is to be determined at only a few points in the region; if temperature is to be determined at a large number of points, the Monte Carlo methods are somewhat slow relative to other methods.

The fixed-random-walk Monte Carlo method described above has also disadvantages. Since the Monte Carlo method is a probability method, the dispersion of Monte Carlo results about the actual temperature becomes smaller as N becomes larger. For the fixed-random-walk procedure described above, the Monte Carlo solution approaches to the corresponding finite-difference solution which is already an approximate solution. For fixed-random-walk Monte Carlo the length of each step and the location of mesh points are fixed by constructing a square network over the region, for arbitrary-shaped bodies there is the difficulty that the mesh points do not necessarily coincide with the boundary of the region.

When the boundary condition is of the second or of the third kind the particle is reflected back into the solid normal to the boundary. For boundaries of irregular shape the point where the particle is reflected may not coincide with the mesh point, hence representation of boundary conditions involving the normal derivative of temperature becomes highly approximate.

By employing a floating-random-walk Monte Carlo method which will be described next, the above disadvantages are eliminated.

9-15. THE FLOATING-RANDOM-WALK MONTE CARLO METHOD IN THE SOLUTION OF STEADY-STATE PROBLEMS

The basis of the floating-random-walk Monte Carlo as applied to the solution of steady temperature in two-dimensional regions is deduced by considering the center temperature T_0 for a circular region of radius r_0 subjected to a prescribed temperature $T(r_0, \omega)$ at the circumference (ω being the angular coordinate). We assume further that there is heat generation in this circular region at a uniform rate of G Btu/hr ft^3.

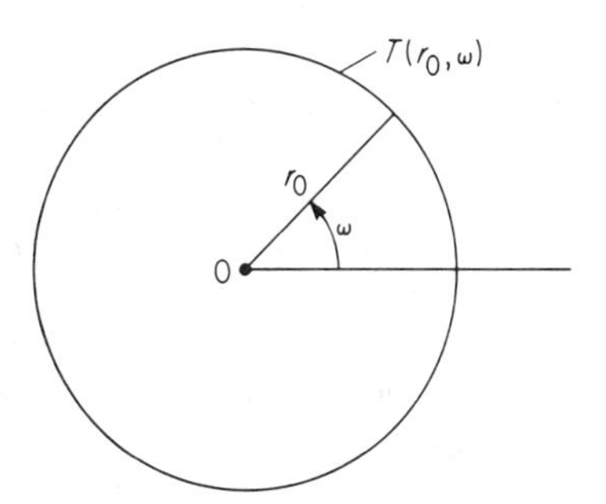

FIG. 9-14. A two-dimensional circular region subjected to prescribed temperature at the circumference.

Figure 9-14 shows the geometry under consideration. The center temperature T_0 can be related to the boundary temperature $T(r, \omega)$ as follows:

$$T_0 = \frac{1}{2\pi} \int_0^{2\pi} \left[T(r_0, \omega) + \frac{Gr_0^2}{4k} \right] d\omega \tag{9-165a}$$

$$= \int_0^1 \left[T(r_0, \omega) + \frac{Gr_0^2}{4k} \right] \cdot dF(\omega) \tag{9-165b}$$

where $F(\omega) = \dfrac{\omega}{2\pi}$

G = heat-generation rate per unit volume
k = thermal conductivity

It is to be noted that Eq. 9-165 follows from the application of the fixed-random-walk procedure with mesh size, $l \to 0$; therefore, it is an exact representation of the center temperature.

$F(\omega)$ is considered as the probability distribution function and its derivative

$$f = \frac{dF(\omega)}{d\omega} = \frac{1}{2\pi}$$

the probability density function. The constant value of the probability density function implies that for a random-walking particle situated at the center of the circle, every angular position ω is equally probable.

The relationship given by Eq. 9-165 is the basis of floating-random-walk Monte Carlo Method for the solution of steady-state heat-conduction problem in two-dimensional finite regions of arbitrary shape.

Consider an arbitrary-shaped, homogeneous, two-dimensional, finite region subjected to a prescribed temperature at the boundaries, and assume that there is heat generation within the solid at a uniform rate of

G Btu/hr ft^3. We wish to determine the steady temperature at a point $o(x_0, y_0)$ in the region using floating random-walk Monte Carlo method. The procedure is as follows.

A circle with center at $o(x_0, y_0)$ and radius r_0 equal to the shortest distance between $o(x_0, y_0)$ and the boundary of the region is constructed as shown in Fig. 9-15. A random-number magnitude between 0 and 1 is supplied by a digital computer. When this number is multiplied by 2π it gives the angular position ω_0 on the circumference of this circle to which the random-moving particle situated at the center $o(x_0, y_0)$ will step. The coordinate of this point on the circumference of the circle is designated by (x_1, y_1) and expressible as

$$x_1 = x_0 + r_0 \cos \omega_0 \qquad y_1 = y_0 + r_0 \sin \omega_0 \tag{9-166}$$

The *first particle* which is set into motion from $o(x_0, y_0)$ takes the first step to the point (x_1, y_1).

Then a circle is drawn with center at (x_1, y_1) and radius r_1 equal to the shortest distance between (x_1, y_1) and the boundary of the region. The computer supplies, upon request, a number of random magnitude between 0 and 1. When this number is multiplied by 2π the angular position ω_1 is determined, and the coordinates of the corresponding point (x_2, y_2) on the circumference of this circle is evaluated from

$$x_2 = x_1 + r_1 \cos \omega_1 \qquad y_2 = y_1 + r_1 \sin \omega_1 \tag{9-167}$$

The random-walk particle takes the second step from (x_1, y_1) to (x_2, y_2). The third, fourth, ..., steps are determined with the same procedure as described above. The *first* random-walking particle wanders from point to point in the region and when it reaches within a strip of preassigned width from the boundary the particle is considered has reached the boundary. Later in the analysis more will be discussed on the choice of the width of such a band around the boundary. Since the boundary con-

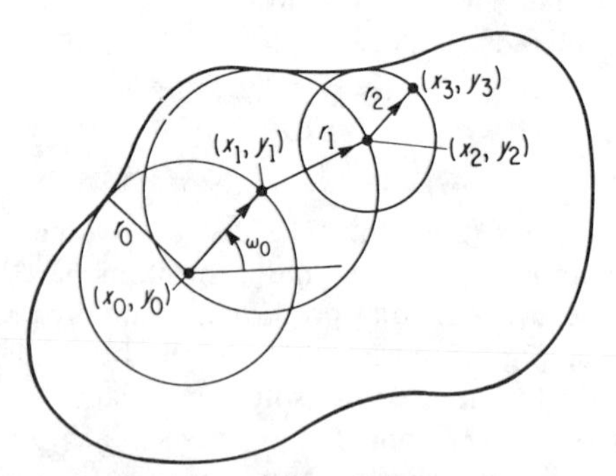

FIG. 9-15. Floating random-walk for arbitrarily shaped, two-dimensional finite region.

dition is assumed to be of the first kind, the boundary is an *absorbing* barrier and the random walk is terminated at the point where the particle reaches a point within this strip; the corresponding *effective random-walk temperature*, for the random-walk of the first particle, $T_E(1)$ is recorded.

A second, third, ..., Nth random-walking particles are successively released from thc point $o(x_0, y_0)$ for a floating-random-walk in the region and upon the termination of each walk at the boundary the corresponding effective random-walk temperatures $T_E(2)$, $T_E(3)$, ..., $T_E(N)$ are recorded.

For the floating-random-walk process described above the *effective random-walk temperature* for the walk of the jth particle is given as

$$T_E(j) = T_b(j) + \sum_{i=0}^{m_j-1} \frac{Gr_{ij}^2}{4k} \tag{9-168}$$

where $T_b(j)$ = temperature of the boundary at the point where the random walk of the jth particle is terminated

m_j = number of steps executed by the jth random-walking particle

r_{ij} = radius of the circle for the $(i+1)$th step of the jth random-walking paarticle

G = heat generation per unit volume

k = thermal conductivity

It is to be noted that the term $(Gr_{ij}^2/4k)$ is the contribution to the temperature due to heat generation at a rate of G Btu/hr ft^3 within the solid.

If N particles are set into motion from the point (x_0, y_0), the *expected temperature* $T_m(x_0, y_0)$ at the point (x_0, y_0) is equal to the mean value of the N effective-random-walk temperatures recorded and is given as

$$\begin{aligned} T_m(x_0, y_0) &= \frac{1}{N}\sum_{j=1}^{N} T_E(j) \\ &= \frac{1}{N}\sum_{j=1}^{N}\left[T_b(j) + \sum_{i=0}^{m_j-1}\frac{Gr_{ij}^2}{4k}\right] \\ &= \frac{1}{N}\sum_{j=1}^{N} T_b(j) + \frac{1}{N}\sum_{j=1}^{N}\left[\sum_{i=0}^{m_j-1}\frac{Gr_{ij}^2}{4k}\right] \end{aligned} \tag{9-169}$$

In the limit as $N \to \infty$, the solution given by Eq. 9-169 approaches the *exact* solution, because the basic equation 9-165 from which the floating random walk is deduced is exact. On the other hand, the solution with fixed-random-walk Monte Carlo method, as $N \to \infty$, approaches the finite-difference solution which is already approximate.

Another advantage of the floating-random-walk Monte Carlo over the fixed-random-walk Monte Carlo is that there is no difficulty in dealing with boundary conditions that involve normal derivatives of temperature at the boundary. This is because the method is indifferent to the shape of the boundary and that the particle approaching the boundary is moving normal to it.

Extension to Three-Dimensional Problems. Extension of the floating-random-walk Monte Carlo to the solution of three-dimensional steady-state problems is done by a similar procedure as outlined above.

Consider an arbitrary-shaped, three-dimensional, homogeneous, finite region with internal heat generation and prescribed temperature distribution at the boundaries. Suppose we wish to determine the steady-state temperature at the point $o(x_0, y_0, z_0)$ in the region. We construct a sphere with center at (x_0, y_0, z_0) and radius r_0 equal to the shortest distance between (x_0, y_0, z_0) and the boundary surface of the region.

Let $T(r_0, \omega, \psi)$ be the temperature distribution at the surface of this sphere. The exact solution of the center temperature $T(x_0, y_0, z_0)$ is expressible in terms of the temperature distribution at the surface of this sphere in the form

$$T(x_0, y_0, z_0) = \frac{1}{4\pi} \int_{\omega=0}^{2\pi} \int_{\psi=0}^{\pi} \left[T(r_0, \omega, \psi) + \frac{Gr_0^2}{6k} \right] \sin\psi \, d\psi \, d\omega \qquad \text{(9-170a)}$$

$$= \int_{F=0}^{1} \int_{U=0}^{1} \left[T(r_0, \omega, \psi) + \frac{Gr_0^2}{6k} \right] \cdot dU(\psi) \, dF(\omega) \qquad \text{(9-170b)}$$

where $F(\omega) = \dfrac{\omega}{2\pi}$

$U(\psi) = \frac{1}{2}(1 - \cos\psi)$

ω, ψ = polar coordinate angle and the "cone" angle respectively in the spherical coordinate system

The functions $F(\omega)$ and $U(\psi)$ are the probability distribution functions for the angular positions ω and ψ respectively; and their derivatives are the probability density functions. Since the derivative $dF(\omega)/d\omega$ is constant, for a particle situated at the center (x_0, y_0, z_0) all angles ω are equally probable; but the derivative $dU(\psi)/d\psi = \frac{1}{2}\sin\psi$ being dependent upon ψ implies that for a particle situated at (x_0, y_0, z_0), all angles ψ are not equally probable.

Now, by making use of the probability distribution functions $F(\omega)$ and $U(\psi)$ as defined above, the *first* particle which is instantaneously situated of the point (x_0, y_0, z_0) takes a first step to a point on the surface of the sphere with the following considerations.

A random number of magnitude between 0 and 1 is supplied by the digital computer for the value of $F(\omega)$. When this number is multiplied by

2π it gives the angular position ω_0 for a step to the surface of the sphere. Another random number of magnitude between 0 and 1 is supplied by the digital computer for the value of $U(\psi)$. Using this number the angular position ψ_0 is obtained from the relation

$$\psi_0 = \cos^{-1}(1 - 2U)$$

The *first* particle takes its first step from the center $o(x_0, y_0, z_0)$ to the point (x_1, y_1, z_1) on the surface of the sphere; the coordinates of this point are given as

$$\begin{aligned} x_1 &= x_0 + r_0 \cdot \sin\psi_0 \cdot \cos\omega_0 \\ y_1 &= y_0 + r_0 \cdot \sin\psi_0 \cdot \sin\omega_0 \\ z_1 &= z_0 + r_0 \cdot \cos\psi_0 \end{aligned} \tag{9-171}$$

The second step of the particle is determined by constructing a sphere with center (x_1, y_1, z_1) and radius r_1 equal to the shortest distance between the point (x_1, y_1, z_1) and the boundary surface of the region. The above procedure is repeated and the particle takes its second step from (x_1, y_1, z_1) to a point (x_2, y_2, z_2) on the surface of the second sphere. The first particle wanders in the region from point to point until it reaches a point within the strip of preassigned width around the boundary. At this point the particle considered has reached the boundary; the walk is terminated and the corresponding effective random-walk temperature $T_E(1)$ for the random walk of the first particle is recorded.

A second, third, ..., and Nth particle are released from the point $O(x_0, y_0, z_0)$ and by proceeding along the lines outlined above effective random-walk temperatures $T_E(2)$, $T_E(3)$, ..., and $T_E(N)$ are recorded.

The effective random-walk temperature for the walk of the jth particle is given as

$$T_E(j) = T_b(j) + \sum_{i=0}^{m_j-1} \frac{Gr_{ij}^2}{6k} \tag{9-172}$$

It is to be noted that Eq. 9-172 is similar to Eq. 9-168 except for the denominator of the second term on the right-hand side.

The *expected temperature*, $T_m(x_0, y_0, z_0)$, at the point (x_0, y_0, z_0) for the three-dimensional region considered above is given as

$$\begin{aligned} T_m(x_0, y_0, z_0) &= \frac{1}{N}\sum_{j=1}^{N} T_E(j) \\ &= \frac{1}{N}\sum_{j=1}^{N} T_b(j) + \frac{1}{N}\sum_{j=1}^{N}\left[\sum_{i=0}^{m_j-1} \frac{G \cdot r_{ij}^2}{6k}\right] \end{aligned} \tag{9-173}$$

Various quantities in this equation as have been defined previously.

Regions Near the Boundary. When a random-walking particle approaches the boundary with the floating random steps, in the immediate vicinity of the boundary the radius of the circle (or sphere) defining the step size may become so small that the time required to carry out the computations increases. To circumvent this difficulty Haji-Sheikh and Sparrow introduced four different types of approximate representations for the motion of the particle in the regions near the boundary. We consider here two of these approximate representations.

(1) When the radius r_i of a circle (or sphere) is smaller than a preassigned value r_{min}, the floating-random-walking particle is regarded as being located at the nearest point on the boundary. The value chosen for r_{min} should be sufficiently small; it may be chosen of the order of $l/4$, where l is the step size of an admissible finite-difference solution. Figure 9-16a shows a strip of width r_{min} along the boundary of a two-dimensional region.

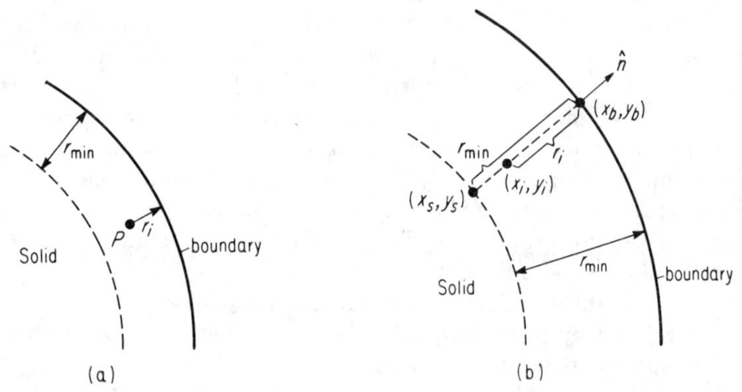

FIG. 9-16. Regions near the boundary.

(2) When the radius r_i of a circle (or sphere) is smaller than a preassigned value r_{min}, the next step of the particle is determined from the solution of the one-dimensional energy equation as applied in the direction normal to the boundary, provided there is no abrupt change in the temperature along the boundary. To illustrate this concept we consider a strip of width r_{min} along the boundary of a two-dimensional region as shown in Fig. 9-16b. Let (x_i, y_i) be the instantaneous position of the random-walking particle within this strip, at a distance r_i from the boundary. Let the normal from this point to the boundary intersect the boundary at a point (x_b, y_b) and the inner edge of the strip at (x_s, y_s). Temperature at the point (x_i, y_i) can be expressed in terms of the temperatures at the points (x_b, y_b) and (x_s, y_s) as

$$T(x_i, y_i) = \frac{r_{min} - r_i}{r_{min}} \cdot T(x_b, y_b) + \frac{r_i}{r_{min}} \cdot T(x_s, y_s) + \frac{G}{2k}(r_i r_{min} - r_i^2) \tag{9-174}$$

This result is also obtainable from Eq. 3-16 of Ref. 31 by letting $h \to \infty$. A probabilistic interpretation of Eq. 9-174 is as follows. When a random-walking particle is instantaneously situated within the strip at a point (x_i, y_i), the probability to step to the point (x_b, y_b) on the boundary is $\frac{r_{\min} - r_i}{r_{\min}}$, and the probability to step to the point (x_s, y_s) at the inner edge of the strip is $r_i/r_{\min}$. These probabilities are calculated as soon as the particle enters the strip and the particle takes the next step accordingly. If the particle goes to the inner edge of the strip the floating random-walk process is continued. The last term in Eq. 9-174 is the quantity to be scored due to heat generation within the strip.

Other Types of Boundary Conditions. In the previous sections we examined the Monte Carlo method for regions subjected to a prescribed temperature boundary condition. Such boundaries are called *absorbing barriers*. When a random-walking particle arrives at such a boundary the walk is terminated and the corresponding effective-random-walk temperature $T_E(j)$ is evaluated.

When a random-walking particle arrives at a boundary subjected to prescribed heat flux (or insulated), the particle is reflected back into the solid. When it arrives at a boundary subjected to convective boundary condition the particle is either reflected or absorbed at the boundary. The walk is terminated when the particle is absorbed, but the random-walk process is continued when the particle is reflected. We now examine the treatment of such boundary conditions.

Boundary Condition of the Second Kind. Consider a boundary subjected to a prescribed heat flux in the form

$$-k \frac{\partial T}{\partial n} = f \quad \text{at the boundary} \tag{9-175}$$

where $\frac{\partial}{\partial n}$ = differentiation along the outward-drawn normal at the boundary. Such a boundary is called a *reflecting barrier*; the physical significance of this term will now be apparent.

Figure 9-17 shows a boundary subjected to a prescribed heat flux as defined by Eq. 9-175. Let B be a point at the boundary and R a point inside the solid at a distance ΔR from B along the normal drawn to the boundary at the point B. Let T_b and T_r be the temperatures at the points B and R respectively. For generality we assume internal heat generation with the solid.

A Taylor series expansion about the point R, according to Eq. 9-5, yields

$$T_r = T_b - \Delta r \left(\frac{\partial T}{\partial x}\right)_B + \frac{(\Delta r)^2}{2} \left(\frac{\partial^2 T}{\partial x^2}\right)_B \tag{9-176}$$

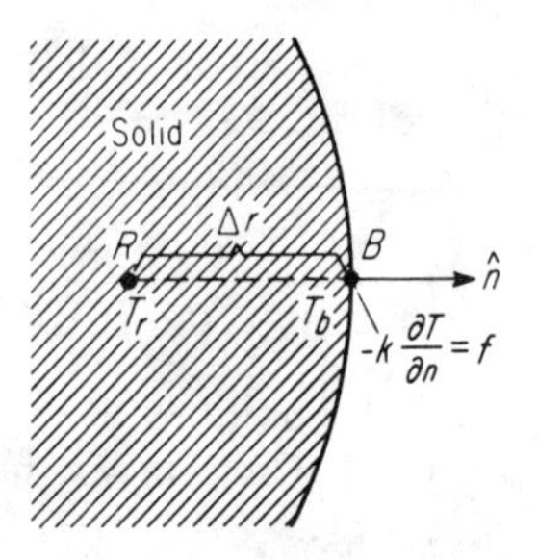

FIG. 9-17. The reflection distance Δr for a boundary subjected to prescribed heat flux.

where

$$\left(\frac{\partial T}{\partial x}\right)_B = -\frac{f}{k} \quad \text{from Eq. 9-175}$$

$$\left(\frac{\partial^2 T}{\partial x^2}\right)_B = -\frac{G}{k} \quad \text{from the heat-conduction equation}$$

Solving Eq. 9-176 for T_b,

$$T_b = T_r + \left[\frac{(\Delta r)^2}{2k} G - \Delta r \cdot \frac{f}{k}\right] \tag{9-177}$$

A probabilistic interpretation of Eq. 9-177 is as follows. When a particle is instantaneously situated at a boundary point B subjected to a prescribed heat flux, the probability of being reflected to the interior point R is unity. That is, the particle is surely reflected back and such a boundary is considered as a reflecting barrier. When the particle is reflected, the following quantity

$$\frac{(\Delta r)^2}{2k} G - \Delta r \cdot \frac{f}{k}$$

should be recorded and added to the *effective random-walk* temperature for the walk under consideration. The particle which is reflected continues the random-walk process until it is absorbed by an absorbing boundary.

Boundary Condition of the Third Kind. Consider a boundary subjected to a convective boundary condition of the form

$$-k \frac{\partial T}{\partial n} = h(T - T_e)$$

or

$$k \frac{\partial T}{\partial n} + hT = hT_e \quad \text{at the boundary} \tag{9-178}$$

where $\dfrac{\partial}{\partial n}$ = differentiation along the outward-drawn normal

T_e = temperature of the environment

Figure 9-18 shows a point B on the boundary which is subjected to convection as defined by Eq. 9-178 and a point R inside the solid at a distance Δr from B along the normal drawn to the boundary at B. Let T_b and T_r

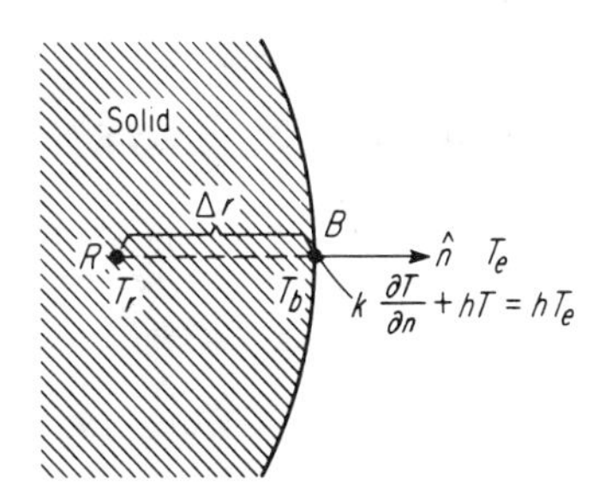

FIG. 9-18. The reflection distance Δr for a boundary subjected to convective boundary condition.

be the temperatures at the points B and R respectively. For generality assume heat generation within the solid at a rate of G Btu/hr ft^3.

Temperature T_b at the boundary is related to the temperatures T_r and T_e by the following relation:

$$T_b = \frac{1}{1 + \dfrac{h\Delta r}{k}} T_r + \frac{h\Delta r/k}{1 + \dfrac{h\Delta r}{k}} T_e + \frac{\Delta r^2/2k}{1 + \dfrac{h\Delta r}{k}} \cdot G \tag{9-179}$$

A probabilistic interpretation of Eq. 9-179 is as follows. When a random-walking particle is instantaneously situated at the boundary point B, the probability to step to the point R inside the region (i.e., being reflected back into the solid) is $1/\left(1 + \dfrac{h\Delta r}{r}\right)$, and the probability for being absorbed is

$$\frac{\dfrac{h\Delta r}{k}}{1 + \dfrac{h\Delta r}{k}}$$

When the particle is reflected back into the solid, the random-walk procedure is continued and the quantity

$$\frac{\Delta r^2/2k}{1 + \dfrac{h\Delta r}{k}} G$$

is recorded in evaluating the effective random-walk temperature. When the particle is absorbed, the walk is terminated and the quantity

$$T_e + \frac{\Delta r^2/2k}{1 + \dfrac{h\Delta r}{k}} G$$

is recorded in evaluating the effective random-walk temperature.

REFERENCES

1. L. Fox, *Numerical Solution of Ordinary and Partial Differential Equations*, Addison-Wesley Publishing Company, Inc., Reading, Mass., 1962.

2. Gordon D. Smith, *Numerical Solution of Partial Differential Equations with Exercises and Worked Solutions*, Oxford University Press, London, 1965.

3. Robert D. Richtmeyer, *Difference Methods for Initial Value Problems*, Interscience Publishers, Inc., New York, 1957.

4. D. D. McCracken, *A Guide to FORTRAN Programming*, John Wiley & Sons, Inc., New York, 1961.

5. H. Bottenbruch, "Structure and Use of ALGOL 60," *J. Assoc. Comp. Mach.*, Vol. 9 (1962), pp. 161–221.

6. J. Crank and P. Nicolson, "A Practical Method for Numerical Evaluation of Solutions of Partial Differential Equations of the Heat Conduction Type," *Proc. Camb. Phil. Soc.*, Vol. 43 (1947), pp. 50–67.

7. W. F. Ames, *Nonlinear Partial Differential Equations in Engineering*, Academic Press, New York, 1965, Chapter 7.

8. S. P. Frankel, "Convergence Rates of Iterative Treatment of Partial Differential Equations," *Math. Tables Aids Comput.*, Vol. 4 (1950), pp. 65–75.

9. D. M. Young, "Iterative Methods for Solving Partial Differential Equations of Elliptic Type," *Trans. Amer. Math. Soc.*, Vol. 76 (1954), pp. 92–111.

10. R. S. Varga, *Matrix Iterative Analysis*, Prentice-Hall, Englewood Cliffs, N. J., 1962.

11. T. B. Fowler and E. R. Volk, "Generalized Heat Conduction Code for the IBM-704 Computer." *ORNL-2734* (October 1959), Oak Ridge National Laboratory, Oak Ridge, Tenn.

12. D. W. Peaceman and H. H. Rachford, "The Numerical Solution of Parabolic and Eliptic Differential Equations," *J. Soc. Indust. Appl. Math.*, Vol. 3, (1955), pp. 28–41.

13. R. B. Smith and J. Spanier, "HOT-1: A Two-Dimensional Steady-State Heat-Conduction Program for the Philco-2000," *WAPD-TM-465*, July (1964), Bettis Atomic Power Laboratory, West Mifflin, Pa.

14. S. R. Allada and D. Quon, "A Stable, Explicit Numerical Solution of the Conduction Equation for Multidimensional Nonhomogeneous Media," *Heat Transfer Los Angeles, Chemical Engineering Progress Symposium Series*, Vol. 62, No. 64 (1966), pp. 151–156.

15. H. Z. Barakat and J. A. Clark, "On the Solution of Diffusion Equation by Numerical Methods," *J. Heat Transfer* (November 1966), pp. 421–427.

16. Frank W. Schmidt, "The Formation of Difference Equations Using

Taylor's Series Expansion," *Mechanical Engineering News*, Vol. 1, No. 2 (May 1964), pp. 15–19.

17. R. F. Thomas and M. D. J. MacRoberts, "RATH Thermal Analysis Programs," LA-3264-MS, UC-32, *Mathematics and Computers* (March 1965), Los Alamos Scientific Laboratory, Los Alamos, N. M.

18. I. S. Berezin and N. P. Zhidkov, *Computing Methods*, Vol. 1, pp. 210–215, Addison-Wesley Publishing Company, Inc., Reading, Mass., 1965.

19. C. A. Foster, "Finite Difference Approach to Some Heat Conduction Problems Involving Change of State," *Report of English Electric Company*, Luton, England, 1954.

20. S. J. Citron, "Heat Conduction in a Melting Slab," *Institute of Aeronautical Sciences, Report No. 59–68*, January 1959.

21. J. E. Sunderland and R. J. Grosh, "Transient Temperature in a Melting Solid," *J. Heat Transfer*, pp. 409–414, November 1961.

22. W. D. Murray and F. Landis, "Numerical and Machine Solutions of Transient Heat Conduction Problems Involving Melting or Freezing," Part I, *Tran. ASME*, Vol. 81 (1959), pp. 106–112.

23. G. S. Springer and D. Olson, "Method of Solution of Axisymmetric Solidification and Melting Problems," ASME Paper 62-WA-246.

24. J. R. Howell and M. Perlmutter, "Monte Carlo Solution of Thermal Transfer Through Radiant Media between Gray Walls," *J. Heat Transfer, Trans. ASME*, Vol. 86 (1964), pp. 116–122.

25. J. R. Howell and M. Perlmutter, "Monte Carlo Solution of Radiant Heat Transfer in a Nongray, Nonisothermal Gas with Temperature-Dependent Properties," *AIChE. J.*, Vol. 10 (1964), pp. 562–567.

26. M. Perlmutter and J. R. Howell, "Radiant Transfer Through a Gray Gas between Concentric Cylinders Using Monte Carlo," *J. Heat Transfer. Trans. ASME*, Vol. 86 (1964), pp. 169–179.

27. J. R. Howell, "Calculation of Radiant Heat Exchange by the Monte Carlo Method," ASME Paper No. 65–WA/HT-54.

28. J. M. Hammersley and D. C. Handscomb, *Monte Carlo Method*, Methuen and Co., Ltd., London, 1964.

29. L. W. Ehrlich, "Monte Carlo Solution of Boundary-Value Problems Involving the Difference Analogue of $\frac{\partial^2 U}{\partial x^2} + \frac{\partial^2 U}{\partial y^2} + \frac{k}{y}\frac{\partial U}{\partial y} = 0$," *J. Assoc. Comp. Mach.*, Vol. 6 (1959), pp. 204–218.

30. N. P. Buslenko, D. I. Golenko, Yu. A. Shreider, I. M. Sobol' and V. G. Sragovich, "Monte Carlo Method," Translated from the Russian by G. J. Tee, Pergamon Press, 1966, pp. 35–47.

31. A. Haji-Sheikh and E. M. Sparrow, "The Solution of Heat Conduction Problems by Probability Methods," ASME Paper No. 66–WA/HT-1.

32. A. Haji-Sheikh, "Application of Monte Carlo Methods to Thermal Conduction Problems," Ph.D. Thesis, Mechanical Engineering, University of Minnesota, Minneapolis, Minn., December 1965.

33. J. H. Curtis, "Sampling Methods Applied to Differential and Difference Equations," *Proceedings of IBM Seminar on Scientific Computation*, November

1949, International Business Machines Corporation, New York (1950), pp. 87–109.

34. V. D. Barnett, "The Behavior of Pseudo-Random Sequence Generated on Computers by Multiplicative Congruential Method," *Math. Comp.*, Vol. 16 (1962), pp. 63–69.

35. IBM Application Program, H20-0205-0. *System/360 Scientific Subroutine Package* [360A-CM-03x] *Programmer's Manual.* International Business Machines Corporation, Technical Publications Department, White Plains, N.Y., 1966, p. 47.

PROBLEMS

1. Consider the following boundary-value problem of heat conduction:

$$\frac{\partial T}{\partial t} = \alpha \frac{\partial^2 T}{\partial x^2} \quad \text{in} \quad 0 \le x \le L,\ t > 0$$

$$-\frac{\partial T}{\partial x} = f_1 \quad \text{at} \quad x = 0,\ t > 0$$

$$k_2 \frac{\partial T}{\partial x} + h_2 T = f_2 \quad \text{at} \quad x = L,\ t > 0$$

$$T = F(x) \quad \text{in} \quad 0 \le x \le L,\ t = 0$$

Write the above system in finite differences by using (i) an explicit method of finite differences, (ii) Crank-Nicolson method of finite differences.

2. Repeat the finite differencing of Prob. 1 when both boundary conditions are of the first kind.

3. Consider the two-dimensional, unsteady heat-conduction equation in the rectangular coordinate system in the form

$$\frac{1}{\alpha}\frac{\partial T}{\partial t} = \frac{\partial^2 T}{\partial x^2} + \frac{\partial^2 T}{\partial y^2}$$

Write this differential equation in finite differences by using (i) a fully implicit method, (ii) the Crank-Nicolson method.

4. By assuming the solution of difference Eq. 9-43 can be represented by a Fourier series in which the time and space variables can be separated with the general term of the expansion in the form

$$e^{\gamma t} \cdot e^{i\beta x}$$

Show that the Crank-Nicolson method of finite differences of the one-dimensional transient heat-conduction equation 9-44 is stable for all values of the time step Δt.

5. Consider two-dimensional, unsteady heat-conduction problem in the rectangular coordinate system in the form

$$\frac{1}{\alpha}\frac{\partial T}{\partial t} = \left(\frac{\partial^2 T}{\partial x^2} + \frac{\partial^2 T}{\partial y^2}\right)$$

subjected to prescribed temperature at the boundaries and an initial condition. Write this differential equation in finite differences by using an explicit method.

By assuming the solution of the difference equation can be represented by a Fourier series such that the general term of the expansion is in the form

$$e^{\gamma t} \cdot e^{i\beta x} \cdot e^{i\delta y}$$

Show that the explicit finite-difference equation is stable if

$$2\alpha \Delta t\left(\frac{1}{(\Delta x)^2} + \frac{1}{(\Delta y)^2}\right) \leq 1$$

6. Barakat and Clark (Ref. 15) examined the finite-difference representation of the following boundary-value problem of heat conduction:

$$\frac{\partial T}{\partial t} = \frac{\partial^2 T}{\partial x^2} + \frac{\partial^2 T}{\partial y^2} \quad \text{in } 0 \leq x \leq 1,\ 0 \leq y \leq 1$$

subjected to prescribed temperature at the boundaries and to a prescribed initial condition.

To solve the above problem with finite differences, they considered two auxiliary functions $U_{j,k}$ and $V_{j,k}$ satisfying the following difference equations:

$$\frac{U_{j,k}^{n+1} - U_{j,k}^{n}}{\Delta t} = \frac{U_{j+1,k}^{n} - U_{j,k}^{n} - U_{j,k}^{n+1} + U_{j-1,k}^{n+1}}{(\Delta x)^2} + \frac{U_{j,k+1}^{n} - U_{j,k}^{n} - U_{j,k}^{n+1} + U_{j,k-1}^{n+1}}{(\Delta y)^2}$$

and a similar expression for the $V_{j,k}$ function with superscripts n and $(n + 1)$ being interchanged on the right-hand side of the above equation.

Assuming $U_{j,k}$ and $V_{j,k}$ functions also satisfy the boundary and initial conditions for the heat conduction problem, the temperature $T_{j,k}$ at any time level $(n + 1)$ may be taken as the arithmetic average of $U_{j,k}$ and $V_{j,k}$ functions, i.e.

$$T_{j,k}^{n+1} = \tfrac{1}{2}[U_{j,k}^{n+1} + V_{j,k}^{n+1}]$$

If finite difference equations for $U_{j,k}$ and $V_{j,k}$ functions are stable the solution of the problem will be stable. To analyze the stability assume that the solution of this finite-difference equation can be represented by a Fourier series with a general term of expansion in the form

$$e^{\gamma t} \cdot e^{i(\beta_1 x + \beta_2 y)}$$

and then show that the above finite-difference equation is unconditionally stable for all values of Δt, Δx, Δy.

7. Show that the finite-difference representation of Laplacian in the cylindrical coordinate system is as given by Eq. 9-82.

8. Consider the two-dimensional, transient heat-conduction equation in the cylindrical coordinate system in the form

$$\frac{1}{\alpha}\frac{\partial T}{\partial t} = \frac{\partial^2 T}{\partial r^2} + \frac{1}{r}\frac{\partial T}{\partial r} + \frac{1}{r^2}\frac{\partial^2 T}{\partial \phi^2}$$

Write this differential equation in the finite-difference form for the nodal point (i, j) shown in Fig. 9-6b.

Repeat the finite-difference representation of this differential equation for the nodal point (i, j) in Fig. 9-6b by assuming both boundary surfaces passing through the point (i, j) are subjected to boundary conditions of the third kind.

9. Write expressions equivalent to Eqs. 9-163 and 9-164 for the effective random-walk temperature and the expected temperature $T(x, y)$ respectively when the internal heat generation rate within the solid varies with position.

10

Heat Conduction in Anisotropic Solids

Solids which have transport properties that do not depend on direction are said to be *isotropic*, and those that have properties which exhibit directional characteristics are said to be *anisotropic*. In the study of heat conduction in solids, anisotropy is associated with the thermal conductivity of the material. In the previous chapters we considered heat conduction in isotropic solids in which the thermal conductivity remained the same in all directions. The heat conduction then obeyed the Fourier law in the form

$$\bar{q} = -k\nabla T \tag{10-1}$$

with the three Cartesian components of the heat flux along the ox_1, ox_2, and ox_3 axes given as

$$q_1 = -k\frac{\partial T}{\partial x_1} \qquad q_2 = -k\frac{\partial T}{\partial x_2} \qquad q_3 = -k\frac{\partial T}{\partial x_3} \tag{10-2}$$

which can be written in a compact form as

$$q_i = -k\frac{\partial T}{\partial x_i} \qquad (i = 1,2,3) \tag{10-2a}$$

For isotropic solids the heat flux at a particular position is directly proportional to the temperature gradient at that position, and that the heat flux vector is normal to the isothermal surface passing through that position.[1]

[1] Although discussed in Chapter 1, it can be shown in general that for isotropic solids the heat flux vector, that is described by Fourier's law, is normal to an isothermal surface. Consider a field point $\bar{r} = \hat{i}x_1 + \hat{j}x_2\ \hat{k}x_3$ and the temperature gradient at this point $\nabla T = \hat{i}\dfrac{\partial T}{\partial x_1} + \hat{j}\dfrac{\partial T}{\partial x_2} + \hat{k}\dfrac{\partial T}{\partial x_3}$. Consider a surface T = constant passing through the point $\bar{r}$. The tangent plane at the point $\bar{r}$ is $d\bar{r} = \hat{i}\,dx_1 + \hat{j}\,dx_2 + \hat{k}\,dx_3$. Since the temperature T at the isothermal surface does not change with the position

$$\begin{aligned} dT &= \frac{\partial T}{\partial x_1}dx_1 + \frac{\partial T}{\partial x_2}dx_2 + \frac{\partial T}{\partial x_3}dx_3 \\ &= \left(\frac{\partial T}{\partial x_1}\hat{i} + \frac{\partial T}{\partial x_2}\hat{j} + \frac{\partial T}{\partial x_3}\hat{k}\right)\cdot(dx_1\hat{i} + dx_2\hat{j} + dx_3\hat{k}) = \nabla T\cdot d\bar{r} = 0 \end{aligned}$$

Thus, ∇T vector is perpendicular to the tangent plane passing through a surface T = constant. Since from Fourier's law $\bar{q}$ is parallel to ∇T, then $\bar{q}$ is always normal to an isothermal surface.

In the case of an anisotropic solid, however, the heat flux vector $\bar{q}$ is not necessarily parallel to the temperature gradient, ∇T, and therefore the direction of the heat flux vector is not necessarily normal to an isothermal surface. As it will be apparent later in this chapter, $\bar{q}$, for anisotropic solids, is not simply proportional to the temperature gradient, ∇T. Each of the components of the temperature gradient are weighted according to the anisotropy of the particular material. Thus, even though ∇T is normal to a given surface $d\bar{r}$ (i.e. $\nabla T \cdot d\bar{r} = 0$) the heat flux vector $\bar{q}$ is not normal to the surface $d\bar{r}$ (i.e. $\bar{q} \cdot d\bar{r} \neq 0$) and $\bar{q}$ is not parallel to ∇T.[2]

There are several engineering materials such as crystals, woods and laminated sheets in which the thermal conductivity varies with direction.

[2]Existence of the obtuse angle relation between the temperature gradient vector, ∇T, and the heat flux vector, $\bar{q}$, can be proved with thermodynamic considerations.

Let V be the volume of a system and V' its environment. Let ds be the change in entropy of system V and ds' be the change in entropy of its environment, V', during the time interval dt. Assuming no expansion and material flux, the change of entropy of the system during the time interval dt is given by Gibbs relation [Ref. 4, p. 23] as

$$ds = dt \int_V \frac{1}{T} \frac{\partial(\rho U)}{\partial t} \, dV$$

and for the environment it is given as

$$ds' = dt \int_{\text{surface}} \hat{n} \cdot \left(\frac{\bar{q}}{T}\right) dA$$

where ρ = the density of the material
U = the specific internal energy
$\hat{n}$ = the unit normal to the surface
$\bar{q}$ = the heat flux vector

From the second law of thermodynamics we have

$$ds + ds' \geq 0$$

Applying the second law to the above relations and changing surface integral to volume integral, we obtain

$$\int_V \left[\frac{1}{T} \frac{\partial(\rho U)}{\partial t} + \nabla \cdot \left(\frac{\bar{q}}{T}\right)\right] dV \geq 0$$

Since V is arbitrary we may choose it so small as to remove the integral sign

$$\frac{1}{T} \frac{\partial(\rho U)}{\partial t} + \nabla \cdot \left(\frac{\bar{q}}{T}\right) \geq 0$$

or

$$\frac{1}{T}\left[\frac{\partial(\rho U)}{\partial t} + \nabla \cdot \bar{q}\right] - \frac{1}{T^2} \bar{q} \cdot \nabla T \geq 0$$

From the general energy equation, for no generation, we have $\dfrac{\partial(\rho U)}{\partial t} + \nabla \cdot \bar{q} = 0$.

Then we obtain $-\bar{q} \cdot \nabla T \geq 0$ or $\bar{q} \cdot \nabla T \leq 0$, which illustrates the obtuse angle relation between $\bar{q}$ and ∇T.

The various types of plywoods and core elements of transformers are examples of anisotropic laminates. For such anisotropic materials the heat conduction laws (Eqs. 10-1 and 10-2) must be modified. The treatment of heat conduction in anisotropic solids may be found in the books by Carslaw and Jaeger [1], Nye [2], Wooster [3], and de Groot and Mazur [4]. In this chapter we examine the basic ideas along with the differential equations of heat conduction for anisotropic solids with some applications to simpler problems.

10-1. THERMAL CONDUCTIVITY TENSOR

Consider conduction of heat in an anisotropic solid referred to a set of rectangular axes ox_1, ox_2, and ox_3. The components of the heat flux vector $\bar{q}$, along a given direction ox_1, will depend not only upon the temperature gradient along the ox_1 axis, but in general upon a linear combination of the temperature gradients along the ox_1, ox_2, and ox_3 directions. Then the general expressions for the three components of the heat flux along the ox_1, ox_2, and ox_3 axes become

$$-q_1 = k_{11}\frac{\partial T}{\partial x_1} + k_{12}\frac{\partial T}{\partial x_2} + k_{13}\frac{\partial T}{\partial x_3} \equiv \sum_{j=1,2,3} k_{1j}\frac{\partial T}{\partial x_j} \tag{10-3a}$$

$$-q_2 = k_{21}\frac{\partial T}{\partial x_1} + k_{22}\frac{\partial T}{\partial x_2} + k_{23}\frac{\partial T}{\partial x_3} \equiv \sum_{j=1,2,3} k_{2j}\frac{\partial T}{\partial x_j} \tag{10-3b}$$

$$-q_3 = k_{31}\frac{\partial T}{\partial x_1} + k_{32}\frac{\partial T}{\partial x_2} + k_{33}\frac{\partial T}{\partial x_3} \equiv \sum_{j=1,2,3} k_{3j}\frac{\partial T}{\partial x_j} \tag{10-3c}$$

These equations can be written in a more compact form by using the *Cartesian tensor* notation as[3]

$$-q_i = k_{ij}\frac{\partial T}{\partial x_j} \qquad i,j = 1,2,3 \tag{10-4}$$

The thermal conductivity of an anisotropic solid k_{ij}, involves nine components called the *conductivity coefficients* that form a second-order tensor called the *thermal conductivity tensor*, which is written in the form

$$k_{ij} = \begin{bmatrix} k_{11} & k_{12} & k_{13} \\ k_{21} & k_{22} & k_{23} \\ k_{31} & k_{32} & k_{33} \end{bmatrix} \tag{10-5}$$

If the thermal conductivity tensor k_{ij} is taken as a symmetrical tensor, i.e.,

$$k_{ij} = k_{ji} \tag{10-6}$$

[3]Whenever an index is repeated twice in the same term, summation with that index is automatically understood (i.e. $j = 1,2,3$); then the equation is repeated for $i = 1,2,3$. The reader may refer to the book by Jeffreys [5] for the use of Cartesian tensor notation.

then the number of unknown conductivity coefficients reduces from nine to six, i.e.,

$$k_{11} \qquad k_{12} = k_{21} \qquad k_{13} = k_{31} \qquad k_{22} \qquad k_{23} = k_{32} \qquad k_{33}$$

The validity of this symmetry assumption is taken on the basis of Onsager's [2, 6] principles of thermodynamics of irreversible processes.[4]

10-2. THERMAL RESISTIVITY TENSOR

In Eqs. 10-3a, b, and c each component of the heat flux vector is expressed as a linear combination of temperature gradients along the ox_1, ox_2, and ox_3 directions. It is sometimes desirable to express the temperature gradient in a given direction as linear combination of the heat flux components along the ox_1, ox_2, and ox_3 axes. Such a relationship is obtainable from Eq. 10-4 by writing it in matrix notation as

$$[-q_i] = [k_{ij}] \left[\frac{\partial T}{\partial x_j}\right] \tag{10-7a}$$

[4]Onsager's phenomenlogical equations for linear processes between the fluxes and forces may be written as

$$J_i = \sum_j L_{ij} Z_j$$

where J_i = Cartesian component of the fluxes
Z_i = Cartesian component of the forces (i.e. for heat conduction the temperature gradient $\frac{\partial T}{\partial x_i}$ is the force)
L_{ij} = Phenomenlogical coefficients

The phenomenlogical coefficients L_{ij} are related by the Onsager relations which state a symmetry between the effect of the jth force on the ith flux, and the effect of the ith force on the jth flux. In the absence of an externally applied magnetic field, the Onsager relation for L_{ij} is

$$L_{ij} = L_{ji}$$

In applying Onsager's principles to heat conduction problems in anisotropic solids the components of the heat flux along the reference axes are considered as the fluxes J_i and the components of the temperature gradient as the forces Z_i; and they do interfere with one another, i.e. the temperature gradient along the ox_1 axis produces heat flow not only along the ox_1 axis but also along ox_2 and ox_3 axes. With this assumption the L_{ij}'s become the conductivity coefficients k_{ij}, hence one writes

$$k_{ij} = k_{ji}$$

On the other hand, in applying Onsager's principles the fluxes and forces must obey certain rules. Casimir [Ref. 2, p. 210 and Ref. 7] pointed out some of the difficulties associated with the application of Onsager's principles to Eq. 10-4. First, the heat flow does not satisfy one of the rules for the proper choice of fluxes, which requires that they shall be the time derivatives of the variables which define the thermodynamical state. Second, in the experiments only the divergence of the heat flux, $\nabla \bar{q}$ (i.e. the difference between the heat entering and leaving a volume) is measured. Hence, there is the possibility that an antisymmetrical tensor k_{ij}^o satisfying the condition $\partial k_{ij}^o/\partial x_i = 0$ exists, but it produces no observable effect on $\nabla \bar{q}$. Since the antisymmetrical part of k_{ij} is not observable, the antisymmetrical part may be taken as zero and we write $k_{ij} = k_{ij}$.

and then taking the inverse,

$$\left[\frac{\partial T}{\partial x_j}\right] = [k_{ij}]^{-1}[-q_i] \tag{10-7b}$$

This equation can then be rewritten as

$$\left[\frac{\partial T}{\partial x_i}\right] = [r_{ij}][-q_j] \tag{10-7c}$$

or in tensor notation as

$$-\frac{\partial T}{\partial x_i} = r_{ij}\, q_j \qquad i,j = 1,2,3 \tag{10-8a}$$

where

$$r_{ij} = (-1)^{i+j}\frac{a_{ij}}{\Delta} \tag{10-8b}$$

$$\Delta \equiv \begin{vmatrix} k_{11} & k_{12} & k_{13} \\ k_{21} & k_{22} & k_{23} \\ k_{31} & k_{32} & k_{33} \end{vmatrix} \tag{10-8c}$$

a_{ij} = the cofactor obtained from Δ by omitting the ith row and the jth column

A new quantity r_{ij} which has been defined in Eq. 10-8a is called the *thermal resistivity tensor*; it involves nine components which are better envisioned when Eq. 10-8a is written in the expanded form as

$$-\frac{\partial T}{\partial x_1} = r_{11}\, q_1 + r_{12}\, q_2 + r_{13}\, q_3 \tag{10-9a}$$

$$-\frac{\partial T}{\partial x_2} = r_{21}\, q_1 + r_{22}\, q_2 + r_{23}\, q_3 \tag{10-9b}$$

$$-\frac{\partial T}{\partial x_3} = r_{31}\, q_1 + r_{32} q_2 + r_{33}\, q_3 \tag{10-9c}$$

The nine components of the thermal resistivity tensor r_{ij} are

$$r_{ij} \equiv \begin{bmatrix} r_{11} & r_{12} & r_{13} \\ r_{21} & r_{22} & r_{23} \\ r_{31} & r_{32} & r_{33} \end{bmatrix} \tag{10-10}$$

As in the case of the thermal conductivity tensor, the thermal resistivity tensor r_{ij} is a symmetrical tensor, in that

$$r_{ij} = r_{ji} \tag{10-11}$$

and the number of unknown resistivity coefficients is reduced from nine to six. The components of the thermal resistivity tensor are related to the conductivity coefficients according to Eq. 10-8b; for example, r_{12} is given as

$$r_{12} = (-1)^3 \frac{\begin{vmatrix} k_{21} & k_{23} \\ k_{31} & k_{33} \end{vmatrix}}{\Delta} = \frac{k_{23} k_{31} - k_{21} k_{33}}{\Delta}$$

10-3. TRANSFORMATION OF AXES

Values of the conductivity coefficients depend on the orientation of the reference coordinate axes. When a set of new coordinate axes are chosen it is important to determine the values of the conductivity coefficients for the new system.

Let k_{ij} represent the conductivity coefficients with respect to a set of *old* rectangular axes ox_1, ox_2, and ox_3. Consider a set of *new* rectangular axes ox_1', ox_2', and ox_3' having the same origin o as the old system as shown in Fig. 10-1. The positions of the new axes with respect to the old axes are defined by the direction cosines c_{ij} (i.e. the cosine of the angle between the new axis ox_i' and the old axis ox_j). For example, c_{23} is the cosine of the angle between the ox_2' axis and the ox_3 axis. It is to be noted that c_{ij} and c_{ji} are not necessarily equal.

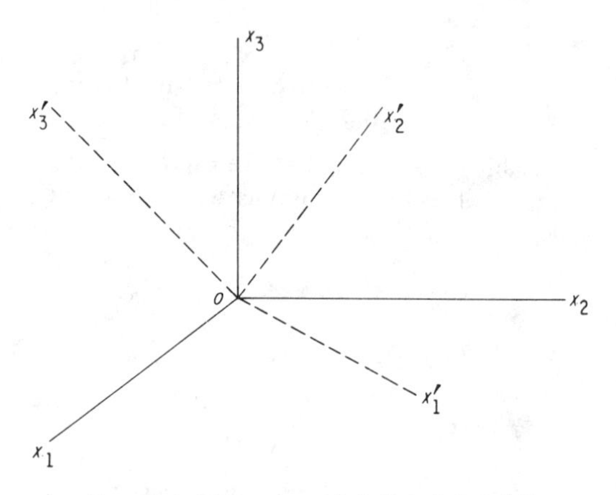

FIG. 10-1. The new axes $0x_1'$, $0x_2'$, $0x_3'$ and the old axes $0x_1$, $0x_2$, $0x_3$.

The conductivity coefficients k_{ij}' referred to the *new* coordinate system are given in terms of those referred to the *old* coordinate system (Ref. 2, Chapter 1) as

$$k_{ij}' = c_{il}c_{jm}k_{lm} \qquad l, m = 1,2,3 \tag{10-12}$$

where k_{lm} denotes the conductivity coefficients referred to the *old* system. Here again the repeated indices, sometimes called dummy indices, indicate summation over a set of values 1, 2, and 3. Equation 10-12 in the expanded form is given as

$$\begin{aligned} k_{ij}' = {} & c_{i1}c_{j1}k_{11} + c_{i1}c_{j2}k_{12} + c_{i1}c_{j3}k_{13} + c_{i2}c_{j1}k_{21} + c_{i2}c_{j2}k_{22} \\ & + c_{i2}c_{j3}k_{23} + c_{i3}c_{j1}k_{31} + c_{i3}c_{j2}k_{32} + c_{i3}c_{j3}k_{33} \end{aligned} \tag{10-13}$$

10-4. SYMMETRY CONSIDERATIONS IN CRYSTALS

The thermal conductivity tensor for an anisotropic solid involves, in general, nine components which are reduced to six because k_{ij} is a symmetrical tensor. On the other hand, crystals possess various forms of structural symmetry that makes some of the coefficients of the thermal conductivity tensor k_{ij} vanish. A brief summary of the types of crystal symmetries and their effects on the conductivity coefficients are given below. For additional information the reader may refer to several references on this subject [2, 3, 8].

A crystal lattice is composed of a repeating array of identical *unit cells* which are parallelepipeds defined by the lengths a, b, and c of three nonparallel sides and the angles α, β, γ between them. Figure 10-2 illustrates three nonparallel sides a, b, c with angles α, β, γ. Some crystals possess a symmetrical array of these units. A list of the symmetries in crystals and their effects on the conductivity coefficients are summarized below.

(1) The crystal system called *triclinic* is the most unsymmetrical system, because all three edges of the parallelepiped are unequal (i.e. $a \neq b \neq c$) and the angles between them are unequal and inclined (i.e., $\alpha \neq \beta \neq \gamma$ and none are at 90°). Therefore, no limitations are imposed on the coefficients of the thermal conductivity tensor k_{ij} and all nine components can be non-zero. That is

$$\begin{bmatrix} k_{11} & k_{12} & k_{13} \\ k_{21} & k_{22} & k_{23} \\ k_{31} & k_{32} & k_{33} \end{bmatrix} \tag{10-14}$$

(2) In a *monoclinic* system all three edges are unequal (i.e. $a \neq b \neq c$) but two sides are at right angles and the third is inclined to them (i.e. $\alpha = \gamma = 90 \neq \beta$). For such a system it can be shown [3] that the conductivity coefficients $k_{13} = k_{31} = k_{23} = k_{32} = 0$, and the remaining five coefficients are nonzero. That is

$$\begin{bmatrix} k_{11} & k_{12} & 0 \\ k_{21} & k_{22} & 0 \\ 0 & 0 & k_{33} \end{bmatrix} \tag{10-15}$$

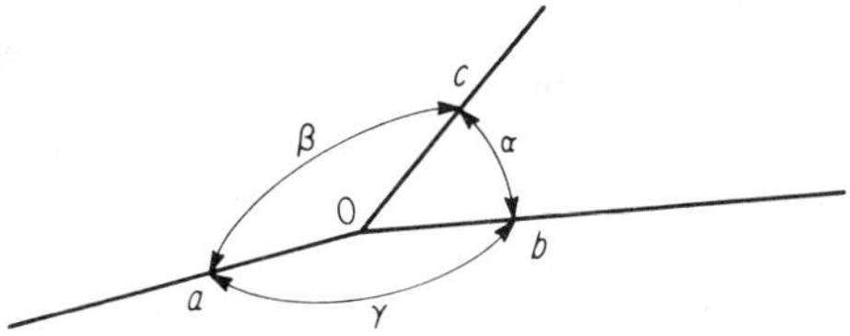

FIG. 10-2. Three sides a, b, c and the angles α, β, γ for a unit cell.

(3) In an *orthorhombic* system (also called *rhombic*) all three edges are unequal (i.e., $a \neq b \neq c$) but all the angles are at right angles (i.e., $\alpha = \beta = \gamma = 90°$). For this case all k_{ij}'s for $i \neq j$ are zero and only the coefficients k_{11}, k_{22}, k_{33} are of interest. That is

$$\begin{bmatrix} k_{11} & 0 & 0 \\ 0 & k_{22} & 0 \\ 0 & 0 & k_{33} \end{bmatrix} \tag{10-16}$$

(4) In the *tetragonal*, *trigonal* and *hexagonal* systems the relations between the sides are $a = b \neq c$. The relation between the angles are: $\alpha = \beta = \gamma = 90$ for the *tetragonal*; and $\alpha = \beta = 90°$, $\gamma = 120°$ for the *trigonal* and *hexagonal* systems. For these systems the thermal conductivity tensor reduces to

$$\begin{bmatrix} k_{11} & k_{12} & 0 \\ -k_{12} & k_{11} & 0 \\ 0 & 0 & k_{33} \end{bmatrix} \tag{10-17}$$

It has been shown that [Ref. 2, p. 207] if a crystal of this type is cut in the form of a circular plate with the principal conductivity axis ox_3 aligned with the cylinder axis, heated at the center and outer edge kept at uniform temperature, a spiral heat flow should result. Since several attempts have failed to detect any spiral heat flow experimentally within the accuracy of measurements, k_{12} may be considered symmetrical; this implies that k_{12} may be taken as zero in Eq. 10-17.

(5) In a *cubic* system all three sides are equal (i.e. $a = b = c$) and all angles are right angles (i.e. $\alpha = \beta = \gamma = 90$). The resulting symmetry considerations require that all k_{ij}'s for $i \neq j$ vanish and that $k_{11} = k_{22} = k_{33}$. Then the thermal conductivity tensor simplifies to

$$\begin{bmatrix} k_{11} & 0 & 0 \\ 0 & k_{11} & 0 \\ 0 & 0 & k_{11} \end{bmatrix} \tag{10-18}$$

10-5. A GEOMETRICAL INTERPRETATION OF THE CONDUCTIVITY TENSOR

In the previous sections we discussed the transformation of axes and evaluation of the conductivity coefficients with reference to a set of new axes. In the problems of heat conduction in anisotropic solids the *principal axes* and *principal conductivities* are terms frequently used. The physical significance of these terms in relation to the transformation of axes is better understood if a geometrical interpretation is made of the transformation of the conductivity coefficients.

Consider a second-degree equation with reference to the rectangular coordinate axes ox_1, ox_2, ox_3 of the form

$$s_{11}x_1^2 + s_{22}x_2^2 + s_{33}x_3^2 + 2s_{12}x_1x_2 + 2s_{13}x_1x_3 + 2s_{23}x_2x_3 = 1 \qquad (10\text{-}19)$$

where the s's represent the coefficients. The relation given by Eq. 10-19 is an equation of a *quadric* (i.e. second-degree surface) provided that all the coefficients of the second-degree terms do not simultaneously vanish. When all the coefficients in Eq. 10-19 are positive it represents an *ellipsoid.* (For detailed discussion of the properties of quadrics the reader may refer to the book by Eisenhart [9].)

Assuming $s_{ij} = s_{ji}$, Eq. 10-19 is written in tensor notation as

$$s_{ij}x_ix_j = 1 \qquad s_{ij} = s_{ji} \qquad (10\text{-}20)$$

We now examine the transformation of the equation of the quadric (Eq. 10-20) from the old rectangular coordinate system ox_1, ox_2, ox_3 to a new rectangular coordinate system ox_1', ox_2', ox_3'. Coordinates of any point in the system ox_1, ox_2, ox_3 are transformed to the system ox_1', ox_2', ox_3' by the relation

$$x_i = c_{li}x_l' \qquad l = 1,2,3 \qquad (10\text{-}21a)$$

$$x_j = c_{mj}x_m' \qquad m = 1,2,3 \qquad (10\text{-}21b)$$

where c_{li} and c_{mj} are the direction cosines. Substituting Eq. 10-21 into Eq. 10-20, the transformation of a *quadric* from the coordinate system ox_1, ox_2, ox_3 into the coordinate system ox_1', ox_2', ox_3' is given by

$$s_{ij}c_{li}x_l'c_{mj}x_m' = 1 \qquad (10\text{-}22)$$

which is written in the form

$$s_{lm}'x_l'x_m' = 1 \qquad (10\text{-}23a)$$

where

$$s_{lm}' = c_{li}c_{mj}s_{ij} \qquad (10\text{-}23b)$$

Equation 10-23b is the relation for the transformation of the coefficients of a quadric. It is interesting to compare Eq. 10-23b with the relation for the transformation of the conductivity coefficients from the rectangular coordinate system ox_1, ox_2, ox_3 to the ox_1', ox_2', ox_3' system as given by Eq. 10-12, i.e.,

$$k_{ij}' = c_{il} \cdot c_{jm} \cdot k_{lm} \qquad (10\text{-}12)$$

The transformations (Eqs. 10-23a and b and 10-12) are of the identical form so far as the relative positions of the indices are concerned.

Therefore, transformation of the conductivity coefficients is similar to the transformation of the coefficients of a quadric. Several of the properties associated with the transformation of a quadric are applicable to the transformation of the conductivity coefficients. One property of the quadric given by Eq. 10-19 is that a new rectangular coordinate system, say, $o\xi_1$, $o\xi_2$, and $o\xi_3$ can be found such that Eq. 10-19 simplifies to the form

$$s_1\xi_1^2 + s_2\xi_2^2 + s_3\xi_3^2 = 1 \tag{10-24a}$$

This is an ellipsoid when the coefficients s_1, s_2, and s_3 are all positive. The new coordinate axes $o\xi_1$, $o\xi_2$, $o\xi_3$ are called the *principal axes* of the ellipsoid. The physical significance of Eq. 10-24 is better envisioned if it is written in the form

$$\frac{\xi_1^2}{A^2} + \frac{\xi_2^2}{B^2} + \frac{\xi_3^2}{C^2} = 1 \tag{10-24b}$$

where

$$A = \frac{1}{\sqrt{s_1}} \qquad B = \frac{1}{\sqrt{s_2}} \qquad C = \frac{1}{\sqrt{s_3}}$$

The coefficients A, B, C represent the lengths of the semi-axes of the ellipsoid along the $o\xi_1$, $o\xi_2$ and $o\xi_3$ axes respectively.

We now examine the geometrical interpretation of Eq. 10-12 for the transformation of the conductivity coefficients. We rewrite Eq. 10-12 for $i = j$,

$$k'_{jj} = c_{jl}c_{jm}k_{l,m} \qquad l, m = 1,2,3 \tag{10-25}$$

This equation is expanded as

$$k'_{jj} = c_{j1}^2 k_{11} + c_{j2}^2 k_{22} + c_{j3}^2 k_{33} + 2c_{j1}c_{j2}k_{12} + 2c_{j1}c_{j3}k_{13} + 2c_{j2}c_{j3}k_{23} \tag{10-26}$$

Equation 10-26 relates the conductivity coefficients k'_{jj} along the *new* axis ox'_j to the conductivity coefficients referred to the old axes ox_1, ox_2, ox_3, and to the direction cosines of the ox'_j axis with the old axes ox_1, ox_2, ox_3. Since, from physical considerations, all the conductivity coefficients are positive, Eq. 10-26 represents an ellipsoid with the length of the radius vector along the ox'_j axis equal to $1/\sqrt{k'_{jj}}$. The ellipsoid as represented by Eq. 10-26 is transformed into a compact form if we define

$$x_1 = \frac{c_{j1}}{\sqrt{k'_{jj}}}$$

$$x_2 = \frac{c_{j2}}{\sqrt{k'_{jj}}}$$

$$x_3 = \frac{c_{j3}}{\sqrt{k'_{jj}}}$$

Substituting these relations into Eq. 10-26 we obtain

$$k_{11}x_1^2 + k_{22}x_2^2 + k_{33}x_3^2 + 2k_{12}x_1x_2 + 2k_{13}x_1x_3 + 2k_{23}x_2x_3 = 1 \tag{10-27a}$$

which is written in tensor notation as

$$k_{ij}x_i x_j = 1 \qquad (i, j = 1, 2, 3, \text{ and } k_{ij} = k_{ji}) \tag{10-27b}$$

The ellipsoid represented by Eq. 10-26b is the locus of the end point of the radius vector $1/\sqrt{k'_{jj}}$; it is called the *conductivity ellipsoid*. When the reference axes coincide with the *principal axes* $o\xi_1, o\xi_2, o\xi_3$ of the ellipsoid, Eq. 10-26 simplifies to

$$k'_{jj} = c_{j1}^2 k_1 + c_{j2}^2 k_2 + c_{j3}^2 k_3 \tag{10-28a}$$

where k_1, k_2, k_3 are called the *principal conductivities*. Defining,

$$\xi_1 = \frac{c_{j1}}{\sqrt{k'_{jj}}} \qquad \xi_2 = \frac{c_{j2}}{\sqrt{k'_{jj}}} \qquad \xi_3 = \frac{c_{j3}}{\sqrt{k'_{jj}}}$$

Eq. 10-28a is written in the form

$$k_1\xi_1^2 + k_2\xi_2^2 + k_3\xi_3^2 = 1 \tag{10-28b}$$

When referred to the principal axes $o\xi_1, o\xi_2, o\xi_3$, the three components of the heat flux given by Eq. 10-3 simplify to

$$q_1 = -k_1 \frac{\partial T}{\partial \xi_1} \tag{10-29a}$$

$$q_2 = -k_2 \frac{\partial T}{\partial \xi_2} \tag{10-29b}$$

$$q_3 = -k_3 \frac{\partial T}{\partial \xi_3} \tag{10-29c}$$

Figure 10-3 shows a *conductivity ellipsoid*. As illustrated in this figure the lengths of the radius vectors in the direction of the semi-axes are related to the principal conductivities by

$$\frac{1}{\sqrt{k_1}} \qquad \frac{1}{\sqrt{k_2}} \qquad \frac{1}{\sqrt{k_3}}$$

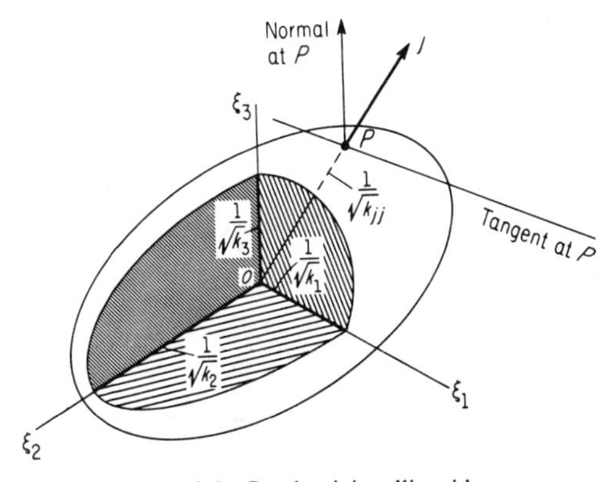

FIG. 10-3. Conductivity ellipsoid.

and the length of the radius vector OP in an arbitrary direction x_j is related to the conductivity coefficient in that direction by

$$\frac{1}{\sqrt{k'_{jj}}}$$

The conductivity ellipsoid can be used to determine geometrically the direction of the heat flux vector when the direction of the temperature gradient vector is prescribed. Let c_1, c_2, c_3 be the direction cosines of the temperature gradient vector ∇T with the principal axes $o\xi_1, o\xi_2, o\xi_3$ of the conductivity ellipsoid. The heat flux vector $\bar{q}$, when referred to the principal axes, is of the form

$$q_i = -k_i \frac{\partial T}{\partial \xi_i} = -k_i c_i G \tag{10-30}$$

where $G \equiv |\nabla T|$. Therefore, the direction cosines of the heat flux vector with the principal axes $o\xi_1, o\xi_2, o\xi_3$ are $k_1 c_1, k_2 c_2, k_3 c_3$ respectively.

Suppose the radius-vector OP is chosen parallel to the temperature gradient vector. Since the point P on the ellipsoid satisfies Eq. 10-28b, the normal to the tangent plane at P has direction cosines $k_1 c_1, k_2 c_2, k_3 c_3$ with respect to the principal axes.[5] Therefore, the normal at P is parallel to the heat flux vector. [i.e. when the radius vector OP is parallel to ∇T, the normal to the tangent plane at P is parallel to the heat flux vector, $\bar{q}$].

The crystal systems, as we have already seen, possess various forms of structural symmetry and because of these symmetry considerations some of the conductivity coefficients vanish and Eq. 10-26 can be simplified further. Simplification of Eq. 10-26 with symmetry considerations is summarized below.

(1) In the *triclinic* crystal system no simplification of Eq. 10-26 is possible because such systems possess no symmetry.

(2) In the *monoclinic* crystal system $k_{13} = k_{23} = 0$; hence Eq. 10-26 simplifies to

$$k'_{jj} = c_{j1}^2 k_{11} + c_{j2}^2 k_{22} + c_{j3}^2 k_{33} + 2c_{j1} c_{j2} k_{12} \tag*{[10-26a]}$$

(3) In the *orthorhombic* crystal system it simplifies to

$$k'_{jj} = c_{j1}^2 k_{11} + c_{j2}^2 k_{22} + c_{j3}^2 k_{33} \tag*{[10-26b]}$$

[5] Since the point P is on the ellipsoid

$$k_1 \xi_1^2 + k_2 \xi_2^2 + k_3 \xi_3^2 = 1$$

the equation of the tangent plane at P is

$$\frac{1}{\sqrt{k'_{jj}}} [c_1 k_1 \xi_1 + c_2 k_2 \xi_2 + c_3 k_3 \xi_3] = 1$$

Hence, the direction cosines of the normal to the tangent plane are $c_1 k_1, c_2 k_2, c_3 k_3$.

(4) In the *tetragonal*, *trigonal* and *hexagonal* crystal systems it becomes

$$k'_{jj} = c_{j1}^2 k_{11} + c_{j2}^2 k_{11} + c_{j3}^2 k_{33} \qquad [10\text{-}26c]$$

(5) In the *cubic* crystal system it becomes

$$k'_{jj} = (c_{j1}^2 + c_{j2}^2 + c_{j3}^2) k_{11} = k_{11} \qquad [10\text{-}26d]$$

The last equation implies that in the cubic crystal system the ellipsoid becomes a sphere.

It is apparent from Eqs. 10-26 through 10-26d that the number of unknown conductivity coefficients involved is six in the *triclinic* crystal system, four in the *monoclinic*, three in the *orthorhombic*, two in the *tetragonal*, *trigonal* and *hexagonal*, and one in the *cubic*. These equations may be used to determine the unknown conductivity coefficients when the coefficient k'_{jj} along the ox'_j axis can be determined from experiments; at least as many experiments are needed to provide different values of k'_{jj} as the number of unknown conductivity coefficients.

10-6. DETERMINATION OF PRINCIPAL CONDUCTIVITIES

The values of the principal conductivities k_1, k_2, k_3 of an anisotropic solid can be determined when the conductivity coefficients k_{ij} referred to a set of rectangular axis ox_1, ox_2 and ox_3 are known.

Consider a *conductivity ellipsoid* whose equation when referred to a set of rectangular axis ox_1, ox_2, and ox_3 is of the form

$$k_{ij} x_i x_j = 1 \qquad k_{ij} = k_{ji}$$

Let P be a point on the ellipsoid. In the previous section we have seen that if the radius vector OP is parallel to the temperature gradient vector, the heat flux vector is parallel to the normal drawn to the ellipsoid at P. When the radius vector OP coincides with one of the principal axes it is normal to the ellipsoid at the intersection, because the principal axes, by definition, are normal to the ellipsoid at their intersection. For such cases the radius vector OP, which is parallel to the temperature gradient vector, is also parallel to the heat flux vector, since the latter is normal to the ellipsoid at P. The condition that the temperature gradient and the heat flux vectors should be parallel requires that the corresponding components along the ox_1, ox_2, and ox_3 axes should be proportional, that is,

$$k_{ij} \frac{\partial T}{\partial x_j} = \lambda \frac{\partial T}{\partial x_i} \qquad i, j = 1, 2, 3 \qquad (10\text{-}31a)$$

where λ is a constant. Defining for convenience

$$\frac{\partial T}{\partial x_i} \equiv G_i$$

Eq. 10-31 is written in the form

$$k_{ij} G_j = \lambda G_i \qquad i, j = 1, 2, 3 \tag{10-31b}$$

or

$$(k_{ij} - \lambda \delta_{ij}) G_i = 0 \tag{10-31c}$$

where δ_{ij} is the kronecker delta. In the expanded form the equations become

$$\begin{aligned} (k_{11} - \lambda) G_1 + \quad k_{12} G_2 + \quad k_{13} G_3 &= 0 \\ k_{21} G_1 + (k_{22} - \lambda) G_2 + \quad k_{23} G_3 &= 0 \\ k_{31} G_1 + \quad k_{32} G_2 + (k_{33} - \lambda) G_3 &= 0 \end{aligned} \tag{10-31d}$$

The homogeneous system of Eqs. 10-31d possesses a nontrivial solution if the determinant of the coefficients vanishes, that is

$$\begin{vmatrix} k_{11} - \lambda & k_{12} & k_{13} \\ k_{21} & k_{22} - \lambda & k_{23} \\ k_{31} & k_{32} & k_{33} - \lambda \end{vmatrix} = 0 \tag{10-32}$$

Equation 10-32 is known as the *characteristic* (or *secular*) equation: It is a cubic equation in λ and has three roots; each of the roots are real numbers because the conductivity coefficients k_{ij} are all real by definition (a proof of this is given in reference 9, p. 259). Each of these roots correspond to the value of the conductivity coefficients in the directions in which the radius vector OP is normal to the ellipsoid; that is, the three roots $\lambda_1, \lambda_2, \lambda_3$ correspond to the values of the three *principle conductivities* k_1, k_2, k_3.

The problem of determining the directions of the three principal axes corresponding to the three principal conductivities is similar to the problem of determining the proper coordinate transformation associated with the *diagonalization* of a symmetric, second-order tensor with real coefficients. The procedure can be summarized as follows.

Let $\hat{n}(1)$ be the unit direction vector of the principal axis $o\xi_1$ corresponding to the principal conductivity $\lambda_1 \equiv k_1$. Let $n(1)_1, n(1)_2, n(1)_3$ be the direction cosines of the unit vector $\hat{n}(1)$ with the reference axes ox_1, ox_2, ox_3 to which the conductivities k_{ij} are referred to. The components $n(1)_1, n(1)_2, n(1)_3$ satisfy the following relations,

$$\begin{bmatrix} k_{11} - \lambda_1 & k_{12} & k_{13} \\ k_{21} & k_{22} - \lambda_1 & k_{23} \\ k_{31} & k_{32} & k_{33} - \lambda_1 \end{bmatrix} \begin{bmatrix} n(1)_1 \\ n(1)_2 \\ n(1)_3 \end{bmatrix} = 0 \tag{10-33a}$$

which provides 3 homogeneous equations for the three unknowns $n(1)_1$, $n(1)_2, n(1)_3$; however, only two of these equations are linearly independent. A third relation is obtained from the requirement that the direction cosines satisfy

$$n(1)_1^2 + n(1)_2^2 + n(1)_3^2 = 1 \tag{10-33b}$$

Thus, the three unknown coefficients $n(1)_1, n(1)_2, n(1)_3$ are determined from Eqs. 10-33a and 10-33b.

This procedure is repeated for λ_2 and λ_3. That is, the components $n(2)_1, n(2)_2, n(2)_3$ corresponding to $\lambda_2 \equiv k_2$ and the components $n(3)_1$, $n(3)_2, n(3)_3$ corresponding to $\lambda_3 \equiv k_3$ are determined. Since $\hat{n}(1)$, $\hat{n}(2)$ and $\hat{n}(3)$ are unit vectors their components $n(1)_1, n(1)_2, n(1)_3$, $n(2)_1$, etc. are the direction cosines. That is,

$$n(1)_1 \equiv c_{11} \qquad n(1)_2 \equiv c_{12} \qquad n(1)_3 \equiv c_{13} \qquad n(2)_1 \equiv c_{21} \text{ etc.}$$

Hence the directions of the principal conductivities $\lambda_1 \equiv k_1$, $\lambda_2 \equiv k_2$ and $\lambda_3 \equiv k_3$ are known.

As a further check on the above procedure for finding the directions of the principal axes we consider Eq. 10-12 for the transformation of conductivity coefficients in matrix notation

$$[k'_{ij}] = Q^T \cdot [\lambda] \cdot Q \tag{10-34}$$

where Q^T = transpose of Q
Q = the matrix of the direction cosines c_{ij}

$$[k'_{ij}] = \begin{bmatrix} k_{11} & k_{12} & k_{13} \\ k_{21} & k_{22} & k_{23} \\ k_{31} & k_{32} & k_{33} \end{bmatrix}$$

$$[\lambda] = \begin{bmatrix} \lambda_1(=k_1) & 0 & 0 \\ 0 & \lambda_2(=k_2) & 0 \\ 0 & 0 & \lambda_3(=k_3) \end{bmatrix}$$

and the convention is that $\lambda_1 \geq \lambda_2 \geq \lambda_3$. Performing the indicated calculations on the right-hand side of Eq. 10-34, it can be verified that the resulting components coincide with those on the left-hand side.

10-7. DIFFERENTIAL EQUATION OF HEAT CONDUCTION

The differential equation of heat conduction for isotropic solids was given in Chapter 1 in the form

$$\rho c_p \frac{\partial T}{\partial t} = -\nabla \cdot \bar{q} + G \tag{10-35}$$

This equation is applicable to anisotropic solids if the heat flux vector $\bar{q}$ is defined as given by Eq. 10-3. Then, the differential equation of heat conduction for an anisotropic solid becomes

$$\rho c_p \frac{\partial T}{\partial t} = \frac{\partial}{\partial x_1}\left[k_{11}\frac{\partial T}{\partial x_1} + k_{12}\frac{\partial T}{\partial x_2} + k_{13}\frac{\partial T}{\partial x_3}\right] + \frac{\partial}{\partial x_2}\left[k_{21}\frac{\partial T}{\partial x_1} + k_{22}\frac{\partial T}{\partial x_2} + k_{23}\frac{\partial T}{\partial x_3}\right] + \frac{\partial}{\partial x_3}\left[k_{31}\frac{\partial T}{\partial x_1} + k_{32}\frac{\partial T}{\partial x_2} + k_{33}\frac{\partial T}{\partial x_3}\right] + G \qquad \text{(10-35a)}$$

For the case of constant thermal conductivity coefficients equation 10-35a reduces to

$$\rho c_p \frac{\partial T}{\partial t} = k_{11}\frac{\partial^2 T}{\partial x_1^2} + k_{22}\frac{\partial^2 T}{\partial x_2^2} + k_{33}\frac{\partial^2 T}{\partial x_3^2} + 2k_{12}\frac{\partial^2 T}{\partial x_1 \partial x_2} + 2k_{13}\frac{\partial^2 T}{\partial x_1 \partial x_3} + 2k_{23}\frac{\partial^2 T}{\partial x_2 \partial x_3} + G \qquad \text{(10-35b)}$$

When the coordinate system is transformed from the rectangular axes ox_1, ox_2 and ox_3 to another set of rectangular axes $o\xi_1$, $o\xi_2$, and $o\xi_3$ which are the *principal axes*, the differential equation of heat conduction 10-35b simplifies to

$$\rho c_p \frac{\partial T}{\partial t} = k_1\frac{\partial^2 T}{\partial \xi_1^2} + k_2\frac{\partial^2 T}{\partial \xi_2^2} + k_3\frac{\partial^2 T}{\partial \xi_3^2} + G \qquad \text{(10-36)}$$

where k_1, k_2, k_3 are the principal conductivities.

Equation 10-36 can be transformed into a more convenient form by defining new independent variables η_1, η_2, η_3 as

$$\eta_1 = \left(\frac{K}{k_1}\right)^{1/2}\xi_1 \qquad \eta_2 = \left(\frac{K}{k_2}\right)^{1/2}\xi_2 \qquad \eta_3 = \left(\frac{K}{k_3}\right)^{1/2}\xi_3$$

where K is a constant which may be chosen arbitrarily. Then Eq. 10-36 simplifies to

$$\rho c_p \frac{\partial T}{\partial t} = K\left[\frac{\partial^2 T}{\partial \eta_1^2} + \frac{\partial^2 T}{\partial \eta_2^2} + \frac{\partial^2 T}{\partial \eta_3^2}\right] + G \qquad \text{(10-37)}$$

If the arbitrary constant K is chosen as

$$K = (k_1 k_2 k_3)^{1/3}$$

then, under the transformation given above the volume of a small element in the region remains fixed, i.e.

$$d\xi_1 d\xi_2 d\xi_3 = \frac{(k_1 k_2 k_3)^{1/2}}{K^{3/2}} d\eta_1 d\eta_2 d\eta_3 = d\eta_1 d\eta_2 d\eta_3$$

That is, the strength of the heat source, G, per unit volume, remains the same during the transformation. Equation 10-37 has the same form as the heat conduction equation for an isotropic solid. However, for most problems, the bounding surfaces of the region are distorted under the transformation considered above. For example a circular region may be distorted into an elliptic region.

For non-crystalline *orthotropic* solids, that is the solids which have three different thermal conductivities k_1, k_2, k_3 along the rectangular axes, say, $o\xi_1, o\xi_2, o\xi_3$ respectively, the differential equation of heat conduction has the form of equation 10-36. In the case of cylindrical and spherical coordinate systems the differential operators should be modified accordingly. For example, a tree trunk is a non-crystalline *orthotropic* solid which has three different thermal conductivities along the axial, radial and circumferential directions. Choosing a cylindrical coordinate system with z-axis lying along the axis, the three thermal conductivity coefficients are k_z along the axial grain, k_r along the radial grain and k_θ along the circumferential θ direction. Then, the differential equation of heat conduction, neglecting heat generation becomes

$$\rho c_p \frac{\partial T}{\partial t} = k_r \frac{1}{r}\frac{\partial T}{\partial r}\left(r\frac{\partial T}{\partial r}\right) + k_\theta \frac{1}{r^2}\frac{\partial^2 T}{\partial \theta^2} + k_z \frac{\partial^2 T}{\partial z^2} \qquad (10\text{-}38)$$

10-8. EXAMPLE—HEAT FLOW ACROSS AN ANISOTROPIC SLAB

To illustrate a method for determining the thermal conductivity coefficients of an anisotropic solid we examine the heat flow across an anisotropic slab, with ox_3' axis chosen normal to the slab, as shown in Fig. 10-4. It is assumed that the bounding surfaces of the slab are maintained at two different constant temperatures. Since the temperature gradient is applied only in the ox_3' direction, temperature gradients in the ox_1' and ox_2' directions are taken as zero, that is

$$\frac{dT}{dx_1'} = \frac{dT}{dx_2'} = 0 \qquad (10\text{-}39)$$

In view of the relations in Eq. 10-39, the components of the heat flux along the ox_1', ox_2', ox_3' directions, as obtained from Eq. 10-3 simplify to

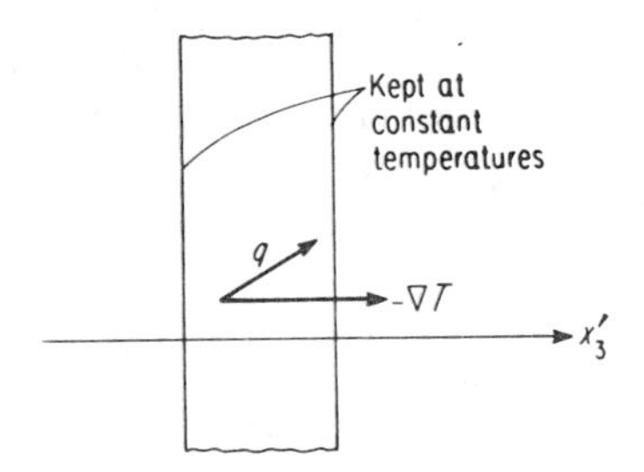

FIG. 10-4. Heat flow across anisotropic slab (temperature gradient is along x_3' axis.)

$$-q_1 = k'_{13}\frac{dT}{dx'_3} \tag{10-40a}$$

$$-q_2 = k'_{23}\frac{dT}{dx'_3} \tag{10-40b}$$

$$-q_3 = k'_{33}\frac{dT}{dx'_3} \tag{10-40c}$$

Equation 10-39 implies that the vector ∇T is along the ox'_3 direction; on the other hand the heat flux vector $\bar{q}$, has 3 components as given by Eq. 10-40, therefore ∇T is not necessarily along the ox'_3 axis but is in an oblique direction.

Since the heat flux components q_1 and q_2 are not directed across the slab, the measured value of the heat flux across the slab is due to the q_3 component. Knowing the heat flux component q_3 and the temperature gradient dT/dx'_3 from measurements, the thermal conductivity coefficient k'_{33} can be determined from Eq. 10-40c, i.e.

$$-q'_3 = k'_{33}\frac{dT}{dx'_3} \qquad [10\text{-}40c]$$

The measured thermal conductivity coefficient k'_{33} is related to the *principle conductivities* by Eq. 10-28a, for $j = 3$, as

$$k'_{33} = c_{31}^2 k_1 + c_{32}^2 k_2 + c_{33}^2 k_3 \tag{10-41}$$

where c_{31}, c_{32}, c_{33} are the direction cosines of the ox'_3 axis with the principal axes of the thermal conductivity.

Knowing the direction cosines of the ox'_3 axis with the principal axes, three different measurements are needed with three differently oriented slabs, to determine the principal conductivities k_1, k_2, k_3.

10-9. EXAMPLE—HEAT FLOW ALONG AN ANISOTROPIC ROD

The thermal resistivity of an anisotropic solid can be determined experimentally using a long, thin rod, with its two ends kept at two different constant temperatures. Figure 10-5 shows a long, thin crystal rod with rod axis chosen in the ox'_3 direction. When the thermal conductivity of the rod is much greater than that of its surrounding (i.e. when the rod is insulated) the heat flow through the rod must be along the ox'_3 direction only, hence we have

$$q_1 = q_2 = 0 \tag{10-42}$$

In view of 10-42, equations 10-9a, b, c simplify to

$$-\frac{dT}{dx'_1} = r'_{13}q_3 \tag{10-43a}$$

$$-\frac{dT}{dx'_2} = r'_{23}q_3 \tag{10-43b}$$

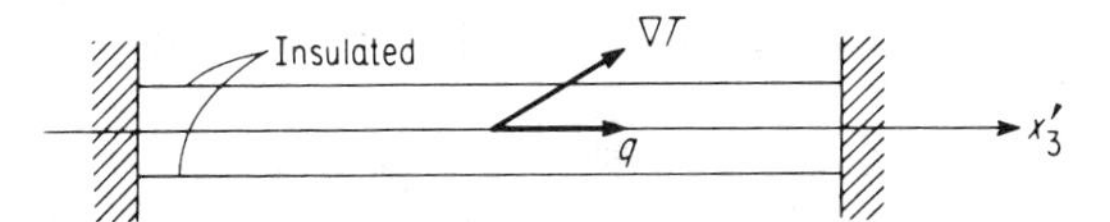

FIG. 10-5. Heat flow along an anisotropic rod (heat flux vector is along x' axis).

$$-\frac{dT}{dx_3'} = r_{33}' q_3 \tag{10-43c}$$

Since heat flows only along the ox_3' axis, the heat flux vector $\bar{q}$ is also in the ox_3' direction. On the other hand, it is apparent from Eq. 10-43 that the temperature gradient, ∇T, is not necessarily along the ox_3' direction but is in an oblique direction.

Knowing the heat flow q_3 and the temperature gradient dT/dx_3' from measurements, the thermal resistivity coefficient r_{33}' can be determined from Eq. 10-43c.

The resistivity coefficient r_{33}' is related to the principal resistivities r_1, r_2, r_3 by

$$r_{33}' = c_{31}^2 r_1 + c_{32}^2 r_2 + c_{33}^2 r_3 \tag{10-44}$$

where c_{31}, c_{32}, c_{33} are the direction cosines of the x_3' direction with the principal axes. Three different measurements with three differently oriented rods are needed to evaluate the principal resistivities r_1, r_2 and r_3.

10-10. EXAMPLE—RECTANGULAR SOLID WITH ORTHOTROPIC THERMAL PROPERTIES

The problem of the temperature distribution in laminated solids subjected to Joulean heating (i.e. heating as a result of current flow or applied voltage across the solid) is of considerable interest in the design of laminated coils. The temperature distribution in a rectangular solid with

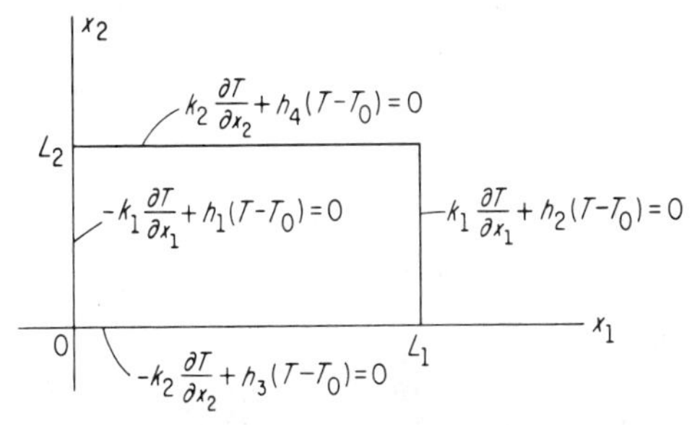

FIG. 10-6. Rectangular solid with orthotropic thermal properties.

orthotropic thermal conductivity and subjected to Joulean heating was examined by Cooper [10, 11].

Consider a rectangular region $0 \leq x_1 \leq L_1$, $0 \leq x_2 \leq L_2$, having orthotropic thermal conductivities k_1 and k_2 in the ox_1 and ox_2 directions, respectively, and shown in Fig. 10-6. Initially the solid is at a uniform temperature T_0. For times $t > 0$ heat is generated within the solid as a result of current flow at constant rate or an applied constant voltage difference.

Since the electrical resistivity of the solid varies with temperature, the rate of heat generation within the solid resulting from current flowing at a constant rate or from an applied constant voltage can be related to the temperature of the solid by[6]

$$g = g_0[1 \mp \mu(T - T_0)] \text{ Btu/hr ft}^3 \tag{10-45}$$

where μ = temperature coefficient of electrical resistance, ohm/ohm °F. The plus sign is chosen for constant current heating, the minus sign for constant voltage heating.

The heat generated within the solid is assumed to be dissipated by convection into a medium at constant temperature T_0. For generality, the heat transfer coefficients at the four boundaries are taken not to be the same.

The boundary value problem of heat conduction for a constant voltage heat source is given as

$$k_1 \frac{\partial^2 T}{\partial x_1^2} + k_2 \frac{\partial^2 T}{\partial x_2^2} + g_0[1 - \mu(T - T_0)] = \rho c_p \frac{\partial T}{\partial t}$$

$$\text{in } 0 \leq x_1 \leq L_1, 0 \leq x_2 \leq L_2, \text{ for } t > 0 \tag{10-46a}$$

$$-k_1 \frac{\partial T}{\partial x_1} + h_1(T - T_0) = 0 \quad \text{at} \quad x_1 = 0, t > 0 \tag{10-46b}$$

$$k_1 \frac{\partial T}{\partial x_1} + h_2(T - T_0) = 0 \quad \text{at} \quad x_1 = L_1, t > 0 \tag{10-46c}$$

$$-k_2 \frac{\partial T}{\partial x_2} + h_3(T - T_0) = 0 \quad \text{at} \quad x_2 = 0, t > 0 \tag{10-46d}$$

$$k_2 \frac{\partial T}{\partial x_2} + h_4(T - T_0) = 0 \quad \text{at} \quad x_2 = L_2, t > 0 \tag{10-46e}$$

[6] For a constant current heat source,

$$g = I_0^2 R_0[1 + \mu(T - T_0)] \equiv g_0[1 + \mu(T - T_0)]$$

For a constant voltage heat source,

$$g = \frac{V_0^2}{R_0} \frac{1}{1 + \mu(T - T_0)} \cong \frac{V_0^2}{R_0}[1 - \mu(T - T_0)] \equiv g_0[1 - \mu(T - T_0)]$$

$$\text{if } \mu(T - T_0) << 1$$

$$T = T_0$$
$$\text{in } 0 \le x_1 \le L_1, 0 \le x_2 \le L_2, \text{ for } t = 0 \qquad (10\text{-}46f)$$

System 10-46 can be expressed in a more convenient form as

$$\frac{\partial^2 \theta}{\partial \xi^2} + \kappa \frac{\partial^2 \theta}{\partial \eta^2} + \mu G_0 \theta + G_0 = \frac{\partial \theta}{\partial \tau} \quad \text{in } 0 \le \xi \le 1, 0 \le \eta \le 1 \text{ for } \tau > 0 \qquad (10\text{-}47a)$$

$$-\frac{\partial \theta}{\partial \xi} + H_1 \theta = 0 \quad \text{at } \xi = 0, \tau > 0 \qquad (10\text{-}47b)$$

$$\frac{\partial \theta}{\partial \xi} + H_2 \theta = 0 \quad \text{at } \xi = 1, \tau > 0 \qquad (10\text{-}47c)$$

$$-\frac{\partial \theta}{\partial \eta} + H_3 \theta = 0 \quad \text{at } \eta = 0, \tau > 0 \qquad (10\text{-}47d)$$

$$\frac{\partial \theta}{\partial \eta} + H_4 \theta = 0 \quad \text{at } \eta = 1, \tau > 0 \qquad (10\text{-}47e)$$

$$\theta = 0 \quad \text{in } 0 \le \xi \le 1, 0 \le \eta \le 1 \text{ for } \tau = 0 \qquad (10\text{-}47f)$$

where various quantities are defined as

$$\theta = T - T_0$$

$$\xi = \frac{x_1}{L_1} \qquad \eta = \frac{x_2}{L_2}$$

$$\tau = \frac{k_1 t}{\rho c_p L_1^2} \qquad \kappa = \frac{k_2 L_1^2}{k_1 L_2^2}$$

$$G_0 = \frac{g_0 L_1^2}{k_1}$$

$$H_1 = \frac{h_1 L_1}{k_1} \qquad H_2 = \frac{h_2 L_1}{k_1} \qquad H_3 = \frac{h_3 L_2}{k_2} \qquad H_4 = \frac{h_4 L_2}{k_2}$$

We solve system 10-47 by applying a finite-integral transform technique described in Chapters 1 and 2. Define the double integral transform and double inversion formula for the temperature function $\theta(\xi, \eta, \tau)$ with respect to the space variables ξ and η as

$$\bar{\theta}(\beta_m, \nu_n, \tau) = \int_{\xi'=0}^{1} \int_{\eta'=0}^{1} K(\beta_m, \xi') \cdot K(\nu_n, \eta') \cdot \theta(\xi', \eta', \tau) \cdot d\xi' \cdot d\eta' \qquad (10\text{-}48a)$$

$$\theta(\xi, \eta, \tau) = \sum_{\beta_m} \sum_{\nu_n} K(\beta_m, \xi) \cdot K(\nu_n, \eta) \cdot \bar{\theta}(\beta_m, \nu_n, \tau) \qquad (10\text{-}48b)$$

where the kernels $K(\beta_m, \xi)$, $K(\nu_n, \eta)$ and the eigenvalues β_m, ν_n are obtained from Table 2-1 (Chapter 2) as

$$K(\beta_m, \xi) = \sqrt{2}\frac{\beta_m \cos \beta_m \xi + H_1 \sin \beta_m \xi}{\left[(\beta_m^2 + H_1^2)\left(1 + \dfrac{H_2}{\beta_m^2 + H_2^2}\right) + H_1\right]^{1/2}}$$

$$K(\nu_n, \eta) = \sqrt{2}\frac{\nu_n \cos \nu_n \eta + H_3 \sin \nu_n \eta}{\left[(\nu_n^2 + H_3^2)\left(1 + \dfrac{H_4}{\nu_n^2 + H_4^2}\right) + H_3\right]^{1/2}}$$

and β_m, ν_n are the positive roots of the following transcendental equations

$$\tan \beta = \frac{\beta(H_1 + H_2)}{\beta^2 - H_1 H_2},$$

$$\tan \nu = \frac{\nu(H_3 + H_4)}{\nu^2 - H_3 H_4}$$

We take the integral transform of the system (Eqs. 10-47a–f) by applying the transform (Eq. 10-48a) and obtain

$$-\beta_m^2 \bar{\theta} - \kappa \nu_n^2 \bar{\theta} + \mu G_0 \bar{\theta} + \bar{G}_0 = \frac{d\bar{\theta}}{d\tau}$$

or

$$\frac{d\bar{\theta}}{d\tau} + \lambda_{mn} \bar{\theta} = \bar{G}_0 \tag{10-49a}$$

subject to the condition $\bar{\theta} = 0$ when $\tau = 0$ (10-49b)

where,

$$\lambda_{mn} \equiv \beta_m^2 + \kappa \nu_n^2 - \mu G_0 \tag{10-49c}$$

Quantities with a bar refer to the integral transform according to the transform Eq. 10-48a. The solution of the ordinary differential Eq. 10-49 gives

$$\bar{\theta} = \bar{G}_0 \cdot \frac{1 - e^{-\lambda_{mn}\tau}}{\lambda_{mn}}. \tag{10-50}$$

Inversion of Eq. 10-50 by the inversion formula Eq. 10-48b gives the solution of system 10-47 as

$$\theta(\xi, \eta, \tau) = \sum_{\beta_m} \sum_{\nu_n} K(\beta_m, \xi) \cdot K(\nu_n, \eta) \cdot \bar{G}_0 \cdot \frac{1 - e^{-\lambda_{mn}\tau}}{\lambda_{mn}} \tag{10-51}$$

where

$$\bar{G}_0 = \int_{\xi=0}^{1} \int_{\eta=0}^{1} G_0 \cdot K(\beta_m, \xi') \cdot K(\nu_n, \eta') \cdot d\xi' \cdot d\eta'$$

$$= 2\frac{G_0\left[\sin \beta_m + \dfrac{H_1}{\beta_m}(1 - \cos \beta_m)\right]\left[\sin \nu_n + \dfrac{H_3}{\nu_n}(1 - \cos \nu_n)\right]}{\left[(\beta_m^2 + H_1^2)\left(1 + \dfrac{H_2}{\beta_m^2 + H_2^2}\right) + H_1\right]^{1/2}\left[(\nu_n^2 + H_3^2)\left(1 + \dfrac{H_2}{\nu_n^2 + H_4^2}\right) + H_3\right]^{1/2}}$$

REFERENCES

1. H. S. Carslaw and J. C. Jaeger, *Conduction of Heat in Solids*, Clarendon Press, Oxford, 1959, pp. 38–49.

2. J. F. Nye, *Physical Properties of Crystals*, Clarendon Press, Oxford, 1957, pp. 195–214.

3. W. A. Wooster, *A Textbook on Crystal Physics*, at the University Press, Cambridge, 1938, pp. 63–84.

4. S. R. DeGroot and P. Mazur, *Non-Equilibrium Thermodynamics*, North-Holland Publishing Company, Amsterdam; Interscience Publishers Inc., New York, 1963, pp. 235–303.

5. H. Jeffreys, *Cartesian Tensors*, University Press, Cambridge, 1963.

6. L. Onsager, Reciprocal Relations in Irreversible Process I," *Physics Review*, Vol. 37, 1931, pp. 405–426.

7. H. B. G. Casimir, "On Onsager's Principle of Microscopic Reversibility," *Review of Modern Physics*, Vol. 17, 1945, pp. 343–50.

8. F. Prutton and S. H. Maron, *Fundamental Principles of Physical Chemistry*, McGraw-Hill Book Co., New York, 1951.

9. L. P. Eisenhart, *Coordinate Geometry*, Dover Publications, Inc., New York, 1960, pp. 252–262.

10. H. F. Cooper, "Transient and Steady-State Temperature Distribution in Foil Wound Solenoids and Other Electric Apparatus of Rectangular Cross Section," *1965 IEEE International Convention Record*, Part 10, March, 1965, pp. 67–75.

11. H. F. Cooper, "Joulean Heating of an Inflnite Rectangular Rod with Orthotropic Thermal Properties," *ASME Paper* No. 66-WA/HT-14.

PROBLEMS

1. Consider the problem of one-dimensional heat flow across an anisotropic slab, the two boundary surfaces of which are kept at constant but different isothermal temperatures. Let ox_3 be the axis normal to the slab. Discuss the effects of the position of the thermal conductivity ellipse in relation to the ox_3 axis on the direction of the heat flux vector. Determine the direction of the heat flux vector when the principal axis of the thermal conductivity ellipse coincides with the ox_3 axis, and when it is inclined to the ox_3 axis.

2. Consider the problem of one-dimensional heat flow along an anisotropic rod which is insulated at the sides and kept at constant but different isothermal temperatures at the ends. Let ox_3 be the axis of the rod. Discuss the effects of the position of the thermal resistivity ellipse in relation to the ox_3 axis on the direction of the temperature gradient vector. What is the direction of the temperature gradient vector when the principal axis of the thermal resistivity ellipse coincides with the ox_3 axis and when it is inclined to it?

3. Let k_{ij} be the nine components of the thermal conductivity tensor referred to an arbitrary cartesian coordinate axis x_i, i.e.

$$k_{ij} = \begin{bmatrix} k_{11} & k_{12} & k_{13} \\ k_{21} & k_{22} & k_{23} \\ k_{31} & k_{32} & k_{33} \end{bmatrix}$$

Write down the relations for each of the thermal resistivity coefficients when the anisotropic solid is (1) a monoclinic crystal system, (2) an orthorombic crystal system, (3) a cubic crystal system. Discuss the physical significance of the relation between the conductivity coefficients and the resistivity coefficients for the orthorhombic system.

4. The conductivity coefficients for a monoclinic crystal system are given with reference to an arbitrary coordinate axes ox_1, ox_2, and ox_3. Consider an arbitrary axis ox_j whose direction cosines with the ox_1, ox_2, ox_3 axes are respectively c_1, c_2, c_3. Write down a relation for the conductivity coefficient k_{jj} along the ox_j axis in terms of the conductivity coefficients referred to the ox_1, ox_2, and ox_3 axis.

Repeat the same problem for an orthorhombic and cubic crystal systems.

5. Consider a laminated material with principal conductivities k_1, k_2, k_3 along the $0\xi_1, 0\xi_2, 0\xi_3$ axes respectively. Let ox_j be an axis lying in the plane $\xi_1 0 \xi_2$ and bisecting the angle $\xi_1 0 \xi_2$. A slab of thickness L is cut out of this material in such a direction that the ox_j axis is perpendicular to the slab.

Write down a relation for the thermal conductivity coefficient of this slab along the ox_j axis, k_{jj}, in terms of the principal conductivities k_1, k_2, k_3. The slab is initially at constant temperature T_i, and for times $t > o$ the two bounding surfaces are kept at constant but different isothermal temperatures T_1 and T_2. Find a relation for the transient temperature distribution in the slab for times $t > o$.

6. The thermal conductivity coefficients of an anisotropic solid are given as

$$k_{ij} = \begin{bmatrix} 1/\sqrt{2} & 1/\sqrt{2} & 0 \\ 1/\sqrt{2} & 1/\sqrt{2} & 0 \\ 0 & 0 & 1 \end{bmatrix}$$

Determine the principal conductivities and their directions in relation to the reference Cartesian axes ox_1, ox_2, ox_3.

7. Consider an anisotropic solid whose thermal conductivity tensor k_{ij} is in the form

$$k_{ij} = \begin{bmatrix} k_{11} & k_{12} & 0 \\ k_{12} & k_{22} & 0 \\ 0 & 0 & k_{33} \end{bmatrix}$$

and each of the thermal conductivity coefficients with respect to the Cartesian reference axes ox_1, ox_2, ox_3 are given.

The negative temperature gradient $-\nabla T$ in the solid is along the ox_1 axis.

(1) Write down the relations for the three components of the heat flux vector, $\bar{q}$.

(2) Using the relations for the components of the heat flux find a relation for the angle between the heat flux vector and the ox_1 axis in terms of the conductivity coefficients.

8. Consider a rectangular region $0 \le x_1 \le L_1$ and $0 \le x_2 \le L_2$ having orthotropic thermal conductivities k_1 and k_2 in the ox_1 and ox_2 directions respectively. The side at $x_2 = 0$ is kept at a constant temperature T_o while the other sides are kept at zero temperature. By means of a suitable transformation of the independent variables the problem for the anisotropic medium can be reduced to the solution of the problem for an isotropic medium. Determine the relation for the steady state temperature distribution in the region.

Appendices

APPENDIX I

ROOTS OF TRANSCENDENTAL EQUATIONS

First six roots β_n of $\beta \tan \beta = c$

c	β_1	β_2	β_3	β_4	β_5	β_6
0	0	3.1416	6.2832	9.4248	12.5664	15.7080
0.001	0.0316	3.1419	6.2833	9.4249	12.5665	15.7080
0.002	0.0447	3.1422	6.2835	9.4250	12.5665	15.7081
0.004	0.0632	3.1429	6.2838	9.4252	12.5667	15.7082
0.006	0.0774	3.1435	6.2841	9.4254	12.5668	15.7083
0.008	0.0893	3.1441	6.2845	9.4256	12.5670	15.7085
0.01	0.0998	3.1448	6.2848	9.4258	12.5672	15.7086
0.02	0.1410	3.1479	6.2864	9.4269	12.5680	15.7092
0.04	0.1987	3.1543	6.2895	9.4290	12.5696	15.7105
0.06	0.2425	3.1606	6.2927	9.4311	12.5711	15.7118
0.08	0.2791	3.1668	6.2959	9.4333	12.5727	15.7131
0.1	0.3111	3.1731	6.2991	9.4354	12.5743	15.7143
0.2	0.4328	3.2039	6.3148	9.4459	12.5823	15.7207
0.3	0.5218	3.2341	6.3305	9.4565	12.5902	15.7270
0.4	0.5932	3.2636	6.3461	9.4670	12.5981	15.7334
0.5	0.6533	3.2923	6.3616	9.4775	12.6060	15.7397
0.6	0.7051	3.3204	6.3770	9.4879	12.6139	15.7460
0.7	0.7506	3.3477	6.3923	9.4983	12.6218	15.7524
0.8	0.7910	3.3744	6.4074	9.5087	12.6296	15.7587
0.9	0.8274	3.4003	6.4224	9.5190	12.6375	15.7650
1.0	0.8603	3.4256	6.4373	9.5293	12.6453	15.7713
1.5	0.9882	3.5422	6.5097	9.5801	12.6841	15.8026
2.0	1.0769	3.6436	6.5783	9.6296	12.7223	15.8336
3.0	1.1925	3.8088	6.7040	9.7240	12.7966	15.8945
4.0	1.2646	3.9352	6.8140	9.8119	12.8678	15.9536
5.0	1.3138	4.0336	6.9096	9.8928	12.9352	16.0107
6.0	1.3496	4.1116	6.9924	9.9667	12.9988	16.0654
7.0	1.3766	4.1746	7.0640	10.0339	13.0584	16.1177
8.0	1.3978	4.2264	7.1263	10.0949	13.1141	16.1675
9.0	1.4149	4.2694	7.1806	10.1502	13.1660	16.2147
10.0	1.4289	4.3058	7.2281	10.2003	13.2142	16.2594
15.0	1.4729	4.4255	7.3959	10.3898	13.4078	16.4474
20.0	1.4961	4.4915	7.4954	10.5117	13.5420	16.5864
30.0	1.5202	4.5615	7.6057	10.6543	13.7085	16.7691

APPENDIX I (*Continued*)

c	β_1	β_2	β_3	β_4	β_5	β_6
40.0	1.5325	4.5979	7.6647	10.7334	13.8048	16.8794
50.0	1.5400	4.6202	7.7012	10.7832	13.8666	16.9519
60.0	1.5451	4.6353	7.7259	10.8172	13.9094	17.0026
80.0	1.5514	4.6543	7.7573	10.8606	13.9644	17.0686
100.0	1.5552	4.6658	7.7764	10.8871	13.9981	17.1093
∞	1.5708	4.7124	7.8540	10.9956	14.1372	17.2788

Roots are all real if $c > 0$.

ROOTS OF TRANSCENDENTAL EQUATIONS

First six roots β_n of $\beta \cot \beta = -c$

c	β_1	β_2	β_3	β_4	β_5	β_6
−1.0	0	4.4934	7.7253	10.9041	14.0662	17.2208
−0.995	0.1224	4.4945	7.7259	10.9046	14.0666	17.2210
−0.99	0.1730	4.4956	7.7265	10.9050	14.0669	17.2213
−0.98	0.2445	4.4979	7.7278	10.9060	14.0676	17.2219
−0.97	0.2991	4.5001	7.7291	10.9069	14.0683	17.2225
−0.96	0.3450	4.5023	7.7304	10.9078	14.0690	17.2231
−0.95	0.3854	4.5045	7.7317	10.9087	14.0697	17.2237
−0.94	0.4217	4.5068	7.7330	10.9096	14.0705	17.2242
−0.93	0.4551	4.5090	7.7343	10.9105	14.0712	17.2248
−0.92	0.4860	4.5112	7.7356	10.9115	14.0719	17.2254
−0.91	0.5150	4.5134	7.7369	10.9124	14.0726	17.2260
−0.90	0.5423	4.5157	7.7382	10.9133	14.0733	17.2266
−0.85	0.6609	4.5268	7.7447	10.9179	14.0769	17.2295
−0.8	0.7593	4.5379	7.7511	10.9225	14.0804	17.2324
−0.7	0.9208	4.5601	7.7641	10.9316	14.0875	17.2382
−0.6	1.0528	4.5822	7.7770	10.9408	14.0946	17.2440
−0.5	1.1656	4.6042	7.7899	10.9499	14.1017	17.2498
−0.4	1.2644	4.6261	7.8028	10.9591	14.1088	17.2556
−0.3	1.3525	4.6479	7.8156	10.9682	14.1159	17.2614
−0.2	1.4320	4.6696	7.8284	10.9774	14.1230	17.2672
−0.1	1.5044	4.6911	7.8412	10.9865	14.1301	17.2730
0	1.5708	4.7124	7.8540	10.9956	14.1372	17.2788
0.1	1.6320	4.7335	7.8667	11.0047	14.1443	17.2845
0.2	1.6887	4.7544	7.8794	11.0137	14.1513	17.2903
0.3	1.7414	4.7751	7.8920	11.0228	14.1584	17.2961
0.4	1.7906	4.7956	7.9046	11.0318	14.1654	17.3019
0.5	1.8366	4.8158	7.9171	11.0409	14.1724	17.3076
0.6	1.8798	4.8358	7.9295	11.0498	14.1795	17.3134
0.7	1.9203	4.8556	7.9419	11.0588	14.1865	17.3192
0.8	1.9586	4.8751	7.9542	11.0677	14.1935	17.3249
0.9	1.9947	4.8943	7.9665	11.0767	14.2005	17.3306
1.0	2.0288	4.9132	7.9787	11.0856	14.2075	17.3364

c	β_1	β_2	β_3	β_4	β_5	β_6
1.5	2.1746	5.0037	8.0385	11.1296	14.2421	17.3649
2.0	2.2889	5.0870	8.0962	11.1727	14.2764	17.3932
3.0	2.4557	5.2329	8.2045	11.2560	14.3434	17.4490
4.0	2.5704	5.3540	8.3029	11.3349	14.4080	17.5034
5.0	2.6537	5.4544	8.3914	11.4086	14.4699	17.5562
6.0	2.7165	5.5378	8.4703	11.4773	14.5288	17.6072
7.0	2.7654	5.6078	8.5406	11.5408	14.5847	17.6562
8.0	2.8044	5.6669	8.6031	11.5994	14.6374	17.7032
9.0	2.8363	5.7172	8.6587	11.6532	14.6870	17.7481
10.0	2.8628	5.7606	8.7083	11.7027	14.7335	17.7908
15.0	2.9476	5.9080	8.8898	11.8959	14.9251	17.9742
20.0	2.9930	5.9921	9.0019	12.0250	15.0625	18.1136
30.0	3.0406	6.0831	9.1294	12.1807	15.2380	18.3018
40.0	3.0651	6.1311	9.1987	12.2688	15.3417	18.4180
50.0	3.0801	6.1606	9.2420	12.3247	15.4090	18.4953
60.0	3.0901	6.1805	9.2715	12.3632	15.4559	18.5497
80.0	3.1028	6.2058	9.3089	12.4124	15.5164	18.6209
100.0	3.1105	6.2211	9.3317	12.4426	15.5537	18.6650
∞	3.1416	6.2832	9.4248	12.5664	15.7080	18.8496

Roots are all real if $c > -1$.

APPENDIX II

Numerical Values of Error Function $\operatorname{erf} z = \dfrac{2}{\sqrt{\pi}} \int_0^z e^{-\xi^2}\, d\xi$

z	erf z	z	erf z	z	erf z	z	erf z	z	erf z
0.00	0.00000	0.50	0.52049	1.00	0.84270	1.50	0.96610	2.00	0.99532
0.01	0.01128	0.51	0.52924	1.01	0.84681	1.51	0.96727	2.20	0.99814
0.02	0.02256	0.52	0.53789	1.02	0.85083	1.52	0.96841	2.40	0.99931
0.03	0.03384	0.53	0.54646	1.03	0.85478	1.53	0.96951	2.60	0.99976
0.04	0.04511	0.54	0.55493	1.04	0.85864	1.54	0.97058	2.80	0.99992
0.05	0.05637	0.55	0.56332	1.05	0.86243	1.55	0.97162	3.00	0.99998
0.06	0.06762	0.56	0.57161	1.06	0.86614	1.56	0.97262		
0.07	0.07885	0.57	0.57981	1.07	0.86977	1.57	0.97360		
0.08	0.09007	0.58	0.58792	1.08	0.87332	1.58	0.97454		
0.09	0.10128	0.59	0.59593	1.09	0.87680	1.59	0.97546		
0.10	0.11246	0.60	0.60385	1.10	0.88020	1.60	0.97634		
0.11	0.12362	0.61	0.61168	1.11	0.88353	1.61	0.97720		
0.12	0.13475	0.62	0.61941	1.12	0.88678	1.62	0.97803		
0.13	0.14586	0.63	0.62704	1.13	0.88997	1.63	0.97884		
0.14	0.15694	0.64	0.63458	1.14	0.89308	1.64	0.97962		

APPENDIX II (*Continued*)

z	erf z	z	erf z	z	erf z	z	erf z	z	erf z
0.15	0.16799	0.65	0.64202	1.15	0.89612	1.65	0.98037		
0.16	0.17901	0.66	0.64937	1.16	0.89909	1.66	0.98110		
0.17	0.18999	0.67	0.65662	1.17	0.90200	1.67	0.98181		
0.18	0.20093	0.68	0.66378	1.18	0.90483	1.68	0.98249		
0.19	0.21183	0.69	0.67084	1.19	0.90760	1.69	0.98315		
0.20	0.22270	0.70	0.67780	1.20	0.91031	1.70	0.98379		
0.21	0.23352	0.71	0.68466	1.21	0.91295	1.71	0.98440		
0.22	0.24429	0.72	0.69143	1.22	0.91553	1.72	0.98500		
0.23	0.25502	0.73	0.69810	1.23	0.91805	1.73	0.98557		
0.24	0.26570	0.74	0.70467	1.24	0.92050	1.74	0.98613		
0.25	0.27632	0.75	0.71115	1.25	0.92290	1.75	0.98667		
0.26	0.28689	0.76	0.71753	1.26	0.92523	1.76	0.98719		
0.27	0.29741	0.77	0.72382	1.27	0.92751	1.77	0.98769		
0.28	0.30788	0.78	0.73001	1.28	0.92973	1.78	0.98817		
0.29	0.31828	0.79	0.73610	1.29	0.93189	1.79	0.98864		
0.30	0.32862	0.80	0.74210	1.30	0.93400	1.80	0.98909		
0.31	0.33890	0.81	0.74800	1.31	0.93606	1.81	0.98952		
0.32	0.34912	0.82	0.75381	1.32	0.93806	1.82	0.98994		
0.33	0.35927	0.83	0.75952	1.33	0.94001	1.83	0.99034		
0.34	0.36936	0.84	0.76514	1.34	0.94191	1.84	0.99073		
0.35	0.37938	0.85	0.77066	1.35	0.94376	1.85	0.99111		
0.36	0.38932	0.86	0.77610	1.36	0.94556	1.86	0.99147		
0.37	0.39920	0.87	0.78143	1.37	0.94731	1.87	0.99182		
0.38	0.40900	0.88	0.78668	1.38	0.94901	1.88	0.99215		
0.39	0.41873	0.89	0.79184	1.39	0.95067	1.89	0.99247		
0.40	0.42839	0.90	0.79690	1.40	0.95228	1.90	0.99279		
0.41	0.43796	0.91	0.80188	1.41	0.95385	1.91	0.99308		
0.42	0.44746	0.92	0.80676	1.42	0.95537	1.92	0.99337		
0.43	0.45688	0.93	0.81156	1.43	0.95685	1.93	0.99365		
0.44	0.46622	0.94	0.81627	1.44	0.95829	1.94	0.99392		
0.45	0.47548	0.95	0.82089	1.45	0.95969	1.95	0.99417		
0.46	0.48465	0.96	0.82542	1.46	0.96105	1.96	0.99442		
0.47	0.49374	0.97	0.82987	1.47	0.96237	1.97	0.99466		
0.48	0.50274	0.98	0.83423	1.48	0.96365	1.98	0.99489		
0.49	0.51166	0.99	0.83850	1.49	0.96489	1.99	0.99511		

APPENDIX III

Numerical Values of Bessel Functions[1]

$J_0(z)$

z	0	0.1	0.2	0.3	0.4	0.5	0.6	0.7	0.8	0.9
0	1.0000	0.9975	0.9900	0.9776	0.9604	0.9385	0.9120	0.8812	0.8463	0.8075
1	0.7652	0.7196	0.6711	0.6201	0.5669	0.5118	0.4554	0.3980	0.3400	0.2818
2	0.2239	0.1666	0.1104	0.0555	0.0025	−0.0484	−0.0968	−0.1424	−0.1850	−0.2243
3	−0.2601	−0.2921	−0.3202	−0.3443	−0.3643	−0.3801	−0.3918	−0.3992	−0.4026	−0.4018
4	−0.3971	−0.3887	−0.3766	−0.3610	−0.3423	−0.3205	−0.2961	−0.2693	−0.2404	−0.2097
5	−0.1776	−0.1443	−0.1103	−0.0758	−0.0412	−0.0068	0.0270	+0.0599	0.0917	0.1220
6	0.1506	0.1773	0.2017	0.2238	0.2433	0.2601	0.2740	0.2851	0.2931	0.2981
7	0.3001	0.2991	0.2951	0.2882	0.2786	0.2663	0.2516	0.2346	0.2154	0.1944
8	0.1717	0.1475	0.1222	0.0960	0.0692	0.0419	0.0146	−0.0125	−0.0392	−0.0653
9	−0.0903	−0.1142	−0.1367	−0.1577	−0.1768	−0.1939	−0.2090	−0.2218	−0.2323	−0.2403
10	−0.2459	−0.2490	−0.2496	−0.2477	−0.2434	−0.2366	−0.2276	−0.2164	−0.2032	−0.1881
11	−0.1712	−0.1528	−0.1330	−0.1121	−0.0902	−0.0677	−0.0446	−0.0213	+0.0020	0.0250
12	0.0477	0.0697	0.0908	0.1108	0.1296	0.1469	0.1626	0.1766	0.1887	0.1988
13	0.2069	0.2129	0.2167	0.2183	0.2177	0.2150	0.2101	0.2032	0.1943	0.1836
14	0.1711	0.1570	0.1414	0.1245	0.1065	0.0875	0.0679	0.0476	0.0271	0.0064
15	−0.0142	−0.0346	−0.0544	−0.0736	−0.0919	−0.1092	−0.1253	−0.1401	−0.1533	−0.1650

When $z > 15.9$,

$$J_0(z) \cong \sqrt{\left(\frac{2}{\pi z}\right)}\left\{\sin\left(z + \frac{1}{4}\pi\right) + \frac{1}{8z}\sin\left(z - \frac{1}{4}\pi\right)\right\}$$

$$\cong \frac{0.7979}{\sqrt{z}}\left\{\sin(57.296z + 45)^\circ + \frac{1}{8z}\sin(57.296z - 45)^\circ\right\}.$$

APPENDIX III (*Continued*)

$J_1(z)$

z	0	0.1	0.2	0.3	0.4	0.5	0.6	0.7	0.8	0.9
0	0.0000	0.0499	0.0995	0.1483	0.1960	0.2423	0.2867	0.3290	0.3688	0.4059
1	0.4401	0.4709	0.4983	0.5220	0.5419	0.5579	0.5699	0.5778	0.5815	0.5812
2	0.5767	0.5683	0.5560	0.5399	0.5202	0.4971	0.4708	0.4416	0.4097	0.3754
3	0.3391	0.3009	0.2613	0.2207	0.1792	0.1374	0.0955	0.0538	0.0128	−0.0272
4	−0.0660	−0.1033	−0.1386	−0.1719	−0.2028	−0.2311	−0.2566	−0.2791	−0.2985	−0.3147
5	−0.3276	−0.3371	−0.3432	−0.3460	−0.3453	−0.3414	−0.3343	−0.3241	−0.3110	−0.2951
6	−0.2767	−0.2559	−0.2329	−0.2081	−0.1816	−0.1538	−0.1250	−0.0953	−0.0652	−0.0349
7	−0.0047	+0.0252	0.0543	0.0826	0.1096	0.1352	0.1592	0.1813	0.2014	0.2192
8	0.2346	0.2476	0.2580	0.2657	0.2708	0.2731	0.2728	0.2697	0.2641	0.2559
9	0.2453	0.2324	0.2174	0.2004	0.1816	0.1613	0.1395	0.1166	0.0928	0.0684
10	0.0435	0.0184	−0.0066	−0.0313	−0.0555	−0.0789	−0.1012	−0.1224	−0.1422	−0.1603
11	−0.1768	−0.1913	−0.2039	−0.2143	−0.2225	−0.2284	−0.2320	−0.2333	−0.2323	−0.2290
12	−0.2234	−0.2157	−0.2060	−0.1943	−0.1807	−0.1655	−0.1487	−0.1307	−0.1114	−0.0912
13	−0.0703	−0.0489	−0.0271	−0.0052	+0.0166	0.0380	0.0590	0.0791	0.0984	0.1165
14	0.1334	0.1488	0.1626	0.1747	0.1850	0.1934	0.1999	0.2043	0.2066	0.2069
15	0.2051	0.2013	0.1955	0.1879	0.1784	0.1672	0.1544	0.1402	0.1247	0.1080

When $z > 15.9$,

$$J_1(z) \cong \sqrt{\left(\frac{2}{\pi z}\right)} \left\{\sin\left(z - \frac{1}{4}\pi\right) + \frac{3}{8z}\sin\left(z + \frac{1}{4}\pi\right)\right\}$$

$$\cong \frac{0.7979}{\sqrt{z}} \left\{\sin(57.296z - 45)^\circ + \frac{3}{8z}\sin(57.296z + 45)^\circ\right\}.$$

$Y_0(z)$

z	0	0.1	0.2	0.3	0.4	0.5	0.6	0.7	0.8	0.9
0	$-\infty$	−1.534	−1.081	−0.8073	−0.6060	−0.4445	−0.3085	−0.1907	−0.0868	+0.0056
1	0.0883	0.1622	0.2281	0.2865	0.3379	0.3824	0.4204	0.4520	0.4774	0.4968
2	0.5104	0.5183	0.5208	0.5181	0.5104	0.4981	0.4813	0.4605	0.4359	0.4079
3	0.3769	0.3431	0.3071	0.2691	0.2296	0.1890	0.1477	0.1061	0.0645	0.0234
4	−0.0169	−0.0561	−0.0938	−0.1296	−0.1633	−0.1947	−0.2235	−0.2494	−0.2723	−0.2921
5	−0.3085	−0.3216	−0.3313	−0.3374	−0.3402	−0.3395	−0.3354	−0.3282	−0.3177	−0.3044
6	−0.2882	−0.2694	−0.2483	−0.2251	−0.1999	−0.1732	−0.1452	−0.1162	−0.0864	−0.0563
7	−0.0259	+0.0042	0.0339	0.0628	0.0907	0.1173	0.1424	0.1658	0.1872	0.2065
8	0.2235	0.2381	0.2501	0.2595	0.2662	0.2702	0.2715	0.2700	0.2659	0.2592
9	0.2499	0.2383	0.2245	0.2086	0.1907	0.1712	0.1502	0.1279	0.1045	0.0804
10	0.0557	0.0307	0.0056	−0.0193	−0.0437	−0.0675	−0.0904	−0.1122	−0.1326	−0.1516
11	−0.1688	−0.1843	−0.1977	−0.2091	−0.2183	−0.2252	−0.2299	−0.2322	−0.2322	−0.2298
12	−0.2252	−0.2184	−0.2095	−0.1986	−0.1858	−0.1712	−0.1551	−0.1375	−0.1187	−0.0989
13	−0.0782	−0.0569	−0.0352	−0.0134	+0.0085	+0.0301	+0.0512	0.0717	0.0913	0.1099
14	0.1272	0.1431	0.1575	0.1703	0.1812	0.1903	0.1974	0.2025	0.2056	0.2065
15	0.2055	0.2023	0.1972	0.1902	0.1813	0.1706	0.1584	0.1446	0.1295	0.1132

When $z > 15.9$,

$$Y_0(z) \cong \sqrt{\left(\frac{2}{\pi z}\right)} \left\{ \sin\left(z - \frac{1}{4}\pi\right) - \frac{1}{8z} \sin\left(z + \frac{1}{4}\pi\right) \right\}$$

$$\cong \frac{0.7979}{\sqrt{z}} \left\{ \sin(57.296z - 45)^\circ - \frac{1}{8z} \sin(57.296z + 45)^\circ \right\}.$$

APPENDIX III (*Continued*)

$Y_1(z)$

z	0	0.1	0.2	0.3	0.4	0.5	0.6	0.7	0.8	0.9
0	$-\infty$	−6.4590	−3.3238	−2.293	−1.7809	−1.4715	−1.2602	−1.103	−0.9781	−0.8731
1	−0.7812	−0.6981	−0.6211	−0.5485	−0.4791	−0.4123	−0.3476	−0.2847	−0.2237	−0.1644
2	−0.1070	−0.0517	+0.0015	+0.0523	0.1005	0.1459	0.1884	0.2276	0.2635	0.2959
3	0.3247	0.3496	0.3707	0.3879	0.4010	0.4102	0.4154	0.4167	0.4141	0.4078
4	0.3979	0.3846	0.3680	0.3484	0.3260	0.3010	0.2737	0.2445	0.2136	0.1812
5	0.1479	0.1137	0.0792	0.0445	0.0101	−0.0238	−0.0568	−0.0887	−0.1192	−0.1481
6	−0.1750	−0.1998	−0.2223	−0.2422	−0.2596	−0.2741	−0.2857	−0.2945	−0.3002	−0.3029
7	−0.3027	−0.2995	−0.2934	−0.2846	−0.2731	−0.2591	−0.2428	−0.2243	−0.2039	−0.1817
8	−0.1581	−0.1331	−0.1072	−0.0806	−0.0535	−0.0262	+0.0011	+0.0280	0.0544	0.0799
9	+0.1043	0.1275	0.1491	0.1691	0.1871	0.2032	0.2171	0.2287	0.2379	0.2447
10	0.2490	0.2508	0.2502	0.2471	0.2416	0.2337	0.2236	0.2114	0.1973	0.1813
11	0.1637	0.1446	0.1243	0.1029	0.0807	0.0579	0.0348	0.0114	−0.0118	−0.0347
12	−0.0571	−0.0787	−0.0994	−0.1189	−0.1371	−0.1538	−0.1689	−0.1821	−0.1935	−0.2028
13	−0.2101	−0.2152	−0.2182	−0.2190	−0.2176	−0.2140	−0.2084	−0.2007	−0.1912	−0.1798
14	−0.1666	−0.1520	−0.1359	−0.1186	−0.1003	−0.0810	−0.0612	−0.0408	−0.0202	+0.0005
15	0.0211	0.0413	0.0609	0.0799	0.0979	0.1148	0.1305	0.1447	0.1575	0.1686

When $z > 15.9$,

$$Y_1(z) \cong \sqrt{\left(\frac{2}{\pi z}\right)}\left\{\sin\left(z - \frac{3}{4}\pi\right) + \frac{3}{8z}\sin\left(z - \frac{1}{4}\pi\right)\right\}$$

$$\cong \frac{0.7979}{\sqrt{z}}\left\{\sin(57.296z - 135)^\circ + \frac{3}{8z}\sin(57.296z - 45)^\circ\right\}.$$

$I_0(z)$

	z	0	0.1	0.2	0.3	0.4	0.5	0.6	0.7	0.8	0.9
	0	1.0000	1.0025	1.0100	1.0226	1.0404	1.0635	1.0920	1.1263	1.1665	1.2130
	1	1.2661	1.3262	1.3937	1.4693	1.5534	1.6467	1.7500	1.8640	1.9896	2.1277
	2	2.2796	2.4463	2.6291	2.8296	3.0493	3.2898	3.5533	3.8417	4.1573	4.5027
	3	4.8808	5.2945	5.7472	6.2426	6.7848	7.3782	8.0277	8.7386	9.5169	10.369
$10\times$	4	1.1302	1.2324	1.3442	1.4668	1.6010	1.7481	1.9093	2.0858	2.2794	2.4915
$10\times$	5	2.7240	2.9789	3.2584	3.5648	3.9009	4.2695	4.6738	5.1173	5.6038	6.1377
$10\times$	6	6.7234	7.3663	8.0718	8.8462	9.6962	10.629	11.654	12.779	14.014	15.370
$10^2\times$	7	1.6859	1.8495	2.0292	2.2266	2.4434	2.6816	2.9433	3.2309	3.5468	3.8941
$10^2\times$	8	4.2756	4.6950	5.1559	5.6626	6.2194	6.8316	7.5046	8.2445	9.0580	9.9524
$10^3\times$	9	1.0936	1.2017	1.3207	1.4514	1.5953	1.7535	1.9275	2.1189	2.3294	2.5610

When $z \geq 10$, $I_0(z) \cong \dfrac{0.3989e^z}{z^{1/2}}\left\{1 + \dfrac{1}{8z} + \dfrac{9}{128z^2} + \dfrac{75}{1024z^3}\right\}.$

$K_0(z)$

	z	0	0.1	0.2	0.3	0.4	0.5	0.6	0.7	0.8	0.9
	0	∞	2.4271	1.7527	1.3725	1.1145	0.9244	0.7775	0.6605	0.5653	0.4867
	1	0.4210	0.3656	0.3185	0.2782	0.2437	0.2138	0.1880	0.1655	0.1459	0.1288
$10^{-1}\times$	2	1.1389	1.0078	0.8926	0.7914	0.7022	0.6235	0.5540	0.4926	0.4382	0.3901
$10^{-1}\times$	3	0.3474	0.3095	0.2759	0.2461	0.2196	0.1960	0.1750	0.1563	0.1397	0.1248
$10^{-2}\times$	4	1.1160	0.9980	0.8927	0.7988	0.7149	0.6400	0.5730	0.5132	0.4597	0.4119
$10^{-2}\times$	5	0.3691	0.3308	0.2966	0.2659	0.2385	0.2139	0.1918	0.1721	0.1544	0.1386
$10^{-3}\times$	6	1.2440	1.1167	1.0025	0.9001	0.8083	0.7259	0.6520	0.5857	0.5262	0.4728
$10^{-3}\times$	7	0.4248	0.3817	0.3431	0.3084	0.2772	0.2492	0.2240	0.2014	0.1811	0.1629
$10^{-4}\times$	8	1.4647	1.3173	1.1849	1.0658	0.9588	0.8626	0.7761	0.6983	0.6283	0.5654
$10^{-4}\times$	9	0.5088	0.4579	0.4121	0.3710	0.3339	0.3006	0.2706	0.2436	0.2193	0.1975

When $z \geq 10$, $K_0(z) \cong \dfrac{1.2533e^{-z}}{z^{1/2}}\left\{1 - \dfrac{1}{8z} + \dfrac{9}{128z^2} - \dfrac{75}{1024z^3}\right\}.$

APPENDIX III (*Continued*)

$I_1(z)$

	z	0	0.1	0.2	0.3	0.4	0.5	0.6	0.7	0.8	0.9
	0	0	0.0501	0.1005	0.1517	0.2040	0.2579	0.3137	0.3719	0.4329	0.4971
	1	0.5652	0.6375	0.7147	0.7973	0.8861	0.9817	1.0848	1.1963	1.3172	1.4482
	2	1.5906	1.7455	1.9141	2.0978	2.2981	2.5167	2.7554	3.0161	3.3011	3.6126
	3	3.9534	4.3262	4.7343	5.1810	5.6701	6.2058	6.7927	7.4357	8.1404	8.9128
$10\times$	4	0.97595	1.0688	1.1706	1.2822	1.4046	1.5389	1.6863	1.8479	2.0253	2.2199
$10\times$	5	2.4336	2.6680	2.9254	3.2080	3.5182	3.8588	4.2328	4.6436	5.0946	5.5900
$10\times$	6	6.1342	6.7319	7.3886	8.1100	8.9026	9.7735	10.730	11.782	12.938	14.208
$10^2\times$	7	1.5604	1.7138	1.8825	2.0679	2.2717	2.4958	2.7422	3.0131	3.3110	3.6385
$10^2\times$	8	3.9987	4.3948	4.8305	5.3096	5.8366	6.4162	7.0538	7.7551	8.5266	9.3754
$10^3\times$	9	1.0309	1.1336	1.2467	1.3710	1.5079	1.6585	1.8241	2.0065	2.2071	2.4280

When $z \geq 10$, $I_1(z) \cong \dfrac{0.3989e^z}{z^{1/2}}\left\{1 - \dfrac{3}{8z} - \dfrac{15}{128z^2} - \dfrac{105}{1024z^3}\right\}$.

$K_1(z)$

	z	0	0.1	0.2	0.3	0.4	0.5	0.6	0.7	0.8	0.9
	0	∞	9.8538	4.7760	3.0560	2.1844	1.6564	1.3028	1.0503	0.8618	0.7165
	1	0.6019	0.5098	0.4346	0.3725	0.3208	0.2774	0.2406	0.2094	0.1826	0.1597
$10^{-1}\times$	2	1.3987	1.2275	1.0790	0.9498	0.8372	0.7389	0.6528	0.5774	0.5111	0.4529
$10^{-1}\times$	3	0.4016	0.3563	0.3164	0.2812	0.2500	0.2224	0.1979	0.1763	0.1571	0.1400
$10^{-2}\times$	4	1.2484	1.1136	0.9938	0.8872	0.7923	0.7078	0.6325	0.5654	0.5055	0.4521
$10^{-2}\times$	5	0.4045	0.3619	0.3239	0.2900	0.2597	0.2326	0.2083	0.1866	0.1673	0.1499
$10^{-3}\times$	6	1.3439	1.2050	1.0805	0.9691	0.8693	0.7799	0.6998	0.6280	0.5636	0.5059
$10^{-3}\times$	7	0.4542	0.4078	0.3662	0.3288	0.2953	0.2653	0.2383	0.2141	0.1924	0.1729
$10^{-4}\times$	8	1.5537	1.3964	1.2552	1.1283	1.0143	0.9120	0.8200	0.7374	0.6631	0.5964
$10^{-4}\times$	9	0.5364	0.4825	0.4340	0.3904	0.3512	0.3160	0.2843	0.2559	0.2302	0.2072

When $z \geq 10$, $K_1(z) \cong \dfrac{1.2533e^{-z}}{z^{1/2}}\left\{1 + \dfrac{3}{8z} - \dfrac{15}{128z^2} + \dfrac{105}{1024z^3}\right\}$.

First Ten Roots of $J_n(z) = 0; n = 0, 1, 2, 3, 4, 5$

	J_0	J_1	J_2	J_3	J_4	J_5
1	2.4048	3.8317	5.1356	6.3802	7.5883	8.7715
2	5.5201	7.0156	8.4172	9.7610	11.0647	12.3386
3	8.6537	10.1735	11.6198	13.0152	14.3725	15.7002
4	11.7915	13.3237	14.7960	16.2235	17.6160	18.9801
5	14.9309	16.4706	17.9598	19.4094	20.8269	22.2178
6	18.0711	19.6159	21.1170	22.5827	24.0190	25.4303
7	21.2116	22.7601	24.2701	25.7482	27.1991	28.6266
8	24.3525	25.9037	27.4206	28.9084	30.3710	31.8117
9	27.4935	29.0468	30.5692	32.0649	33.5371	34.9888
10	30.6346	32.1897	33.7165	35.2187	36.6990	38.1599

First Six Roots β_n of $\beta J_1(\beta) - cJ_0(\beta) = 0$

(From Carslaw and Jaeger, Ref. 2)

c	β_1	β_2	β_3	β_4	β_5	β_6
0	0	3.8317	7.0156	10.1735	13.3237	16.4706
0.01	0.1412	3.8343	7.0170	10.1745	13.3244	16.4712
0.02	0.1995	3.8369	7.0184	10.1754	13.3252	16.4718
0.04	0.2814	3.8421	7.0213	10.1774	13.3267	16.4731
0.06	0.3438	3.8473	7.0241	10.1794	13.3282	16.4743
0.08	0.3960	3.8525	7.0270	10.1813	13.3297	16.4755
0.1	0.4417	3.8577	7.0298	10.1833	13.3312	16.4767
0.15	0.5376	3.8706	7.0369	10.1882	13.3349	16.4797
0.2	0.6170	3.8835	7.0440	10.1931	13.3387	16.4828
0.3	0.7465	3.9091	7.0582	10.2029	13.3462	16.4888
0.4	0.8516	3.9344	7.0723	10.2127	13.3537	16.4949
0.5	0.9408	3.9594	7.0864	10.2225	13.3611	16.5010
0.6	1.0184	3.9841	7.1004	10.2322	13.3686	16.5070
0.7	1.0873	4.0085	7.1143	10.2419	13.3761	16.5131
0.8	1.1490	4.0325	7.1282	10.2516	13.3835	16.5191
0.9	1.2048	4.0562	7.1421	10.2613	13.3910	16.5251
1.0	1.2558	4.0795	7.1558	10.2710	13.3984	16.5312
1.5	1.4569	4.1902	7.2233	10.3188	13.4353	16.5612
2.0	1.5994	4.2910	7.2884	10.3658	13.4719	16.5910
3.0	1.7887	4.4634	7.4103	10.4566	13.5434	16.6499
4.0	1.9081	4.6018	7.5201	10.5423	13.6125	16.7073
5.0	1.9898	4.7131	7.6177	10.6223	13.6786	16.7630
6.0	2.0490	4.8033	7.7039	10.6964	13.7414	16.8168
7.0	2.0937	4.8772	7.7797	10.7646	13.8008	16.8684
8.0	2.1286	4.9384	7.8464	10.8271	13.8566	16.9179
9.0	2.1566	4.9897	7.9051	10.8842	13.9090	16.9650
10.0	2.1795	5.0332	7.9569	10.9363	13.9580	17.0099

APPENDIX III (*Continued*)

c	β_1	β_2	β_3	β_4	β_5	β_6
15.0	2.2509	5.1773	8.1422	11.1367	14.1576	17.2008
20.0	2.2880	5.2568	8.2534	11.2677	14.2983	17.3442
30.0	2.3261	5.3410	8.3771	11.4221	14.4748	17.5348
40.0	2.3455	5.3846	8.4432	11.5081	14.5774	17.6508
50.0	2.3572	5.4112	8.4840	11.5621	14.6433	17.7272
60.0	2.3651	5.4291	8.5116	11.5990	14.6889	17.7807
80.0	2.3750	5.4516	8.5466	11.6461	14.7475	17.8502
100.0	2.3809	5.4652	8.5678	11.6747	14.7834	17.8931
∞	2.4048	5.5201	8.6537	11.7915	14.9309	18.0711

First Five Roots β_n of $$\frac{J_0(\beta)}{J_0(c\beta)} - \frac{Y_0(\beta)}{Y_0(c\beta)} = 0$$

(From Carslaw and Jaeger, Ref. 2.)

c	β_1	β_2	β_3	β_4	β_5
1.2	15.7014	31.4126	47.1217	62.8302	78.5385
1.5	6.2702	12.5598	18.8451	25.1294	31.4133
2.0	3.1230	6.2734	9.4182	12.5614	15.7040
2.5	2.0732	4.1773	6.2754	8.3717	10.4672
3.0	1.5485	3.1291	4.7038	6.2767	7.8487
3.5	1.2339	2.5002	3.7608	5.0196	6.2776
4.0	1.0244	2.0809	3.1322	4.1816	5.2301

APPENDIX IV
Some Properties of Bessel Functions

VALUES OF BESSEL FUNCTIONS FOR SMALL z:[3]

$$J_n(z) \cong \frac{1}{2^n \cdot n!} z^n$$

$$Y_n(z) \cong \begin{cases} \dfrac{2}{\pi} \ln z & \text{for } n = 0 \\ -\dfrac{2^n(n-1)!}{\pi} \dfrac{1}{z^n} & \text{for } n \neq 0 \end{cases}$$

$$I_n(z) \cong \frac{1}{2^n \cdot n!} z^n$$

$$K_n(z) \cong \begin{cases} -\ln z & \text{for } n = 0 \\ \dfrac{2^{n-1} \cdot (n-1)!}{z^n} & \text{for } n \neq 0 \end{cases}$$

VALUES OF BESSEL FUNCTIONS FOR LARGE z:[3]

$$J_n(z) \cong \sqrt{\frac{2}{\pi z}} \cdot \cos\left(z - \frac{\pi}{4} - \frac{n\pi}{2}\right)$$

$$Y_n(z) \cong \sqrt{\frac{2}{\pi z}} \cdot \sin\left(z - \frac{\pi}{4} - \frac{n\pi}{2}\right)$$

$$I_n(z) \cong \frac{e^z}{\sqrt{2\pi z}}$$

$$K_n(z) \cong \sqrt{\frac{\pi}{2z}} \cdot e^{-z}$$

DERIVATIVES OF $J_n(z)$ AND $Y_n(z)$ FUNCTIONS:[3]

$$\frac{d}{dz}[z^n W_n(\beta z)] = \beta z^n W_{n-1}(\beta z)$$

$$\frac{d}{dz}\left[\frac{W_n(\beta z)}{z^n}\right] = -\beta \frac{W_{n+1}(\beta z)}{z^n}$$

where $W \equiv J$ or Y.

APPENDIX IV (*Continued*)

DERIVATIVES OF $I_n(z)$ AND $K_n(z)$ FUNCTIONS:[3]

$$\frac{d}{dz}[z^n I_n(\beta z)] = \beta z^n I_{n-1}(\beta z)$$

$$\frac{d}{dz}\left[\frac{I_n(\beta z)}{z^n}\right] = \beta\frac{I_{n+1}(\beta z)}{z^n}$$

$$\frac{d}{dz}[z^n K_n(\beta z)] = -\beta z^n K_{n-1}(\beta z)$$

$$\frac{d}{dz}\left[\frac{K_n(\beta z)}{z^n}\right] = -\beta\frac{K_{n+1}(\beta z)}{z^n}$$

INTEGRATION OF BESSEL FUNCTIONS:[1,4]

$$\int z^n W_{n-1}(\beta z)\cdot dz = \frac{1}{\beta}z^n W_n(\beta z)$$

$$\int \frac{W_{n+1}(\beta z)}{z^n}dz = -\frac{1}{\beta}\frac{W_n(\beta z)}{z^n}$$

where $W = J$ or Y.

$$\int zW_n^2(\beta z)\,dz = \frac{1}{2}z^2[W_n^2(\beta z) - W_{n-1}(\beta z)\cdot W_{n+1}(\beta z)]$$

$$= \frac{1}{2}z^2\left[\left(1-\frac{n^2}{\beta^2 z^2}\right)\cdot W_n^2(\beta z) + W_n'^2(\beta z)\right]$$

where $W = J$ or Y.

$$\int_0^\infty e^{-Pz^2}\cdot z\cdot J_n(az)\cdot J_n(bz)\cdot dz = \frac{1}{2P}\cdot e^{-\frac{a^2+b^2}{4P}} I_n\left(\frac{ab}{2P}\right)$$

$$\int_0^\infty z\cdot e^{-Pz^2}\cdot J_o(\alpha z)\cdot dz = \frac{1}{2P}\cdot e^{-\frac{\alpha^2}{4P}}$$

Systematic tabulations of various integrals of Bessel functions are given in Refs. 5 and 6.

APPENDIX V

Numerical Values of Legendre Polynomials of the First Kind.[7]

x	$P_1(x)$	$P_2(x)$	$P_3(x)$	$P_4(x)$	$P_5(x)$	$P_6(x)$	$P_7(x)$
0.00	0.0000	−.5000	0.0000	0.3750	0.0000	−.3125	0.0000
.01	.0100	−.4998	−.0150	.3746	.0187	−.3118	−.0219
.02	.0200	−.4994	−.0300	.3735	.0374	−.3099	−.0436
.03	.0300	−.4986	−.0449	.3716	.0560	−.3066	−.0651
.04	.0400	−.4976	−.0598	.3690	.0744	−.3021	−.0862
.05	.0500	−.4962	−.0747	.3657	.0927	−.2962	−.1069
.06	.0600	−.4946	−.0895	.3616	.1106	−.2891	−.1270
.07	.0700	−.4926	−.1041	.3567	.1283	−.2808	−.1464
.08	.0800	−.4904	−.1187	.3512	.1455	−.2713	−.1651
.09	.0900	−.4878	−.1332	.3449	.1624	−.2606	−.1828
.10	.1000	−.4850	−.1475	.3379	.1788	−.2488	−.1995
.11	.1100	−.4818	−.1617	.3303	.1947	−.2360	−.2151
.12	.1200	−.4784	−.1757	.3219	.2101	−.2220	−.2295
.13	.1300	−.4746	−.1895	.3129	.2248	−.2071	−.2427
.14	.1400	−.4706	−.2031	.3032	.2389	−.1913	−.2545
.15	.1500	−.4662	−.2166	.2928	.2523	−.1746	−.2649
.16	.1600	−.4616	−.2298	.2819	.2650	−.1572	−.2738
.17	.1700	−.4566	−.2427	.2703	.2769	−.1389	−.2812
.18	.1800	−.4514	−.2554	.2581	.2880	−.1201	−.2870
.19	.1900	−.4458	−.2679	.2453	.2982	−.1006	−.2911
.20	.2000	−.4400	−.2800	.2320	.3075	−.0806	−.2935
.21	.2100	−.4338	−.2918	.2181	.3159	−.0601	−.2943
.22	.2200	−.4274	−.3034	.2037	.3234	−.0394	−.2933
.23	.2300	−.4206	−.3146	.1889	.3299	−.0183	−.2906
.24	.2400	−.4136	−.3254	.1735	.3353	.0029	−.2861
.25	.2500	−.4062	−.3359	.1577	.3397	.0243	−.2799
.26	.2600	−.3986	−.3461	.1415	.3431	.0456	−.2720
.27	.2700	−.3906	−.3558	.1249	.3453	.0669	−.2625
.28	.2800	−.3824	−.3651	.1079	.3465	.0879	−.2512
.29	.2900	−.3738	−.3740	.0906	.3465	.1087	−.2384
.30	.3000	−.3650	−.3825	.0729	.3454	.1292	−.2241
.31	.3100	−.3558	−.3905	.0550	.3431	.1492	−.2082
.32	.3200	−.3464	−.3981	.0369	.3397	.1686	−.1910
.33	.3300	−.3366	−.4052	.0185	.3351	.1873	−.1724
.34	.3400	−.3266	−.4117	−.0000	.3294	.2053	−.1527
.35	.3500	−.3162	−.4178	−.0187	.3225	.2225	−.1318
.36	.3600	−.3056	−.4234	−.0375	.3144	.2388	−.1098
.37	.3700	−.2946	−.4284	−.0564	.3051	.2540	−.0870
.38	.3800	−.2834	−.4328	−.0753	.2948	.2681	−.0635
.39	.3900	−.2718	−.4367	−.0942	.2833	.2810	−.0393

APPENDIX V (*Continued*)

x	$P_1(x)$	$P_2(x)$	$P_3(x)$	$P_4(x)$	$P_5(x)$	$P_6(x)$	$P_7(x)$
.40	.4000	−.2600	−.4400	−.1130	.2706	.2926	−.0146
.41	.4100	−.2478	−.4427	−.1317	.2569	.3029	.0104
.42	.4200	−.2354	−.4448	−.1504	.2421	.3118	.0356
.43	.4300	−.2226	−.4462	−.1688	.2263	.3191	.0608
.44	.4400	−.2096	−.4470	−.1870	.2095	.3249	.0859
.45	.4500	−.1962	−.4472	−.2050	.1917	.3290	.1106
.46	.4600	−.1826	−.4467	−.2226	.1730	.3314	.1348
.47	.4700	−.1686	−.4454	−.2399	.1534	.3321	.1584
.48	.4800	−.1544	−.4435	−.2568	.1330	.3310	.1811
.49	.4900	−.1398	−.4409	−.2732	.1118	.3280	.2027
.50	.5000	−.1250	−.4375	−.2891	.0898	.3232	.2231
.51	.5100	−.1098	−.4334	−.3044	.0673	.3166	.2422
.52	.5200	−.0944	−.4285	−.3191	.0441	.3080	.2596
.53	.5300	−.0786	−.4228	−.3332	.0204	.2975	.2753
.54	.5400	−.0626	−.4163	−.3465	−.0037	.2851	.2891
.55	.5500	−.0462	−.4091	−.3590	−.0282	.2708	.3007
.56	.5600	−.0296	−.4010	−.3707	−.0529	.2546	.3102
.57	.5700	−.0126	−.3920	−.3815	−.0779	.2366	.3172
.58	.5800	.0046	−.3822	−.3914	−.1028	.2168	.3217
.59	.5900	.0222	−.3716	−.4002	−.1278	.1953	.3235
.60	.6000	.0400	−.3600	−.4080	−.1526	.1721	.3226
.61	.6100	.0582	−.3475	−.4146	−.1772	.1473	.3188
.62	.6200	.0766	−.3342	−.4200	−.2014	.1211	.3121
.63	.6300	.0954	−.3199	−.4242	−.2251	.0935	.3023
.64	.6400	.1144	−.3046	−.4270	−.2482	.0646	.2895
.65	.6500	.1338	−.2884	−.4284	−.2705	.0347	.2737
.66	.6600	.1534	−.2713	−.4284	−.2919	.0038	.2548
.67	.6700	.1734	−.2531	−.4268	−.3122	−.0278	.2329
.68	.6800	.1936	−.2339	−.4236	−.3313	−.0601	.2081
.69	.6900	.2142	−.2137	−.4187	−.3490	−.0926	.1805
.70	.7000	.2350	−.1925	−.4121	−.3652	−.1253	.1502
.71	.7100	.2562	−.1702	−.4036	−.3796	−.1578	.1173
.72	.7200	.2776	−.1469	−.3933	−.3922	−.1899	.0822
.73	.7300	.2994	−.1225	−.3810	−.4026	−.2214	.0450
.74	.7400	.3214	−.0969	−.3666	−.4107	−.2518	.0061
.75	.7500	.3438	−.0703	−.3501	−.4164	−.2808	−.0342
.76	.7600	.3664	−.0426	−.3314	−.4193	−.3081	−.0754
.77	.7700	.3894	−.0137	−.3104	−.4193	−.3333	−.1171
.78	.7800	.4126	.0164	−.2871	−.4162	−.3559	−.1588
.79	.7900	.4362	.0476	−.2613	−.4097	−.3756	−.1999
.80	.8000	.4600	.0800	−.2330	−.3995	−.3918	−.2397
.81	.8100	.4842	.1136	−.2021	−.3855	−.4041	−.2774

x	$P_1(x)$	$P_2(x)$	$P_3(x)$	$P_4(x)$	$P_5(x)$	$P_6(x)$	$P_7(x)$
.82	.8200	.5086	.1484	−.1685	−.3674	−.4119	−.3124
.83	.8300	.5334	.1845	−.1321	−.3449	−.4147	−.3437
.84	.8400	.5584	.2218	−.0928	−.3177	−.4120	−.3703
.85	.8500	.5838	.2603	−.0506	−.2857	−.4030	−.3913
.86	.8600	.6094	.3001	−.0053	−.2484	−.3872	−.4055
.87	.8700	.6354	.3413	.0431	−.2056	−.3638	−.4116
.88	.8800	.6616	.3837	.0947	−.1570	−.3322	−.4083
.89	.8900	.6882	.4274	.1496	−.1023	−.2916	−.3942
.90	.9000	.7150	.4725	.2079	−.0411	−.2412	−.3678
.91	.9100	.7422	.5189	.2698	.0268	−.1802	−.3274
.92	.9200	.7696	.5667	.3352	.1017	−.1077	−.2713
.93	.9300	.7974	.6159	.4044	.1842	−.0229	−.1975
.94	.9400	.8254	.6665	.4773	.2744	.0751	−.1040
.95	.9500	.8538	.7184	.5541	.3727	.1875	.0112
.96	.9600	.8824	.7718	.6349	.4796	.3151	.1506
.97	.9700	.9114	.8267	.7198	.5954	.4590	.3165
.98	.9800	.9406	.8830	.8089	.7204	.6204	.5115
.99	.9900	.9702	.9407	.9022	.8552	.8003	.7384
1.00	1.0000	1.0000	1.0000	1.0000	1.0000	1.0000	1.0000

REFERENCES

1. N. W. McLachlan, *Bessel Functions for Engineers*, Clarendon Press, London, 1961, pp. 215–221, 110,192.

2. H. S. Carslaw and J. C. Jaeger, *Conduction of Heat in Solids*, Clarendon Press, London, 1959, p. 493.

3. F. B. Hildebrand, *Advanced Calculus for Engineers*, Prentice-Hall, Inc., Englewood Cliffs, N. J., 1949, pp. 161–163.

4. G. N. Watson, *A Treatise on the Theory of Bessel Functions*, Cambridge University Press, London, 1944, pp. 132, 393, 395.

5. I. S. Gradshteyn and I. M. Ryzhik, *Table of Integrals, Series, and Products*. Trans. from the Russian and edited by Alan Jeffrey. Academic Press, New York, 1965.

6. Y. L. Luke, *Integrals of Bessel Functions*, McGraw-Hill Book Company, New York, 1962.

7. W. E. Byerly, *Fourier Series and Spherical, Cylindrical, and Ellipsoidal Harmonics*, Dover Publications, Inc., New York, 1959, pp. 280–281.

Index

A CATALOG OF SELECTED

DOVER BOOKS

IN ALL FIELDS OF INTEREST

THE BOOK OF BEASTS: Being a Translation from a Latin Bestiary of the Twelfth Century, T. H. White. Wonderful catalog real and fanciful beasts: manticore, griffin, phoenix, amphivius, jaculus, many more. White's witty erudite commentary on scientific, historical aspects. Fascinating glimpse of medieval mind. Illustrated. 296pp. 5⅜ × 8¼. (Available in U.S. only) 24609-4 Pa. $5.95

FRANK LLOYD WRIGHT: ARCHITECTURE AND NATURE With 160 Illustrations, Donald Hoffmann. Profusely illustrated study of influence of nature—especially prairie—on Wright's designs for Fallingwater, Robie House, Guggenheim Museum, other masterpieces. 96pp. 9¼ × 10¾. 25098-9 Pa. $7.95

FRANK LLOYD WRIGHT'S FALLINGWATER, Donald Hoffmann. Wright's famous waterfall house: planning and construction of organic idea. History of site, owners, Wright's personal involvement. Photographs of various stages of building. Preface by Edgar Kaufmann, Jr. 100 illustrations. 112pp. 9¼ × 10.
23671-4 Pa. $7.95

YEARS WITH FRANK LLOYD WRIGHT: Apprentice to Genius, Edgar Tafel. Insightful memoir by a former apprentice presents a revealing portrait of Wright the man, the inspired teacher, the greatest American architect. 372 black-and-white illustrations. Preface. Index. vi + 228pp. 8¼ × 11. 24801-1 Pa. $9.95

THE STORY OF KING ARTHUR AND HIS KNIGHTS, Howard Pyle. Enchanting version of King Arthur fable has delighted generations with imaginative narratives of exciting adventures and unforgettable illustrations by the author. 41 illustrations. xviii + 313pp. 6⅛ × 9¼. 21445-1 Pa. $6.50

THE GODS OF THE EGYPTIANS, E. A. Wallis Budge. Thorough coverage of numerous gods of ancient Egypt by foremost Egyptologist. Information on evolution of cults, rites and gods; the cult of Osiris; the Book of the Dead and its rites; the sacred animals and birds; Heaven and Hell; and more. 956pp. 6⅛ × 9¼.
22055-9, 22056-7 Pa., Two-vol. set $20.00

A THEOLOGICO-POLITICAL TREATISE, Benedict Spinoza. Also contains unfinished *Political Treatise*. Great classic on religious liberty, theory of government on common consent. R. Elwes translation. Total of 421pp. 5⅜ × 8½.
20249-6 Pa. $6.95

INCIDENTS OF TRAVEL IN CENTRAL AMERICA, CHIAPAS, AND YUCATAN, John L. Stephens. Almost single-handed discovery of Maya culture; exploration of ruined cities, monuments, temples; customs of Indians. 115 drawings. 892pp. 5⅜ × 8½. 22404-X, 22405-8 Pa., Two-vol. set $15.90

LOS CAPRICHOS, Francisco Goya. 80 plates of wild, grotesque monsters and caricatures. Prado manuscript included. 183pp. 6⅜ × 9⅝. 22384-1 Pa. $4.95

AUTOBIOGRAPHY: The Story of My Experiments with Truth, Mohandas K. Gandhi. Not hagiography, but Gandhi in his own words. Boyhood, legal studies, purification, the growth of the Satyagraha (nonviolent protest) movement. Critical, inspiring work of the man who freed India. 480pp. 5⅜ × 8½. (Available in U.S. only)
24593-4 Pa. $6.95

HOW TO WRITE, Gertrude Stein. Gertrude Stein claimed anyone could understand her unconventional writing—here are clues to help. Fascinating improvisations, language experiments, explanations illuminate Stein's craft and the art of writing. Total of 414pp. 4⅜ × 6⅜. 23144-5 Pa. $5.95

ADVENTURES AT SEA IN THE GREAT AGE OF SAIL: Five Firsthand Narratives, edited by Elliot Snow. Rare true accounts of exploration, whaling, shipwreck, fierce natives, trade, shipboard life, more. 33 illustrations. Introduction. 353pp. 5⅜ × 8½. 25177-2 Pa. $7.95

THE HERBAL OR GENERAL HISTORY OF PLANTS, John Gerard. Classic descriptions of about 2,850 plants—with over 2,700 illustrations—includes Latin and English names, physical descriptions, varieties, time and place of growth, more. 2,706 illustrations. xlv + 1,678pp. 8½ × 12¼. 23147-X Cloth. $75.00

DOROTHY AND THE WIZARD IN OZ, L. Frank Baum. Dorothy and the Wizard visit the center of the Earth, where people are vegetables, glass houses grow and Oz characters reappear. Classic sequel to *Wizard of Oz*. 256pp. 5⅜ × 8. 24714-7 Pa. $4.95

SONGS OF EXPERIENCE: Facsimile Reproduction with 26 Plates in Full Color, William Blake. This facsimile of Blake's original "Illuminated Book" reproduces 26 full-color plates from a rare 1826 edition. Includes "The Tyger," "London," "Holy Thursday," and other immortal poems. 26 color plates. Printed text of poems. 48pp. 5¼ × 7. 24636-1 Pa. $3.50

SONGS OF INNOCENCE, William Blake. The first and most popular of Blake's famous "Illuminated Books," in a facsimile edition reproducing all 31 brightly colored plates. Additional printed text of each poem. 64pp. 5¼ × 7. 22764-2 Pa. $3.50

PRECIOUS STONES, Max Bauer. Classic, thorough study of diamonds, rubies, emeralds, garnets, etc.: physical character, occurrence, properties, use, similar topics. 20 plates, 8 in color. 94 figures. 659pp. 6⅛ × 9¼. 21910-0, 21911-9 Pa., Two-vol. set $15.90

ENCYCLOPEDIA OF VICTORIAN NEEDLEWORK, S. F. A. Caulfeild and Blanche Saward. Full, precise descriptions of stitches, techniques for dozens of needlecrafts—most exhaustive reference of its kind. Over 800 figures. Total of 679pp. 8⅛ × 11. Two volumes. Vol. 1 22800-2 Pa. $11.95
Vol. 2 22801-0 Pa. $11.95

THE MARVELOUS LAND OF OZ, L. Frank Baum. Second Oz book, the Scarecrow and Tin Woodman are back with hero named Tip, Oz magic. 136 illustrations. 287pp. 5⅜ × 8½. 20692-0 Pa. $5.95

WILD FOWL DECOYS, Joel Barber. Basic book on the subject, by foremost authority and collector. Reveals history of decoy making and rigging, place in American culture, different kinds of decoys, how to make them, and how to use them. 140 plates. 156pp. 7⅞ × 10¾. 20011-6 Pa. $8.95

HISTORY OF LACE, Mrs. Bury Palliser. Definitive, profusely illustrated chronicle of lace from earliest times to late 19th century. Laces of Italy, Greece, England, France, Belgium, etc. Landmark of needlework scholarship. 266 illustrations. 672pp. 6⅛ × 9¼. 24742-2 Pa. $14.95

ILLUSTRATED GUIDE TO SHAKER FURNITURE, Robert Meader. All furniture and appurtenances, with much on unknown local styles. 235 photos. 146pp. 9 × 12. 22819-3 Pa. $7.95

WHALE SHIPS AND WHALING: A Pictorial Survey, George Francis Dow. Over 200 vintage engravings, drawings, photographs of barks, brigs, cutters, other vessels. Also harpoons, lances, whaling guns, many other artifacts. Comprehensive text by foremost authority. 207 black-and-white illustrations. 288pp. 6 × 9. 24808-9 Pa. $8.95

THE BERTRAMS, Anthony Trollope. Powerful portrayal of blind self-will and thwarted ambition includes one of Trollope's most heartrending love stories. 497pp. 5⅜ × 8½. 25119-5 Pa. $8.95

ADVENTURES WITH A HAND LENS, Richard Headstrom. Clearly written guide to observing and studying flowers and grasses, fish scales, moth and insect wings, egg cases, buds, feathers, seeds, leaf scars, moss, molds, ferns, common crystals, etc.—all with an ordinary, inexpensive magnifying glass. 209 exact line drawings aid in your discoveries. 220pp. 5⅜ × 8½. 23330-8 Pa. $3.95

RODIN ON ART AND ARTISTS, Auguste Rodin. Great sculptor's candid, wide-ranging comments on meaning of art; great artists; relation of sculpture to poetry, painting, music; philosophy of life, more. 76 superb black-and-white illustrations of Rodin's sculpture, drawings and prints. 119pp. 8⅜ × 11¼. 24487-3 Pa. $6.95

FIFTY CLASSIC FRENCH FILMS, 1912–1982: A Pictorial Record, Anthony Slide. Memorable stills from Grand Illusion, Beauty and the Beast, Hiroshima, Mon Amour, many more. Credits, plot synopses, reviews, etc. 160pp. 8¼ × 11. 25256-6 Pa. $11.95

THE PRINCIPLES OF PSYCHOLOGY, William James. Famous long course complete, unabridged. Stream of thought, time perception, memory, experimental methods; great work decades ahead of its time. 94 figures. 1,391pp. 5⅜ × 8½. 20381-6, 20382-4 Pa., Two-vol. set $19.90

BODIES IN A BOOKSHOP, R. T. Campbell. Challenging mystery of blackmail and murder with ingenious plot and superbly drawn characters. In the best tradition of British suspense fiction. 192pp. 5⅜ × 8½. 24720-1 Pa. $3.95

CALLAS: PORTRAIT OF A PRIMA DONNA, George Jellinek. Renowned commentator on the musical scene chronicles incredible career and life of the most controversial, fascinating, influential operatic personality of our time. 64 black-and-white photographs. 416pp. 5⅜ × 8¼. 25047-4 Pa. $7.95

GEOMETRY, RELATIVITY AND THE FOURTH DIMENSION, Rudolph Rucker. Exposition of fourth dimension, concepts of relativity as Flatland characters continue adventures. Popular, easily followed yet accurate, profound. 141 illustrations. 133pp. 5⅜ × 8½. 23400-2 Pa. $3.95

HOUSEHOLD STORIES BY THE BROTHERS GRIMM, with pictures by Walter Crane. 53 classic stories—Rumpelstiltskin, Rapunzel, Hansel and Gretel, the Fisherman and his Wife, Snow White, Tom Thumb, Sleeping Beauty, Cinderella, and so much more—lavishly illustrated with original 19th century drawings. 114 illustrations. x + 269pp. 5⅜ × 8½. 21080-4 Pa. $4.50

SUNDIALS, Albert Waugh. Far and away the best, most thorough coverage of ideas, mathematics concerned, types, construction, adjusting anywhere. Over 100 illustrations. 230pp. 5⅜ × 8½. 22947-5 Pa. $4.50

PICTURE HISTORY OF THE NORMANDIE: With 190 Illustrations, Frank O. Braynard. Full story of legendary French ocean liner: Art Deco interiors, design innovations, furnishings, celebrities, maiden voyage, tragic fire, much more. Extensive text. 144pp. 8⅞ × 11¾. 25257-4 Pa. $9.95

THE FIRST AMERICAN COOKBOOK: A Facsimile of "American Cookery," 1796, Amelia Simmons. Facsimile of the first American-written cookbook published in the United States contains authentic recipes for colonial favorites—pumpkin pudding, winter squash pudding, spruce beer, Indian slapjacks, and more. Introductory Essay and Glossary of colonial cooking terms. 80pp. 5⅜ × 8½. 24710-4 Pa. $3.50

101 PUZZLES IN THOUGHT AND LOGIC, C. R. Wylie, Jr. Solve murders and robberies, find out which fishermen are liars, how a blind man could possibly identify a color—purely by your own reasoning! 107pp. 5⅜ × 8½. 20367-0 Pa. $2.50

THE BOOK OF WORLD-FAMOUS MUSIC—CLASSICAL, POPULAR AND FOLK, James J. Fuld. Revised and enlarged republication of landmark work in musico-bibliography. Full information about nearly 1,000 songs and compositions including first lines of music and lyrics. New supplement. Index. 800pp. 5⅜ × 8¼. 24857-7 Pa. $14.95

ANTHROPOLOGY AND MODERN LIFE, Franz Boas. Great anthropologist's classic treatise on race and culture. Introduction by Ruth Bunzel. Only inexpensive paperback edition. 255pp. 5⅜ × 8½. 25245-0 Pa. $5.95

THE TALE OF PETER RABBIT, Beatrix Potter. The inimitable Peter's terrifying adventure in Mr. McGregor's garden, with all 27 wonderful, full-color Potter illustrations. 55pp. 4¼ × 5½. (Available in U.S. only) 22827-4 Pa. $1.75

THREE PROPHETIC SCIENCE FICTION NOVELS, H. G. Wells. *When the Sleeper Wakes, A Story of the Days to Come* and *The Time Machine* (full version). 335pp. 5⅜ × 8½. (Available in U.S. only) 20605-X Pa. $5.95

APICIUS COOKERY AND DINING IN IMPERIAL ROME, edited and translated by Joseph Dommers Vehling. Oldest known cookbook in existence offers readers a clear picture of what foods Romans ate, how they prepared them, etc. 49 illustrations. 301pp. 6⅛ × 9¼. 23563-7 Pa. $6.50

SHAKESPEARE LEXICON AND QUOTATION DICTIONARY, Alexander Schmidt. Full definitions, locations, shades of meaning of every word in plays and poems. More than 50,000 exact quotations. 1,485pp. 6½ × 9¼. 22726-X, 22727-8 Pa., Two-vol. set $27.90

THE WORLD'S GREAT SPEECHES, edited by Lewis Copeland and Lawrence W. Lamm. Vast collection of 278 speeches from Greeks to 1970. Powerful and effective models; unique look at history. 842pp. 5⅜ × 8½. 20468-5 Pa. $11.95

CHRISTMAS CUSTOMS AND TRADITIONS, Clement A. Miles. Origin, evolution, significance of religious, secular practices. Caroling, gifts, yule logs, much more. Full, scholarly yet fascinating; non-sectarian. 400pp. 5⅜ × 8½.
23354-5 Pa. $6.50

THE HUMAN FIGURE IN MOTION, Eadweard Muybridge. More than 4,500 stopped-action photos, in action series, showing undraped men, women, children jumping, lying down, throwing, sitting, wrestling, carrying, etc. 390pp. 7⅞ × 10⅝.
20204-6 Cloth. $21.95

THE MAN WHO WAS THURSDAY, Gilbert Keith Chesterton. Witty, fast-paced novel about a club of anarchists in turn-of-the-century London. Brilliant social, religious, philosophical speculations. 128pp. 5⅜ × 8½. 25121-7 Pa. $3.95

A CEZANNE SKETCHBOOK: Figures, Portraits, Landscapes and Still Lifes, Paul Cezanne. Great artist experiments with tonal effects, light, mass, other qualities in over 100 drawings. A revealing view of developing master painter, precursor of Cubism. 102 black-and-white illustrations. 144pp. 8⅜ × 6⅜. 24790-2 Pa. $5.95

AN ENCYCLOPEDIA OF BATTLES: Accounts of Over 1,560 Battles from 1479 B.C. to the Present, David Eggenberger. Presents essential details of every major battle in recorded history, from the first battle of Megiddo in 1479 B.C. to Grenada in 1984. List of Battle Maps. New Appendix covering the years 1967–1984. Index. 99 illustrations. 544pp. 6½ × 9¼. 24913-1 Pa. $14.95

AN ETYMOLOGICAL DICTIONARY OF MODERN ENGLISH, Ernest Weekley. Richest, fullest work, by foremost British lexicographer. Detailed word histories. Inexhaustible. Total of 856pp. 6½ × 9¼.
21873-2, 21874-0 Pa., Two-vol. set $17.00

WEBSTER'S AMERICAN MILITARY BIOGRAPHIES, edited by Robert McHenry. Over 1,000 figures who shaped 3 centuries of American military history. Detailed biographies of Nathan Hale, Douglas MacArthur, Mary Hallaren, others. Chronologies of engagements, more. Introduction. Addenda. 1,033 entries in alphabetical order. xi + 548pp. 6½ × 9¼. (Available in U.S. only)
24758-9 Pa. $11.95

LIFE IN ANCIENT EGYPT, Adolf Erman. Detailed older account, with much not in more recent books: domestic life, religion, magic, medicine, commerce, and whatever else needed for complete picture. Many illustrations. 597pp. 5⅜ × 8½.
22632-8 Pa. $8.50

HISTORIC COSTUME IN PICTURES, Braun & Schneider. Over 1,450 costumed figures shown, covering a wide variety of peoples: kings, emperors, nobles, priests, servants, soldiers, scholars, townsfolk, peasants, merchants, courtiers, cavaliers, and more. 256pp. 8⅜ × 11¼. 23150-X Pa. $7.95

THE NOTEBOOKS OF LEONARDO DA VINCI, edited by J. P. Richter. Extracts from manuscripts reveal great genius; on painting, sculpture, anatomy, sciences, geography, etc. Both Italian and English. 186 ms. pages reproduced, plus 500 additional drawings, including studies for *Last Supper, Sforza* monument, etc. 860pp. 7⅞ × 10¾. (Available in U.S. only) 22572-0, 22573-9 Pa., Two-vol. set $25.90

THE ART NOUVEAU STYLE BOOK OF ALPHONSE MUCHA: All 72 Plates from "Documents Decoratifs" in Original Color, Alphonse Mucha. Rare copyright-free design portfolio by high priest of Art Nouveau. Jewelry, wallpaper, stained glass, furniture, figure studies, plant and animal motifs, etc. Only complete one-volume edition. 80pp. 9⅜ × 12¼. 24044-4 Pa. $8.95

ANIMALS: 1,419 COPYRIGHT-FREE ILLUSTRATIONS OF MAMMALS, BIRDS, FISH, INSECTS, ETC., edited by Jim Harter. Clear wood engravings present, in extremely lifelike poses, over 1,000 species of animals. One of the most extensive pictorial sourcebooks of its kind. Captions. Index. 284pp. 9 × 12. 23766-4 Pa. $9.95

OBELISTS FLY HIGH, C. Daly King. Masterpiece of American detective fiction, long out of print, involves murder on a 1935 transcontinental flight—"a very thrilling story"—NY Times. Unabridged and unaltered republication of the edition published by William Collins Sons & Co. Ltd., London, 1935. 288pp. 5⅜ × 8½. (Available in U.S. only) 25036-9 Pa. $4.95

VICTORIAN AND EDWARDIAN FASHION: A Photographic Survey, Alison Gernsheim. First fashion history completely illustrated by contemporary photographs. Full text plus 235 photos, 1840–1914, in which many celebrities appear. 240pp. 6½ × 9¼. 24205-6 Pa. $6.00

THE ART OF THE FRENCH ILLUSTRATED BOOK, 1700–1914, Gordon N. Ray. Over 630 superb book illustrations by Fragonard, Delacroix, Daumier, Doré, Grandville, Manet, Mucha, Steinlen, Toulouse-Lautrec and many others. Preface. Introduction. 633 halftones. Indices of artists, authors & titles, binders and provenances. Appendices. Bibliography. 608pp. 8⅜ × 11¼. 25086-5 Pa. $24.95

THE WONDERFUL WIZARD OF OZ, L. Frank Baum. Facsimile in full color of America's finest children's classic. 143 illustrations by W. W. Denslow. 267pp. 5⅜ × 8½. 20691-2 Pa. $5.95

FRONTIERS OF MODERN PHYSICS: New Perspectives on Cosmology, Relativity, Black Holes and Extraterrestrial Intelligence, Tony Rothman, et al. For the intelligent layman. Subjects include: cosmological models of the universe; black holes; the neutrino; the search for extraterrestrial intelligence. Introduction. 46 black-and-white illustrations. 192pp. 5⅜ × 8½. 24587-X Pa. $6.95

THE FRIENDLY STARS, Martha Evans Martin & Donald Howard Menzel. Classic text marshalls the stars together in an engaging, non-technical survey, presenting them as sources of beauty in night sky. 23 illustrations. Foreword. 2 star charts. Index. 147pp. 5⅜ × 8½. 21099-5 Pa. $3.50

FADS AND FALLACIES IN THE NAME OF SCIENCE, Martin Gardner. Fair, witty appraisal of cranks, quacks, and quackeries of science and pseudoscience: hollow earth, Velikovsky, orgone energy, Dianetics, flying saucers, Bridey Murphy, food and medical fads, etc. Revised, expanded In the Name of Science. "A very able and even-tempered presentation."—The New Yorker. 363pp. 5⅜ × 8. 20394-8 Pa. $6.50

ANCIENT EGYPT: ITS CULTURE AND HISTORY, J. E Manchip White. From pre-dynastics through Ptolemies: society, history, political structure, religion, daily life, literature, cultural heritage. 48 plates. 217pp. 5⅜ × 8½. 22548-8 Pa. $4.95

SIR HARRY HOTSPUR OF HUMBLETHWAITE, Anthony Trollope. Incisive, unconventional psychological study of a conflict between a wealthy baronet, his idealistic daughter, and their scapegrace cousin. The 1870 novel in its first inexpensive edition in years. 250pp. 5⅜ × 8½. 24953-0 Pa. $5.95

LASERS AND HOLOGRAPHY, Winston E. Kock. Sound introduction to burgeoning field, expanded (1981) for second edition. Wave patterns, coherence, lasers, diffraction, zone plates, properties of holograms, recent advances. 84 illustrations. 160pp. 5⅜ × 8¼. (Except in United Kingdom) 24041-X Pa. $3.50

INTRODUCTION TO ARTIFICIAL INTELLIGENCE: SECOND, ENLARGED EDITION, Philip C. Jackson, Jr. Comprehensive survey of artificial intelligence—the study of how machines (computers) can be made to act intelligently. Includes introductory and advanced material. Extensive notes updating the main text. 132 black-and-white illustrations. 512pp. 5⅜ × 8½. 24864-X Pa. $8.95

HISTORY OF INDIAN AND INDONESIAN ART, Ananda K. Coomaraswamy. Over 400 illustrations illuminate classic study of Indian art from earliest Harappa finds to early 20th century. Provides philosophical, religious and social insights. 304pp. 6⅝ × 9⅝. 25005-9 Pa. $8.95

THE GOLEM, Gustav Meyrink. Most famous supernatural novel in modern European literature, set in Ghetto of Old Prague around 1890. Compelling story of mystical experiences, strange transformations, profound terror. 13 black-and-white illustrations. 224pp. 5⅜ × 8½. (Available in U.S. only) 25025-3 Pa. $5.95

ARMADALE, Wilkie Collins. Third great mystery novel by the author of *The Woman in White* and *The Moonstone*. Original magazine version with 40 illustrations. 597pp. 5⅜ × 8½. 23429-0 Pa. $9.95

PICTORIAL ENCYCLOPEDIA OF HISTORIC ARCHITECTURAL PLANS, DETAILS AND ELEMENTS: With 1,880 Line Drawings of Arches, Domes, Doorways, Facades, Gables, Windows, etc., John Theodore Haneman. Sourcebook of inspiration for architects, designers, others. Bibliography. Captions. 141pp. 9 × 12. 24605-1 Pa. $6.95

BENCHLEY LOST AND FOUND, Robert Benchley. Finest humor from early 30's, about pet peeves, child psychologists, post office and others. Mostly unavailable elsewhere. 73 illustrations by Peter Arno and others. 183pp. 5⅜ × 8½. 22410-4 Pa. $3.95

ERTÉ GRAPHICS, Erté. Collection of striking color graphics: *Seasons, Alphabet, Numerals, Aces* and *Precious Stones*. 50 plates, including 4 on covers. 48pp. 9⅜ × 12¼. 23580-7 Pa. $6.95

THE JOURNAL OF HENRY D. THOREAU, edited by Bradford Torrey, F. H. Allen. Complete reprinting of 14 volumes, 1837-61, over two million words; the sourcebooks for *Walden*, etc. Definitive. All original sketches, plus 75 photographs. 1,804pp. 8½ × 12¼. 20312-3, 20313-1 Cloth., Two-vol. set $80.00

CASTLES: THEIR CONSTRUCTION AND HISTORY, Sidney Toy. Traces castle development from ancient roots. Nearly 200 photographs and drawings illustrate moats, keeps, baileys, many other features. Caernarvon, Dover Castles, Hadrian's Wall, Tower of London, dozens more. 256pp. 5⅜ × 8¼. 24898-4 Pa. $5.95

CATALOG OF DOVER BOOKS

AMERICAN CLIPPER SHIPS: 1833–1858, Octavius T. Howe & Frederick C. Matthews. Fully-illustrated, encyclopedic review of 352 clipper ships from the period of America's greatest maritime supremacy. Introduction. 109 halftones. 5 black-and-white line illustrations. Index. Total of 928pp. 5⅜ × 8½.
25115-2, 25116-0 Pa., Two-vol. set $17.90

TOWARDS A NEW ARCHITECTURE, Le Corbusier. Pioneering manifesto by great architect, near legendary founder of "International School." Technical and aesthetic theories, views on industry, economics, relation of form to function, "mass-production spirit," much more. Profusely illustrated. Unabridged translation of 13th French edition. Introduction by Frederick Etchells. 320pp. 6⅛ × 9¼. (Available in U.S. only) 25023-7 Pa. $8.95

THE BOOK OF KELLS, edited by Blanche Cirker. Inexpensive collection of 32 full-color, full-page plates from the greatest illuminated manuscript of the Middle Ages, painstakingly reproduced from rare facsimile edition. Publisher's Note. Captions. 32pp. 9⅜ × 12¼. 24345-1 Pa. $4.95

BEST SCIENCE FICTION STORIES OF H. G. WELLS, H. G. Wells. Full novel *The Invisible Man,* plus 17 short stories: "The Crystal Egg," "Aepyornis Island," "The Strange Orchid," etc. 303pp. 5⅜ × 8½. (Available in U.S. only)
21531-8 Pa. $4.95

AMERICAN SAILING SHIPS: Their Plans and History, Charles G. Davis. Photos, construction details of schooners, frigates, clippers, other sailcraft of 18th to early 20th centuries—plus entertaining discourse on design, rigging, nautical lore, much more. 137 black-and-white illustrations. 240pp. 6⅛ × 9¼.
24658-2 Pa. $5.95

ENTERTAINING MATHEMATICAL PUZZLES, Martin Gardner. Selection of author's favorite conundrums involving arithmetic, money, speed, etc., with lively commentary. Complete solutions. 112pp. 5⅜ × 8½. 25211-6 Pa. $2.95

THE WILL TO BELIEVE, HUMAN IMMORTALITY, William James. Two books bound together. Effect of irrational on logical, and arguments for human immortality. 402pp. 5⅜ × 8½. 20291-7 Pa. $7.50

THE HAUNTED MONASTERY and THE CHINESE MAZE MURDERS, Robert Van Gulik. 2 full novels by Van Gulik continue adventures of Judge Dee and his companions. An evil Taoist monastery, seemingly supernatural events; overgrown topiary maze that hides strange crimes. Set in 7th-century China. 27 illustrations. 328pp. 5⅜ × 8½. 23502-5 Pa. $5.95

CELEBRATED CASES OF JUDGE DEE (DEE GOONG AN), translated by Robert Van Gulik. Authentic 18th-century Chinese detective novel; Dee and associates solve three interlocked cases. Led to Van Gulik's own stories with same characters. Extensive introduction. 9 illustrations. 237pp. 5⅜ × 8½.
23337-5 Pa. $4.95

Prices subject to change without notice.

Available at your book dealer or write for free catalog to Dept. GI, Dover Publications, Inc., 31 East 2nd St., Mineola, N.Y. 11501. Dover publishes more than 175 books each year on science, elementary and advanced mathematics, biology, music, art, literary history, social sciences and other areas.